Asymptotics of Linear Differential Equations

Mathematics and Its Applications

Managing Editor:

M. HAZEWINKEL

Centre for Mathematics and Computer Science, Amsterdam, The Netherlands

Asymptotics of Linear Differential Equations

by

M.H. Lantsman

KLUWER ACADEMIC PUBLISHERS
DORDRECHT / BOSTON / LONDON

A C.I.P. Catalogue record for this book is available from the Library of Congress.

ISBN 978-90-481-5773-0

Published by Kluwer Academic Publishers,
P.O. Box 17, 3300 AA Dordrecht, The Netherlands.

Sold and distributed in North, Central and South America
by Kluwer Academic Publishers,
101 Philip Drive, Norwell, MA 02061, U.S.A.

In all other countries, sold and distributed
by Kluwer Academic Publishers,
P.O. Box 322, 3300 AH Dordrecht, The Netherlands.

Printed on acid-free paper

Contents

Preface

The asymptotic theory deals with the problem of determining the behaviour of a function in a neighborhood of its singular point. The function is replaced by another known function (named the asymptotic function) close (in a sense) to the function under consideration. Many problems of mathematics, physics, and other divisions of natural science bring out the necessity of solving such problems. At the present time asymptotic theory has become an important and independent branch of mathematical analysis.

The present consideration is mainly based on the theory of asymptotic spaces. Each asymptotic space is a collection of asymptotics united by an associated real function which determines their growth near the given point and (perhaps) some other analytic properties.

The main contents of this book is the asymptotic theory of ordinary linear differential equations with variable coefficients. The equations with power order growth coefficients are considered in detail.

As the application of the theory of differential asymptotic fields, we also consider the following asymptotic problems: the behaviour of explicit and implicit functions, improper integrals, integrals dependent on a large parameter, linear differential and difference equations, etc.. The obtained results have an independent meaning.

The reader is assumed to be familiar with a comprehensive course of the mathematical analysis studied, for instance at mathematical departments of universities. Further necessary information is given in this book in summarized form with proofs of the main aspects.

The book can be of use to readers of various professions dealing with the application of asymptotic methods in their current work.

Chapter 1

INTRODUCTION

1. INTRODUCTION

In an asymptotic problem we look for an approximation of a function in at least one neighborhood of its (as a rule, singular) point. More precisely, we replace the function by another function supposed to be close to the first one. It is important to notice that the dimension of the neighborhood is not indicated and no matter how small it is. Such problems are called *local*. Thus, every asymptotic problem is a local problem of a function estimate. First, the asymptotic relations obtained give qualitative characteristics of the function. Then, we may obtain some numerical characteristics of its behaviour or approximate formulae. The fields of application may be as follows: computation of some limits and obtaining some approximate formulae in order to estimate the function under consideration (the approximate equality $\sqrt{1+t} \approx 1 + t/2$ for $t \ll 1$ is a simple example of such relations); outline about zeros distribution (for instance, for the Airy equation $x'' + tx = 0$, the distance between nearest zeros of a solution is approximately equal to $\pi/\sqrt{t}$ for $t \gg 1$); criteria of asymptotic stability and instability etc. Solutions of asymptotic problems may be included as an important stage for solving global problems [for example, it is impossible to compute many functions on a sufficiently large interval without asymptotic relations (see the Supplement)]. Asymptotic problems are found in many physical problems.

At the present time the asymptotic theory of functions becomes as an integrate and independent part of the theory of mathematical analysis.

One of the main problems of the asymptotic theory is to solve an operator equation of the form $x = A(x)$. But instead of exact solutions, the final aim

1

of any such asymptotic problem is to obtain so called asymptotic solutions in an appropriate asymptotic space.

Certain asymptotic problems may be solved by means of the classical methods of calculus. The asymptotic theory began to be developed as an independent part of the mathematical analysis at the close of the XIX century in connection with the need to understand the behavior of solutions of a differential equation possessing singular points. This was connected with requirements of many problems in pure and applied mathematics.

TH. STILTJES (1886) and H. POINCARÉ (1886) introduced asymptotic series. In this method of investigation, an asymptotic series is represented as a series in already known functions with constant coefficients such that the general (mth) term was an $o-$small function of the $m-1$th term of the series. For example, let us try to obtain some asymptotic formulae to an equation (say $F(t, x) = 0$) in a neighborhood of the point $t = a$. This method implies the following procedure. We replace the unknown by an appropriate asymptotic series with undetermined coefficients and compute the coefficients annihilating the sum of coefficients in like terms. The result may lead to a divergent series (for any t in the region considered). However, partial sums (say $s_m(t)$) may form asymptotic approximations to the considered function. The asymptotic approximation becomes 'better' when m increases. Such consideration (using divergent series) was the first and an important step towards creating the asymptotic theory. This method permits us to solve several problems, and we believe this as an extension of methods of the classical analysis.

The notion of asymptotic series was somewhat extended by A. ERDÉL-YI. In addition, there exist many partial methods, each of which requires an individual approach and individual technique to obtain the desired result. Evidently, an unified method of all asymptotic problems does not exist. But some special methods (having sense analogous to the general methods of investigation in the classical analysis) must be elaborated. The theory presented in this book is a step in this direction.

The attempts to solve asymptotic problems by means of the classical methods and by divergent asymptotic series with intermediate coefficients, even if they allow to obtain the desired result, as a rule, lead to complicated transformations, and, as a consequence, the answer obscures the connection with the original problem.

In the theoretical part of this book we introduce the main notions and prove several theorems having adequate meaning for asymptotic problems as the corresponding parts of the classical and functional analysis. First, we define some numerical estimates and sets of functions satisfying such estimates in a neighborhood of their singular point.

The main method considered in this book is based on the theory of so called asymptotic spaces [introduced by the author (see [**131**])]. Asymptotic space is an abelian group with an operation of summation, and it is assigned a real function $F(x)$ (estimate of growth) to any its element x. $F(x)$ possesses the following property: $F(x + y) \leq \max[F(x), F(y)]$ (the exact definition is given in Chapter 3).

For example, let $f(t)$ be a complex function of a real argument defined for $t \gg 1$ and comparable with a real power function t^p such that $|f(t)| < t^p$ for $t \gg 1$. Clearly, the infimum of p is a characteristics of the function growth. If the growth characteristics of functions $f_1(t)$ and $f_2(t)$ are equal to p_1 and p_2, respectively, then the characteristics of their sum $|f_1(t) + f_2(t)|$ no more than $\max(p_1, p_2)$.

On the basis of power growth functions, we may consider other sets, for instance, with exponential growth, etc..

Some simple asymptotic information about the investigate function may be the clue to obtain a more detailed asymptotic information.

Example 1.1. Given the equation $\ln x + (1/t)x = 0$. We look for asymptotic representation for real solutions of the equation for $t \to +\infty$. First, it is easy to prove that the equation (for any positive t) has a unique solution $x(t)$ bounded for $t > 1$. In the meanwhile, we have very rough information about the asymptotic behavior of the solution $x(t)$. However, evidently, $(1/t)x(t) \to 0$. Hence (from the original equation) $\ln x(t) \to 0$. Hence $x(t) \to 1$ and $\ln x(t) \sim -1/t$. Thus, $x(t) - 1 \sim -1/t$. Taking into account that $\ln(1 + u) - u \sim -u^2/2$ for $u \to 0$, we have

$$x(t) - 1 + 1/t \sim -(x(t) - 1)/t + (x(t) - 1)^2/2 \sim 3/(2t^2) \quad (t \to +\infty),$$

and so on.

Many sets of functions whose elements possess estimates of growth are used to formulate problem states, and they form sets of the desired asymptotic approximations. Function sets, where it is possible to estimate simultaneously the function and its derivatives of any order, play a considerable role. They are called analytic estimates. The spaces may be useful applied for investigation of different kinds of differential and difference equations.

Various estimates of function growth are known in literature. A wide set of functions is given by the J. KARAMATA class of regularly varying functions. Estimates of holomorphic functions in a complex plane are known, etc.

There exist other more complicated estimates of function growth which characterizes behavior of a function more in detail. For example, the growth of functions of the type

$$f(t) \sim ct^{k_0}(\ln_1 t)^{k_1}...(\ln_m t)^{k_m} \text{ for } t \to +\infty, \qquad (1.1)$$

gives the vector $(k_0, k_1, ..., k_m)$. Here $k_0, k_1, ..., k_m$ are real numbers, $c \neq 0$ is a complex number. Complex numbers may be used as characteristics of behavior of a function.

For the goal of investigating equations (where the operation of differentiation is applied), so called asymptotic differential rings and fields play a considerable role. The fields are closed about the algebraic operations and operation of differentiation.

An interesting attempt to construct general asymptotic differential fields for analytic functions was undertaken by W.STRODT. Functions of this type (roughly speaking) are holomorphic and equivalent to a monomial of the form (1.1) in a sector of the complex plane.

The notion of general asymptotic field is given in this book, and several asymptotic differential fields of functions of a real and complex argument were considered. They are named fields of type M and N on a positive semi-axis and fields of type N_S and normal fields of type N_S in a sector of the complex plane. The fields of type N_S seem to be more convenient for this goal than the Strodt fields, because they are given in a simple axiomatic form, and some of them contains functions which do not belong to any Strodt field. For example, $e^{\sqrt{\ln z}}$ is such a function. It is not equivalent to any monomial of the considered form and $e^{\sqrt{\ln z}} = o(z)$ for $z \to \infty$ in any sector of the complex plane. Moreover, the fields of type N and M convenient for the goal of investigating numerous wide types of differential and difference equations.

Solution of the equation $x = A(x)$ frequently is split up into two stages. In asymptotic problems (as a rule) operator $A(x)$ is singular, and to solve this problem we have to obtain its formal solutions (as the first and may be the most difficult stage). By a formal asymptotic solution we usually understand a function obtained by means of some no correct method, but it is supposedly an approximation to the required solution. In asymptotic spaces, formal solutions find an exact meaning. In short, a formal solution is an element x of the asymptotic space such that $x \asymp A(x)$ (definition of the symbol $\asymp$ is given in this book). So that the problem of obtaining formal solutions becomes as an independent problem of the considered equation. Formal solution may be found much easier than asymptotic solution. For the formal solutions, it is possible to construct a so called factor space (which corresponds to the asymptotic space considered). This is a special kind of metric spaces, and the original singular problem turns into a regular problem in the factor space. For instance, the principle of the contractive mappings or some other principles of a fixed point may be applied to the singular equation $x = A(x)$.

The process of solving this problem is called the *formal procedure*.

In this book the main theorems for the formal procedure are given. And the technique of its application to various important problems are shown.

Proof that the required asymptotic solutions are found among the obtain formal solutions is the second stage of solving the asymptotic problem.

In the subsequent chapters we consider several applications of the elaborated asymptotic theory. The greatest part of them is devoted to linear differential equations on the positive semi-axis with power order growth coefficients which considered in detail. In the regular case (when its characteristic equation has a complete set of roots possessing the property of asymptotic separability) the asymptotic representations of solutions are given in a simple explicit form. In particular, they permit to obtain the main parts of asymptotic representation of a fundamental set of solutions to the considered equation, and on this basis, to formulate the main its qualitative characteristics such as criteria of asymptotic stability, instability, and oscillation, etc. For other cases, there are given a standard procedure to obtain the desired properties of the solutions. In a sector of the complex plane, the linear differential equation is only considered in the regular case. The formulae (like in the regular case for the equation of a real argument) are given, and the so called Stokes phenomenon is considered in general form. Many results obtained for the equation of real and complex argument are new and of independent interest.

In Chapter 7 we mainly deal with problems to estimate integrals containing a large parameter. The Laplace and Saddle Point methods are considered. We give several examples of application of these methods and consider some problems about *Euler's gamma function*. In division 10 linear difference equations with power order growth coefficients are considered. The asymptotic solutions on the positive semi-axis (in the considered case) are given in a simple explicit form. In conclusion of Chapter 15 we look for formal solutions of a simple differential difference equation of the form $x'(t + 1) = a(t)x(t)$.

2. CONVENTIONS AND NOTATION

(1) t is a *real value*, which means that t is either a real number or $t = \infty$ (more precisely t may be equal to $-\infty$ or $+\infty$); J_+ is a *positive semi-axis*, consists of all numbers $t \geq 0$ ($t = +\infty$ does not belong to J_+). J_- is a *negative semi-axis*, it consists of all numbers $t \leq 0$ ($t = -\infty \notin J_-$). $\mathbf{R} = J_- \cup J_+$ is the set of all real numbers or a *real axis*.

z is a *complex value*. That is, z is a complex number or $z = \infty$. We also say that ∞ is a point of an (*extended*) complex plane. Complex plane **Z** (the *extended plane*) is the set of all complex numbers and of the point $z = \infty$. **C** (the *finite plane*) is the set of all complex numbers, that is, **C** = **Z**\∞.

Let Γ be a continuous line placed entirely within the complex plane. We denote the part of the curve which connects the points a and b by $\Gamma(a, b)$. The parenthesis near a point means that the corresponding point does not belong to the considered part. If the point belongs to the considered part, we write a square bracket. We have three more cases $\Gamma(a, b]$, $\Gamma[a, b)$, and $\Gamma[a, b]$.

We denote a central sector lying in the complex plane by S, $z = \infty \notin S$. The angles of its boundary rays are denoted by φ_1, φ_2 ($\varphi_1 < \varphi_2$). We denote a closed central subsector of the sector S whose any point $z \neq 0$ is an interior point of S by S^* :

$$S^* = \{z : \varphi_1 < \varphi_1^* \leq \arg z \leq \varphi_2^* < \varphi_2, z \neq \infty\},$$

where φ_1^* and φ_2^* are the angles of the boundary rays of S^*.

(2) $t \gg 1$ ($t \ll 1$) reads *sufficiently large* (*sufficiently small*) t.

Let a function $f(t)$ possesses a property P. The expression *the function possesses the property P for $t \gg 1$* means that *in every (fixed) infinite interval of the positive semi axis there exists a number T (which, generally speaking, depends on the interval) such that P takes place in the domain* $\{t : t > T, \ t \neq \infty\}$.

The statement *the function possesses the property P for $t \ll 1$* means that there is a number $\varepsilon > 0$ such that the property P takes place in the domain $\{t : 0 < t < \varepsilon\}$.

Examples: 1° The statement *a function $f(t)$ is determined for $t \gg 1$* means that there is a number $T > 0$ (no matter how large) such that the function $f(t)$ is determined for $t > T$, $t \neq +\infty$.

2° The statement *a function $\widehat{f}(t)$ has zeros for $t \gg 1$* means that there is a sequence of positive numbers $\{t_m\}$ ($n = 1, 2, ...,$) such that $t_1 < t_2 < ... < t_m < ..., t_m \to +\infty$ for $n \to \infty$ and $f(t_n) = 0$. Indeed, take a number $T_1 > 0$. Hence there is a zero $t_1 \neq +\infty$ in the interval $]T_1, +\infty[$. In the same way, there is another zero in the interval $]t_1, +\infty[$. That is, $t_2 > t_1$, $t_2 \neq +\infty$, $f(t_2) = 0$. Clearly, the required result is obtained by induction with respect to $m = 1, 2, ...$

The concept of sufficiently small (large) values is convenient for many local problems and widely used in many asymptotic definitions. For example, the definition of a limit of a function may be formulated as follows: The relation $f(t) \to A$ *for* $t \to a$ *means that* $|f(t) - A| \ll 1$ *for* $|t - a| \ll 1$. Indeed, the last means the following: Let $\varepsilon > 0$ be a fixed positive number (no matter how small). Then $|f(t) - A| < \varepsilon$ is fulfilled for $|t - a| \ll 1$. This means that there is a number $\delta > 0$ such that $|f(t) - A| < \varepsilon$ for $|t - a| < \delta$. Moreover, the last implies that $f(t)$ is determined for $|t - a| < \delta$.

Here (which is important) the relation $|t - a| \ll 1$ implies the inequality $|t - a| > 0$. That is, the function is not considered at the point $t = a$, and $f(a)$ may not exist.

The relation $f(t) \to \infty$ for $t \to a$ means that $|f(t) - A| \gg 1$ for $|t-a| \ll 1$.

(3) The statement *A property* P *occurs for* $z \in [S]$ means that the property P takes place in any fixed subsector S^*.

(4) A function $f(z)$ is said to be *holomorphic* in a domain D if it is analytic at any point $z \in D$ and single valued on D.

Examples 1.2.
1° The expression *A function* $f(z)$ *is holomorphic for* $|z| \gg 1, z \in [S]$ means that *for any fixed subsector* S^* *there is a positive number* T *(which, generally speaking, depends on* S^**) such that the function* ($f(z)$ *is holomorphic in the domain* $S^* \cap \{z : |z| > T\}$).

(5) The expression (limit)

$$\lim_{z \to \infty, z \in [S]} f(z) = q \tag{1.2}$$

means that for any (fixed) subsector S^*

$$\lim_{T \to +\infty} \sup_{|z|=T, z \in S^*} |f(z) - q| = 0. \tag{1.3}$$

We denote a *sufficiently small sector* by S_ε,

$$S_\varepsilon = \{z : |\arg z| < \varepsilon, \varepsilon \ll 1, z \neq \infty\}.$$

The expression *A property* P *occurs in* S_ε means that *the property* P *takes place in the sector* S_ε *for at least one* $\varepsilon \neq 0$.

(6) We suppose *for any function under consideration (if it is not mentioned apart) to be single-valued without removable singularities. Moreover, we suppose that the operation of canceling the removable singularities is always added to the operations under the functions.* For example, the fraction

z/z is considered to be equal to unity in the entire complex plane (including the points $z = 0$ and $z = \infty$).

(7) Let $f(z)$ be a holomorphic function in $[S]$ for $|z| \gg 1$; D is said to be a *sectional* or (which is the same) *permissible domain* of the function $f(z)$ in S if

(i) D is a domain in the complex plane;
(ii) for any subsector S^* there is a positive number T (generally speaking, depending on S^*) such that $S^* \cap \{z : |z| > T\} \subset D$;
(iii) $f(z)$ is holomorphic in D.

(8) Let $f(z)$ be a holomorphic function for $|z| \gg 1, z \in [S]$. Let z_0 be a limit point in S. The expression (integral)

$$\int_{z_0}^{z} f(s)ds \tag{1.4}$$

in $[S]$ means the following: if z_0 belongs to at least one permissible domain of the function $f(z)$ in S, then z is an arbitrary point which belongs to any permissible domain containing the point z_0, and (1.4) means an integral along any piecewise smooth rectangable curve $\Gamma(z_0, z)$ in the considered domain;

if z_0 does not belong to any permissible domain of $f(z)$ in S, we suppose that it is chosen a point z_1 in a permissible domain of the function and a rectangable curve $\Gamma(z_0, z_1)$ in S such that the integral $\int_{\Gamma(z_0,z_1)} f(z)dz$ is convergent, and

$$\int_{z_0}^{z} f(s)ds = \int_{\Gamma(z_0,z_1)} f(z)dz + \int_{z_1}^{z} f(s)ds. \tag{1.5}$$

If the integral $\int_{\Gamma(z_0,z_1)} f(z)dz$ is divergent for any point $z_1 \in S$ and for any curve $\Gamma(z_0, z_1)$ in $[S]$, we say that integral (1.4) is divergent in $[S]$.

(9) We denote the difference $f(z+1) - f(z)$ by $f^-(z)$ or $[f(z)]^-$; $f^=(z)$ or $[f(z)]^=$ means $[f^-(z)]^-$. We define the expression $f^{[m]}(z)$ by induction with respect to $m = 1, 2, \dots$ We set $f^{[0]}(z) = f(z)$ and $f^{[m]}(z) = [f^{[m-1]}(z)]^-$. So that $f^{[1]}(z) = f^-(z)$ and $f^{[2]}(z) = f^=(z)$.

(10) We define the expression $\ln_m z$ by induction with respect to $m = 0, 1, \dots$: $\ln_0 z = z$, and $\ln_m z = \ln \ln_{m-1} z$, where $\ln$ is the sign of the natural logarithm.

(11) $\square$ means the end of Proof.

3. ASYMPTOTIC RELATIONS

3.1 *DEFINITION OF ASYMPTOTICS*

There are problems when the function under consideration is too complicated or cannot be found, while we are able to describe its certain qualitative characteristics by means of another function which is close (in some sense) to the first one. Given a function $f(z)$ in a region D and a is its limit point. We say that a function $g(z)$ is an *asymptotics* or an *asymptotic approximation* of the function $f(z)$ for $z \to a, z \in D$ if the difference $|f(z) - g(z)|$ is small for $|z - a| \ll 1, z \in D$. Thus, asymptotics is a local characteristic of a function. The statement will obtain a strict sense when we define what the 'smallness of the difference' means for the considered problem. A problem of how correctly to determine the state of an asymptotic approximation resembles the problem "what does it mean that a number b is an approximate value for a number a"? The solution of these questions lies beyond general consideration. The exact answer may be given only on the basis of a specific problem. Here we consider some features of asymptotic function approximation.

Let $g_1(z)$ and $g_2(z)$ be functions considered to be asymptotic approximations to a function $f(z)$. We say that the asymptotic approximation $g_1(z)$ is *better* or *more accurate* than $g_2(z)$ for $z \to a, z \in D$, with respect to the function $f(z)$, which is written

$$g_1(z) \prec g_2(z) \text{ or } g_2(z) \succ g_1(z) \text{ for } z \to a, \ z \in D \qquad (1.6)$$

if

$$|f(z) - g_1(z)| \le |f(z) - g_2(z)| \text{ for } |z - a| \ll 1, z \in D. \qquad (1.7)$$

In certain problems, a function $g(z)$ is said to be an *asymptotic approximation* of a function $f(z)$ only if

$$\lim_{z \to a, z \in D} \frac{f(z)}{g(z)} = 1. \qquad (1.8)$$

Such a relation is exemplified by Stirling's formula

$$n! \sim n^n e^{-n} \sqrt{2\pi n} \text{ for } n \to +\infty$$

which means that

$$\lim_{n \to \infty} \frac{n!}{n^n e^{-n} \sqrt{2\pi n}} = 1.$$

An asymptotic approximation may be formulated as follows: For a positive function $h(z)$ or for a set of positive functions $\{h(z)\}$ supposed to be small, a function $g(z)$ is called an *asymptotic approximation* of the function $f(z)$ (for $z \to a, z \in D$) if the following inequality holds:

$$|f(z) - g(z)| \le h(z) \text{ for } |z - a| \ll 1, z \in D. \qquad (1.9)$$

The set $\{h(z)\}$ is said to be a *scale of growth*. If $g(z) = 0$, the function $h(z)$ gives an estimate of the function growth for $z \to a, z \in D$.

Examples 1.3.   Here we consider function growth scales where $a = +\infty$ and $D = J_+$.

1° The scale consists of power functions of the form $h(t) = t^r$ where r are real numbers or, in particular, r may be integers (in the general case when a is a finite point of the complex plane, the scale consists of the functions $|z - a|^r$).

Instead of a power function, we may set a number (functional) in the form

$$r = \frac{\ln h(t)}{\ln t},$$

which becomes a numerical estimate of the asymptotic approximation.

2° *Power-logarithmic scale.* The scale consists of functions of the form

$$h(t) = t^{k_0}(\ln_1 t)^{k_1}...(\ln_m t)^{k_m}, \tag{1.10}$$

where $m \in \{0, , ...\}$, $k_0, k_1, ..., k_m$ are real numbers. The vector $(k_0, k_1, ..., k_m)$ becomes an estimate of growth. The scale can be put in order as follows: consider two functions $h_1(t)$ and $h_2(t)$ of type (1.10) with vectors $(k_{01}, k_{11}, ..., k_{m1})$ and $(k_{02}, k_{12}, ..., k_{m2})$, respectively. The relation

$$(k_{01}, k_{11}, ..., k_{m1}) \prec (k_{02}, k_{12}, ..., k_{m2}), \tag{1.11}$$

means that

$$\lim_{t \to +\infty} \frac{h_1(t)}{h_2(t)} = 0,$$

that is, the first non-zero difference $k_{i1} - k_{i2}$ $(i \in \{0, 1, ..., m\})$ is negative.

3° Given the function

$$e(t) = \int_t^{+\infty} \frac{e^{t-\tau}}{\tau} d\tau. \tag{1.12}$$

Let us take integral (1.12) $n - 1$ times by parts. We have

$$e(t) = e_n(t) + r_n(t),$$

where

$$e_n(t) = \sum_{j=1}^{n} (-1)^{j+1} \frac{(j-1)!}{t^j}, \tag{1.13}$$

and

$$r_n(t) = (-1)^n n! \int_t^{+\infty} \frac{e^{t-\tau}}{\tau^{n+1}} d\tau. \tag{1.14}$$

Clearly,

$$\lim_{t \to +\infty} r_n(t)t^{n+1} = (-1)^n n!,$$

and hence we may suppose that $e_n(t)$ are asymptotic approximations for $e(t)$ as $t \to +\infty$. In the considered case, the scale of growth may consist of the set $\{t^{-n}\}$ $(n = 1, 2, ...)$.

3.2 *MAIN ASYMPTOTIC RELATIONS*

In this subsection we give some definitions which somewhat simplify the consideration of asymptotic behavior of functions. We shall consider functions $f(z), g(z), ...$ defined on a region D, and a is its limit point.

Definition 1.4. We say that $f(z)$ is of the *higher order of smallness* than $g(z)$ (or, which is the same, $f(z)$ is *o-small* of $g(z)$) for $z \to a, z \in D$, which is written

$$f(z) = o(g(z)) \text{ for } z \to a, z \in D, \tag{1.15}$$

if for *any* (fixed) number $\varepsilon > 0$ the following relation holds:

$$|f(z)| \leq \varepsilon|g(z)| \text{ for } |z - a| \ll 1, z \in D. \tag{1.16}$$

If $g(z) \neq 0$ for $|z - a| \ll 1, z \in D$, then the relation

$$f(z) = o(g(z)) \text{ for } z \to a, z \in D$$

is equivalent to the limit

$$\lim_{z \to a, z \in D} \frac{f(z)}{g(z)} = 0.$$

If $f(z) = o(g(z))$ and $g(z) = o(h(z))$, then $f(z) = o(h(z))$ for $z \to a, z \in D$.

If $f(z) = o(g(z))$ and $g(z)$ possesses zeros in $D \cap \{z : |z - a| \ll 1\}$, then the function $f(z)$ possesses zeros in $D \cap \{z : |z - a| \ll 1\}$. Moreover, there exists a number $\delta > 0$ such that if z^* is a zero of the function $g(z)$, $z^* \in D$, and $|z^* - a| < \delta$, then $f(z^*) = 0$.

Examples 1.5. 1° For any a and D, $0 = o(0)$.

2° Suppose $a = \infty$, then $1 = o(z)$ for $|z| \to \infty$; $e^{-t} = o(t^\sigma)$ as $t \to +\infty$ for any number σ.

Definition 1.6. We say that $f(z)$ and $g(z)$ are *equivalent* for $z \to a$, $z \in D$, which is written

$$f(z) \sim g(z) \text{ for } z \to a, z \in D, \tag{1.17}$$

if

$$f(z) = g(z) + o(g(z)) \text{ for } z \to a, z \in D. \tag{1.18}$$

Let $g(z) \neq 0$ for $|z - a| \ll 1, z \in D$, then (1.17) holds if and only if

$$\lim_{z \to a, z \in D} \frac{f(z)}{g(z)} = 1.$$

If $f(z) \sim g(z)$ and $g(z) \sim h(z)$, then $f(z) \sim h(z)$ for $z \to a, z \in D$. If $f(z) \sim g(z)$ and $g(z)$ possesses zeros in $D \cap \{z : |z - a| \ll 1\}$, then the function $f(z)$ possesses zeros in $D \cap \{z : |z - a| \ll 1\}$. Moreover, there exists a number $\varepsilon > 0$ such that if z^* is a zero of the function $g(z)$, $z^* \in D$, and $|z^* - a| < \varepsilon$, then $f(z^*) = 0$.

Examples 1.7. $1°$ Suppose $a = \infty$, then $1 \sim 1 + 1/z$ for $z \to \infty$.

$2°$ $t^\sigma \sim t^\sigma + e^{-t}$ as $t \to +\infty$ for any number σ.

Proposition 1.8. *If $f(z) \sim g(z)$ for $z \to a, z \in D$, then*

$$g(z) \sim f(z) \text{ for } z \to a, z \in D.$$

PROOF (1.18) implies the relation

$$|f(z) - g(z)| \leq \varepsilon |g(z)| \text{ for } |z - a| \ll 1, z \in D.$$

Then

$$|g(z)| \leq \frac{|f(z)|}{1 - \varepsilon} \text{ and } |g(z) - f(z)| \leq \frac{\varepsilon |f(z)|}{1 - \varepsilon}$$

for $\varepsilon < 1$ in the same domain. Since the function $\varepsilon/(1 - \varepsilon)$ is continuous on the interval $[0, 1[$ and takes on any positive values, then, for any (fixed) $c \in]0, +\infty[$ (no matter how small), there exists a number $\varepsilon > 0$ such that $c = \varepsilon/(1 - \varepsilon)$. Thus,

$$|g(z) - f(z)| \leq c |f(z)| \text{ for } |z - a| \ll 1, z \in D.$$

(for any fixed $c > 0$). That is, $g(z) \sim f(z)$ $(z \to a, z \in D)$. $\square$

Definition 1.9. We say that $f(z)$ is of the *order* of $g(z)$ (or, which is the same, $f(z)$ is *O-large* of $g(z)$) for $z \to a, z \in D$, which is written

$$f(z) = O(g(z)) \text{ for } z \to a, z \in D, \tag{1.19}$$

if the following relation holds:

$$|f(z)| \leq C |g(z)| \text{ for } |z - a| \ll 1, z \in D, \tag{1.20}$$

for *at least one number* $C > 0$.

If $g(z) \neq 0$ for $|z - a| \ll 1, z \in D$ the relation $f(z) = O(g(z))$ for $z \to a, z \in D$ is equivalent to the assertion that $f(z)/g(z)$ is a bounded function for $|z - a| \ll 1, z \in D$. In particular the relation

$$f(z) = O(1) \text{ for } z \to a, z \in D$$

means that $f(z)$ is a bounded function for $|z - a| \ll 1, z \in D$.

If $f(z) = O(g(z))$ and $g(z) = O(h(z))$ then $f(z) = O(h(z))$ for $z \to a, z \in D$. If $f(z) = O(g(z))$ and $g(z)$ possesses zeros in $D \cap \{z : |z - a| \ll 1\}$, then the function $f(z)$ possesses zeros in $D \cap \{z : |z - a| \ll 1\}$. Moreover there exists a number $\varepsilon > 0$ such that if z^* is a zero of the function $g(z)$, $z^* \in D$ and $|z^* - a| < \varepsilon$ then $f(z^*) = 0$.

If $f(z) = o(g(z))$ then $f(z) = O(g(z))$ (for any a and D). But the inverse assertion is obviously not fulfilled. That is if $f(z) = O(g(z))$ the relation $f(z) = o(g(z))$ generally speaking is not true.

Examples 1.10. $1°$ $0 = O(0)$ for any a and D,

$2°$ $\sin t = O(1)$ for any real a and real set D,

$3°$ Let $a = \infty$ then $2z + 1 = O(z)$ for $z \to \infty$; $\sin t - t = O(t^3)$ for $t \to 0$.

Definition 1.11. $f(z)$ and $g(z)$ are said to be *comparable* for $z \to a, z \in D$ if there exists at least one finite limit

$$\lim_{z \to a, z \in D} \frac{f(z)}{g(z)} \text{ or } \lim_{z \to a, z \in D} \frac{g(z)}{f(z)}. \tag{1.21}$$

Example 1.12. The functions t and $\sin t$ are comparable for $t \to +\infty$ because $\lim_{t \to +\infty} (\sin t)/t = 0$. The estimate $\sin t = O(t)$ for $t \to +\infty$ is also true but it is somewhat more rough than the previous relation.

Definition 1.13. Consider a set of functions

$$F = \{f_1(z), ..., f_n(z)\}.$$

We say that $f(z)$ is a *function of the greatest growth* of the set F for $z \to a, z \in D$ if $f(z) \in F$ and all the relations hold:

$$f_i(z) = O(f(z)) \text{ for } z \to a, z \in D \ (i = 1, 2, ..., n). \tag{1.22}$$

For example the set

$$\{\sin 2t, \ 3t^2 - e^{-t}, \ 2t, \ t^2\},$$

$3t^2 - e^{-t}$ and t^2 are functions of the greatest growth for $t \to +\infty$.

Chapter 2

METRIC SPACES

1.   MAIN DEFINITIONS

A space is a non-empty set (no matter how the nature of its elements). Thus, the concept of a space is the most general in mathematics. Instead of the term *space*, it is possible to use another synonyms as follows: *class, system, collection, family* etc. In a space we may introduce the concept of an operator.

Definition 2.1. Let $X = \{x\}$ and $Y = \{y\}$ be two spaces. Consider a correspondence: for every fixed $x \in X$, there exist an element $y \in Y$, and it is unique for the element x, let Y^* be the subset of Y consisting of all elements each of which corresponds to at least one element $x \in X$. Then we say that there is defined an *operator* with the *domain of definition* X and the *domain of values* (or *range*) Y^* located in Y.

So that we have a collection of pairs (x, y) where all the elements x are different and form the set X. All y of these pairs form the set Y^*, $Y^* \subset Y$. Arbitrary elements $x \in X$ and $y \in Y^*$ are called *variables*. More precisely, x is an *independent variable* or *argument*, and y is a *dependent* variable. In the special case when Y is a set of real numbers, we also say that it is defined a *functional*.

The operator can be written by different symbols, for example, as follows:

$$y = A(x), \quad y = f(x), \quad y = Ax, \quad x \to y, \quad (x, y) \text{ etc.}$$

There are used another names of an operator such as a *function, correspondence, transformation, mapping*, etc.

Let a and b be elements of a space E. The relation $a = b$ means that a and b is the same element. So that if $a = b$ and $b = c$, then $b = a$ and $a = c$.

14

Metric spaces generalize the important notion of distance in the Euclidean plane. They play an extraordinary role in many divisions of mathematics.

Definition 2.2. A non-empty set **M** is said to be a *metric space* if for every fixed pair of its elements x and y, it is assigned a non-negative number $\rho(x, y)$ satisfying the following conditions (axioms):

(1) $\rho(x, y) = 0$ if and only if $x = y$ (axiom of identity).

(2) $\rho(x, y) = \rho(y, x)$ (axiom of symmetry).

(3) $\rho(x, z) \leq \rho(x, y) + \rho(y, z)$ (triangle axiom) (z is an arbitrary element belonging to **M**).

$\rho(x, y)$ is called the *distance* between x and y (or *distance function*) of the space.

Consider a sequence of the form

$$\{s_m\} \quad (m = 0, 1, ...) \tag{2.1}$$

of elements (or *points*) of the metric space **M**.

Definition 2.3. Let there exist an element $s \in \mathbf{M}$ such that $\rho(s_m, s) \to 0$ for $m \to \infty$, then s is called a *limit* of the sequence, and we write $s_m \to s$ (for $m \to \infty$) or $\lim_{m \to \infty} s_m = s$. If the sequence has a limit, then we also say that (1.1) *converges* or is *convergent*. Otherwise (1.1) *diverges* or is *divergent*.

If $\{s_m\}$ is a convergent sequence, then it is bounded in the following sense: for any (fixed) point $x \in \mathbf{M}$, there exists a number $K > 0$ (generally speaking, depending on x) such that $\rho(s_m, x) < K$ for any m.

Proposition 2.4. *The sequence $\{s_m\}$ may have only one limit.*

PROOF. Let s and s^* be two limits of the sequence. We have for $m \gg 1$ (on the basis of the triangle axiom),

$$\rho(s, s^*) \leq \rho(s, s_m) + \rho(s_m, s^*) \ll 1.$$

Since the left side is constant, we conclude that $\rho(s, s^*) = 0$. By axiom (1) of definition 2.2, $s = s^*$. $\qquad\square$

Clearly, *any subsequence $\{s_{m_k}\}$, where $m_k \to \infty$ for $k \to \infty$, of a convergent sequence $\{s_m\}$ possesses the same limit (for $k \to \infty$).*

Introduce several useful notations. We denote an *open sphere* or an $h - neighborhood$ (or simply *neighborhood*) of radius r with center at the

point $a \in \mathbf{M}$ by
$$S_{ar} = \{x : x \in \mathbf{M}, \rho(x, a) < r\}.$$

A closed sphere $\overline{S}_{ar} = \{x : x \in \mathbf{M}, \rho(x, a) \le r\}$. An open (closed) punctured sphere $S'_{ar} = S_{ar}\backslash a$ $(\overline{S}'_{ar} = \overline{S}_{ar}\backslash a)$ that is, $S'_{ar} = \{x : x \in \mathbf{M}, \rho(x, a) < r, x \ne a\}$ $(\overline{S}'_{ar} = \{x : x \in \mathbf{M}, \rho(x, a) \le r, x \ne a\})$. A *sufficiently small sphere* or *sufficiently small neighborhood*:
$$S_{a\varepsilon} = \{x : x \in \mathbf{M}, \rho(x, a) \ll 1, x = a\}$$

and a *punctured sufficiently small sphere (neighborhood)* $S'_{a\varepsilon} = S_{a\varepsilon}\backslash a$.

The notion of a limit of the sequence $\{s_m\}$ may be formulate as follows: The sequence $\{s_m\}$ has a limit $a \in M$ if any punctured infinitesimal sphere $S'_{a\varepsilon}$ contains all the points s_m $(m = N, N + 1, ...)$ for $N \gg 1$.

2.   SEVERAL EXAMPLES OF METRIC SPACES

$1°$ The real line $\mathbf{R}$ (consisting of all real numbers, $+\infty$ and $-\infty$ do not belong to $\mathbf{R}$) where $\rho(x, y) = |x - y|$. $\mathbf{R}$ is also denoted by $\mathbf{R}_1$. Check the axioms of a metric space. If $\rho(x, y) \equiv |x - y| = 0$ then clearly $x = y$. The relation $\rho(x, y) = \rho(y, x)$ is obvious and

$$\rho(x, z) \equiv |x - z|$$
$$= |(x - y) + (y - z)|$$
$$\le |x - y| + |x - z|$$
$$= \rho(x, y) + \rho(y, z)$$

for any $z \in \mathbf{M}$. Hence $\mathbf{R}$ is a metric space.

$2°$ The *n-dimensional Euclidean space* $\mathbf{R}_n$ is a space consisting of elements (points) each of which is an ordered collection of n real numbers $\xi_1, \xi_2, ..., \xi_n$. The numbers are called the *coordinates* of the element $x = (\xi_1, \xi_2, ..., \xi_n) \in \mathbf{R}_n$. The distance between the points x and $y = (\eta_1, \eta_2, ..., \eta_n) \in \mathbf{R}_n$ is given by the relation

$$\rho(x, y) = \sqrt{(\xi_1 - \eta_1)^2 + (\xi_2 - \eta_2)^2 + (\xi_n - \eta_n)^2}.$$

It is easy to check that $\mathbf{R}_n$ is a metric space.

For a sequence of the form
$$\{x_m = (\xi_{1m}, \xi_{2m}, ..., \xi_{nm})\} \subset \mathbf{R}_n \quad m = 1, 2, ...$$

Let
$$\rho(x, x_m) \to 0, \text{ i.e. } \sqrt{\sum_{j=1}^{n}(\xi_j - \xi_{jm})^2} \to 0 \text{ for } m \to \infty.$$

This is equivalent to the conditions $\xi_{jm} \to \xi_j$ as $m \to \infty$ $(j = 1, 2, ..., n)$.

$3°$ In the same way we introduce the (numerical) $n-dimensional$ *complex metric space* denoted by $\mathbf{C}_n$. Here any $x \in \mathbf{C}_n$ is in the form $x = (\xi_1, \xi_2, ..., \xi_n)$, where the components ξ_j are complex numbers and

$$\rho(x, y) = \sqrt{|\xi_1 - \eta_1|^2 + |\xi_2 - \eta_2|^2 + |\xi_n - \eta_n)|^2}$$

$(y = (\eta_1, \eta_2, ..., \eta_n)$ is an arbitrary element of the space).

$4°$ We denote by $\mathbf{C}(a, b) = \{x(t)\}$ the *space of all continuous complex-valued functions* of the real argument t, defined on the segment $[a, b]$ with

$$\rho(x(t), y(t)) = \sup_{t \in [a,b]} |x(t) - y(t)|.$$

Clearly $\rho(x(t), y(t)) = 0$ if and only if $x(t) = y(t)$ at any point $t \in [a, b]$. That is, $x(t)$ and $y(t)$ is the same element of the space. Besides $\rho(x(t), y(t)) = \rho(y(t), x(t))$ and

$$\rho(x(t), z(t)) = \sup_{t \in [a,b]} |x(t) - z(t)|$$

$$= \sup_{t \in [a,b]} |(x(t) - y(t)) + (y(t) - z(t))|$$

$$\leq \sup_{t \in [a,b]} |(x(t) - y(t))| + \sup_{t \in [a,b]} |(y(t) - z(t))|$$

$$= \rho(x(t), y(t)) + \rho(y(t), z(t))$$

for any $z(t) \in \mathbf{C}(a, b)$. Consequently $\mathbf{C}(a, b)$ is a metric space.

$5°$ The space $\mathbf{A}(S_{ar})$ of all holomorphic functions in a circle $S_{ar} = \{z : |z - a| < r\}$ and continuous on the boundary of the circle $\partial S_{ar} = \{z : |z - a| = r\}$ where

$$\rho(x(z), y(z)) = \sup_{z \in \partial S_{ar}} |x(z) - y(z)| \text{ for any } x(z), y(z) \in \mathbf{A}(S_{ar}).$$

The metric space axioms can be easily checked.

3.   SEQUENCES IN A METRIC SPACE

The first and the main important problem which is considered for sequence (2.1) is to test a sequence for convergence. Several the most important tests may already be formulated for a general metric space.

Proposition 2.5 (Necessary test of convergence). *If sequence* (2.1) *is convergent in* **M** *then*

$$\rho(s_{m-1}, s_m) \to 0 \text{ for } m \to \infty. \tag{2.2}$$

PROOF. Let s be the limit of the sequence (2.1). That is, $\rho(s_m, s) \to 0$ $(m \to \infty)$. Hence

$$\rho(s_{m-1}, s_m) \leq \rho(s_{m-1}, s) + \rho(s, s_m) \to 0.$$

That is $\rho(s_{m-1}, s_m) \to 0$ $(m \to \infty)$.   □

Examples 2.6. 1° Consider the sequence

$$\left\{ s_m = 1 + \frac{1}{1!} + \frac{1}{2!} + \dots + \frac{1}{m!} \right\},$$

in the space **R**. The sequence $\{s_m\}$ has a limit which is equal to the number e. And $\rho(s_m, s_{m+1}) = 1/(m+1)! \to 0$ for $m \to \infty$. But in the metric space of all rational numbers (with the same metric function), a limit does not exist because the number e is irrational.

2° It is obvious that the sequence $\{s_m = \ln m\}$, considered in the space **R**, is divergent. But

$$\rho(s_{m-1}, s_m) \equiv \ln m - \ln(m-1) = \ln\left(1 - \frac{1}{m}\right) \to 0$$

for $m \to \infty$. That is, the necessary test of convergence is fulfilled, but the sequence diverges.

The second example shows that the considered necessary test may be insufficient for convergence in any metric space. The first example shows that the following situation is possible: for two metric spaces $\mathbf{M}_1$ and $\mathbf{M}_2$ such that $\mathbf{M}_1 \subset \mathbf{M}_2$, a sequence may be convergent in $\mathbf{M}_2$ but divergent in $\mathbf{M}_1$.

In the space **R**, Cauchy's test gives the necessary and sufficient conditions of convergence for $\{s_m\}$. We formulate the test without proof.

Proposition 2.7. *Let s_m be complex numbers. Sequence (2.1) is convergent in **R** if and only if $|s_m - s_n| \to 0$ for $\min[n, m] \to \infty$ where m, n are arbitrary natural numbers.*

The Cauchy test reflects a fundamental property of the space **R** leading to many properties composed the basis of the classical mathematical analysis. There are many other metric spaces possessing the same property. They form so called *complete metric spaces*. They possess many properties similar to the space **R**.

Definition 2.8. A sequence $\{s_m\} \subset \mathbf{M}$ is said to be *fundamental* (or a *Cauchy's sequence*) if $\rho(s_m, s_n) \to 0$ for $\min[m, n] \to \infty$.

Clearly, any convergent sequence is fundamental. Indeed if s is its limit, then $\rho(s_m, s_n) \leq \rho(s_m, s) + \rho(s, s_n) \to 0$ for $\min[m, n] \to \infty$, which leads to the required property. But the converse theorem, generally speaking, is not valid (see Example 2.6 1°).

Definition 2.9. If any fundamental sequence is convergent in **M**, then **M** is called a *complete metric space*.

There are many tests giving (only) sufficient conditions for a sequence to be fundamental. First, we consider the comparison tests. Let

$$a_0 + a_1 + \ldots + a_m + \ldots \tag{2.3}$$

be a positive number series (a_m are positive numbers). Mark, (2.3) is either convergent or properly divergent (that is, divergent to infinity). Indeed, its partial sums $s_m = a_0 + a_1 + \ldots + a_m$ form an increasing sequence. If it is bounded, then it has a finite positive limit s which is the sum of the series. If $\{s_m\}$ is unbounded, then $s = +\infty$.

Let series (2.3) be convergent, and let the positive number series

$$b_1 + b_2 + \ldots + b_m + \ldots \tag{2.4}$$

be divergent.

Proposition 2.10. *Let, for any $m \gg 1$, if*

(1) $\rho(s_{m-1}, s_m) \leq a_m$, *then (2.1) is fundamental;*

(2) $\rho(s_m, s_{m+1}) \geq b_m$, *then (2.1) is not fundamental;*

(3) *there exist a finite limit $\lim_{m \to \infty} \rho(s_m, s_{m+1})/a_m$, then (2.1) is fundamental;*

(4) *there exist a finite limit $\lim_{m \to \infty} b_m/\rho(s_m, s_{m+1})$, then (2.1) is not fundamental.*

PROOF. Prove property (1). Let (for definiteness) $n > m$. Clearly,

$$\rho(s_m, s_n) \leq \rho(s_m, s_{m+1}) + \rho(s_{m+1}, s_{m+2}) + \ldots + \rho(s_{n-1}, s_n)$$

$$\leq a_m + a_{m+1} + \ldots + a_n \to 0$$

for $m \to \infty$, which leads to the required property.

Prove property (3). Let $\lim_{m \to \infty} \rho(s_{m-1}, s_m)/a_m = l < +\infty$. Then there exists a number $\varepsilon > 0$ such that $\rho(s_{m-1}, s_m)/a_m \leq l + \varepsilon$ for $m \gg 1$. That is, $\rho(s_{m-1}, s_m) \leq (l + \varepsilon)a_m$. Obviously, the series with the general term $(l + \varepsilon)a_m$ is convergent. Consequently, the considered sequence is fundamental.

We prove property (2) by contradiction. Suppose $\rho(s_m, s_n) \to 0$ for $\min[m, n] \to \infty$. Since $b_m + b_{m+1} + ... + b_n \leq \rho(s_m, s_n)$, hence $b_m + b_{m+1} + ... + b_n \to 0$. Consequently, series (2.4) is convergent. The obtained contradiction proves our assertion. In the same way, if in (4) the limit is finite and the considered sequence is fundamental, we can easily prove that series (2.4) is convergent which is impossible. $\qquad\square$

The following proposition is a simple consequence of the comparison tests.

Lemma 2.11. *Let there be a positive number $q < 1$ such that for all $m \gg 1$, the following inequalities hold:*

$$\rho(s_m, s_{m+1}) \leq q\rho(s_{m-1}, s_m). \qquad (2.5)$$

Then the sequence $\{s_m\}$ is fundamental.

PROOF. We may suppose (for simplicity, without loss of generality) that inequality (2.5) holds for all $m \geq 1$. We have $\rho(s_1, s_2) \leq q\rho(s_0, s_1)$,

$$\rho(s_2, s_3) \leq q\rho(s_1, s_2) \leq q^2\rho(s_0, s_1).$$

It is easy to show (by induction with respect to m) that

$$\rho(s_{m-1}, s_m) \leq q^m\rho(s_0, s_1) \ (m = 1, 2, ...).$$

Clearly, the series with the general term $q^m\rho(s_0, s_1)$ is convergent (it is a geometric progression and its sum is equal to $\rho(s_0, s_1)/(1-q)$). Consequently, the required property is a consequence of Proposition 2.10 (1). $\qquad\square$

The following proposition is a direct extension of the well known D'Alembert's test.

Proposition 2.12. *If there exists a (finite or infinite) limit*

$$\lim_{m\to\infty} \frac{\rho(s_m, s_{m+1})}{\rho(s_{m-1}, s_m)} = r \begin{cases} < 1, \text{ then } (2.1) \text{ is fundamental,} \\ > 1, \text{ then } (2.1) \text{ is not fundamental,} \\ = 1, \text{ then the problem remains unsolved.} \end{cases}$$

$$(2.6)$$

PROOF. Let $r < 1$, then it is possible to choose a positive number ε such that $r + \varepsilon < 1$, and for sufficiently large m, the following inequality holds:

$$\frac{\rho(s_m, s_{m+1})}{\rho(s_{m-1}, s_m)} \leq r + \varepsilon \equiv q < 1.$$

Because of Lemma 2.11, the considered sequence is fundamental.

Let $r > 1$. In the same way as in the previous proof, we may prove that $\rho(s_{m-1}, s_m) > cq^m$ for $m \gg 1$ where $q > 1$ and c is a positive number. We have $\rho(s_m, s_n) \geq \rho(s_m, s_{m+1}) > cq^m \to \infty$ for $m \to \infty$, which leads to the required property.

Let $r = 1$. We have to show that for some fundamental and not fundamental sequences, the limit in (2.6) may be equal to 1.

Clearly, the sequence $\{s_m = m\}$ is not fundamental. We have $\rho(s_m, s_{m+1})/\rho(s_{m-1}, s_m) = m/(m - 1) \to 1$. Hence $r = 1$.

The sequence $\{s_m = 1 - 1/m\}$ is fundamental because $|s_n - s_m| = |1/m - 1/n| \to 0$ for $\min[m, n] \to \infty$. And $\rho(s_{m-1}, s_m) = 1/m - 1/(m-1) = 1/[(m-1)m]$. Hence

$$r = \lim_{m \to \infty} \frac{(m-1)m}{m(m+1)} = 1. \qquad \square$$

Proposition 2.13 (Integral test). *Let there be a positive monotonic continuous function $f(t)$ (for $t \geq 1$) such that*

$$f(m) = \rho(s_{m-1}, s_m) \quad for \ m \gg 1.$$

Then if

$$\int_1^{+\infty} f(t)dt = \begin{cases} A \neq \infty, \ then \ sequence \ (2.1) \ is \ fundamental, \\ \infty, \ then \ sequence \ (2.1) \ is \ not \ fundamental. \end{cases} \tag{2.7}$$

PROOF. We may suppose that the function $f(t)$ is not increase for $t \gg 1$ (otherwise, the considered series and the integral are simultaneously divergent). Hence for $m \gg 1$,

$$f(m + 1) \leq \int_m^{m+1} f(t)dt \leq f(m). \tag{2.8}$$

Integral (2.7) is equal to the sum of the series

$$\sum_{m=1}^{\infty} \int_m^{m+1} f(t)dt. \tag{2.9}$$

Let (2.7) be convergent. This implies the convergence of series (2.9). Taking into account the inequality (2.8) for $f(m + 1)$ and Proposition 2.10 (1), we conclude that sequence (2.1) is fundamental.

Let the integral be divergent. That is series (2.8) is divergent. Taking into account the inequality (2.8) for $f(m)$ and Proposition 2.10 (2), we conclude that sequence (2.1) is not fundamental. $\qquad \square$

4.   EQUATIONS IN A METRIC SPACE

In this section, we consider an operator $A(y)$ which operates from a metric space $\mathbf{M}$ to $\mathbf{M}$, that is, $A(\mathbf{M}) \subset \mathbf{M}$.

Definition 2.14. We say that the operator $A(y)$ has a limit $b \in \mathbf{M}$ for $y \to a \in \mathbf{M}, y \in \mathbf{M}$, which is denoted by

$$A(y) \to b \text{ for } y \to a, y \in \mathbf{M}, \text{ or } \lim_{y \to a, y \in \mathbf{M}} A(y) = b, \qquad (2.10)$$

if

$$\rho(A(y), b) \ll 1 \text{ for } \rho(y, a) \ll 1 \ (y \in \mathbf{M}). \qquad (2.11)$$

Definition 2.15. We say that $A(y)$ is continuous at the point a in $\mathbf{M}$ if $\lim_{y \to a, y \in \mathbf{M}} A(y) = A(a)$.

Consider an equation of the form

$$y = A(y). \qquad (2.12)$$

Since $A(M) \subset M$, we may suppose that $A(\mathbf{M})$ is a mapping $\mathbf{M}$ onto $\mathbf{M}$. Thus, a solution $y \in \mathbf{M}$ of the equation is a *fixed point* of the mapping.

The required solution to (2.12) frequently may be obtained as a limit of a sequence of the form

$$s_{m+1} = A(s_m) \ (m = 0, 1, ...), \qquad (2.13)$$

where s_0 an appropriate element belonging to $\mathbf{M}$. The following conditions for a fixed point existence are obvious.

Lemma 2.16. *Let* $\mathbf{M}$ *be complete, and* $A(\mathbf{M}) \subset \mathbf{M}$. *Let* $A(y)$ *be continuous at any point* $y \in \mathbf{M}$, *and let there exist a limit* $y = \lim_{m \to \infty} A(s_m) \in \mathbf{M}$. *Then* $y = A(y)$.

Clearly, the solution y may depend on the initial point s_0. The following proposition gives a sufficient condition of uniqueness of the solution.

Lemma 2.17. *Let for any* $y_1, y_2 \in \mathbf{M}$ *and* $y_1 \neq y_2$,

$$\rho(A(y_1), A(y_2)) < \rho(y_1, y_2). \qquad (2.14)$$

Then equation (2.13) possesses no more than one solution belonging to $\mathbf{M}$.

PROOF. Let there be two solutions $y, y^* \in \mathbf{M}$ of the equation, and $y \neq y^*$. Then $\rho(y, y^*) = \rho(A(y), A(y^*)) < \rho(y, y^*)$, which is impossible. $\square$

The following theorem is called the *Principle of contractive mappings*. It gives sufficient conditions of existence and uniqueness of a solution of equation (2.12). First, consider the following:

Definition 2.18. Let there be a positive number q, called a *Lipschitz constant*, such that for any $y_1, y_2 \in \mathbf{M}$, the following inequality is valid

$$\rho(A(y_1, y_2)) \leq q\rho(y_1, y_2). \tag{2.15}$$

Then we say that $A(y)$ possesses the *Lipschitz condition* with a Lipschitz constant q in $\mathbf{M}$.

Theorem 2.19. *Let* $\mathbf{M}$ *be complete. Let* $A(\mathbf{M}) \subset \mathbf{M}$. *Let* $A(y)$ *possesses the Lipschitz condition in* $\mathbf{M}$ *with a Lipschitz constant* $q < 1$. *Then equation* (2.12) *has a solution* $y \in \mathbf{M}$. *It is unique and* y *is a limit of sequence* (2.13) *where* s_0 *is an arbitrary point belonging to* $\mathbf{M}$.

PROOF. It is enough to prove that, for the chosen initial element $s_0 \in \mathbf{M}$, there exists a limit of the sequence $\{s_m\}$. The other properties follow from lemmata 2.16 and 2.17. Since all s_m belong to $\mathbf{M}$, we have

$$\rho(s_m, s_{m+1}) = \rho(A(s_{m-1}), A(s_m)) \leq q\rho(s_{m-1}, s_m).$$

So that (see Lemma 2.11) the sequence is fundamental. And since the space is complete, the sequence possesses a limit belonging to $\mathbf{M}$. $\qquad \square$

Mark, that some other sufficient conditions for existence of a solution to equation (2.12) may be obtained on the basis of Propositions 2.12 and 2.13.

5. ABSTRACT FUNCTIONS

Let $\mathbf{M} = \{x\}$ be a metric space. Let U be a bounded domain of the complex plane. In particular it can be a bounded interval in $\mathbf{R}$. Let $x(z)$ be a function defined on U and $x(U) \subset \mathbf{M}$. Such a function is referred as an *abstract function*.

A limit and continuity of a function are defined as usual. For example, the function $x(z)$ is continuous at the point $z_0 \in U$ if $\lim_{z \in U, z \to z_0} x(z) = x(z_0)$, that is, $\lim_{z \in U, \Delta z \to 0} \rho(x(z_0), x(z_0 + \Delta z)) = 0$.

Proposition 2.20. *Let* U *be a closed bounded domain. If the function* $x(z)$ *is continuous on* U, *then it is uniformly continuous on* U. *That is,* $\rho(x(z), x(z + \Delta z)) \ll 1$ *as* $|\Delta z| \ll 1$ *for any* $z, z + \Delta z \in U$.

PROOF. Let ξ be an arbitrary point of the space $\mathbf{M}$. Let $z, z + \Delta z \in U$, and let $\rho(x(z), \xi) \leq \rho(x(z + \Delta z), \xi)$. We have

$$\rho(\xi, x(z)) + \rho(x(z), x(z + \Delta z)) \geq \rho(\xi, x(z + \Delta z)).$$

Hence

$$\rho(x(z + \Delta z), \xi) - \rho(x(z), \xi) \le \rho(x(z), x(z + \Delta z)).$$

In the same way, if $\rho(x(z + \Delta z), \xi) \le \rho(x(z), \xi)$, then

$$\rho(\xi, x(z + \Delta z)) + \rho(x(z + \Delta z), x(z)) \ge \rho(\xi, x(z)).$$

Hence

$$\rho(x(z), \xi) - \rho(x(z + \Delta z), \xi) \le \rho(x(z), x(z + \Delta z)).$$

Consequently,

$$|\rho(x(z + \Delta z), \xi) - \rho(x(z), \xi)| \le \rho(x(z), x(z + \Delta z))$$

for any $\xi \in \mathbf{M}$. This means (taking into account that the region U is closed and bounded) that the function $\rho(x(z), \xi)$ is uniformly continuous on U and on $\xi \in \mathbf{M}$. That is, for $z, z + \Delta z \in U$, $\xi \in \mathbf{M}$, and for any ϵ there is a number δ such that $|\rho(x(z + \Delta z), \xi) - \rho(x(z), \xi)| < \epsilon$ if $|\Delta z| < \delta$. Put $\xi = x(z + \Delta z)$. Since $\rho(x(z + \Delta z), x(z + \Delta z)) = 0$, we have $\rho(x(z), x(z + \Delta z)) < \epsilon$ for any $z \in U$. That is, $x(z)$ is uniformly continuous on U. $\qquad\square$

The following proposition is evidently true.

Proposition 2.21. *Let U be a closed bounded domain. If the function $x(z)$ is continuous on U, then it is bounded on U.*

Definition 2.22. We denote by $\mathbf{MU}$ the space consisting of all continuous functions specified on a closed bounded domain U. Let $\rho^*(x(z), y(z)) = \sup_{z \in U} \rho(x(z), y(z))$.

The following proposition is obvious.

Proposition 2.23. $\mathbf{MU}$ *is a complete metric space with metric function $\rho^*(x(z), y(z))$.*

6.   LINEAR SETS

General metric spaces may be very wide in their scope. As the matter of fact, we frequently deal with more restricted sets. In particular, we frequently consider spaces with arithmetic operations determined between their elements. We prefer to deal with sets closed to the given operations because we have to perform the expressions under consideration in process of their investigation. The main mathematical operations are addition and multiplication. First, we introduce sets where one operation is defined. For definiteness, let it be *addition*.

Definition 2.24. A non-empty set $\mathbf{G}$ is said to be an *additive abelian group* if the following properties (axioms) are fulfilled:

(1) The *addition law* is given: for any pair of elements $a, b \in \mathbf{G}$, there is given an element $a + b \in \mathbf{G}$ which is called the *sum* of elements a and b.

(2) The *commutative law*:

$$a + b = b + a.$$

(3) The *associative law*: for any $a, b, c \in \mathbf{G}$

$$(a + b) + c = a + (b + c).$$

(4) There exists at least one element $\theta \in \mathbf{G}$ such that, for any $a \in \mathbf{G}$, $a + \theta = a$. The element θ is called a *zero* or *null* of the group.

(5) For any $a \in G$, there exists at least one element $b = -a$ such that $a + b = \theta$. The element $-a$ is said to be an *opposite element* of a.

Remark 2.25. In (3), we may omit the brackets and write (instead of both the sides of the equality) $a + b + c$.

Remark 2.26. *The group has only one element* θ. Indeed, let θ_1 and θ_2 be two zeros of the group. We have on the one hand $\theta_1 + \theta_2 = \theta_1$, and on the other hand $\theta_1 + \theta_2 = \theta_2 + \theta_1 = \theta_2$. Hence $\theta_1 = \theta_2$.

Remark 2.27. *There exists only one opposite element for a fixed element* $a \in \mathbf{G}$. Indeed if b_1 and b_2 are two opposite elements, then $a + b_1 = \theta$ and $a + b_2 = \theta$. Hence $b_1 = b_1 + (a + b_2) = b_2 + (a + b_1) = b_2$.

Remark 2.28. *The equation* $a + x = b$ *has one and only one solution.* Clearly, $x = b + (-a)$ is a solution. If there is another solution x_1, then $a + x = a + x_1$. Hence $(-a) + a + x_1 = (-a) + a + x$ Consequently, $\theta + x = \theta + x_1$. That is, $x = x_1$. For simplicity, we write the solution by $x = b - a$ (instead of $x = b + (-a)$).

Definition 2.29. *Multiplicative groups.* In the definition of the notion of a group, we can change the sign '+' of the operation and set it as multiplication. In this case in the axioms (1)–(4), we must throughout read *multiplication* instead of *addition*, and $a \cdot b$ or ab instead of $a + b$. Then $\mathbf{G}$ is called a *multiplicative abelian group*. Instead of the zero element θ, we have the unity element I with the property $aI = a$ *for any* $a \in \mathbf{G}$. Instead of the opposite element $-a$, we have the inverse element a^{-1} such that $a^{-1}a = I$.

7.   METRIC GROUPS. MAIN ASYMPTOTIC RELATIONS IN A METRIC GROUP

Definition 2.30. Let **G** be an additive abelian group. We say that **G** is a *metric group* if for any (fixed) $a \in$ **G** it is assigned a non-negative number $\rho(a)$ satisfying the following conditions (axioms):

(1) $\rho(a) = 0$ if and only if $a = \theta$;

(2) for any $a, b \in$ **G**, $\rho(a + b) \le \rho(a) + \rho(b)$ (triangle axiom).

Clearly the metric group **MG** is a metric space with a metric function $\rho(x - y)$, $\rho(x)$ means the distance between x and θ and (by analogy with the length of a central vector in the three-dimensional space) $\rho(x)$ is called the *length* (or a *length function*) of the elements $x \in$ **G**. The set **G** with the length function $\rho(x)$ is denoted by **MG** and called a *metric group*.

In a metric group it is possible to consider functions of a real and complex argument. Let functions $f(z)$ and $g(z)$ be defined in a domain D of the complex plane (or in particular on the real axis **R**), let z_0 be a limit point of the domain. Let $f(D)$ and $g(D)$ be subsets of **MG**.

Definition 2.31. We say that the function $f(z)$ is of *higher order of smallness* than the function $g(z)$ (or, which is the same, $f(z)$ is *o-small* of $g(z)$) for $z \to a, z \in D$ which is written

$$f(z) = o(g(z)) \text{ for } z \to a, z \in D \qquad (2.16)$$

if for *any* (fixed) number $\varepsilon > 0$ the following relation holds:

$$\rho(f(z)) \le \varepsilon \rho(g(z)) \text{ for } |z - a| \ll 1, z \in D. \qquad (2.17)$$

Definition 2.32. We say that two functions $f(z)$ and $g(z)$ are *equivalent* for $z \to a, z \in D$ which is written

$$f(z) \sim g(z) \text{ for } z \to a, z \in D, \qquad (2.18)$$

if

$$f(z) = g(z) + o(g(z)) \text{ for } z \to a, z \in D. \qquad (2.19)$$

Definition 2.33. We say that the function $f(z)$ is of the *order* of the function $g(z)$ (or which is the same $f(z)$ is *O-large* of $g(z)$) for $z \to a, z \in D$ which is written

$$f(z) = O(g(z)) \text{ for } z \to a, z \in D \qquad (2.20)$$

if for *at least one number* $C > 0$, the following relation holds:

$$\rho(f(z)) \leq C\rho(g(z)) \quad \text{for } |z - a| \ll 1, z \in D. \qquad (2.21)$$

8. LINEAR SYSTEMS

First, we consider an additive abelian group where it is defined an operation of multiplication by a number.

Definition 2.34. We say that a set N of complex (real) numbers is an *number field* if

(1) N is an additive abelian group with the operation of addition (0 is its zero element);

(2) $N \backslash 0$ is a multiplicative abelian group (1 is its unite element).

Note that the set N is closed over the operations of addition and multiplication.

Definition 2.35. A set $\mathbf{E}$ is said to be a *linear system over a field of complex (or real) numbers,* or, simply *linear system* if the following properties (axioms) are fulfilled:

(1) $\mathbf{E}$ is an additive abelian group,

(2) there is defined a commutative operation of multiplication between the elements of $\mathbf{E}$ and elements of a number field N such that

(1) for any $a, b \in \mathbf{E}$ and $\lambda \in N$, $\lambda a \in \mathbf{E}$;

(2) for any $a \in \mathbf{E}$, and $\lambda, \mu \in N$,

(i) $\lambda(\mu a) = (\lambda\mu)a$;

(ii) $(\lambda + \mu)a = \lambda a + \mu a$ and $\lambda(a + b) = \lambda a + \lambda b$;

(iii) $1a = a$;

(iv) $0a = \theta$;

(v) if $\lambda a = \theta$ and $a \neq \theta$, then $\lambda = 0$.

Definition 2.36. Let $\mathbf{ME}$ be a metric group and let (in addition) $\mathbf{E}$ be a linear system (determined in Definition 2.35). Then $\mathbf{ME}$ is called a *linear metric system.*

In a linear metric system $\mathbf{ME}$ it is possible to define the notions of the derivative, a primitive function, and of the definite integral. We give the definitions for functions of a real argument. In the same way, they can be defined for functions of a complex argument. Let $t, t + \Delta t$ belong to a segment J, and let $x(t)$ be a function such that $x(J) \subset \mathbf{ME}$.

Definition 2.37. If the limit

$$x'(t) = \lim_{\Delta t \to 0} \frac{x(t + \Delta t) - x(t)}{\Delta t}$$

exists it is named the *derivative* of the function $x(t)$. We use an alternate designations of the derivative y', $dx(t)/dt$ or $\frac{dx(t)}{dt}$ instead of $x'(t)$.

If the function has the derivative at the point t (in a region R) we say that the function is *differentiable* at the point t (in R).

It is easy to obtain the following properties of the operation of differentiation: If $x'(t)$ and $y'(t)$ exist then

(1) $[x(t) + y(t)]' = x'(t) + y'(t)$,

(2) $[\lambda x(t)]' = \lambda x'(t)$ for any number λ.

Let $I(t)$ be a function in **ME** such that $I(J) \subset$ **ME**.

Definition 2.38. The function $I(t)$ is said to be an *antiderivative* (or *primitive*) function of a function $f(t) \in ME$ if $I'(t) = f(t)$.

The definite integral is defined by the following procedure. Consider a *partition* of the segment $J = [\alpha, \beta]$ by points $\alpha = t_0 < t_1 < \ldots < t_m < \ldots < t_n = \beta$, which we denote by $P = [t_0, t_1, \ldots, t_n]$; the value $\tau = \max_{j=1,\ldots,n}(t_{j+1} - t_j)$, is called the *diameter* of the partition, we also call P the $\tau - partition$. A partition $P' = [t'_0, t'_1, \ldots, t'_n]$ is said to be a *subpartition* of the partition P if $\{t_0, t_1, \ldots, t_n\} \subset \{t'_0, t'_1, \ldots, t'_n\}$ and we write $P' \subset P$ or $P \supset P'$. It means that P' contain all the points of the partition P, and may be there are in addition several points in P' which does not belong to P. So that any segment $[t'_s, t'_{s+1}]$ of P' belongs to a segment of P. Clearly diameter τ' of the subpartition is $\leq \tau$. A sum of the form

$$S(P, f(t)) = \sum_{j=1}^{n} f(t_j)(t_{j+1} - t_j) \qquad (2.22)$$

is said to be the *integral sum* corresponding to the partition P. Consider a sequence of partitions $\sigma = \{P_1 \supset P_2 \supset \ldots \supset P_n \ldots\}$ and a sequence of their integral sums $\{S(P_n, f(t)\}$. Let us denote the limit of the last sum (if it exists) as S_σ.

Definition 2.39. If any limit of type S_σ exists and the same for any system of partitions then it is denoted by symbols

$$\int_{\alpha}^{\beta} f(t)dt \text{ or } \int_{J} f(t)dt \qquad (2.23)$$

and called the *definite integral* of the function $f(t)$ taken over the interval $J = [\alpha, \beta]$. The numbers α and β are called the *lower limit* and *upper limit* of integration, respectively, $f(t)$ is called the *integrand*, and $f(t)dt$ is the *element of integration*.

9.   NORMED SPACES

An important particular case of a linear metric system is a linear normed space. In such spaces, we denote the length function by $\|x\|$ which is called the *norm* of x.

Definition 2.40. Linear metric system **ME** is called a *linear normed space* if for any $x \in ME$ and any complex (real) number $\lambda \in N$, the following inequality holds: $\|\lambda x\| \equiv |\lambda| \|x\|$.

So that a linear normed space satisfy the following properties: Let $E = \{x\}$ be a linear system. Let, for any x, it be assigned a non-negative number $\|x\|$ satisfying the following conditions (axioms):

(1) $\|x\| = 0$ if and only if $x = \theta$;

(2) for any $x, y \in E$, $\|x + y\| \leq \|x\| + \|z\|$ (triangle axiom);

(3) for any complex (real) number λ, $\|\lambda x\| = |\lambda| \|x\|$.

A complete normed space is also called a *Banach space,* or a *space of type B.*

Let us mark. If $\lambda_m, \lambda \in N$, $x_m, x \in E$, and $\lambda_m \to \lambda$, $x_m \to x$ for $m \to \infty$, then $\lambda_m x_m \to \lambda x$. Indeed,

$$\|\lambda x - \lambda_m x_m\| = \|(\lambda x - \lambda_m x) + (\lambda_m x - \lambda_m x_m)\|$$

$$\leq \|\lambda x - \lambda_m x\| + \|\lambda_m x - \lambda_m x_m\|$$

$$= |\lambda - \lambda_m| \|x\| + |\lambda_m| \|x - x_m\| \to 0 \text{ for } m \to \infty,$$

which leads to the required property.

Theorem 2.41. *Let B be a complete linear normed space. Let $f(t)$ be a function such that $f(J) \subset B$, where $J = [\alpha, \beta]$ is a finite segment. Let $f(t)$ be continuous on J. Then the definite integral (2.23) exists.*

PROOF. Since $f(t)$ is continuous, it is uniformly continuous on J. This means that for any ε there is a positive number τ such that $\|f(t + \Delta t)) - f(t)\| \leq \varepsilon$ if $t, t + \Delta t \in J$ and $|\Delta t| < \tau$. So that there exists a positive function (in τ) $\varepsilon_\tau \to 0$ for $0 < \tau \to 0$ such that $\|f(t + \Delta t)) - f(t)\| \leq \varepsilon_\tau$ if $|\Delta t| < \tau$.

To prove this Theorem, we have to prove several lemmata.

Lemma 2.42. *If a partition Q is a subpartition of the τ-partition P (of the segment J) then (see (2.22))*

$$\|S(P, f(t)) - S(Q, f(t))\| \leq \varepsilon_\tau(\beta - \alpha)$$

where ε_τ is determined above.

PROOF. Let $P = [t_0, t_1, ..., t_n]$ and let the partition Q_1 consist of all the points of the partition P and of the points $t_{s_1}, t_{s_2}, ..., t_{s_m}$, ordered on the segment $[t_s, t_{s+1}]$. Let for definiteness $t_s < t_{s_1} < ... < t_{s_m} < t_{s+1}$. Then

$$
\begin{aligned}
S(P, f(t)) - S(Q_1, f(t)) &= f(t_s)(t_{s+1} - t_s) - f(t_s)(t_{t_{s_1}} - t_s) \\
&+ f(t_{s_1})(t_{s_2} - t_{s_1}) + ... + f(t_{s_m})(t_{s+1} - t_{s_m}) \\
&= [f(t_{s_1}) - f(t_s)](t_{s_2} - t_{s_1}) + ... \\
&+ [f(t_{s_m}) - f(t_s)](t_{s+1} - t_{s_m}).
\end{aligned}
$$

Consequently

$$S\|(P, f(t)) - S(Q_1, f(t))\| \leq \varepsilon_\tau[(t_{s_2} - t_{s_1}) + (t_{s_3} - t_{s_2}) + ... + (t_{s+1} - t_{s_m})]$$

$$\leq \varepsilon_\tau(t_{s+1} - t_s).$$

Clearly for an arbitrary subpartition Q we have

$$S\|(P, f(t)) - S(Q, f(t))\| \leq \varepsilon_\tau[(t_1 - t_0) + (t_2 - t_1) + ... + (t_n - t_{n-1})]$$

$$= \varepsilon_\tau(\beta - \alpha). \qquad \square$$

Lemma 2.43. *Let P_τ and $P_{\tau'}$ be an arbitrary $\tau-$ and $\tau'-$partitions of the segment $J = [\alpha, \beta]$ respectively. Then*

$$\|S(P_\tau, f(t)) - S(P_{\tau'}, f(t))\| \leq (\varepsilon_\tau + \varepsilon_{\tau'})(\beta - \alpha).$$

PROOF. Indeed let P' be the partition which contains the union of the points of the partitions P_τ and $P_{\tau'}$. Hence it is a subpartition of P_τ and $P_{\tau'}$ simultaneously. Due to Lemma 2.42 we have

$$\|S(P_\tau, f(t)) - S(P', f(t))\| \leq \varepsilon_\tau(\beta - \alpha),$$

$$\|S(P_{\tau'}, f(t)) - S(P', f(t))\| \leq \varepsilon_{\tau'}(\beta - \alpha).$$

Consequently

$$
\begin{aligned}
\|S(P_\tau, f(t)) - S(P_{\tau'}, f(t))\| &\leq \|S(P_\tau, f(t)) - S(P', f(t))\| \\
&+ \|S(P_{\tau'}, f(t)) - S(P', f(t))\| \\
&\leq (\varepsilon_\tau + \varepsilon_{\tau'})(\beta - \alpha). \qquad \square
\end{aligned}
$$

Continuation of the proof of Theorem 2.41. Consider a sequence of τ_n-partitions $Q = \{P_{\tau_n}\}$ of the segment $J = [\alpha, \beta]$ such that $\tau_n \to 0$ for $n \to \infty$. For the corresponding integral sums $S(P_{\tau_n}, f(t))$, owing to Lemma 2.42, we have, for any natural m and p,

$$\|S(P_{\tau_n}, f(t)) - S(P_{\tau_{n+p}}, f(t))\| \le (\varepsilon_{\tau_n} + \varepsilon_{\tau_{n+p}})(b - a) \to 0 \text{ for } m \to \infty.$$

Consequently, the sequence is fundamental. Since the space is complete, there exists a limit $S(Q)$ of the sequence for $m \to \infty$. Now, we have to prove the limits of any such sequences are the same. Indeed, consider another such sequence $Q' = \{P_{\tau'_n}\}$ of partitions where $\tau'_n \to 0$ for $n \to \infty$. Let its limit is equal to $S(Q')$. Unite the sequences such that the new sequence is in the form $\tilde{Q} = \{P_{\tau_1}, P_{\tau'_1}, P_{\tau_2}, P_{\tau'_2}, ..., P_{\tau_n}, P_{\tau'_n}, ...\} \equiv \{P_{\tau''_1}, P_{\tau''_2}, .., P_{\tau''_n}, ...\}$. Clearly, their common value τ''_n is equal to $\tau_{(n+1)/2}$ if n is odd, and $\tau''_n = \tau'_{n/2+1}$ if n is even. Hence $\tau''_n \to 0$ for $n \to \infty$, and the corresponding integral sums of the sequence have a limit which must be equal to $S(Q)$ and $S(Q')$ simultaneously. Consequently, $S(Q) = S(Q')$. $\qquad\square$

10. BASIC PROPERTIES OF DEFINITE INTEGRALS

Here we consider functions $f(t)$ and $g(t)$ in a complete linear normed space **B** and suppose that they are continuous on $J = [\alpha, b]$.

Proposition 2.44.

(1)
$$\int_\alpha^\beta [f(t) + g(t)]dt = \int_\alpha^\beta f(t)dt + \int_\alpha^\beta g(t)dt.$$

(2) *For any number* λ

$$\int_\alpha^\beta \lambda f(t)dt = \lambda \int_\alpha^\beta f(t)dt.$$

(3) *For* $C = $ const, $C \in$ **B**,

$$\int_\alpha^\beta C dt = C(\beta - \alpha).$$

These properties are obvious.

Proposition 2.45.

$$\left\| \int_\alpha^\beta f(t)dt \right\| \le \int_\alpha^\beta \|f(t)\|dt. \tag{2.24}$$

PROOF.   The function $\|f(t)\|$ is also continuous on $[\alpha, \beta]$, and for a τ−partition $P = [t_0, t_1, ..., t_n]$, we have

$$\|S(P, f(t))\| = \left\|\sum_{j=0}^{n-1} f(t_j)(t_{j+1} - t_j)\right\|$$

$$\leq \sum_{j=0}^{n-1} \|f(t_j)\|(t_{j+1} - t_j)$$

$$= S(P, \|f(t)\|). \tag{2.25}$$

For $\tau \to 0$, the integral sums tend, respectively, to the integrals

$$\int_\alpha^\beta f(t)dt \text{ and } \int_\alpha^\beta \|f(t)\|dt,$$

and inequality (2.24) directly follows from (2.25).   □

Proposition 2.46. *Let $c \in [\alpha, \beta]$, then*

$$\int_\alpha^\beta f(s)ds = \int_\alpha^c f(t)dt + \int_c^\beta f(t)dt. \tag{2.26}$$

PROOF.   Choose a sequence of τ_n−partitions $S(P_n, f(t))$ on the segment $[\alpha, \beta]$ such that any partition P_n contains the point c. Then $P_n = P_n' + P_n''$ where P_n' and P_n'' are τ_n−partitions of the segments $[\alpha, c]$ and $[c, \beta]$ respectively. Consequently

$$S(P_n, f(t)) = S(P_n', f(t)) + S(P_n'', f(t)).$$

Clearly

$$\lim_{n\to\infty} S(P_n, f(t)) = \int_\alpha^\beta f(t)dt,$$

$$\lim_{n\to\infty} S(P_n', f(t)) = \int_\alpha^c f(t)dt, \text{ and } \lim_{n\to\infty} S(P_n'', f(t)) = \int_c^\beta f(t)dt,$$

which leads to (2.27).   □

Proposition 2.47. *Let*

$$I(t) = \int_\alpha^t f(s)ds, \text{ then } I'(t) = f(t).$$

PROOF. Indeed

$$\Delta I(t) \equiv I(t + \Delta t) - I(t) = \int_t^{t+\Delta t} f(s)ds.$$

Hence

$$\frac{\Delta I(t)}{\Delta t} - f(t) = \frac{1}{\Delta t}\int_t^{t+\Delta t} [f(s) - f(t)]ds.$$

Since $f(t)$ is continuous, $\lim_{\Delta t \to 0} \sup_{s \in [t, t+\Delta t]} \|f(s) - f(t)\| = 0$. Consequently,

$$\left\|\frac{1}{\Delta t}\int_t^{t+\Delta t} [f(s) - f(t)]ds\right\| \le \sup_{s \in [t, t+\Delta t]} \|f(s) - f(t)\| \to 0$$

for $\Delta t \to 0$ which leads to the required relation. $\square$

Proposition 2.48 (The Newton–Leibnitz formula). *Let $I'(t) = f(t)$ at any point $t \in J$. Then*

$$\int_\alpha^\beta f(t)dt = I(\beta) - I(\alpha). \qquad (2.27)$$

PROOF. Since the function $f(t)$ is continuous, integral (2.27) exists, and the function is uniformly continuous on J.

Let P be a $\tau-$partition of the segment J, and

$$S(P, f(t)) = f(\alpha)(t_1 - \alpha) + f(t_1)(t_2 - t_1) + ... + f(t_{n-1}(\beta - t_{n-1})$$

be the corresponding integral sum. We have

$$\|I(t_{m+1}) - I(t_m) - f(t_m)(t_{m+1} - t_m)\| \le \varepsilon_m(t_{m+1} - t_m),$$

where $(m = 0, 1, ..., n - 1,\ t_0 = \alpha)$. Clearly, $\varepsilon = \max_j \varepsilon_j \to 0$ for $\tau \to 0$. Consequently,

$$\begin{aligned}\|S(P, f(t)) &- [I(t_1) - I(\alpha)] - [I(t_2) - I(t_1)] - ... - [I(\beta) - I(t_{n-1})]\|\\ &\le \varepsilon_0(t_1 - t_0) + \varepsilon_1(t_2 - t_1) + ... + \varepsilon_{n-1}(\beta - t_{n-1})\\ &\le \varepsilon_\tau(\beta - a) \to 0.\end{aligned}$$

That is, $\|S(P, f(t)) - I(\beta) + I(\alpha)\| \to 0$ for $\tau \to 0$, which leads to the required property. $\square$

Proposition 2.49. *Let $I_1(t)$ and $I_2(t)$ be two primitive functions of $f(t)$ over the interval $J = [\alpha, \beta]$. Then there is a constant $C \in \mathbf{B}$ such that $I_1(t) - I_2(t) = C$ at any point $t \in J$.*

PROOF. By Proposition 2.48 we have $I_1(t) - I_1(\alpha) = I_2(t) - I_2(\alpha)$ that is, $I_1(t) - I_2(t) = C$, where $C = I_1(\alpha) - I_2(\beta)$ is independent of t.   □

The Propositions 2.44 (3) and 2.48 permit us to introduce the concept of indefinite integral.

Definition 2.50. The collection of all primitives of the function $f(t)$ (on J) is called the *indefinite integral* of the function $f(t)$ and denoted by the symbol $\int f(t)dt$. If $F'(t) = f(t)$ we write the set of all primitives in the form $F(t) + C$ where C is called an *arbitrary constant* or a *constant of integration*. Thus

$$\int f(t)dt = F(t) + C. \qquad (2.28)$$

We also say that the indefinite integral is an expression of the form (2.28).

11.   SOLUTION OF DIFFERENTIAL EQUATIONS

Consider a differential equation of the form

$$\frac{dx}{dt} = f(t, x), \qquad (2.29)$$

where $x = x(t)$ and $f(t, x)$ are elements of a complete linear normed space B for any $t \in J = [\alpha, \beta]$; α, β are numbers.

Let $f(t, x)$, as a function of x, satisfy the *Lipschitz condition*

$$\|f(t, x) - f(t, y)\| \leq p\|x - y\| \qquad (2.30)$$

for any $x, y \in B$ and $t \in J$, where p is a positive number called a *Lipschitz constant*.

Let $t_0 \in J$ and $t_0 < \beta$. Let us take a number $\delta > 0$ and $\delta < \max[\beta - t_0, 1/p]$, and consider equation (2.29) in the space $C^B(t_0, t_0 + \delta)$ of all continuous abstract functions $x(t) \in B$ for $t \in [t_0, t_0 + \delta]$. Introduce a norm in the space by

$$\|x\|_\delta = \max_{t \in [t_0, t_0+\delta]} \|x(t)\| \qquad (2.31)$$

normed space.

First, consider an integral equation of the form

$$x(t) = x_0 + \int_{t_0}^{t} f(\tau, x(\tau))d\tau, \qquad (2.32)$$

where $t_0 \leq t \leq t_0 + \delta$. and $x_0 \in B$. Designate the right side of (2.32) by $A(x)$. $A(x)$ is an operator which transforms $x = x(t)$ from $C^B(t_0, t_0 + \delta)$ to an element of the same space. We have, due to (2.32),

$$\|A(x) - A(y)\|_\delta = \left\| \int_{t_0}^t \|f(x(\tau)) - f(y(\tau))\| d\tau \right\|_C$$

$$\leq \int_{t_0}^{t_0+\delta} \|f(t, x(t)) - f(t, y(t))\| dt.$$

$$\leq p \int_a^{t_0+\delta} \|x(t) - y(t)\| dt$$

$$\leq p\delta \max_{t_0 \leq t \leq t_0+\delta} \|x(t) - y(t)\|$$

$$= q\|x(t) - y(t)\|_\delta,$$

where $q = p\delta < 1$. Then on the basis of the principle of contractive mappings, we may conclude that there is a solution $x(t) \in C^B(t_0, t_0 + \delta)$ such that $x(t_0) = x_0$, and the solution is unique.

Equation (2.29) with initial condition $x(t_0) = x_0$ is equivalent to equation (2.32) in the following sense: Any such solution of (2.29) is a solution of equation (2.32), and vice versa. Consequently, equation (2.29) possesses a unique solution $x(t) \in C^B(t_0, t_0+\delta)$ such that $x(t_0) = x_0$. Denote $x(t_0+\delta) = x_1$. Reasoning in the same way as in the case $x(t_0) = x_0$, we obtain a unique solution $x(t)$ of equation (2.29) on the segment $t_0 + \delta, t_0 + 2\delta$ with the new initial condition. Clearly, we obtain a unique solution $x(t)$ of the equation on the segment $[t_0, t_0+2\delta]$, and so we obtain a unique solution $x(t)$ satisfying the condition $x(t_0) = x_0$ on the segment $[t_0, b]$. In the same way, (if $a < t_0$) considering the equation on the interval $[a, t_0]$, it is easy to continue this solution also on the segment $[a, t_0]$. Thus, we proved the following:

Theorem 2.51. *Let $f(t, x) \in C^E(\alpha, \beta)$ for any $x \in B$ and $t \in J$, and let it satisfy the Lipschitz condition (2.30). Then for any initial condition $x(t_0) = x_0$ ($x_0 \in B$, and $t_0 \in J$) there exists a solution $x(t) \in C^E(\alpha, \beta)$ (such that $x(t_0) = x_0$), and this solution is unique.*

Example 2.52. Consider a linear equation of the form

$$\frac{dx}{dt} = A(t)x + F(t), \quad x(\alpha) = x_0, \tag{2.33}$$

where $x \in B$ and $F(t) \in B$ for any $t \in J = [\alpha, \beta]$ and continuous as an abstract function on J; for any $t \in J$, $A(t)$ belongs to a complete metric space **E**. For any elements $a \in \mathbf{E}$ and $x \in \mathbf{B}$, there is defined a multiplicative operation $ax \in \mathbf{B}$ such that and $\|ax\| \leq \|a\|\|x\|$. Besides, $A(t)$ belongs to $\mathbf{E}$ and continuous for any $t \in J$. Let us denote the right side of (2.33) by $A(t, x)$, and let $\|A(t)\|_C < q$. Then

$$\|A(t, x) - A(t, y)\| \leq \|A(t)\|_C \|x - y\| \leq q\|x - y\|.$$

And hence equation (2.33) satisfies all the condition of Theorem 2.51. Consequently, for any fixed $x_0 \in \mathbf{B}$ and $t_0 \in J$, there exists a unique solution $x(t) \in C^B(\alpha, \beta)$ satisfying the condition $x(t_0) = x_0$.

Remark 2.53. Instead of (2.31), any other equivalent norm may be taken. We say that the norms $\|x\|_1$ and $\|x\|_2$ are equivalent in the spaces $\mathbf{M}$ if for any $x \in M$, there are positive numbers p_1 and p_2 such that $\|x\|_1 \leq p_1\|x\|_2$ and $\|x\|_2 \leq p_2\|x_1\|$. For example: Let $M = \{x(t)\}$ be a space of all n-ordered complex (real) functions continuous on the interval $J = [\alpha, \beta]$. Introduce the following norms:

$$\|x(t)\|_1 = \sup_{t=J}[|x_1(t)| + ... + |x_n(t)|],$$

$$\|x(t)\|_2 = \sup_{t=J}|x_1(t)| + ... + \sup_{t=J}|x_n(t)|].$$

Clearly, $\|x(t)\|_1 \leq \|x(t)\|_1$. On the other hand,

$$\sup_{t=J}|x_i(t)| \leq \sup_{t=J}[|x_1(t)| + ... + |x_n(t)|].$$

Hence $\|x(t)\|_2 \leq n\|x(t)\|_2$. That is, the norms are equivalent.

Example 2.54. Consider an important particular case of equation (2.33). Let $A(t) = (a_{ij}(t))_n$ be a square matrix of n-order; $a_{ij}(t)$ are continuous (scalar) functions on J, that is, $a_{ij}(t)$ are complex (real) functions on J. $F(t) = (f_1(t), ..., f_n(t))^T$ is column-matrix, where $f_j(t)$ are scalar continuous functions on J $(j = 1, 2, ..., n)$. Let $x = (x_1, ..., x_n)^T$ be n-dimensional column-matrix where the coordinates are complex (real) numbers. Let B be a space consisting of column-matrices of the form $x = (x_1, ..., x_n)^T$ and $\|x\| = |x_1| + ... + |x_n|$. We set $C^B(\alpha, \beta)$ to be a space of any continuous (column-matrices) functions $x(t) = (x_1(t), ..., x_n(t))^T$ in J, and

$$\|x(t)\|_C = \max_{t \in J}|x_1(t)| + ... + \max_{t \in J}|x_n(t)|.$$

Clearly, $C^B(\alpha, \beta)$ is a complete linear normed space. B is a space of all square complex (real) numerical matrices. For any $A = (a_{ij})_n$

$$\|A\| = \sum_{i,j=1}^{n}|a_{ij}| \quad \text{and} \quad \|A(t)\|_C = \sum_{i,j=1}^{n}\max_{t \in J}|a_{ij}(t)|.$$

Then for any (fixed) column-matrix $(x_{10}, ..., x_{n0})$ and $t_0 \in J$, there exists a unique solution $x(t) = (x_1(t), ..., x_n(t))^T \in C^B(\alpha, \beta)$ such that $x(t_0) = (x_1, ..., x_n)^T$.

Counterexample 2.55. Given the equation

$$\frac{dx}{dt} = x^2$$

in the space $C(-\infty, +\infty)$ of all real continuous functions. Denote $A(x) = x^2$. We have $|A(x) - A(y)| = |x + y||x - y|$. Since x and y may admit any large value, x^2 does not possess the Lipschitz condition (in the entire space).

Clearly all the solutions to the equation consist of the functions $x = 0$ and $x = -1/(t + a)$ where a is a (real) arbitrary constant. For any (fixed) a the corresponding solution has a pole at the point $t = -a$. But for any fixed t_0 and $\xi \neq 0$ we have a solution $x(t)$ such that $x(t_0) = \xi$ (here $a = -1/\xi - t_0$). It is continuous for any t except $t = -a$. And we may assert that for any t_0 and ξ there is at least one neighborhood of the point t_0 such that the equation has a unique continuous solution satisfying the condition $x(t_0) = \xi$.

Remark 2.56. Theorem 2.51 is valid if the function $f(t, x)$ satisfies the Lipschitz condition in the entire space **B**. But in many important cases property (2.32) is fulfilled only in a bounded region of x, say $\|x\| \leq K$ (K is a positive number). $D = \{(t, x) : t \in J = [\alpha, \beta], \|x\| \leq K\}$. K is a positive it is possible to obtain a unique solution in a (may be sufficiently small) neighborhood of the point t_0 (which is proved in the same way as Theorem 2.51). The solution can be continued on an subinterval of J as far as the solution will satisfy the inequality $\|x(t)\| \leq K$.

Example 2.57. We assume the hypothesis and notation of Example 2.54. Consider a scalar system of differential equations in the form

$$\frac{dx_j}{dt} = f_j(t, x_1, ..., x_n) \ j = 1, 2, ..., n \tag{2.34}$$

or in the equivalent form (2.29).

Let $f(t, x)$ be defined and continuous in J in the domain $D = \{(t, x) : t \in J = [\alpha, \beta], \|x\| \leq K\}$ (K is a positive number). Moreover let $f(t, x)$ satisfy the Lipschitz condition (2.12). Let t_0 be an interior point belonging to J, let $\xi \in B$ and $\|\xi\| < p$. Then in a sufficiently small neighborhood $U_\varepsilon = \{t : |t - t_0| \leq \varepsilon\}$ there is a unique solution $x(t)$ such that $x(t) \in C^B(t_0 - \varepsilon, t_0 + \varepsilon)$ and $x(t_0) = \xi$.

Chapter 3

ASYMPTOTIC SPACES

1. DEFINITION OF AN ASYMPTOTIC SPACE

The theory of asymptotic spaces is used for formal solution of an operator equation of the form $x = A(x)$. Formal solution is the first (and may be very difficult) step of the total asymptotic solution of the equation. Since many different spaces are used for different problems, we consider this notion in an axiomatic form.

Definition 3.1. An additive abelian group E is said to be an *asymptotic group* with its characteristic function $F(x)$, which is denote by E_F, if to any element $\varphi \in E$, there is assigned a definite value $F(\varphi)$, which is either a real number or $-\infty$, such that the following conditions (axioms of an asymptotic group) are fulfilled:

(1) $F(\varphi) = F(-\varphi)$ (property of symmetry);

(2) $F(\varphi + \psi) \leq \max[F(\varphi), F(\psi)]$ for any $\varphi, \psi \in E$ (characteristic property);

(3) $F(\theta) = -\infty$, where θ is the zero of the group.

By Θ we denote the set of all elements $\varphi \in E_F$ such that $F(\varphi) = -\infty$, we also say that φ is an element with *negligible asymptotics*. If for $x, y \in E_F$, the relation $F(x - y) = -\infty$ holds, then we say that x, y are *asymptotically equal in E_F and write $x \asymp y$.*

Clearly, if $x \asymp y$ and $y \asymp z$, then $y \asymp x$ and $x \asymp z$.

Definition 3.2. A space E is called a *commutative ring* if there are given two operations – summation and multiplication: for any $x, y, z \in E$,

1. Summation: E is an additive abelian group;

2. Multiplication:

(i) $\qquad\qquad\qquad xy = yx$ (commutative law);

(ii) $\qquad\qquad\qquad x(yz) = (xy)z$ (associative law);

38

3. The operations of summation and multiplication satisfy the following relation:

$$x(y + z) = xy + xz \quad \text{(distributive law)}.$$

Owing to the associative law of multiplication, we may write xyz instead of both sides of (ii).

A commutative ring may have a unit element, which we denote by the symbol I. That is, $Ix = x$ for any $x \in E$. A commutative ring may have only one unit element. Indeed, if there is another unit element I^*, then $I^*I = I$, and in the same way $I^*I = II^* = I^*$, hence $I^* = I$. We have $x\theta = \theta$. Indeed, $x\theta = x(x - x) = xx - xx = \theta$.

Definition 3.3. Let E be a commutative ring. We say that E_F is an *asymptotic ring* if it is an asymptotic group, and for any $x, y \in E$,

$$F(xy) \le F(x) + F(y).$$

The common name of an asymptotic group and asymptotic ring is an *asymptotic space* (or simply, *A-space*). The function $F(x)$ is said to be the *characteristic function* or simply *characteristics* of the space.

2.   EXAMPLES OF ASYMPTOTIC SPACES

The trivial asymptotic group consists of one element θ (which is the zero of the group), and, clearly, $F(\theta) = -\infty$.

Example 3.4. 1° Let E be the space consisting of all formal Laurent series of the form

$$x = a_n z^n + a_{n-1} z^{n-1} + \ldots + a_j z^j + \ldots, \tag{3.1}$$

where n is an arbitrary integer, a_j are complex numbers (the set $\{j\}$ contains infinitely many integers, $j = n, n-1, \ldots$). Let a_k be the first coefficient which does not equal to zero (that is, $a_n = a_{n-1} = \ldots = a_{k+1} = 0$, and $a_k \ne 0$). We set $F(x) = k$, and k is said to be the *degree* of the sequence. We denote an element with all zero coefficients by θ and set $F(\theta) = -\infty$.

We define (as usual) the operations of addition and multiplication. Let $y = b_n z^n + b_{n-1} z^{n-1} + \ldots + b_j z^j + \ldots$, then

$$x + y = (a_n + b_n) z^n + \ldots + (a_j + b_j) z^j + \ldots$$

and

$$xy = c_{2n} z^{2n} + c_{2n-1} z^{2n-1} + \ldots + c_j z^j + \ldots$$

where $c_{2n} = a_n b_n$, $c_{2n-1} = a_{2n} b_{2n-1} + a_{2n-1} b_{2n}$, and so on. Clearly, the considered set $\{x\}$ with characteristics $F(x)$ is an asymptotic ring.

2° We denote by RF the set of all rational functions of the form $u = P(z)/Q(z)$, where $P(z)$ and $Q(z)$ are polynomials in z with complex coefficients and $Q(z) \not\equiv 0$. Its characteristic function is equal to $F(u) = p - q$, where p and q are the degrees of $P(z)$ and $Q(z)$, respectively. Clearly, RF is an asymptotic ring.

3.   SPACES OF TYPE H_D

Definition 3.5. Let D be a region of the complex plane such that $z = \infty$ is its boundary point (we suppose that $z = \infty \neq D$). Let $f(z)$ be defined in D. Consider a limit of the form

$$k = \varlimsup_{z \to \infty, z \in D} \frac{\ln |f(z)|}{\ln |z|}. \tag{3.2}$$

Here $\varlimsup$ means an upper limit. Limit (3.2) can be rewritten in the form

$$k = \lim_{T \to +\infty} \sup_{|z| \geq T, z \in D} \frac{\ln |f(z)|}{\ln |z|}.$$

So that either k is a real number, or $k = -\infty$, $k = +\infty$. We denote by H_D a space of all complex functions such that if $f(z) \in H_D$, then $f(z)$ is defined for $|z| \gg 1, z \in D$, and $k < +\infty$. The value k is called a *simple estimate of power growth* (or simply a *simple estimate*). The relation

$$h_D\{f(z)\} = k < +\infty \tag{3.3}$$

means that $f(z) \in H_D$ and its simple estimate is equal to k. If $k = -\infty$, we say that $f(z)$ has a *negligible small simple asymptotics* in D and also write $f(z) = O_D(z^{-\infty})$. If $D = \{t : t \gg 1\}$ (t is a real argument), then we write H, t, $h\{f(t)\} = k$, and $O(t^{-\infty})$ instead of H_D, z, $h_D\{f(z)\} = k$, and $O_D(z^{-\infty})$, respectively.

Proposition 3.6. *Let* $h_D\{f(z)\} = k$ ($k < +\infty$). *Then, for any (fixed) number* $\varepsilon > 0$, *there exists a number* T_ε *such that*

$$|f(z)| < |z|^{k+\varepsilon} \text{ for any } |z| \geq T_\varepsilon, z \in D \tag{3.4}$$

Besides, for any number $T > 0$ *(no matter how large), there exists at least one number* $|z| > T, z \in D$ *such that*

$$|f(z)| > |z|^{k-\varepsilon}. \tag{3.5}$$

Proposition 3.6 easily follows from relation (3.2).

Proposition 3.7. *The function* $f(z)$ *possesses the estimate* $h_D\{f(z)\} = k < +\infty$ *if and only if there exists at least one real number* r *such that*

$f(z) = o(z^r)$ *for* $z \to \infty, z \in D$, *and* $k = \inf\{r\}$ *for any* r *satisfying the indicated property.*

PROOF. We distinguish the following two cases: $k > -\infty$ and $k = -\infty$. We prove the case $k > -\infty$ (the case $k = -\infty$ is also easily proved). Let relation (3.2) hold. Then $f(z) = o(z^{k+\varepsilon})$ as $z \to \infty, z \in D$ for any number $\varepsilon > 0$. Hence a number r (which satisfy the required property) exists. Moreover (taking into account the arbitrariness of ε), we conclude that $\inf\{r\} \leq k$. Let $\inf\{r\} < k$. Then there exists a number $\delta > 0$ such that $f(z) = o(z^{k-\delta})$ for $z \to \infty, z \in D$. Then the limit (in (3.2)) is less than k. This contradiction proves the necessity of the required property.

Let $\inf\{r\} = k$ then obviously in (3.2) the limit $\leq k$. Hence there exists a number $\delta > 0$ such that $f(z) = o(z^{k-\delta})$ for $z \to \infty, z \in D$. This leads to the inequality $\inf\{r\} < k$ which is impossible. The obtained contradiction proves this proposition.   □

Let us give some properties of simple estimates which easily follow from Definition 3.5.

Proposition 3.8. *Let* $h_D\{f(z)\} = p$ *and* $h_D\{g(z)\} = q$. *Then*
(1) $h_D\{[f(z)]^\sigma\} = \sigma h_D\{f(z)\}$, *where* σ *is a real non-negative number;*
(2) $h_D\{f(z)+g(z)\} \leq \max\{p,q\}$. *If in addition there exists a finite limit*

$$\lim_{z \to \infty, z \in D} \frac{g(z)}{f(z)} \neq -1$$

then $h_D\{f(z) + g(z)\} = \max\{p,q\}$.
(3) $h_D\{f(z)g(z)\} \leq h_D\{f(z)\} + h_D\{g(z)\}$.
So that H_D *is an asymptotic ring.*

Example 3.9. Consider an algebraic equation of the form

$$y^n = a_1(z)y^{n-1} + a_2(z)y^{n-2} + \ldots + a_n(z), \qquad (3.6)$$

where $a_j(z) \in H_D$ $(j = 1,2,...,n)$. Then any solution (say $\lambda(z)$) of the equation belongs to H_D. Moreover

$$h_D\{\lambda(z)\} \leq \max_{j=1,...,n} \left[\frac{1}{j} h_D\{a_j(z)\} \right]. \qquad (3.7)$$

PROOF. Prove that $k < +\infty$ in (3.2). Suppose the contrary. Then there exists a set of numbers (say $z_1, z_2, ..., z_m, ... \in D$, and $z_m \to \infty$ for $m \to \infty$) such that $|\lambda(z_m)| > |z_m|^N$, where N is a sufficiently large positive number. Take a positive number $\varepsilon \ll 1$ such that $|a_j(z_m)| < |z_m|^{k_j+\varepsilon}$, $k_j = h_D\{a_j(z)\}$ $(j = 1,2,...,n)$. Hence

$$|\lambda(z_m)|^n < |z_m|^{\max_{j=1,...,n}[k_j+\varepsilon+(n-j)N]}.$$

That is,

$$|\lambda(z_m)z_m^{-N}|^n < |z_m|^{\max_{j=1,\ldots,n}[k_j+\varepsilon-jN]},$$

which leads to a contradiction because $|\lambda(z_m)z_m^{-N}|^n \geq 1$ and $\max_{j=1,\ldots,n}$ $[k_j + (n+1)\varepsilon - jN] \to -\infty$ for $N \to +\infty$. Thus, the required property is proved. We have (see (3.6))

$$h_D\{\lambda^n(z)\} \leq \max_{j=1,\ldots,n}[h_D\{a_j(z)\lambda^{n-j}(z)\}].$$

Clearly, $h_D\{\lambda^n(z)\} = nh\{\lambda(z)\}$, and $h_D\{a_j(z)\lambda^{n-j}(z)\} \leq h_D\{a_j(z)\} + (n-j)h_D\{\lambda(z)\}$. Consequently,

$$\max_{j=1,\ldots,n}[h_D\{a_j(z)\} - jh_D\{\lambda(z)\}] \geq 0.$$

Let the maximum is attended for $j = j_0 \in \{1, 2, .., n\}$. Then

$$h_D\{a_{j_0}(z)\} - j_0 h_D\{\lambda(z)\} \geq 0.$$

That is,

$$h_D\{\lambda(z)\} \leq \frac{1}{j_0}h_D\{a_{j_0}(z)\} \leq \max_{j=1,\ldots,n}\left[\frac{1}{j}h_D\{a_j(z)\}\right]. \qquad \square$$

In some cases it is possible to estimate a (convergent) integral. Consider an integral of the form

$$\gamma(t) = \int_{t_0}^t f(s)ds. \qquad (3.8)$$

Proposition 3.10.   *Let integral (3.8) be convergent for any* $t \in (t_0, +\infty)$. *Let* $h\{f(t)\} = k$. *Here* $t_o = +\infty$ *if* $k < -1$, *and* t_o *is a sufficiently large positive number if* $k \geq -1$. *Then* $h\{\gamma(t)\} \leq k + 1$.

PROOF. The case $k = -\infty$ is obvious. Let $-1 \leq k < +\infty$ (t_o is a finite point). We have $f(t) = o(t^{k+\varepsilon})$ as $t \to +\infty$ for any positive number ε. Hence $\gamma(t) = C + \varphi(t)$, where $\varphi(t) = O(t^{k+1+\varepsilon})$. Therefore

$$\sup_{R<|t|<+\infty} \frac{\ln|\gamma(t)|}{\ln t} \leq k + 1 + \varepsilon.$$

Taking into account the arbitrariness of ε, we obtain the relation

$$\overline{\lim}_{t\to+\infty} \frac{\ln|f(t)|}{\ln t} \leq k + 1.$$

Let $k < -1$ $(t_o = \infty)$. We have

$$|\gamma(t)| \leq \int_t^{+\infty} \tau^{k+\varepsilon} d\tau = O(z^{k+1+\varepsilon})$$

as $t \gg 1$ for any number ε, $0 < \varepsilon < -1 - k$ which leads to the required inequality.

□

We may formulate a similar proposition for functions of the complex argument. First, introduce the following definition:

Definition 3.11. Let Γ be a curve of the complex plane such that $z = \infty$ is its limiting point ($\infty \notin \Gamma$). Let z_1, z_2 be arbitrary points of Γ. We denote the length of $\Gamma(z_1, z_2)$ by $l(\Gamma(z_1, z_2))$. We say that the curve is *normal* if there exists a number $C > 0$ such the Lipschitz condition is fulfilled:

$$l\{\Gamma(z_1, z_2)\} \leq C|z_2 - z_1|. \tag{3.9}$$

The following proposition is a simple consequence of Proposition 3.10. Consider an integral of the form

$$I(z) = \int_{\Gamma(z_0,z)} f(s) ds. \tag{3.10}$$

Proposition 3.12. *Let Γ be a normal curve in the complex plane. Let integral (3.10) converge for any $z \in \Gamma$. Let $h_\Gamma\{f(z)\} = k$. Here $z_0 = \infty$ if $k < -1$, and z_0 is a sufficiently large positive number if $k \geq -1$. Then $h_\Gamma\{I(z)\} \leq k + 1$.*

Some other examples of asymptotic spaces will be given in the subsequent chapters.

4. SERIES AND SEQUENCES IN ASYMPTOTIC SPACES

Consider a series of the form

$$\delta_0 + \delta_1 + ... + \delta_m + ..., \tag{3.11}$$

where all $\delta_m \in E_F$ (E_F is an asymptotic space; $m = 0, 1, ...$). We set

$$s_m = \delta_0 + \delta_1 + ... + \delta_m. \tag{3.12}$$

Definition 3.13.   We say that $s \in E_F$ is an *asymptotic sum* of series (3.11) and an *asymptotic limit* of the sequence $\{s_m\}$ in E_F $(m = 0, 1, ...)$, which is written

$$s \asymp \delta_0 + \delta_1 + ... + \delta_m + ..., \tag{3.13}$$

if $\lim_{m \to \infty} F(s - s_m) = -\infty$.

Definition 3.14. We say that series (3.11) is *asymptotically fundamental* if $\lim_{m \to \infty} F(\delta_m) = -\infty$.

Definition 3.15. We say that E_F is *asymptotically complete* if every asymptotically fundamental sequence in E_F has at least one asymptotic limit.

5.   FORMAL AND ASYMPTOTIC SOLUTIONS

Here we consider an equation of the form

$$y = A(y), \tag{3.14}$$

where $A(y)$ is an operator which operates from the asymptotic space E_F to E_F.

Definition 3.16. We say that an element s is a *formal solution* to equation (3.14) in E_F if $s \in E_F$ and

$$s \asymp A(s) \text{ or, which is the same, } F(s - A(s)) = -\infty. \tag{3.15}$$

The following condition plays analogous role for formal solution of equation (3.14) as the Lipschitz condition for an equation in a metric space. Therefore we named it the *L–condition*.

Definition 3.17. Let D be a region in the space E_F. We say that the operator $A(y)$ possesses the *L–condition* with constant σ (σ is a real number) in D if, for any $y_1, y_2 \in D$, the following inequality holds:

$$F(A(y_1) - A(y_2)) \le F(y_1 - y_2) - \sigma. \tag{3.16}$$

Let $\lambda \in E_F$ and $F(\lambda) > -\infty$. Consider a ball in E_F of the form

$$U_{\lambda h} = \{y : y \in E_F, F(y - \lambda) \le h\}. \tag{3.17}$$

Here h is a real number.

Lemma 3.18. *Let E_F be asymptotically complete. Let $A(U_{\lambda h}) \subset U_{\lambda h}$. Let $A(y)$ possess the L–condition with constant $\sigma > 0$ in $U_{\lambda h}$. Then there* *exists a formal solution $s \in U_{\lambda h}$ to equation (3.14); s is an asymptotic limit of a sequence $\{s_m\}$ such that s_0 is an arbitrary (fixed) point of $U_{\lambda h}$,*

$$s_m = A(s_{m-1}) \ for \ m = 1, 2, \ldots \tag{3.18}$$

and $F(s_m - s_{m-1}) \le F(s_1 - s_0) - (m - 1)\sigma$. Moreover an element $\tilde{s} \in U_{\lambda h}$ is a formal solution to equation (3.14) if and only if $\tilde{s} \asymp s$.

PROOF. Choose an element $s_0 \in U_{\lambda h}$. Clearly all the members s_m belong to U_{lah}. Let us prove the existence of at least one solution s with the desired properties.

If $s_m \asymp A(s_m)$ for at least one m then the element $s = s_m$ is the desired formal solution. Indeed we have $s_{m+1} - s_m = A(s_m) - s_m \asymp 0$. Hence $s_{m+1} \asymp s$ for all $i = 1, 2, \ldots$ Consequently s is an asymptotic limit of the sequence.

Let $F(s_m - A(s_m)) > -\infty$ for any $m = 0, 1, \ldots$, that is, $F(s_m - s_{m-1}) > -\infty$. For $m = 2, \ldots$ we have $s_m - s_{m-1} = A(s_{m-1}) - A(s_{m-2})$. Hence

$$F(s_m - s_{m-1}) = F(A(s_{m-1}) - A(s_{m-2})) \le F(s_{m-1} - s_{m-2}) - \sigma.$$

Consequently $F(s_m - s_{m-1}) \le F(s_1 - s_0) - (m - 1)\sigma \to -\infty$ for $m \to \infty$. Hence $F(s_m - s_{m-1}) \to -\infty$. Since E_F is complete there exists an element $s \in E_F$ such that $F(s - s_m) \to -\infty \ (m \to \infty)$.

Prove that s is the required solution. If $F(s_m - \lambda) \to -\infty$ for $m \to +\infty$ and $F(s_m - \lambda) > -\infty$ for any $m = 1, 2, \ldots$, then $s = \lambda$ is the required solution. Indeed $\lambda \in U_{\lambda h}$ and

$$F(\lambda - A(\lambda)) = F((\lambda - s_m) + (A(s_m) - A(\lambda)))$$

$$\le \max[F(\lambda - s_m), F(s_{m-1} - s) - \sigma] \to -\infty$$

as $m \to \infty$. Since the left side of the relation does depend on m we have $F(\lambda - A(\lambda)) = -\infty$. Consequently λ is the desired solution.

If $F(s - \lambda) > -\infty$ then there is a sufficiently large number m such that

$$F(s - \lambda) = F((s_m - \lambda) + (s - s_m)) \le F(s_m - \lambda) \le h.$$

That is, $s \in U_{\lambda h}$ and clearly

$$F(s - A(s)) = F(s - s_m + A(s_{m-1}) - A(s))$$

$$\le \max[F(s - s_m), F(s_{m-1} - s) - \sigma] \to -\infty$$

$(m \to \infty)$. Hence $s \asymp A(s)$. That is, s is the desired formal solution to equation (3.14).

Let $\tilde{s} \in U_{\lambda h}$ be another formal solution to the equation, and let $F(s-\tilde{s}) > -\infty$. Then $s-\tilde{s} = A(s)-A(\tilde{s})+\theta$, where $\theta \asymp 0$. Hence $F(s-\tilde{s}) \leq F(s-\tilde{s})-\sigma$, which is possible only if $F(s - \tilde{s}) = -\infty$.

Let $\tilde{s}$ be an arbitrary element of E_F such that $\tilde{s} \asymp s$. Then obviously $\tilde{s} \in U_{\lambda h}$, and $F(\tilde{s} - A(\tilde{s})) = F(\tilde{s} - s + A(s) - A(\tilde{s}) + \theta)$, where $\theta \asymp 0$. Hence

$$F(\tilde{s} - A(\tilde{s})) \leq \max[F(s - \tilde{s}), F(A(s) - A(\tilde{s})), F(\theta)]$$

$$= \max[F(s - \tilde{s}), F(\theta)]$$

$$= -\infty.$$

So that $F(\tilde{s} - A(\tilde{s})) = -\infty$. □

Remark 3.19. If the condition of the space completeness is absent in the conditions of Lemma 3.18, then we may only conclude that *the sequence $\{s_m\}$ obtained in 3.18 is fundamental.*

Definition 3.20. An element $g \in E_F$ is said to be an *asymptotic solution* to equation (3.14) if it is a formal solution to the equation, and there exists an exact solution y to the equation such that $g \asymp y$.

6.   FACTOR SPACES

The equation $F(x) = -\infty$ may have many solutions in E_F. Therefore we introduce some more simple spaces which (in definite sense) are equivalent to the corresponding asymptotic spaces. Let E_F be an asymptotic space.

Definition 3.21.   Let x be an arbitrary element of E_F. Let $F(x+\theta) = F(x)$ for any $\theta \asymp 0$. We denote by X the set consisting of all elements $x^* \in E_F$ such that $x^* \asymp x$. We say that X *corresponds to* x and x is a *representative element* to X. We also use the following designation $X = x + \Theta$ where Θ is the set of all functions $\theta \asymp 0$, we set $F(X) = F(x)$.

The set $\mathbf{E}_F = \{X\}$ corresponding to all the elements belonging to E_F is said to be the *factor space* corresponding to E_F. Element Θ is the *zero* of the space $\mathbf{E}_F$.

Definition 3.22. Let $x, y \in E_F$ and let $X, Y \in \mathbf{E}_F$ be their corresponding elements, respectively. Then we denote the element of $\mathbf{E}_F$ corresponding to $x + y$ by $X + Y$. And if E_F is an asymptotic ring, we denote the element corresponding to xy by XY.

It is easy to show that $\mathbf{E}_F$ is an asymptotic group with its characteristic function $F(X)$. Moreover, if E_F is an asymptotic ring, then $\mathbf{E}_F$ is also an asymptotic ring. The characteristic function $F(X)$ possesses the following properties: Let $X, Y \in \mathbf{E}_F$ then

(1) $F(X - Y) = -\infty$ if and only if $X = Y$, $\Theta \in \mathbf{E}_F$ is a unique element corresponding to θ and $F(\Theta) = -\infty$;

(2) $F(X) = F(-X)$;

(3) $F(X + Y) \le \max[F(X), F(Y)]$;

(3) If E_F is an asymptotic ring then $F(XY) \le F(X) + F(Y)$.

The following proposition is obvious.

Proposition 3.23. *Let Δ_i correspond to δ_i $(i = 0, 1, ...)$ and let series* (3.11) *have an asymptotic sum s. Then*

$$S = \Delta_0 + \Delta_1 + ... + \tag{3.19}$$

in $\mathbf{E}_F$ where S corresponds to s.

Let all the conditions of Lemma 3.18 be fulfilled. Then we have $A(y) \asymp A(\tilde{y})$ for any $y \in U_{\lambda h}$ and $\tilde{y} \asymp y$. Consequently, equation (3.14) may be considered in the factor space $\mathbf{E}_F$. That is, we have the equation

$$Y = A(Y) \tag{3.20}$$

where $A(Y) = A(y) + \Theta$. Equation (3.20) is equivalent to equation (3.14) in the following sense. If equation (3.20) has a solution $Y \in \mathbf{E}_F$ then equation (3.14) has a formal solution y where y is any representative function to Y. And if equation (3.14) has any solution belonging to the set $y + \Theta$, where $y \in E_F$, then equation (3.20) has a solution Y corresponding to y.

Let us mark that if E_F is asymptotically complete then $\mathbf{E}_F$ is also asymptotically complete, and it means that any asymptotically fundamental sequence in $\mathbf{E}_F$ converges in $\mathbf{E}_F$. Set

$$\mathbf{U}_{\lambda h} = \{Y : Y \in \mathbf{E}_F, F(Y - \lambda) \le h\}. \tag{3.21}$$

For equation (3.20) Lemma 3.18 may be rewritten in the following form.

Lemma 3.24. *Let E_F be asymptotically complete. Let $A(\mathbf{U}_{\lambda h}) \subset \mathbf{U}_{\lambda h}$. Let $A(Y)$ possess the $L-$ condition with constant $\sigma > 0$ in $U_{\lambda h}$. Then there exists a unique solution $S \in \mathbf{U}_{\lambda h}$ of equation (3.20); S is a limit of a sequence $\{S_m\}$ such that S_0 is an arbitrary (fixed) point of $\mathbf{U}_{\lambda h}$,*

$$S_m = A(S_{m-1}) \ for \ m = 1, 2, ... \tag{3.22}$$

and $F(S_m - S_{m-1}) \le F(S_1 - S_0) - (m-1)\sigma$.

For a factor space, we may form a metric space, and (in definite sense) they are equivalent. Let E_F be an asymptotic space and $\mathbf{E}_F$ is its corresponding factor space.

Let $r > 1$ be an arbitrary (fixed) number, let X, Y be arbitrary elements of the factor space $\mathbf{E}_F$. Set $\rho^*(X) = r^{F(X)}$.

Proposition 3.25.   *The functions $\rho^*(X)$ possesses the following properties*

(1) $\rho^*(X) = 0$ *if and only if* $X = \Theta$, *of the space* $\mathbf{E}_F$;

(2) $\rho^*(X + Y) \le r^{\max[\rho(X), \rho(Y)]} \le \rho^*(X) + \rho^*(Y)$;

(3) $\rho^*(X) = \rho^*(-X)$.

All properties (1)-(3) are easily verified.

Thus, the function $\rho(X, Y) = \rho^*(X - Y)$ possesses the following properties:

(1) $\rho(X, Y)$ is determined for all $X, Y \in \mathbf{E}_F$, $\rho(X, Y) \ge 0$, and $\rho(X, Y) = 0$ if and only if $X = Y$;

(2) $\rho(X, Y) = \rho(Y, X)$;

(3) $\rho(X, Y) \le \max[\rho(X, Z), \rho(Z, Y)] \le \rho(X, Z) + \rho(Z, Y)$ for any $X, Y, Z \in \mathbf{E}_F$.

So that $\rho(X, Y)$ satisfies all the axioms of a metric function. Thus, the space $\mathbf{E}_F$ (with its characteristic function $\rho(X, Y)$) is a metric space.

Chapter 4

ASYMPTOTIC BEHAVIOR
OF FUNCTIONS

Here we consider the behavior of scalar complex-valued functions for
sufficiently large real and complex arguments in an unbounded domain
D. We look for asymptotic approximations of the functions which may be
written by the relations of type $O, o,$ $\sim$, and we estimate their growth
(when the argument tends to infinity) by means of real numbers. Simple
estimates were regarded in subsection 3.2.1.

For many asymptotic problems including the operation of differentiation,
it is important to have estimates which give simple estimates for a functions and its derivatives simultaneously. Such estimates are studied in this
chapter. For functions of the reel argument $t : f(t)$ has an analytic estimate
$\Pi\{f(t)\} \leq p$ if (and only if) it has all the simple estimates $h\{f^{(m)}(t) \leq p-m$
$(m = 0, 1, ...)$ (p is a real number).

A priori knowledge that the unknowns possesses an analytic estimate
may be the key of the problem solution.

In this chapter, we investigate the property of functions having analytic
estimates. For example, if a function $f(z)$ is holomorphic in a sufficiently
small sector S_ε containing the positive semi-axis and there is a number r
such that $|f(z)| < |z|^r$ for $|z| \gg 1, z \in S_\varepsilon$, then (on the positive semi-axis)
$\Pi\{f(t)\}$ exists and $\leq r$.

First we examine infinitesimal functions of real argument closely connecting with functions possessing analytic estimates. They also have independent meaning for solution some asymptotic problems.

1. INFINITESIMAL FUNCTIONS OF A REAL ARGUMENT

Definition 4.1. *Class C_t is a set of all functions defined on the positive semi-axes such that if $\alpha(t) \in C_t$, then, for any (fixed) $n = 0, 1, ...,$ there exists a derivative*

$$\alpha^{(n)}(t) \text{ for } t \gg 1 \text{ and } \alpha^{(n)}(t) = o(t^{-n}) \text{ for } t \to +\infty.$$

It is easy to show that the following functions belong to C_t : $t^{-\sigma}$ where σ is a positive number; e^{-t}; $(\sin \ln t)/t$; the function of the form

$$ct^{k_0}(\ln_1 t)^{k_1}...(\ln_m t)^{k_m},$$

where c is a complex number and $k_0, k_1, ..., k_m$ are real numbers such that $(k_0, k_1, ..., k_m) \prec (0, 0, ..., 0)$ that is, the non-zero number k_i with minimal i is negative (see it Conventions and Notation relation (0.10)).

C_t is a sufficiently wide set of functions that the following proposition shows.

Proposition 4.2. *Let a function $\alpha(z)$ be holomorphic in a sufficiently small sector $S_\varepsilon = \{z : |\arg z| \ll 1, |z| \gg 1, z \neq \infty\}$ and $\alpha(z) \to 0$ for $z \to \infty, z \in S_\varepsilon$. Then (on the positive semi-axis) $\alpha(t) \in C_t$.*

PROOF. Consider a circle of the form $\gamma(t) = \{z : |z - t| \leq t \tan \varepsilon\}$, where the number $\varepsilon \ll 1$. Clearly, $\gamma(t) \subset S_\varepsilon$, $t(1 - \sin \varepsilon) \leq |z| \leq t(1 + \sin \varepsilon)$. Set

$$M(t) = \max_{z \in \gamma(t)} |\alpha(z)|.$$

Since $z \in S_\varepsilon$ for $t \gg 1$ and $z \to \infty$ for $t \to +\infty$, we have $M(t) \to 0$ for $t \to +\infty$. Apply the Cauchy inequality for the derivative $\alpha^{(n)}(t)$. We have

$$|\alpha^{(n)}(t)| \leq \frac{n! M(t)}{(t \sin \varepsilon)^n}, \tag{4.1}$$

where $t \sin \varepsilon$ is equal to the radius of the circle. Consequently, $\alpha^{(n)}(t) t^n \to 0$ for $t \to +\infty$ □

Proposition 4.3. *Let $\alpha(t) \in C_t$.*

(1) *Linear combination of finitely many functions belonging to C_t is a function belonging to C_t.*

(2) *Let $f(x)$ be a holomorphic function in a neighborhood of the point $x = 0$, and $f(0) = 0$. Then $f(\alpha(t)) \in C_t$.*

(3) *Let $F(x,y)$ be a holomorphic function in a neighborhood $U = \{(x,y) : |x| < p, |y| < q\}$, where x and y are complex variables, p and q are positive numbers; let $F(0,0) = 0$ and $\partial F(0,0)/\partial y \neq 0$. Then the equation $F(x,y) = 0$ possesses a unique infinitesimal solution $y = f(x)$ for $x \to 0$. Moreover, the function $f(x)$ is holomorphic in a neighborhood of the point $x = 0$ and, consequently, $f(\alpha(t)) \in C_t$.*

PROOF. Property (1) is obvious. To prove property (2), it is enough to show that $d^n f(\alpha(t))/dt^n = o(t^{-n})$ for $t \to +\infty$ for any (fixed) number $n = 0, 1, \dots$ This proof is made by induction on n, being trivial for $n = 0$. Assume the statement for $n-1$. The derivative $d^n f(\alpha(t))/dt^n$ may be represented as a finite sum of distinguished members of the form

$$a f^{(m)}(\alpha(t))[\alpha'(t)]^{k_1} \dots [\alpha^{(n)}(t)]^{k_n},$$

where $a, k_1, \dots, k_m$ are non-negative integers such that

$$k_1 + 2k_2 + \dots + nk_n = n.$$

Since $f^{(m)}(\alpha(t))$ is bounded for $t \gg 1$, and

$$\alpha^{(j)}(t) = o(t^{-j}), \quad j = 1, 2, \dots, n, \quad \text{for } t \to +\infty,$$

we have $d^n f(\alpha(t))/dt^n = o(t^{-n})$, and the proof is complete.

Property (3) can be proved by the principle of contractive mappings, and we must perform $F(x,y) = 0$ to the required form. Due to point (2), it is enough to prove that the last equation has a unique infinitesimal solution $y = f(x)$ for $x \to 0$ holomorphic in a circle $c_\varepsilon = \{x : |x| \ll 1, x = 0\}$. We may write the function $F(x,y)$ in the form

$$F(x,y) = a(x) + yb(x) + y^2\psi(x,y),$$

where $a(x), b(x)$, and $\psi(x,y)$ are holomorphic functions. Since $F(0.0) = 0$ and $\partial F(0,0)/\partial y \neq 0$, we have $a(0) = 0$, $b(0) \neq 0$. Besides, there are numbers $N > 0$ and $\delta > 0$ such that $|\psi(x,y)| < N$, $|b(x)| > \delta$. Let $M = N/\delta$. Below we consider the equation in the domain

$$U_{\varphi h} = \{(x,y) : |x| \leq \varepsilon, |y| \leq h\},$$

where ε and h are sufficiently small positive numbers. The equation $F(x,y) = 0$ may be rewritten in the form $y = A(x,y)$, where $A(x,y) = \lambda(x) + y^2\varphi(x,y)$. Here $\lambda(x) = -a(x)/b(x)$, $\varphi(x,y) = -\psi(x,y)/b(x)$. All the functions are holomorphic, and $\lambda(0) = 0$. Consider a sequence of the form

$$y_{m+1}(x) = A(x, y_m(x)) \quad \text{for } m = 0, 1, \dots, \quad y_0(x) = 0.$$

We take the number h so small that $h \leq \min[1/(2M), q/2]$, and take ε so small that $|\lambda(x)| < h/2$. Hence $|y^2 \varphi(x, y)| < h^2 M < h/2$. So that $|A(x.y)| < h$. Consequently, all the members $y_m(x)$ belong to $U_{\varphi h}$. Let $|y_1| < h$, and $|y_2| < h$. We have

$$\begin{aligned} A(x, y_1) - A(x, y_2) &= y_1^2 \varphi(x, y_1) - y_2^2 \varphi(x, y_2) \\ &= (y_1^2 - y_2^2)\varphi(x, y_1) + y_2^2[\varphi(x, y_1) - \varphi(x, y_2)]. \end{aligned}$$

The function $\varphi(x, y)$ is holomorphic in y. Consequently, there exists a positive continuous function $r(x)$ such that $|\varphi(x, y_2) - \varphi(x, y_1)| \leq r(x)|y_1 - y_2|$. There exists a positive number R such that $|r(x)| < R$ (in U). So that (in addition), we may choose the number h so small that the following inequality holds: $|A(x, y_1) - A(x_2)| \leq k|y_1 - y_2|$ in $U_{\varphi h}$, where $k < 1$. On the basis of the principle of contractive mappings, there exists a unique solution $y = f(x)$ in $U_{\varphi h}$ such that it is a sum of an uniformly convergent series

$$y(x) = y_0(x) + (y_2(x) - y_1(x)) + \ldots + (y_{m+1}(x) - y_m(x)) + \ldots$$

Clearly, $y(x)$ is holomorphic because all the members of the uniformly convergent series are holomorphic, and $y(0) = 0$ because all the members of the series vanish when $x = 0$. □

Proposition 4.3 gives us a possibility to obtain asymptotic solutions for some equations.

Example 4.4. Consider an equation of the form

$$xe^x = \alpha(t) \text{ where } \alpha(t) \in C_t.$$

On the basis of Proposition 4.3 (3), the equation possesses a unique infinitesimal solution $x(t) \in C_t$. Rewrite the equation in the form $x = \alpha(t)e^{-x}$. Consequently,

$$x(z) = \alpha(t)(1 + o(1)).$$

Hence

$$x(t) = \alpha(t)e^{-\alpha(t)(1+o(1))} = \alpha(t) - \alpha^2(t)(1 + o(1)) \text{ for } t \to +\infty,$$

and so forth.

Moreover, the following iterates obviously make the asymptotic approximations more accurate:

$$x_n(t) = \alpha(t)e^{-x_{n-1}(t)}, \quad x_0(t) = 0, \quad n = 1, 2, \ldots$$

Let us remark that if $\alpha(t)$ is a real function, then the solution $x(t)$ is also real.

Example 4.5. Consider an equation of the form

$$y^{1/\alpha(t)} e^y = 1, \quad \text{where } \alpha(t) \in C_t, \ \alpha(t) \neq 0 \text{ for } t \gg 1.$$

Show that the equation possesses a solution

$$y(t) \sim 1 \text{ for } z \to \infty, z \in [S].$$

To this end, put $y = 1 + x$ and take the logarithm. Then (by simple performances) we obtain the equation

$$\frac{\ln(1 + x)}{1 + x} + \alpha(t) = 0.$$

By Proposition 4.3 (3), the equation has a unique infinitesimal solution $x(t) \in C_t$. To obtain its asymptotic approximation, rewrite the equation in the form

$$x = -\alpha(t)(1 + x) + x - \ln(1 + x),$$

and hence

$$x = -\alpha(t)(1 + o(1)) \text{ for } t \to +\infty.$$

The following iterates give the asymptotic representations more accurate:

$$x_n(t) = -\alpha(t)(1 + x_{n-1}(t)) + x_{n-1}(t) - \ln(1 + x_{n-1}(t)),$$

$x_0(t) = 0$, $n = 1, 2, \dots$ We have

$$x_2(t) \sim -\alpha(t) + \frac{3}{2}\alpha^2(t),$$

and so forth. Thus,

$$x(t) = -\alpha(t) + \frac{3}{2}\alpha^2(t) + O(\alpha^3(t)).$$

Clearly,

$$y(t) = 1 - \alpha(t) + \frac{3}{2}\alpha^2(t) + O(\alpha^3(t)),$$

for $z \to \infty, z \in [S]$.

Example 4.6. Consider a more complicated case. Show that the equation $\sin x = 1/\ln x$ has solutions

$$x_n \sim \pi n \text{ for } n \to \infty, \ n \in \{1, 2, \dots, \}$$

and let us determine asymptotic approximations for the solutions. It is easy to show that the function $F(x) = \sin x - 1/\ln x$ is continuous for $x > 0$ and changes its sign in any interval $[\pi n - \varepsilon_n, \pi n + \varepsilon_n]$ for $n \gg 1$ where $\varepsilon_n = 2/\ln(\pi n)$. This means that the considered equation has solutions $x_n \sim \pi n$ for $n \to \infty$. Let us put $x = u + t$ where $t = \pi n$. We have

$$(-1)^n \sin u = \frac{1 + (\ln(1 + u/t))/\ln t}{\ln t}.$$

Consider this equation on the positive semi-axis. On the basis of Proposition 4.3 (3) the last equation possesses a unique infinitesimal solution $u(t) \in C_t$. Consequently,

$$\sin u(t) = \frac{(-1)^n}{\ln t} + O\left(\frac{1}{t \ln^3 t}\right) \quad \text{for } t \to +\infty,$$

and hence

$$u(t) = (-1)^n \arcsin \frac{1}{\ln t} + O\left(\frac{1}{t \ln^3 t}\right).$$

Thus, the equation has solutions of the form

$$x_n = \pi n + (-1)^n \arcsin \frac{1}{\ln(\pi n)} + O\left(\frac{1}{n \ln^3 n}\right) \quad \text{for } n \to \infty.$$

2.   ANALYTIC ESTIMATES

Definition 4.7. Let $f(t)$ be a complex-valued function defined on J_+ for $t \gg 1$. If there exists the infimum $\inf\{r\} = p$ for all real numbers r such that $f(t)t^{-r} \in C_t$, where p is a number or $p = -\infty$, then we say that $f(t)$ possesses an *analytic estimate* for $t \to +\infty$.

The designation $\Pi\{f(t)\} = p$ means that the function $f(t)$ possesses an analytic estimate for $t \to +\infty$ which is equal to p.

The designation $f(t) \in O_t$ means that $\Pi\{f(t)\} = -\infty$. We also write $f(t) \asymp g(t)$ if $\Pi\{f(t) - g(t)\} = -\infty$. In particular, $f(t) \asymp 0$ means that $\Pi\{f(t)\} = -\infty$, and we say that $f(t)$ has a negligible small asymptotics. If it does not lead to misunderstanding we will write the last relation in the form $f(t) \asymp 0$, and it is called an *asymptotic zero*.

We designate the set of all function which possess analytic estimates for $t \to +\infty$ by Π. It is obvious that Π is a linear space.

Proposition 4.8. *Let* $\Pi\{f(t)\} = p$ *and* $\Pi\{g(t)\} = q$. *Then*

(1) $$\Pi\{f(t) + g(t)\} \le \max(p, q); \tag{4.2}$$

(2) $$\Pi\{f(t)g(t)\} \le p + q; \tag{4.3}$$

(3) $$\Pi\{f'(t)\} \le p - 1, \tag{4.4}$$

and if $p \ne 0$, then $\Pi\{f'(t)\} = p - 1$;

    (4) *Let*

$$\gamma(t) = \int_{t_o}^t f(s)ds, \tag{4.5}$$

where $t_o = +\infty$ if $p < -1$, and t_o is a sufficiently large positive number if $p \ge -1$. The value t_o may be equal to infinity if the considered integral is convergent. Then

$$\Pi\{\gamma(t)\} = p + 1.$$

    (5) *for the estimate $\Pi\{f(t)\} = p$ to be valid, it is necessary and sufficient that $h\{f^{(m)}(t)\} \le p - m$, where $m = 0, 1, \ldots$ and $\sup_m [h\{f^{(m)}(t)\} + m] = p$.*

PROOF. Properties (1) and (2) are obvious. Prove the inequality $\Pi\{f'(t)\} \le p - 1$. Choose a number $r > p$. Clearly, $f(t)t^{-r} \in C_t$. Therefore (by the definition)

$$(f(t)t^{-r})'t = f'(t)t^{-r+1} - rf(t)t^{-r} \in C_t.$$

Hence $f'(t)t^{-r+1} \in C_t$. This leads to the relation $\Pi\{f'(t)\} \le r - 1$. Taking into account the arbitrariness of r, we obtain $\Pi\{f'(t)\} \le p - 1$.

Prove property (4). Let $p \ge -1$. Then $f(t) = o(t^{p+\varepsilon})$ for $t \to +\infty$. Hence

$$\left| \int_{t_0}^t f(s)ds \right| \le \int_{t_0}^t s^{p+\varepsilon}ds = O(t^{p+1+\varepsilon}) \text{ for } t \to +\infty.$$

If $p < -1$, in the same way (for any $\varepsilon < -p - 1$)

$$\left| \int_{+\infty}^t f(s)ds \right| \le \int_t^\infty s^{p+\varepsilon}ds = O(t^{p+1+\varepsilon}).$$

Taking into account the arbitrariness of ε, we conclude that $\Pi\{\gamma(t)\} \le p + 1$. If $\Pi\{\gamma(t)\} < p + 1$ then (by already proved) $\Pi\{f(t)\} = \Pi\{\gamma'(t)\} < p$, which is impossible.

Prove property (3). It was already proved that $\Pi\{f'(t)\} \le p - 1$. Let $p < 0$. Hence $\lim_{t \to +\infty} f(t) = 0$. Suppose the contrary, that is $\Pi\{f'(t)\} < p - 1 < -1$. Clearly, the integral $\int_t^{+\infty} f'(t)dt$ is absolutely convergent, and $\int_t^{+\infty} f'(t)dt = -f(t)$. Consequently, $\Pi\{f(t)\} \le \Pi\{f'(t)\} + 1 < p$. The obtained contradiction proves the considered case. Let $p > 0$. Hence $t_0 \ne \infty$, $\lim_{t \to +\infty} f(t) = \infty$, and $\Pi\{f(t) - f(t_0)\} = p$. Let $\Pi\{f'(t)\} < p - 1$. We have $\int_{t_0}^t f'(t)dt = f(t) - f(t_0)$, which leads to a contradiction. Thus, property (3) is proved.

It remains to prove property (5). Prove the *necessity of property* (5). Let $\Pi\{f(t)\} = p$. Then, as it was proved, $\Pi\{f'(t)\} \le p - 1$, (by induction with respect to $m = 0, 1...$), we obtain $\Pi\{f^{(m)}(t)\} \le p - m$. That is, $h\{f^{(m)}(t)\} \le p - m$. In particular, $\sup_m[h\{f^{(m)}(t)\} + m] \le p$. Suppose (on the contrary) that $\sup_m[h\{f^{(m)}(t)\} + m] < p$, then there is a positive number ε such that $\sup_m[h\{f^{(m)}(t)\} + m] < p - \varepsilon$, for any m. This means that $f^{(m)}(t)t^{p-m+\varepsilon} = o(1)$ for $t \to +\infty$. That is, $\Pi\{f(t)\} < p$. The obtained contradiction proves our assertion.

Sufficiency: Let $h\{f^{(m)}(t)\} \le p - m$, and $\sup_m[h\{f^{(m)}(t)\} + m] = p$. From the first relation it is obvious that $\Pi\{f(t)\} \le p$. Suppose (the contrary) that $\Pi\{f(t)\} < p$. Then there is a number $\varepsilon > 0$ such that $\Pi\{f(t)\} < p - \varepsilon$. That means that $h\{f^{(m)}(t)\} + m \le p - \varepsilon$. That is, $\sup_m[h\{f^{(m)}(t)\} + m] < p$. The obtained contradiction proves our assertion. □

Proposition 4.8 implies

Proposition 4.9. *The space* Π *is an asymptotic ring.*
We denote the asymptotic factor space corresponding to Π *by* $\mathbf{\Pi}$.

Definition 4.10. The space $\mathbf{\Pi}$ consists of all elements X corresponding to all $x \in \Pi$. The corresponding element X is the set $X(t) = x(t) + \Theta$, where Θ is the set of all function $\theta(t) \asymp 0$. We set $\Pi\{X(t)\} = \Pi\{x(t)\}$.

3. SERIES AND SEQUENCES IN THE SPACE *Π*

Here we investigate so called *asymptotic series* when their members belong to the space Π. We analyze the properties of their *asymptotic sums*, and we look for the conditions when the particular sums are close to the corresponding asymptotic sum. On this basis we formulate the notion of asymptotically fundamental series and deduce the necessary and sufficient conditions for the members of series in the space Π. For asymptotic problems in the space Π, the elaborate theory of asymptotic spaces has like meaning as the theory of converges of series in a Banach space.

Here we consider complex-valued functions of the real argument t on the positive semi-axis J_+. But the consideration is valid if we shall write z, $z \in D$, $|z| \gg 1$, and $z \in D, z \to \infty$ instead of $t \in J_+$, $t \gg 1$, and $t \to +\infty$, respectively. Here D is an unbounded domain of the complex plane, $z = \infty \notin D$.

Consider a series of the form

$$\delta_0(t) + \delta_1(t) + ... + \delta_m(t) + ..., \qquad (4.6)$$

where any $\delta_m(t)$ is a function defined for $t \gg 1$ $(m = 0, 1, ...)$. We denote its partial sum by

$$s_m(t) = \delta_0(t) + \delta_1(t) + ... + \delta_m(t),$$

and consider the sequence

$$\{s_m(t)\} \quad m = 0, 1, ... \tag{4.7}$$

Definition 4.11. We say that (4.6) is an *asymptotic series* and (4.7) is an *asymptotic sequence* for $t \to +\infty$ if

$$\delta_{m+1}(t) = o(\delta_m(t)) \text{ for any } m \gg 1 \text{ as } t \to +\infty.$$

Definition 4.12. We say that a function $s(t)$ is a *simple asymptotic sum* of series (4.6) and a *simple asymptotic limit* of sequence (4.7) for $t \to \infty$ if $s(t)$ is defined for $t \gg 1$ and

$$s(t) - s_m(t) = o(\delta_m(t)) \text{ for any } m \gg 1 \text{ as } t \to +\infty.$$

We also say that (4.6) is a *simple asymptotic expansion* of the function $s(t)$ for $t \to +\infty$, and the function $\delta_0(t)$ is the *main part* (*main term* or *main member*) of the function $s(t)$ for $t \to +\infty$.

Frequently it is convenient to consider the following simple asymptotic series (and, accordingly, simple asymptotic sequences): let there be a sequence of functions $\{\varphi_m(t)\}$ such that

$$\varphi_{m+1}(t) = o(\varphi_m(t)) \text{ for } m = 0, 1, ... \text{ for } t \to +\infty.$$

Any series of the form

$$\sum_{m=0}^{\infty} c_m \varphi_m(t) \tag{4.8}$$

is a simple asymptotic series where $c_0, c_1, ..., c_m, ...$ are non-zero numbers.

Examples 4.13.
1° If $\varphi_m(t) = t^{-m}$, we have a power series. Thus, a power series is a series of the form

$$\sum_{m=0}^{\infty} c_m t^{-m}.$$

2^0 If $\varphi_m(t) = t^{r_m}$, where r_m are real numbers such that

$$r_0 > r_1 > ... > r_m > ..., r_m \to -\infty \text{ for } m \to \infty,$$

series (4.8) is called a *generalized power series*. Thus, a generalized power series is a series of the form

$$\sum_{m=0}^{\infty} c_m t^{r_m}. \tag{4.9}$$

3^0 Let

$$\varphi_m(t) = t^{r_{0m}} (\ln_1 t)^{r_{1m}} ... (\ln_s t)^{r_{sm}}, \tag{4.10}$$

where $r_{0m}, r_{1m}, ..., r_{sm}$ are real numbers such that $\varphi_{m+1}(t) = o(\varphi_m(t))$ for $m = 0, 1, ...$ as $t \to +\infty$. Then we say that (4.8) is a *power-logarithmic series*.

Proposition 4.14. *Let any $\delta_m(t)$ be a function defined for $t \gg 1$ ($m = 0, 1, ...$). Let*

$$\delta_{m+1}(t) = o(\delta_m(t)) \text{ for any } m \gg 1 \text{ as } t \to +\infty.$$

Then there exists a function $g(t)$ defined for $t \gg 1$ such that (see (4.7))

$$g(t) - s_m(t) = o(\delta_m(t)) \text{ for } t \to \infty, t \in D.$$

PROOF. If series (4.6) consists of finitely many members, then, clearly, in the same time it is its asymptotic sum. Therefore it is enough to prove this proposition for infinite series. That is, $\delta_m(t) \not\equiv 0$ for any $m = 0, 1, ...$ Since $\delta_{m+1}(t) = o(\delta_m(t)$ for $t \to +\infty$, there exist numbers T_m such that

$$|\delta_{m+1}(t)| \le \frac{1}{2}|\delta_m(t)| \text{ for } t \ge T_m.$$

We choose the numbers T_m such that $T_1 < T_2 < ... < T_m < ...,$ $T_m \to +\infty$ for $m \to \infty$. Let $\mu_m(t)$ be a continuous function such that $\mu_m(t) = 0$ for $t \le T_m,$ $0 \le \mu_m(t) \le 1$ for $T_m \le t \le T_{m+1},$ and $\mu_m(t) = 1$ for $t \ge T_{m+1}.$ Then, for any $t > T_m$ and $p = 1, 2, ...,$

$$\mu_{m+p}(t)\delta_{m+p}(t) \le 2^{-p}|\delta_m(t)|. \tag{4.11}$$

Clearly, the considered inequality takes place if $t \ge T_{m+p}$, and if $t < T_{m+p}$, then the left side vanishes. Let

$$g(t) = \sum_{m=0}^{\infty} \mu_m(t)\delta_m(t). \tag{4.12}$$

Mark that $g(t)$ is a finite sum for any fixed t. Consequently, series (4.12) is convergent for any $t \in J_+$. We have, for $t \geq T_m$,

$$\left| g(t) - \sum_{j=0}^{m} \mu_j(t)\delta_j(t) \right| \leq \sum_{j=m+1}^{\infty} |\mu_j \delta_j(t)|$$

$$\leq |\delta_{m+1}(t)| \sum_{j=m+1}^{\infty} 2^{m+1-j}$$

$$= 2|\delta_{m+1}(t)|$$

$$= o(\delta_m(t)). \qquad \square$$

Remark 4.15. Series (4.12) consists of finitely many terms for any fixed t. Moreover, if any function $\delta_m(t)$ is $(p-1)$ times differentiable ($p \leq \infty$) then the functions $\mu_m(t)$ may be also chosen p times differentiable. In this case, consequently, $g(t)$ is p times differentiable.

Definition 4.16. Let any function $\delta_m(t)$ belong to the space Π ($m = 0, 1, ...$). A function $s(t)$ is said to be an *analytic asymptotic sum* of series (4.6) and an *analytic asymptotic limit* of sequence (4.7) for $t \to +\infty$, which is written

$$s(t) \asymp \delta_0(t) + \delta_1(t) + ... + \delta_m(t) + ..., \qquad (4.13)$$

if

$$\Pi\{s(t) - s_m(t)\} \to -\infty \text{ for } m \to \infty.$$

We also say that series (4.6) is an *analytic asymptotic representation* of the function $s(t)$ for $t \to +\infty$ in the *class of power growth functions*.

In the next proposition we give the conditions of existence of analytic asymptotic sums of series (4.6) on the positive semi-axis.

Theorem 4.17. *Let all the functions $\delta_m(t)$ possess the estimates $\Pi\{\delta_m(t)\}$ ($m = 0, 1, ...$) and*

$$\Pi\{\delta_m(t)\} \to -\infty \text{ for } m \to \infty.$$

Then there exists a function $s(t)$ such that

$$\Pi\{s(t) - s_m(t)\} \to -\infty \ (m \to \infty).$$

To prove this proposition we shall consider the following lemmas.

Lemma 4.18. *For any positive number* k, *the following inequality holds:*

$$1 - e^{-kt} < kt \ for \ t > 0.$$

PROOF. Indeed, for the function $f(t) = kt + e^{-kt} - 1$, we have $f(0) = 0$ and $f'(t) = k(1 - e^{-kt}) > 0$ for $t > 0$. Hence $f(t)$ is an increasing function, i.e. $f(t) > 0$ for $t > 0$ which implies the required inequality.   □

Lemma 4.19. *Given a sequence of functions of the form*

$$\{\gamma_m(t)\} \ (m = 0, 1, ...),$$

where the functions $\gamma_m(t)$ *are continuous on the interval* $[1, +\infty[$ *and*

$$|\gamma_m(t)| < C(2t)^{-m} \ for \ t \geq 1,$$

where C *is a positive number independent of* m. *Let* $\{k_0, k_1, ..., k_m, ...\}$ *be a sequence of positive numbers. Then the series*

$$\gamma_0(t)e^{-k_0 t} + \gamma_1(t)e^{-k_1 t} + ... + \gamma_m(t)e^{-k_m t} + ... \qquad (4.14)$$

converges on $[1, +\infty[$ *and its sum* $S(t)$ *possesses the estimate* $h\{S(t)\} = -\infty$.

PROOF. Since $|\gamma_m(t)e^{-kt}| < C(2t)^{-m}$ for $t \geq 1$, series (4) converges for $t \geq 1$ to a continuous function $S(t)$. Clearly,

$$|S(t)| \leq |\gamma_0(t)|e^{-k_0 t} + ... + |\gamma_m(t)|e^{-k_m t} + \sum_{j=m+1}^{\infty} C(2t)^{-j} = O(t^{-m-1}).$$

Hence $h\{S(t)\} \leq -m - 1$. Taking into account the arbitrariness of m, we conclude that $S(t) = O(t^{-\infty})$.   □

PROOF of Theorem 4.17. On the basis of Proposition 4.8 (5), it suffices to prove that

$$h\{s^{(n)}(t) - s_m^{(n)}(t)\} \to -\infty \ as \ m \to \infty$$

for any (fixed) $n = 0, 1, ...$
    Let

$$\delta_m(t), \delta_m'(t), ..., \delta_m^{(m)}(t)$$

be continuous functions on an interval $[T_m, +\infty[$ $(T_m = \text{const}, T_m \geq 1)$. Consider a modifiable series

$$\delta_0^*(t) + \delta_1^*(t) + ... + \delta_m^*(t) + ...,$$

and the corresponding sequence $\{s_m^*(t)\}$ $(m = 0, 1, ...)$, where

$$s_m^*(t) = \delta_0^*(t) + \delta_1^*(t) + ... + \delta_m^*(t).$$

Here

$$\delta_m^*(t) = \delta_m(t + e^{T_m - t}).$$

The functions $\delta_m^*(t)$ possesses the following properties:

(1) Each function $\delta_m^{*(n)}(t)$ (n=0,1,...,m) is continuous on the entire real axis (because $t + e^{T_m - t} \geq T_m + 1$ for any t);

(2) $h\{d^p \delta_m^*(t)/dt^p - \delta_m^{(p)}(t)\} = -\infty$ for $p = 0, 1, ...$
Indeed, for $p = 0$,

$$|\delta_m^*(t) - \delta_m(t)| \leq e^{T_m - t} \sup_{t \leq \tau \leq t + e^{T_m - t}} |\delta'(\tau)| = O(t^{-\infty}).$$

On differentiating the difference $\delta_m^*(t) - \delta_m(t)$ p times, and taking into account that any function of the form $q(t)e^{-t} = O(t^{-\infty})$, where $q(t)$ is a function with an estimate $h\{q(t)\} < +\infty$, we easily obtain the required relation. It is enough to prove this Proposition for the modifiable series because if it has the required asymptotic sum $s^*(t)$, then, for any $p = 0, 1, ...$,

$$h\{s^{*(p)}(t) - s_m^{(p)}(t)\}$$

$$\leq \max\left[h\{s^{*(p)}(t) - s_m^{*(p)}(t)\}, h\left\{\frac{d^p s_m^*(t)}{dt^p} - s_m^{(p)}(t)\right\}\right] \to -\infty.$$

Thus, we suppose that series (4.6) possesses all the properties of the modifiable series. Moreover, we may suppose without loss of generality that $\Pi\{\delta_m(t)\} < -m - 1$ for any m, and hence there are positive number A_m such that

$$|\delta_m^{(n)}(t)| < A_m t^{-m-n-1} \text{ for } t \geq 1 \text{ and any } n = 0, 1, ..., m.$$

Consider a sequence of numbers

$$\{R_m = 2^m(1 + A_m)\}, \quad m = 0, 1, ...$$

We prove that the required asymptotic sum may be represented as a sum of the series

$$\delta_0(t)(1 - e^{-t/R_0}) + \delta_1(t)(1 - e^{-t/R_1}) + ... + \delta_m(t)(1 - e^{-t/R_m}) + ... \quad (4.15)$$

Indeed, on the basis of Lemma 4.18

$$|\delta_m(t)(1 - e^{-t/R_m})| \leq (2t)^{-m} \text{ for } t \geq 1, m = 0, 1, ...$$

Hence the series converges to a continuous function $s(t)$ for $t \geq 1$. On the basis of Lemma 4.19,

$$h\{s(t) - s_m(t)\} \to -\infty \text{ for } m \to \infty.$$

On differentiating (4.15) n times we obtain the following series

$$\sum_{m=0}^{\infty} \delta_m^{(n)}(t)(1 - e^{-t/R_m}) + \gamma_m(t)e^{-t/R_m},$$

where

$$\gamma_m(t) = \frac{n\delta_m^{(n-1)}(t)}{R_m} - \frac{n(n-1)}{2}\delta_m^{(n-2)}(t)R_m^{-2} + ... + (-1)^{n-1}\delta_m(t)R_m^{-n}.$$

For $m \geq n$, we have

$$|\delta_m^{(n)}(t)(1 - e^{-t/R_m})| \leq (2t)^{-m}$$

and $|\gamma_m(t)| < 2^n(2t)^{-m}$. Consequently, the series converges for $t \geq 1$ to $s^{(n)}(t)$ which is a continuous function for $t \geq \max(1, T_n)$. Here T_n is a number such that all the functions $\delta_q^{(p)}(t)$ are continuous on the interval $[T_n, +\infty[$. Here $p = 0, 1, ..., n$, and any $q < p$. On the basis of Lemma 4.19 it is easy to show,

$$h\{s^{(n)}(t) - s_m^{(n)}(t)\} \to -\infty \text{ for } m \to \infty,$$

which proves this Proposition.   □

Let us note that a sum of a convergent series may not be its asymptotic sum, as the following example shows. Consider the series

$$u(t) = \sum_{m=0}^{\infty} \frac{e^{-t}}{(1 + me^{-t})[1 + (m+1)e^{-t}]}. \qquad (4.16)$$

On the one hand, each members $\delta_m(t)$ of the series is equivalent to e^{-t} for $t \to +\infty$. Hence it has the estimate $\Pi\{\delta_m(t)\} = -\infty$. Therefore the series satisfies all the conditions of Proposition 4.17, and, clearly, each function $f(t)$ with the estimate $\Pi\{f(t)\} = -\infty$ is an asymptotic sum of the series.
On the other hand, we have

$$\frac{e^{-t}}{(1 + me^{-t})[1 + (m+1)e^{-t}]} = \frac{1}{1 + me^{-t}} - \frac{1}{1 + (m+1)e^{-t}}.$$

Hence

$$u_n(t) \equiv \sum_{m=0}^{n} \frac{e^{-t}}{(1+me^{-t})[1+(m+1)e^{-t}]}$$

$$= \left[1 - \frac{1}{1+e^{-t}}\right] + \left[\frac{1}{1+e^{-t}} - \frac{1}{1+2e^{-t}}\right] + \ldots$$

$$+ \left[\frac{1}{1+ne^{-t}} - \frac{1}{1+(n+1)e^{-t}}\right]$$

$$= 1 - \frac{1}{1+(n+1)e^{-t}}.$$

Clearly, $u(t) \equiv \lim_{n\to\infty} u_n(t) = 1$, and hence $\Pi\{u(t)\} = 0$.

The following proposition is a direct consequence of Theorem 4.17.

Proposition 4.20. *Let all the functions $\delta_m(t)$ possess the estimates $\Pi\{\delta_m(t)\}$ ($m = 0, 1, \ldots$) and $\Pi\{\delta_m(t)\} \to -\infty$ for $m \to \infty$. Then series (4.6) is fundamental. Consequently, the space Π is asymptotically complete (see Definitions 3.14 and 3.15).*

4.   OPERATORS IN THE SPACE Π

Many problems are reduced to solution operator equations which we consider in the form

$$y = A(y) \tag{4.17}$$

We consider (4.17) in the space Π. The general aspect of such equations in an asymptotic space E_F was given in chapter 2. The main aim of any asymptotic problem (reduced to (4.17)) is to obtain so called *asymptotic solutions of the equation* (instead of exact solution in regular problems). Roughly speaking, we look for a function (or functions) $f(t) \in \Pi$ such that there exists a precise solution $y(t)$ of the equation and $f(t) - y(t) = o(t^{-N})$ for any positive N (no matter haw large) as $t \to +\infty$. As a rule, the solution amounts to two dependent stages. The first one is called a *formal procedure* which consists of finding a sequence $\{x_m(t)\}$ ($m = 0, 1, \ldots$) such that $x_m(t) - A(x_m(t)) = o(t^{k_m})$ where the numbers $k_m \to -\infty$ as $m \to \infty$. An asymptotic sum $s(t)$ of the formal series

$$x_0(t) + [x_1(t) - x_0(t)] + \ldots + [x_{m+1}(t) - x_m(t)] + \ldots$$

or (which is the same) asymptotic limit of the sequence $\{x_m(t)\}$ is set two be a *formal solution* to (4.17). The second stage consists of the proof that the formal solution is an asymptotic solution to the equation. We mainly analyze the case when the formal sequence may be obtained as an iterate sequence of the form $x_{m+1}(t) = A(x_m(t))$, in a suitable sphere of the space Π. Let us give the exact formulations.

Definition 4.21. A function $s(t) \in \Pi$ is said to be a *formal solution* to equation (4.17) if $s(t) \asymp A(s(t))$ (that is, $\Pi(s(t) - A(s(t))) = -\infty$); $s(t)$ is said to be an *asymptotic solution* to the equation if there is an exact solution $y(t)$ of the equation such that $y(t) \asymp s(t)$.

Example 4.22. Find all the real asymptotic solutions to the equation

$$y' + y^3 - t^3 = 0, \tag{4.18}$$

First, we obtain all its real formal solutions belonging to Π. If $s(t)$ is a formal solution to the equation, then $s'(t) + s^3(t) - t^3 \asymp 0$. This means that at least two members in the left-hand side of the last expression must have the same estimate. If $\Pi\{s'(t)\} = \Pi\{s^3(t)\}$, then taking into account that $\Pi\{s'(t)\} \leq \Pi\{s(t)\} - 1$, we obtain $\Pi\{s(t)\} \leq -1/2$. But it is impossible because of the relation $s'(t) + s^3(t) \asymp t^3$ implies $\Pi\{s^3(t)\} \geq 3$. If $\Pi\{s'(t)\} = \Pi\{t^3\}$, then $\Pi\{s(t)\} = 4$, which is also impossible because the rest term $s^3(t)$ will have the estimate $\Pi\{s^3(t)\} = 12$. It is possible only if $\Pi\{s^3(t)\} = \Pi\{t^3\}$. That is $\Pi\{s(t)\} = 1$. More precisely, since $\Pi\{s'(t)\} = 0$ in this case, we have $s^3(t) \sim t^3$ for $t \to +\infty$. Thus, any real formal solution have to be equal to t ($t \to +\infty$). Moreover, $s'(t) + (s(t) - t)(s^2(t) + s(t)t + t^2) \asymp 0$. Hence

$$s(t) = t - \frac{s'(t)}{s^2(t) + s(t) + t^3} + \theta(t),$$

where $\theta(t) \asymp 0$. This means that the required formal solutions are the formal solutions of the differential equation

$$s = t - \frac{s'}{s^2 + st + t^2}.$$

This enable us to consider the iteration sequence $\{s_m(t)\}$, where $s_0(t) = t$, and

$$s_m(t) = t - \frac{s'_{m-1}(t)}{s^2_{m-1}(t) + s_{m-1}(t)t + t^2} \quad m = 1, 2, \dots$$

We have $s_1(t) = t - 1/(3t^2)$, $s_2(t) = t - 1/(3t^2) - 1/(6t^5)$, etc. Finally we obtain a series of the form

$$t + a_1 t^{-2} + \dots + a_m t^{-3m+2} + \dots,$$

where a_m are numbers, $a_1 = -1/3$, $a_2 = -1/6$ and so on. Since $\Pi\{a_m t^{-3m+2}\} = -3m + 2 \to -\infty$ for $m \to \infty$, there exists an asymptotic sum $s(t)$ of this series

$$s(t) \asymp t + a_1 t^{-3} + \dots + a_m t^{-2m-1} + \dots$$

Clearly, it is a formal solution to the considered equation. Moreover, if there is another real formal solution $s^*(t)$, then it has to be equivalent to t $(t \to +\infty)$. Since it has the same asymptotic representation just as the function $s(t)$, we conclude that $s^*(t) \asymp s(t)$.

Let us prove that equation (4.18) has a solution $y(t) \asymp s(t)$. To this end, substitute $y = u + s(t)$ in (4.18). Since $s(t)$ is a formal solution, we obtain the equation

$$u' + 3s^2(t)u + 3s(t)u^2 + u^3 = \alpha(t). \tag{4.19}$$

Here $\alpha(t) \equiv s'(t) - s^3(t) - t^3 \asymp 0$. We pass to an integral equation putting $u = ve^{-3\int s^2(t)dt}$. We have

$$v' = \alpha(t)e^{-3\int s^2(t)dt} - [3s(t)v^2 - v^3]e^{-3\int s^2(t)dt}, \tag{4.20}$$

and for any arbitrary constant C and sufficiently large number $T > 0$, we obtain the integral equation

$$
\begin{aligned}
u(t) =& \\
& Ce^{-3\int_T^t u^2(t)dt} + \int_T^{+\infty} \alpha(\tau)\exp\left[-3\int_\tau^t s^2(s)ds\right]d\tau \\
& +e^{-3\int_T^t s^2(t)dt}\int_t^{+\infty}\left[3s(\tau)u^2(\tau)e^{-6\int_T^\tau s^2(s)ds} + u^3(\tau)e^{-9\int_T^t s^2(s)ds}\right]d\tau.
\end{aligned}
$$

Clearly,

$$q(t) = Ce^{-3\int_T^t s^2(t)dt} + \int_T^{+\infty}\alpha(\tau)\exp\left[-3\int_\tau^t s^2(s)ds\right]d\tau \asymp 0,$$

and the last equation has (for any fixed C) a solution $u(t,C) \asymp 0$. It is easy to see that any solution of the considered equation is a solution to equation (4.19), so that the original equation has a family of solutions (depending on the parameter C) of the form

$$y = s(t) + u(t,C).$$

Let us remark that equation (4.18) has two more complex solutions equivalent to the roots of the equation $y^2 + yt + t^2 = 0$ for $t \to +\infty$.

Definition 4.23. Owing to Definition 3.16, we say that $A(y)$ possesses the *L–condition* with constant σ in a region $D \subset \Pi$ if for any $y_1(t), y_2(t) \in D$, the following inequality holds:

$$\Pi\{A(y_1(t)) - A(y_2(t))\} \leq \Pi\{y_1(t) - y_2(t)\} - \sigma. \tag{4.21}$$

Introduce the following designations

$$U_\lambda = \{y(t) : y(t) \in \Pi, y(t) \sim \lambda(t) \text{ for } t \to +\infty\}. \qquad (4.22)$$

$$U_{\lambda h} = \{y(t) : y(t) \in \Pi, \Pi\{y(t) - \lambda(t)\} \le h\}, \qquad (4.23)$$

where h is a real number, $h < \Pi\{\lambda(t)\}$.

For the space Π, Lemma 3.18 is rewritten in the following form:

Lemma 4.24. *Let $\lambda(t) \in \Pi$ and $\Pi\{\lambda(t)\} > -\infty$. Let $A(U_\lambda) \subset U_\lambda$. Let $A(y)$ possess the L-condition with constant $\sigma > 0$ in U_λ. Then there exists a formal solution $s(t) \in U_\lambda$ of equation (4.17). Besides, $s(t)$ is an analytic asymptotic limit of any sequence of the form $\{s_m(t)\} \subset U_\lambda$ where $s_0(t)$ is an arbitrary (fixed) function belonging to U_λ,*

$$s_m(t) = A(s_{m-1}(t)) \; m = 1, 2, \ldots \qquad (4.24)$$

and $\Pi\{s_m(t) - s_{m-1}(t)\} \le -\sigma$, $s_m(t) \in U_\lambda$. Moreover, an element $\tilde{s}(t) \in U_\lambda$ is a formal solution to equation (4.17) if and only if $\tilde{s}(t) \asymp s(t)$.

Remark 4.25. *Lemma 4.24 is valid if we consider equation (4.17) in the region $U_{\lambda h}$ ($h < \Pi\{\lambda(t)\}$). That is, we may replace U_λ by $U_{\lambda h}$ everywhere in this Lemma.*

Apply this Lemma to Example 4.22. We represent equation (4.18) in the form

$$y = A(y) \equiv t - \frac{y'}{y^2 + yt + t^2},$$

and consider in the region

$$U_t = \{y(t) : y(t) \in \Pi, y(t) \sim t \text{ for } t \to +\infty\}.$$

Check the conditions of Lemma 4.24. For any $y(t) \in U_t$, clearly,

$$\Pi\{y'(t)\} = 1, \quad \Pi\{y^2(t) + y(t)t + t^2\} = \Pi\{3t^2\} = 2,$$

and

$$\Pi\{y'(t)/(y^2(t) + y(t)t + t^2)\} = \Pi\{1/(3t^2)\} = -2.$$

Hence $A(U_t) \subset U_t$.

For any $y_1(t), y_2(t) \in U_t = \{y(t) : y(t) \in \Pi, y(t) \sim t$ for $t \to +\infty$,

$$A(y_1(t)) - A(y_2(t)) = \frac{y_1'(t)}{y_1^2(t) + ty_1(t) + t^2} - \frac{y_2'(t)}{y_2^2(t) + ty_2(t) + t^2}$$

$$= \frac{(y_1(t) - y_2(t))'}{y_1^2(t) + 2y_1(t)t + t^2}$$

$$- \frac{(y_1(t) - y_2(t))(y_1(t) + y_2(t) + t)}{(y_1^2(t) + 2y_1(t)t + t^2)(y_2^2(t) + 2y_2(t)t + t^2)}.$$

Since $y_{1,2}(t) \sim t$, $y_{1,2}'(t) \sim 1$ for $t \to +\infty$, and $\Pi\{(y_1(t) - y_2(t))'\} = \Pi\{y_1(t) - y_2(t)\} - 1$, we have

$$\Pi\{A(y_1(t)) - A(y_2(t))\} \leq \Pi\{y_1(t) - y_2(t)\} - 3.$$

Hence $A(y)$ possesses the L-property with constant $\sigma = 3$ in U_t. Thus, all the conditions of Lemma 4.24 are fulfilled which leads to the desired result.

Definition 4.26. Let $\lambda(t) \in \Pi$ and $\Pi\{\lambda(t)\} > -\infty$. The operator $A(y)$ is said to be *continuous* at the point $\lambda(t)$ [*(conditionally) continuous on* $U_{\lambda h}$] if $A(y(t)) \in \Pi$ and $A(y(t)) \sim A(\lambda(t))$ as $t \to +\infty$ for any $y(t) \in U_\lambda$ $(y(t) \in U_{\lambda h})$.

5. DERIVATIVE ESTIMATES

Derivative estimates considerably facilitate the testing of the conditions of Lemma 4.24. Let $\lambda(t) \in \Pi$, and $\Pi\{\lambda(t)\} > -\infty$.

Definition 4.27. The value

$$\Pi'\{A(\lambda(t))\} = \sup_{y_1(t), y_2(t) \in U_\lambda} [\Pi\{A(y_1(t)) - A(y_2(t))\} - \Pi\{y_1(t) - y_2(t)\}],$$

(4.25)

is said to be the *derivative estimate* of the operator $A(y)$ with respect to the variable y at the point $\lambda(t)$. The notation $\Pi'\{A(\lambda(t))\} = d$ (where d is either a real number or $d = -\infty$) means that the derivative estimate exists at the point $\lambda(t)$ and it is equal to d.

In $U_{\lambda h}$ we also introduce (in the same way) the notion of a conditional derivative estimate.

Definition 4.28. The value

$$\Pi_h'\{A(\lambda(t))\} = \sup_{y_1(t), y_2(t) \in U_{\lambda h}} [\Pi\{A(y_1(t)) - A(y_2(t))\} - \Pi\{y_1(t) - y_2(t)\}],$$

(4.26)

is said to be the *derivative estimate* of the operator $A(y))$ with respect to the variable y in $U_{\lambda h}$ (or *conditional derivative estimate*). The notation $\Pi'_h\{A(\lambda(t))\} = d$ (where d is either a real number or $d = -\infty$) means that the conditional derivative estimate exists and equals to d.

The following proposition is obvious. Consider an equation of the form

$$y = \lambda(t) + R(y). \tag{4.27}$$

Proposition 4.29. *Let for any* $y_1(t), y_2(t) \in U_\lambda$ $(U_{\lambda h})$ *and* $y_1(t) \asymp y_2(t)$, *the relation* $R(y_1(t)) \asymp R(y_2(t))$ *holds. Let* $\Pi'\{R(\lambda(t))\} = -\sigma < 0$ $(\Pi'_h\{A(\lambda(t)))\} = -\sigma < 0)$. *Then* $R(y)$ *possesses the L-property with constant* σ *in* U_λ $(U_{\lambda h})$.

Consider several important properties of derivative estimates.

Proposition 4.30. *Let* $y(t), y_1(t), y_2(t)$ *are arbitrary functions belonging to* U_λ $(U_{\lambda h})$ *and* $\Pi\{y_1(t) - y_2(t)\} > -\infty$; *let operators* $R_j(y)$ *operate from* U_λ $(U_{\lambda h})$ *to* Π *and possess the estimates* $\Pi\{R_i(y(t))\} \le p_j$ *and* $\Pi'\{R_j(\lambda(t))\} \le d_j$, $j = 1, 2, ..., n$. *Let*

$$R(y) = R_1(y) + ... + R_n(y) \ and \ Q(y) = R_1(y)...R_n(y).$$

Then
(i)
$$\Pi\{R(y(t))\} \le \max_{j=1,...,n} p_j \ \Pi'\{R(\lambda)\} \le \max_{j=1,...,n} d_j. \tag{4.28}$$

(ii)
$$\Pi\{Q(y(t))\} \le p_1 + ... + p_n; \ \Pi'\{Q(\lambda(t))\}$$
$$\le p_1 + ... + p_n + \max[d_1 - p_1, ..., d_n - p_n]. \tag{4.29}$$

PROOF. Clearly,

$$\Pi\{R(y(t))\} \le \max_{j=1,...,n} \Pi\{R_j(y(t))\} \le \max_{j=1,...,n} p_j$$

and

$$\Pi\{Q(y(t))\} \le \sum_{j=1}^{n} \Pi\{R_j(y(t))\} \le \sum_{j=1}^{n} p_j.$$

Besides, we have

$$\Pi\{R(y_1(t) \quad - \quad R(y_2(t)\} - \Pi\{y_1(t) - y_2(t)\}$$

$$\leq \quad \max_{j=1,2,\ldots,n}[\Pi\{R_j(y_1(t) - R_j(y_2(t)\} - \Pi\{y_1(t) - y_2(t)\}]$$

$$\leq \quad \max_{j=1,\ldots,n} d_j.$$

Thus, property (i) is proved.

For $n = 1$, the relation $\Pi'\{R_1(\lambda(t)\} = d_1$ is trivial. For $n = 2$,

$$R_1(y_1(t))R_2(y_1(t)) - R_1(y_2(t))R_2(y_2(t))$$

$$= R_1(y_1(t))[R_2(y_1(t)) - R_2(y_2(t))]$$

$$+ R_2(y_2(t))[R_1(y_1(t)) - R_1(y_2(t))].$$

Hence

$$\Pi\{R_1(y_1(t))R_2(y_1(t)) - R_1(y_2(t))R_2(y_2(t))\}$$

$$\leq \max[\Pi\{R_1(y(t)\} + \Pi\{R_2(y_1(t)) - R_2(y_2(t))\},$$

$$\Pi\{R_2(y(t)\} + \Pi\{R_1(y_1(t)) - R_1(y_2(t))\}].$$

Consequently,

$$\Pi'\{R_1(\lambda(t))R_2(\lambda(t))\} \quad \leq \quad \max[p_1 + d_2, p_2 + d_1]$$

$$= \quad p_1 + p_2 + \max[d_1 - p_1, d_2 - p_2].$$

Let the formula of the required derivative estimate for product be true for $n - 1$. That is, for $Q^*(y) = R_1(y)...R_{n-1}(y)$, we have

$$d^* = \Pi'\{Q^*(\lambda(t))\} \leq p_1 + ... + p_{n-1} + \max_{j=1,\ldots,n-1}[d_j - p_j].$$

Consequently,

$$\Pi'\{Q(\lambda(t))\} \quad \leq \quad (p_1 + ... + p_{n-1}) + p_n + \max[d^* - p_n, d_n - p_n]$$

$$= \quad p_1 + ... + p_n + \max_{j=1,\ldots,n}[d_j - p_j]. \qquad \square$$

Examples 4.31. 1^0 Let $R(y) = a(t) \in \Pi$. Then $\Pi'\{R(\lambda(t))\} = -\infty$. Indeed, $R(y_1(t)) - R(y_2(t)) = a(t) - a(t) = 0$ Hence $\Pi\{R(y_1(t)) - R(y_2(t))\} - \Pi\{y_1(t) - y_2(t)\} = -\infty$, which leads to the required property.

2^0 $\Pi'\{\lambda^k(t)\} \le (k-1)\Pi\{\lambda(t)\}$, where k is a positive number. Indeed,

$$y_1^k(t) - y_2^k(t) = [y_1(t) - y_2(t)][ky_1^{k-1}(t) + O(y_1^{k-2}(t)(y_1(t) - y_2(t)))]$$

for $t \to +\infty$. This means that $\Pi'\{\lambda^k(t)\} \le \Pi\{k\lambda^{k-1}(t)\} = (k-1)\Pi\{\lambda(t)\}$.

3^0 $\Pi'\{\lambda^{(m)}(t)\} \le -m$, where m is a natural number. Indeed $y_1^{(m)}(t) - y_2^{(m)}(t) = (y_1(t) - y_2(t))^{(m)}$. Hence $\Pi\{y_1^{(m)}(t) - y_2^{(m)}(t)\} \le \Pi\{y_1(t) - y_2(t)\} - m$, That is, $\Pi'\{\lambda^{(m)}(t)\} \le -m$.

4^0 Let $R(y)) = [y^{(m)}]^k$, where k, m are the same just as in Examples 2^0 and 3^0. Then $\Pi'\{[\lambda^{(m)}(t)]^k\} \le k[\Pi\{\lambda(t)\} - m] - \Pi\{\lambda(t)\}$. Indeed, $R(y_1(t)) - R(y_2(t)) = [y_1^{(m)}(t) - y_2^{(m)}(t)][k[y_1^{(m)}(t)]^{(k-1)} + O([y_1^{(m)}(t)]^{k-2})(y_1^{(m)}(t) - y_2^{(m)}(t)]$. Hence

$$\Pi\{R(y_1(t)) \quad - \quad R(y_2(t))\}$$

$$\le \quad \Pi\{[y_1(t) - y_2(t)]^{(m)}\} + (k-1)\Pi\{y_1^{(m)}(t)\}$$

$$\le \quad \Pi\{y_1(t) - y_2(t)\} - m + (k-1)[\Pi\{\lambda(t)\} - m]$$

$$= \quad \Pi\{y_1(t) - y_2(t)\} + k[\Pi\{\lambda(t)\} - m] - \Pi\{\lambda(t)\}.$$

Definition 4.32. Operator $R(y)$ is said to be of *power type* with a majorant r at the point $\lambda(t)$ if $\Pi\{R(y(t))\} \le r$ and $\Pi'\{A(\lambda(t))\} \le r - \Pi\{\lambda(t)\}$.

Remark 4.33. *All the operators considered in Examples 4.28 are of power type at $\lambda(t)$. More precisely, $R(y(t)) = a(t)$ has the majorant $\Pi\{a(t)\}$, and $R(y(t)) = [y^{(m)}(t)]^k$ has the majorant $k[\Pi\{y(t)\} - m]$.*

Proposition 4.34. *Let all the conditions of Proposition 4.30 be fulfilled, and (in addition) $R_j(y)$ are of power type with the majorants r_j, respectively, at the point $\lambda(t)$ $(j = 1, 2, ..., n)$. Then at $\lambda(t)$*

(i) *$R(y)$ is of power type with the majorant $r = \max_{j=1,...,n} r_j$;*

(ii) *$Q(y)$ is of power type with the majorant $r = r_1 + ... + r_n$.*

PROOF. Case (i). Since $R_j(y)$ is of power type, then $d_j \le r_j - \Pi\{\lambda(t)\}$, which leads to the required property. Case (ii). Let (for definiteness) the maximum of $d_j - p_j$ be equal to $j = 1$ (otherwise, we can change the numeration of the operators $R_j(y)$). We have $d_1 \le p_1 - \Pi\{\lambda(t)\}$. Consequently, $d = p_2 + ... + p_n + d_1 \le r - \Pi\{\lambda(t)\}$.   □

As a consequence of this proposition we have:

Proposition 4.35. *Assume the notation and hypothesis of Proposition 4.30. Let*

$$R(y) = \sum_{j=1}^{n} a_j(t) y^{k_{0j}} [y']^{k_{1j}} ... [y^{(m)}]^{k_{mj}},$$

where $a_j(t) \in \Pi$ *and at least one* $\Pi\{a_j(t)\} > -\infty$, $k_{0j}, k_{1j}, ..., k_{mj}$ *are non-negative numbers* $(j = 1, 2, ..., n)$. *Denote by*

$$\Sigma_j = k_{0j} + k_{1j} + ...k_{mj} \text{ and } I_j = k_{1j} + 2k_{2j} + ... + mk_{m_j}.$$

Then $R(y)$ *is an operator of power type with the majorant* $r = \max_{j=1,...,n} [\Pi\{a_j(t)\} + \Pi\{\lambda(t)\}\Sigma_j - I_j]$.

PROOF. Indeed, let $\Pi\{a_j(t)\} > -\infty$. Since $a_j(t)$ and $[y^{(s)}]^{k_{sj}}$ are of power type $(s = 0, 1, ..., m)$, then $A_j(y) = a_j(t) y^{k_{0j}} [y']^{k_{1j}} ... [y^{(m)}]^{k_{mj}}$ is of power type with the majorant $\Sigma_j - I_j$. If $a_j(t) \asymp 0$, then we set that $A_j(y)$ is of power type with the majorant $-\infty$. $R(y)$ is of required type as a sum of operators of power type.

Return once again to Example 4.22. We examined the equation $y = t - y'/(y^2 + yt + t^2)$. Here $\lambda(t) = t$ and $R(y) = -y'/(y^2 + yt + t^2)$. $\Pi\{t\} = 1$. It remains to show that $R(y)$ is of power type with a majorant $r < 0$ at the point t. We have $R(y) = f(y)/H(y)$, where $F(y) = y'$ and $H(y) = y^2 + yt + t^2$. Clearly, they are of power type with the majorants $f = 0$ and $H = 2$, respectively. It is easy to see that $\Pi\{f(y(t))/H(y(t))\} = -2$ for $y(t) \in U_t$ and $\Pi'\{f(t)/H(t)\} \leq \max[\Pi'\{f(t)\} - \Pi\{H(t)\}, \Pi\{f(t)\} + \Pi'\{H(t)\} - 2\Pi\{H(t)\}] = -3$, which leads to the required property.

6. ASYMPTOTIC BEHAVIOR OF ANALYTIC FUNCTIONS

In this paragraph we mainly consider the asymptotic behavior of analytic functions specified in a central sector S of a complex plane. It is convenient to formulate their main asymptotic properties in $[S]$ which is a collection of all closed central sectors $\{S^*\}$ such that $S^* \subset S$ and its each point $z \neq 0$ is an interior point of S (see the section *Conventions and Notation*). The point is that many asymptotic properties of an analytic function determined in S essentially vary if $z \to \infty$, $z \in S$, and infinitely tends to a limiting ray of the sector. Therefore to formulate such properties, we have to choose an exhaustive system $\{D_m\}$ of domains where every D_m is a subset of S, D_m preserves the considered asymptotic property, and any interior point $z \in S$

belongs to at least one of the domains D_m. In this book, we restrict our consideration by the very simple exhaustive system $[S]$.

As a rule, asymptotic properties of analytic functions depend on the domain of their analysis. The estimates given in this section naturally follow from the notion of an infinitesimal analytic function in $[S]$.

First, we consider several general asymptotic properties of analytic functions. Consider an integral of the form

$$\gamma(z) = \int_{z_o}^{z} f(s)ds \text{ in } D, \tag{4.30}$$

Lemma 4.36. *Let $f(z)$ be a holomorphic function for $|z| \gg 1$ in a normal domain D (see Definition 3.11). Let $h_D\{f(z)\} = p$. Let $z_o = \infty$ if $p < -1$, and z_o is a finite point in D if $p \geq -1$. Then $h_D\{\gamma(z)\} \leq p + 1$.*

PROOF The cases $p = -\infty$ and $p = +\infty$ are obvious. Let $-1 \leq p < +\infty$ (z_o is a finite point). Consequently, for any positive number ε,

$$f(z) = o(z^{p+\varepsilon}) \text{ for } z \to \infty, z \in D,$$

and $\gamma(z) = C + \varphi(z)$, where $\varphi(z) = O(z^{p+1+\varepsilon})$. Therefore

$$\sup_{R<|z|<+\infty} \frac{\ln|\gamma(z)|}{\ln|z|} \leq p + 1 + \varepsilon.$$

Because of the arbitrariness of ε, we obtain the required relation.

Let $p < -1$ ($z_o = \infty$). For any number ε such that $0 < \varepsilon < -1 - p$ and for $|z| \gg 1$, we have

$$|\gamma(z)| \leq \int_{\Gamma(z,\infty)} \tau^{p+\varepsilon} d\tau = O(z^{p+1+\varepsilon}),$$

which leads to the required inequality.   □

Lemma 4.37. *Let $f(z)$ be a holomorphic function for $|z| \gg 1$ in a normal domain D and $h_D\{f(z)\} = p \neq 0$. Then $h_D\{f'(z)\} \geq p - 1$.*

PROOF. This proposition is a simple consequence of Lemma 4.36. Indeed,

$$f(z) = f(z_o) + \int_{z_o}^{z} f'(s)ds.$$

Suppose that $p > 0$ ($z_o \neq \infty$). Since $h_D\{f(z_o)\} = 0$ if $f(z_o) \neq 0$, and $h_D\{f(z_o)\} = -\infty$ if $f(z_o)\} = 0$, on the basis of Proposition 4.36, we obtain

$$p = h_D\{f(z)\} \leq h_D\{f'(z)\} + 1,$$

which means the required property. Suppose $p < 0$. Let

$$h_D\{f'(z)\} < p - 1.$$

Then there exists a positive number ε such that

$$f'(z) = o(z^{-1-\varepsilon}) \text{ for } z \to \infty, z \in D,$$

hence

$$f(z) = -\int_z^{+\infty} f'(s)ds,$$

and on the basis of Proposition 4.36,

$$p = h_D\{f(z)\} \le h_D\{f'(z)\} + 1 < p.$$

The obtained contradiction proves this Proposition. $\qquad\square$

Proposition 4.38. *Let $f(z)$ be a holomorphic bounded function for $|z| \gg 1$ in the complex plane. Then*

$$f'(z) = O(z^{-2}) \text{ for } |z| \to \infty.$$

Moreover, if $f(z) \to 0$ for $z \to \infty$, then there exists a natural number k such that

$$f'(z) \sim -k\frac{f(z)}{z}, \text{ and } h_D\{f'(z)\} = h_D\{f(z)\} - 1.$$

PROOF. Clearly, $f(z)$ is a sum of a convergent power series in z^{-1}. That is

$$f(z) = a_0 + a_1 z^{-1} + \ldots + a_n z^{-n} + \ldots,$$

and $f(z) - s_n(z) = O(z^{-n-1})$ for $|z| \to \infty$. Here

$$s_n(z) = a_0 + a_1 z^{-1} + \ldots + a_n z^{-n}.$$

Besides,

$$f'(z) = -a_1 z^{-2} - \ldots - n a_n z^{-n-1} + \ldots$$

Hence if $f'(z) \not\equiv 0$, then there exists an integer $k \ge 1$ such that $f'(z) \sim -k a_k z^{-k-1}$. If in addition $f(z) \to 0$, then $f(z) \sim a_k z^{-k}$. This implies the relation $f'(z) \sim -k f(z)/z$ for $z \to \infty$. $\qquad\square$

The following proposition is important for the further investigation.

Lemma 4.39. *Let $f(z)$ be a holomorphic and bounded function for $|z| \gg 1, z \in [S]$. Then*

$$f^{(n)}(z) = O\left(\sup_{z\in[S],|z|=R} \frac{|f(z)|}{R^n}\right) \quad for \ R \to \infty. \qquad (4.31)$$

Hence if, in addition, $f(z) \to 0$ for $z \to \infty, z \in [S]$, then

$$f^{(n)}(z) = o(z^{-n}) \ for \ z \to \infty, z \in [S].$$

PROOF. Choose a sector S^*. Let φ_1, φ_2, and φ_1^*, φ_2^*, be the boundary angles of the sectors S and S^*, respectively, such that $\varphi_1 < \varphi_1^* < \varphi_2^* < \varphi_2$, and let us choose a number $\varepsilon > 0$ such that $\varphi_1 < \varphi_1^* - \varepsilon$ and $\varphi_2 > \varphi_2^* + \varepsilon$ Consider the domain

$$U_{R\varepsilon} = \{z : R(1 - \varepsilon) \le |z| \le R(1 + \varepsilon), \varphi_1^* - \varepsilon \le \arg z \le \varphi_2^* + \varepsilon\}.$$

(see Fig. 1). The function $f(z)$ is holomorphic in $U_{R\varepsilon}$ and

$$M_R = \max_{z\in U_{R\varepsilon}} |f(z)|$$

bounded for $R \gg 1$. Apply the Cauchy inequality for the derivative $f^{(n)}(z)$ to the points of the curve

$$\Gamma_R = \{z : |z| = R, z \in S^*\}.$$

We have

$$|f^{(n)}(z)| \le \frac{n! M_R l}{2\pi[r(z)]^{n+1}}, \qquad (4.32)$$

where l is the perimeter of the $U_{R\varepsilon}$ boundary, $l = 2R(\varphi_2^* - \varphi_1^* + 4\varepsilon)$, $r(z)$ is the distance from the point $z \in \Gamma_R$ to the boundary of the domain $U_{R\varepsilon}$, $r(z) \ge R \sin \varepsilon$. Hence, taking into account that $|z| = R$ on Γ_R, we obtain the estimate

$$f^{(n)}(z) = O\left(\sup_{z\in\Gamma_R} \frac{|f(z)|}{|z|^n}\right), \quad for \ z \to \infty, z \in \Gamma_R,$$

which implies (4.31).   □

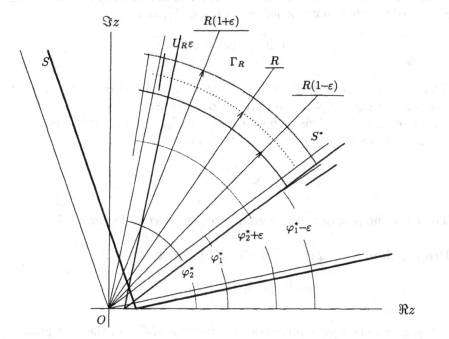

Fig. 1.

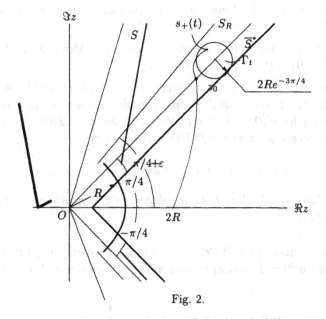

Fig. 2.

Definition 4.40. *Class* C_S *is a set of all functions such that if* $\alpha(z) \in C_S$, *then* $\alpha(z)$ *is holomorphic for* $|z| \gg 1, z \in [S]$, *and*

$$\lim_{z \to \infty, z \in [S]} \alpha(z) = 0.$$

That is, the limit is valid when $z \to \infty$ in any closed central sector S^* where each its point $z \neq 0$ is an interior point of S.

C_ε is a set of all functions such that if $\alpha(z) \in C_\varepsilon$, then $\alpha(z)$ is holomorphic for $|z| \gg 1$ in a sufficiently small sector S_ε and

$$\lim_{z \to \infty, z \in S_\varepsilon} \alpha(z) = 0.$$

The following proposition is a simple consequence of Lemma 4.39.

Proposition 4.41. *Let* $\alpha(z) \in C_S$ (C_ε). *Then*

$$\alpha^{(n)}(z) z^n \in C_S \ (C_\varepsilon).$$

Let us give some simple properties of functions of C_S. In the next proposition *we suppose that* $\alpha(z) \in C_S$.

Proposition 4.42. (1) *Linear combination of finitely many functions belonging to* C_S *is a function belonging to* C_S.

(2) *Let* $F(x)$ *be a holomorphic function in a neighborhood of the point* $x = 0$, *and let* $F(0) = 0$. *Then* $f(\alpha(z)) \in C_S$;

(3) *Let* $F(x,y)$ *be a holomorphic function in a neighborhood* $U = \{(x,y) : |x| < p, \ |y| < q\}$, *where* x *and* y *are complex variables,* p *and* q *are positive numbers; let* $F(0,0) = 0$ *and* $\partial F(0,0)/\partial y \neq 0$. *Then the equation* $F(x,y) = 0$ *possesses a unique infinitesimal solution*

$$y = f(\alpha(z)) \text{ for } z \to \infty, z \in [S].$$

Moreover the function $f(x)$ *is holomorphic in a neighborhood at the point* $x = 0$ *and* $f(\alpha(z)) \in C_S$.

PROOF. Properties (1) and (2) are obvious. Property (3) follows from the theorem of implicit function existence (see Proposition 4.3 (3)). $\square$

Remark 4.43. Similar properties take place for C_ε.

Examples 4.44. 1° The function $M(z) = z^{k_0}(\ln_1 z)^{k_1}...(\ln_m z)^{k_m}$, belongs to C_S if $(k_0, k_1, ..., k_m) \prec (0, 0, ..., 0)$. Here $k_0, k_1, ..., k_m$ are real numbers. The last relation means that the non-zero number k_i with minimal i is negative (see it Conventions and Notation).

2° $(\sin \ln z)/\ln z \in C_S$.

In these examples S is any central sector of the complex plane

3° $e^{-z} \in C_S$, where $S = \{z : |\arg z| \leq \frac{\pi}{2}, z \neq \infty\}$.

4° $\sin \alpha(z) \in C_S$, $e^{\alpha(z)} - 1 \in C_S$, where $\alpha(z) \in C_S$.

Definition 4.45. The designation $\Pi_S\{f(z)\} = p$, $(\Pi_\varepsilon\{f(z)\} = p)$, where p is a real number or $p = -\infty$, means that the function $f(z)$ is holomorphic for $|z| \gg 1, z \in [S]$ $(|z| \gg 1, z \in S_\varepsilon)$ and $h_{[S]}\{f(z)\} = p$ $(h_{S_\varepsilon}\{f(z)\} = p)$.

The designations $f(z) \in O_S$ $(f(z) \in O_\varepsilon)$ means that $\Pi_S\{f(z)\} = -\infty$ $(\Pi_\varepsilon\{f(z)\} = -\infty)$. Instead of $\Pi_S\{f(z) - g(z)\} = -\infty$ or $\Pi_\varepsilon\{f(z) - g(z)\} = -\infty$ (if it does not lead to misunderstanding) we may write $f(z) \asymp g(z)$. In particular, instead of $\Pi_S\{f(z)\} = -\infty$ or $\Pi_\varepsilon\{f(z)\} = -\infty$, we may write $f(z) \asymp 0$.

The value $\Pi_S\{f(z)\}$ $(\Pi_\varepsilon\{f(z)\})$ is said to be the *analytic estimate* of the function $f(z)$ in the sector S (S_ε).

We denote the set of all functions which possess analytic estimates in S (S_ε) by Π_S (Π_ε). Clearly, Π_S and Π_ε are linear spaces.

In the next proposition we consider some simple properties of the analytic estimates. An analytic estimate estimates the growth of function and its derivatives of any order.

The following proposition is proved analogously to Proposition 4.8.

Proposition 4.46. *Let $\Pi_S\{f(z)\} = p$ and $\Pi_S\{g(z)\} = q$. Then*

(1) $\Pi_S\{f(z) + g(z)\} \leq \max(p, q)$;

(2) $\Pi_S\{f(z)g(z)\} \leq p + q$;

(3) $\Pi_S\{f'(z)\} \leq p - 1$, *and if $p \neq 0$ then $\Pi_S\{f'(z)\} = p - 1$;*

(4) *Let*

$$\gamma(z) = \int_{z_o}^{z} f(s)ds, \qquad (4.33)$$

where $z_o = \infty$ if $p < -1$, and z_o belongs to at least one permissible domain of the function $f(z)$ in S if $p \geq -1$. The value z_o may be equal to infinity if the considered integral is convergent. Then $\Pi_S\{\gamma(z)\} = p + 1$.

Remark 4.47. Due to Proposition 4.46, Π_S *is an asymptotic ring with its characteristic function* $\Pi_S\{x(z)\}$. We denote the corresponding factor space by $\mathbf{\Pi}_S$, and $\mathbf{\Pi}_S$ is a metric space with its length function $\rho(X(z)) = r^{\overline{\Pi}_S\{X(z)\}}$ (r is a real number $r > 1$).

Consider an equation of the form

$$y^n = P(z, y, y', ..., y^{(s)}), \qquad (4.34)$$

where n and s are non-negative integers, $P(z, y, y', ..., y^{(s)})$ is a finite sum of different members of the form:

$$a(z)y^{k_0}(y')^{k_1}...(y^{(s)})^{k_s}, \qquad (4.35)$$

where $k_0, k_1, ..., k_s$, are real non-negative numbers, $a(z)$ is a holomorphic function for $|z| \gg 1$, $z \in [S]$.

We say that the number $\Sigma = k_0 + k_1 + ... + k_s$ is the *power* of expression (4.35). For instance, the power of a function $a(z) \in \Pi_S$ is equal to zero.

Proposition 4.48. *Let all the coefficients $a(z)$ of the members in (4.34) possess analytic estimates of type Π_S, and let the powers of all the members $\leq n - 1$. Then any holomorphic solution of equation (4.34) possesses an analytic estimate of type Π_S.*

PROOF. Fix a subsector S^* and let $y(z)$ be an analytic solution of equation (4.34) in S^*. Suppose that $M(|z|) = \sup_{z \in S^*, |z| \gg 1} |y(z)| > |z|^N$ for a sufficiently large numbers N. Since $y(z)$ is a holomorphic function in S^*, we have $\sup_{z \in S^*, |z| \gg 1} |y^{(m)}(z)| \leq M(|z|)|z|^{-m+\varepsilon}$ where $\varepsilon \gg 1$ ($m = 1, 2, ..., s$). Let $p = h_{S^*}\{a(z)\}$. Hence

$$|a(z)y^{k_0}(y')^{k_1}...(y^{(s)})^{k_s}| \leq |z|^{p+\varepsilon}M(|z|)^{\Sigma}|z|^{-k_1-2k_2-...-sk_s+\Sigma\varepsilon}.$$

Consequently (taking into account that $k_1 + 2k_2 + ... + sk_s \geq 0$ and $\Sigma \leq n-1$) we obtain the inequality $M(|z|) \leq |z|^{\max[p+(1+\Sigma)\varepsilon]}$, which contradicts the considered supposition which proves that $\lambda(z) \in \Pi_S$. $\square$

Example 4.49. Consider the equation

$$y' + y^2 + z = 0. \qquad (4.36)$$

On the basis of Proposition 4.46, we have $\Pi_S\{y^2(z)\} \leq \max(\Pi_S\{y'(z)\}, 1)$ for any holomorphic solution $y(z)$ for $|z| \gg 1$, $z \in [S]$. If $\Pi_S\{y^2(z)\} \leq$

$\Pi_S\{y'(z)\}$, taking into account that $\Pi_S\{y'(z)\} \le \Pi_S\{y(z)\} - 1$, we obtain the estimate $2\Pi_S\{y(z)\} \le \Pi_S\{y(z)\} - 1$ and thus $\Pi_S\{y(z)\} \le -1$. Consequently, $\Pi_S\{y(z)\} \le 1$. Moreover, since $y(z)$ is a solution of the considered equation, the set $\{y^2(z), y'(z), z\}$ possess no less than two functions of the greatest growth. Hence it is easy to see

$$y^2(z) \sim -z \quad \text{for} \quad z \to \infty, z \in [S]$$

and $\Pi_S\{y(z)\} = 1/2$. Clearly, $\Pi_S\{y'(z)\} = \Pi_S\{y(z)\} - 1 = -1/2$, which leads to the estimate $y^2(z) = -z + g(z)$, where $\Pi_S\{g(z)\} = -1/2$.

Consider a series of the form

$$\delta_0(z) + \delta_1(z) + \dots + \delta_n(z) + \dots \tag{4.37}$$

Lemma 4.50. *Let all the functions $\delta_m(z)$ possess the estimates $\Pi_S\{\delta_m(z)\}$, and let*

$$\Pi_S\{\delta_m(z)\} \to -\infty \ for \ m \to \infty \ (m = 0, 1, \dots).$$

Then, for any (fixed) closed central subsector $\tilde{S}$ whose any point $z \ne 0$ is an interior point of the sector S, there exists a function $\tilde{g}(z)$ (generally speaking, $\tilde{g}(z)$ depends on $\tilde{S}$) holomorphic for $z \in \tilde{S}$ and

$$\Pi_{\tilde{S}}\{\tilde{g}(z) - s_m(z)\} \to -\infty \ for \ m \to \infty$$

To prove this Proposition we will consider the following lemmata:

Lemma 4.51. *Let $\Pi_S\{\delta(z)\} = p$ and $\theta(z) \asymp 0$. Then*

$$\Pi_S\{\delta(z + \theta(z)) - \delta(z)\} = -\infty.$$

PROOF. Let us choose a subsector S^*. Since $\theta(z) \asymp 0$, the curves $te^{i\varphi_1^*} + \theta(te^{i\varphi_1^*})$ and $te^{i\varphi_2^*} + \theta(te^{i\varphi_2^*})$ asymptotically tend to the boundary rays $te^{i\varphi_1^*}, te^{i\varphi_2^*}$ of the sector S^* for $t \to +\infty$ (φ_1^*, φ_2^* are the angles of the boundary rays). Hence $\delta(z + \theta(z))$ is holomorphic for $|z| \gg 1$ in S^*. Clearly

$$\delta(z + \theta(z)) - \delta(z) = \int_z^{z+\theta(z)} \delta'(s)ds.$$

Since $\Pi\{\delta'(z)\} \le p - 1$ then

$$\Pi\left\{\int_z^{z+\theta(z)} \delta'(s)ds\right\} \le \Pi\{\delta'(z)\} + \Pi\{\theta(t)\} = -\infty$$

which leads to the required property.   □

Lemma 4.52. *Let all the conditions of Proposition 4.51 be fulfilled. Let $\tilde{S}$ be the sector defined in Lemma 4.50. Then there exists a function $\tilde{\delta}(z)$ which is holomorphic in $\tilde{S}$ for $|z| > 0$ and*

$$h_{\tilde{S}}\{\delta(z) - \tilde{\delta}(z)\} = -\infty.$$

PROOF. The case $p = -\infty$ is obvious. Let $p > -\infty$. The idea of the proof is simple. Let $\tilde{S} = \{z : \tilde{\varphi}_1 \leq \arg z \leq \tilde{\varphi}_2, z \neq \infty.\}$ Choose a (fixed) sector $S^* = \{z : \varphi_1^* \leq \arg z \leq \varphi_2^*, z \neq \infty\}$ such that $\varphi_1 < \tilde{\varphi}_1^*$ and $\tilde{\varphi}_2^* < \varphi_2$. There is a number $R > 0$ that $\delta(z)$ is holomorphic in the domain $S_R = S^* \cap \{z : |z| > R\}$. Then we have to choose a function $\theta(z) \asymp 0$ such that $z + \theta(z) \subset S_R$ for any $z \in \tilde{S}$. Hence the function $\tilde{\delta}(z) = \delta(z + \theta(z))$ will be holomorphic in $\tilde{S}$ (for any z) which (due to Lemma 4.51) proves this Lemma.

First consider the case when the boundary angles of the sector $\tilde{S}$ are equal to $\pm\pi/4$ (see Fig. 2).

We put $\theta(z) = 2Re^{-\mu z}$, where μ is a positive number. Clearly, $2Re^{-\mu z} \asymp 0$. It is enough to prove that it is possible to choose a positive number μ so small that the function $\tilde{\delta}(z) = \delta(z + 2Re^{-\mu z})$ will be holomorphic in

$$S_R = \left\{z : |\arg z| \leq \frac{\pi}{4}, \ z \neq 0, \ z \neq \infty\right\}.$$

To this end, it suffices to show that the curves

$$s_+(t) = te^{i\pi/4} + 2Re^{-t\mu\exp(i\pi/4)} \text{ and } s_-(t) = te^{-i\pi/4} + 2Re^{-t\mu\exp(-i\pi/4)}$$

belong to S_R for $t > 0$ ($i = \sqrt{-1}$). Let us prove the first inclusion (the second one is proved in the same way). Since $\mu \ll 1$, the curve $s_+(t)$ may leave the region S_R if it intersects the boundary ray $z = te^{i\pi/4}$ of the sector $\tilde{S}$. Let t_0 be the value corresponding to the first point of intersection $(z_0 = s_+(t_0))$. Then

$$\arg(2R\exp(-t_0\mu e^{i\pi/4})) = -\frac{3\pi}{4}.$$

Hence $t_0 = \frac{3\pi\sqrt{2}}{4\mu}$. If $t > t_0$, then the points $s_+(t)$ remain inside the circle $C(t)$ with center at the point $te^{i\pi/4}$ and radius $r = 2Re^{-3\pi/4}$. Since $t_0 \gg 1$ for $\mu \ll 1$, the circle G_t belongs to S_R for $t > t_0$. Hence $s_+(t) \subset S_R$. Thus, the considered case is proved.

The case, when the boundary angles $\tilde{\varphi}_1$ and $\tilde{\varphi}_2$ are an arbitrary (fixed) numbers $(\tilde{\varphi}_1 < \tilde{\varphi}_2)$, is reduced to the previous case by means of the following mapping: $z = y^s e^{ia}$, where

$$s = 2\frac{\tilde{\varphi}_2 - \tilde{\varphi}_1}{\pi} \quad \text{and} \quad a = \frac{\tilde{\varphi}_1 + \tilde{\varphi}_2}{2}.$$

Indeed, the sector S is mapped onto a sector S' and $\tilde{S}$ - onto $\tilde{S}' = \{y : |\arg y| \leq \frac{\pi}{4}, y \neq \infty\}$, whose any point $y \neq 0$ is an interior point of the sector S'. The function $u(y) = \delta(y^s e^{ia})$ is holomorphic for $y \neq 0$ in S' and has the estimate $\Pi\{u(y)\} = sp$. Consequently (as it was proved above), there exists a function $\tilde{u}(y)$ which is holomorphic in $\tilde{S}'$ for $|y| > 0$, and

$$h_{S'*}\{u^*(y) - u(y)\} = -\infty.$$

Thus, $\delta^*(z) = u^*(z^{1/s}e^{-ia/s})$ is the required function. $\qquad\square$

PROOF of Lemma 4.50. Without loss of generality, we may suppose that
(1) $\Pi_S\{\delta_m(z)\} \leq -m - 1$ $(m = 0, 1, ...)$.

On the basis of Lemmas 4.51 and 4.52, we may suppose that
(2) each function $\delta_m(z)$ is holomorphic for $|z| \geq 1, z \in S^*$;
(3) $S^* = \{z : |\arg z| \leq \frac{\pi}{4}, z \neq \infty\}$.

By properties (1) and (2) there is a sequence of numbers A_m such that

$$|\delta_m(z)| < A_m|z|^{-m-1} \text{ for } |z| \geq 1, z \in S^*.$$

Then we consider the modifiable series

$$\delta_0^*(z) + \delta_1^*(z) + ... + \delta_m^*(z) + ..., \tag{4.38}$$

$$\delta_m^*(z) = \delta_m(z)\left[1 - \exp\left(-\frac{z}{2^m A_m}\right)\right].$$

The inequality

$$\left|1 - \exp\left[-\frac{z}{2^m A_m}\right]\right| \leq \frac{|z|}{2^m A_m}$$

holds in S^*. Hence

$$|\delta_m^*(z)| \leq |2z|^{-m} \quad \text{for} \quad |z| \geq 1, z \in S^*.$$

Consequently, the series is convergent to a function $g(z)$ holomorphic for $|z| \geq 1, z \in S^*$. Moreover,

$$|g(z) - s_m^*(z)| \leq |r_m(z)|,$$

where $r_m(z)$ is a sum of the series

$$\delta_m^*(z) + \delta_{m+1}^*(z) + \cdots,$$

$$s_m^*(z) = \delta_0^*(z) + \delta_1^*(z) + \cdots + \delta_m^*(z).$$

We have

$$|r_m(z)| \leq 1/(2^m |z|^{m+1}).$$

Therefore

$$h_{S^*}\{g(z) - s_m^*(z)|\} \leq -m - 1.$$

Consequently,

$$\Pi_{S^*}\{g(z) - s_m(z)\} = h_{S^*}\{g(z) - s_m^*(z)\} \to -\infty$$

for $m \to \infty$.   □

Lemma 4.52 shows that the asymptotic space Π_S possesses a property which is similar to the property of asymptotic space completeness, but it does not absolutely coincide with the last one.

Definition 4.53. Let $G \subset \Pi_S$. We say that G is almost complete if for any series (4.37) such that $\delta_m(z) \in G$ and $\Pi_S\{\delta_m(z) - \delta_{m-1}(z)\} \to 0$ for $m \to \infty$ and for any (fixed) sector S^*, there exists a function $s(z)$ (generally speaking, depending on S^*) such that $\Pi_{S^*}\{s(z) - s_m(z)\} \to -\infty$ for $m \to \infty$.

On the basis of Lemma 4.50, the space Π_S is almost complete.

Consider an operator equations of the form

$$y = A(y) \tag{4.39}$$

The general properties of the equation may be obtained in the same way as in the case of the spaces Π. The main difference of the spaces Π and Π_S resides in the fact that Π_S is only almost complete. We give several propositions which will be used in the further consideration.

Definition 4.54. The function $s(z)$ is said to be a *formal solution* to equation (4.39) in S if $s(z) \asymp A(s(z))$; $s(z)$ is said to be an *asymptotic solution* to the equation if there is an exact solution $y(z)$ of the equation such that $y(z) \asymp s(z)$.

Definition 4.55. We say that $A(y)$ possesses the *L-condition* with constant σ in a region $D \subset \Pi_S$ if the following inequality holds:

$$\Pi_S\{A(y_1(z)) - A(y_2(z))\} \leq \Pi_S\{y_1(z) - y_2(z)\} - \sigma \qquad (4.40)$$

for any $y_1(z), y_2(z) \in D$.

Introduce two domains of the following form

$$V_\lambda = \{x(z) : x(z) \in \Pi_S, x(z) \sim \lambda(z) \text{ for } z \to \infty, z \in [S]\}, \qquad (4.41)$$

fix a sector S^*

$$V_\lambda^* = \{x(z) : x(z) \in \Pi_{S^*}, x(z) \sim \lambda(z) \text{ for } z \in S^*, z \to \infty\}. \qquad (4.42)$$

Lemma 4.56. *Let* $\lambda(z) \in \Pi_S$ *and* $\Pi_S\{\lambda(z)\} > -\infty$. *Let* $A(V_\lambda) \subset V_\lambda$. *Let* $A(y)$ *possess the L-condition with constant* $\sigma > 0$ *in* V_λ. *Then there exists a formal solution* $s(z) \in V_\lambda^*$ *of equation* (4.39) *in* S^*. *Besides,* $s(z)$ *is an analytic asymptotic limit of any sequence of the form* $\{s_m(z)\} \subset U_\lambda$ *where* $s_0(z)$ *is an arbitrary (fixed) function belonging to* V_λ,

$$s_m(z) = A(s_{m-1}(z)) \quad m = 1, 2, \dots \qquad (4.43)$$

and $\Pi_S\{s_m(z) - s_{m-1}(z)\} \leq -\sigma$. *Moreover, an element* $\tilde{s}(z) \in V_\lambda^*$ *is a formal solution to equation* (4.39) *in* V_λ^* *if and only if* $\Pi_{S^*}\{\tilde{s}(z) - s(z)\} = -\infty$.

Definition 4.57. The value

$$\Pi_S'\{A(\lambda(z))\} = \sup_{y_1(z), y_2(z) \in V_\lambda} [\Pi_S\{A(y_1(z)) - A(y_2(z))\} - \Pi\{y_1(z) - y_2(tz)\}],$$
$$(4.44)$$

is said to be the *derivative estimate of the operator* $A(y))$ *with respect to the variable* y *at the point* $\lambda(z)$. The notation $\Pi_S'\{A(\lambda(z))\} = d$ (where d is either a real number or $d = -\infty$) means that the derivative estimate exists at the point $\lambda(z)$ and it is equal to d.

The following proposition is obvious. Consider an equation of the form

$$y = \lambda(z) + R(y). \qquad (4.45)$$

Proposition 4.58. *Let the relation* $R(y_1(z)) \asymp R(y_2(z))$ *hold for any* $y_1(z), y_2(z) \in V_\lambda$ *and* $y_1(z) \asymp y_2(z)$. *Let* $\Pi'\{R(\lambda(z))\} = -\sigma < 0$. *Then* $R(y)$ *possesses the L-property with constant* σ *in* V_λ.

7.   IMPLICIT FUNCTIONS

In this paragraph we consider functions given by an equation in the form

$$F(x, y) = 0, \tag{4.46}$$

and we look for asymptotic representation for solutions to the equation for $x \to 0$. This problem is directly connected with the equation $F(x, \alpha(z)) = 0$ where $\alpha(z) \in C_S$.

7.1   *SOME GENERAL METHODS*

Let $F(x, y)$ be a holomorphic function as a function of two complex variables x and y for $|x| + |y| \ll 1$. Let $F(0,0) = 0$ and $\partial F(0,0)/\partial y \neq 0$. As it was proved in Proposition 4.3 (3), there exists a unique holomorphic infinitesimal solution $y = f(x)$ in the domain $|x| + |y| \ll 1$. So that the solution is a sum of a power series of the form

$$f(x) = c_1 x + c_1 x^2 + \dots + c_n x^n + \dots \tag{4.47}$$

Two methods of calculation of the coefficients c_n are well known. One of them consists of the following procedure: The numbers c_n are unknowns and they called the *indeterminate coefficients*. Substitute series (4.47) in (4.46). We obtain the identity $F(x, f(x)) = 0$ for $|x| \ll 1$ which is equivalent to an identity of the form

$$\Phi(x, c_1, c_2, \dots, c_n, \dots) = 0$$

for $|x| < \varepsilon$ (ε is a sufficiently small positive number). Thus, the left side have to be independent of the constants. We expand the function into a power series in x. All the coefficients of the series must vanish which makes possible to find the numbers $c_1, c_2, \dots, c_n, \dots$

The other method is based on the solution representation in the form of its MacLaurin series

$$f(x) = \frac{f'(0)}{1!} x + \frac{f''(0)}{2!} x^2 + \dots + \frac{f^{(n)}(0)}{n!} x^2 + \dots \tag{4.48}$$

We have to find $f'(0), f''(0), \dots$ To this end, let us differentiate (4.46) using the initial condition $y(0) = 0$. Clearly,

$$\frac{\partial F(x, y)}{\partial y} \frac{dy}{dx} + \frac{\partial F(x, y)}{\partial x} = 0.$$

Hence

$$f'(0) = -\frac{\partial F(0,0)}{\partial x} \bigg/ \frac{\partial F(0,0)}{\partial y}.$$

Next

$$\frac{\partial F(x,y)}{\partial y}y'' + \frac{\partial^2 F(x,y)}{\partial y^2}y'^2 + 2\frac{\partial^2 F(x,y)}{\partial x\partial y}y' + \frac{\partial^2 F(x,y)}{\partial x^2} = 0.$$

Hence

$$f''(0) =$$

$$-\left\{\frac{\partial^2 F(0,0)}{\partial y^2}[f'(0)]^2 + 2\frac{\partial^2 F(0,0)}{\partial x\partial y}f'(0) + \frac{\partial^2 F(0,0)}{\partial x^2}\right\} \bigg/ \frac{\partial F(0,0)}{\partial y}$$

In the same way, on differentiating (4.46) in succession n times, we may determine all the derivatives $f'(0), f''(0), ..., f^{(n)}(0)$.

In some cases the procedure to calculate the coefficients c_n may be some simpler.

7.2   *LAGRANGE'S FORMULA*

Here we consider the case when equation (4.46) is given in the form

$$x = \frac{y}{g(y)} \tag{4.49}$$

in the region $\{x,y : |x| < p, |y| < q\}$, where x and y are complex variables, p and q are positive numbers.

Proposition 4.59. *Let $g(y)$ be a holomorphic function in the considered region, $g(0) = 0$ and $g'(0) \neq 0$. Then there exists a unique holomorphic solution $y = f(x)$ for $|x| \ll 1$ infinitesimal for $x \to 0$. Let us rewrite it as a sum of a power series in the form*

$$\varphi(x) = c_1 x + c_2 x^2 + ... + c_n x^n + ... \tag{4.50}$$

Then

$$c_n = \frac{1}{n!}\{[g^n(y)]_{y=0}^{(n-1)}. \tag{4.51}$$

PROOF. The uniqueness of the solution $\varphi(z)$ and the relation $\varphi(0) = 0$ is proved in Proposition 4.3 (3). It remains to prove relations (4.51). To this end, we will require some properties of holomorphic functions.

Let $\varphi(x)$ be a holomorphic function in a circle

$$C_\delta = \{x : |x| < \delta, \delta = \text{const} > 0, x \neq 0\}$$

and possess a pole of m order at the point $z = 0$. In the other words in the circle C_δ the function is expanded into a series of the form

$$\varphi(x) = a_{-m}x^{-m} + \ldots + a_{-1}x^{-1} + a_0 + a_1 x + \ldots + a_n x^n + \ldots, \quad (4.52)$$

where a_n are constants, $n = -m, -m+1, \ldots$

Definition 4.60. In (4.52), the number a_{-1} is said to be the *residue* of the function $\varphi(x)$ in C.

Lemma 4.61. *The residue is equal to*

$$a_{-1} = \frac{1}{2\pi i} \int_\gamma \varphi(x)dx, \quad (4.53)$$

where the path of integration γ is a circumference with center at the point $x = 0$ and with a sufficiently small radius.

On multiplying (4.52) through by x^m, we obtain

$$x^m \varphi(x) = a_{-m} + a_{-m+1}x + \ldots + a_{-1}x^{m-1} + \ldots$$

On taking a derivative of $m-1$ order of both sides of this equality and substituting $x = 0$, we obtain the following formula:

$$\text{Res}_{x=0}\varphi(x) = \frac{1}{(m-1)!}\{[x^m\varphi(x)]^{(m-1)}\}_{x=0}.$$

Continuation of the PROOF of Proposition 4.59. Clearly

$$c_n = \frac{1}{2\pi i} \int_\gamma \frac{y(x)}{x^{n+1}}dx.$$

Substitute $x = y/g(y)$ in the last integral. We have

$$c_n = \frac{1}{2\pi i} \int_{\tilde\gamma} y \frac{g^{n+1}(y)}{y^{n+1}} d\frac{y}{g(y)}$$

$$= \frac{1}{2\pi i} \int_{\tilde\gamma} \frac{g^n(y)}{y^n}dy - \frac{1}{2\pi i} \int_{\tilde\gamma} \frac{g^{n-1}(y)g'(y)dy}{y^{n-1}}$$

$$= \frac{1}{2\pi i} \int_{\tilde\gamma} \frac{g^n(y)}{y^n}dy - \frac{1}{2\pi i} \int_{\tilde\gamma} \frac{[g^n(y)]'dy}{ny^{n-1}}, \quad (4.54)$$

where $\tilde\gamma$ is a circumference with center at the point $x = 0$ and with a sufficiently small radius.

Since $g^n(y)$ is holomorphic in the circle, the function is expanded into a convergent series of the form

$$g^n(y) = b_0 + b_1 y + \dots + b_m y^m + \dots \tag{4.55}$$

By (4.54) and (4.55) we have

$$c_n = b_{n-1} - \frac{n-1}{n} b_{n-1} = \frac{1}{n} b_{n-1}.$$

In the same time on differentiating (4.55) $(n-1)$ times and putting $y = 0$, we obtain

$$b_{n-1} = \left\{ \frac{1}{(n-1)!} [g^n(x)]^{(n-1)} \right\}_{x=0},$$

which leads to (4.51). $\square$

Example 4.62. Given the equation

$$y e^y = t^{-1}. \tag{4.56}$$

We look for an infinitesimal positive solution $y(t)$ to the equation for $t \to +\infty$. On putting $x = t^{-1}$, we rewrite equation (4.56) in the form $x = y/e^{-y}$ and apply Lagrange's formula (4.51). Here $g(y) = e^{-y}$. We have

$$[g^n(y)]^{(n-1)} \equiv (e^{-ny})^{(n-1)} = (-1)^{n-1} n^{n-1} e^{-ny}.$$

Hence

$$y(t) = \sum_{n=1}^{\infty} (-1)^{n-1} \frac{n^{n-1} t^{-n}}{n!}. \tag{4.57}$$

The series converges for $t > e$ and in the same time it is an asymptotic expansion of the solution for $t \to +\infty$.

Consider an equation of the form

$$y = x + R(y), \tag{4.58}$$

where $R(y)$ is a holomorphic function at $y = 0$, and $R(0) = R'(0) = 0$. Consequently, $R(y) = \frac{1}{2} R''(0) y^2 + O(y^3)$ for $y \to 0$.

Clearly, there exists a unique infinitesimal solution $y(x)$ of the equation for $x \to 0$. We have

$$y' = 1 + R'(y) y', \text{ hence } y'(0) = 1.$$

$$y'' = 1 + R''(y)(y')^2 + R'(y) y'', \text{ hence } y''(0) = R''(0).$$

$$y''' = R'''(y)(y')^3 + 3R''(y)y'y'' + R'(y)y''',$$

hence $y'''(0) = R'''(0) + 3[R''(0)]^2$. Consequently,

$$y(x) = x + \frac{1}{2}R''(0)x^2 + \frac{1}{6}R'''(0)x^3 + \frac{1}{2}[R''(0)]^2x^3 + O(x^4). \qquad (4.59)$$

Clearly, $\frac{1}{2}R''(0)x^2 = R(x) + O(x^3)$, and

$$\frac{1}{2}R''(0)x^2 + \frac{1}{6}R'''(0)x^3 = R(x) + O(x^4).$$

On substituting the obtained relations in (4.59), we obtain the following approximate formulae

$$y(x) = x + R(x) + O(x^3), \qquad (4.60)$$

$$y(x) = x + R(x) + \frac{2R^2(x)}{x} + O(x^4). \qquad (4.61)$$

For instance, find the asymptotic approximations of an infinitesimal solution (for $x \to 0$) to the equation

$$y = x + 1 - \cos 2y.$$

Clearly, $R(y) = 1 - \cos 2y = y^2 + O(y^4)$ for $y \to 0$. Hence (see (4.60) and (4.61)) we have the first and second approximations:

$$y(x) = x + 1 - \cos 2x + O(x^3) = x + 2x^2 + O(x^3).$$

$$y(x) = x + 1 - \cos 2x + \frac{2(1 - \cos 2x)^2}{x} + O(x^4) = x + 2x^2 + 8x^3 + O(x^4).$$

In conclusion of this paragraph, we consider an algebraic equation of the form

$$x = \alpha(t) + (q_2 + \alpha_2(t))x^2 + ... + (q_n + \alpha_n(t))x^n, \qquad (4.62)$$

where $\alpha(t), \alpha_2(t), ..., \alpha_n(t)$ belong to C_t, $q_2, .., q_n$ are (complex) numbers.

Proposition 4.63. *Equation (4.62) has a simple infinitesimal root $x(t)$ for $t \gg 1$, and $x(t) \in C_t$.*

PROOF Equation (4.62) turns into an algebraic equation of n-th order with constant coefficients for any fixed t, and it has a unique sufficiently small root (or the root is identically equal to zero). Hence (because of the continuous dependence of the roots on the parameter t of the equation) (4.62) has a unique infinitesimal solution $x(t)$ for $t \to +\infty$ which is continuous for $t \gg 1$. We shall prove that $x(t) \in C_t$. To this end, it remains to prove that each derivative $x^{(m)}(t)$ exists for $t \gg 1$ and $x^{(m)}(t)t^m \to 0$ as $t \to +\infty$ $(m = 1, 2, ...)$. Substitute $x(t)$ in (4.62). We have

$$x'(t) = \frac{\alpha'(t) + \alpha'_2(t)x^2(t) + ... + \alpha'_n(t)x^n(t)}{1 - x(t)[2(q_2 + \alpha_2(t)) + 3(q_3 + \alpha_3(t))x(t) + ... + n(q_n + \alpha_n(t))x^{n-2}(t)]} \tag{4.63}$$

Since $x(t)$ and $\alpha'(t)t, a'_2(t), ..., a'_n(t)$ are infinitesimal functions for $t \to +\infty$, we conclude that $x'(t)$ exists for $t \gg 1$ and $x'(t)t \to 0$ $(t \to +\infty)$. On differentiating in succession (4.63) $m-1$ times, we easily obtain that $x^{(m)}(t)$ exists for $t \gg 1$ and $x^m(t)t^m \to 0$ for $t \to +\infty$. $\square$

Chapter 5

POWER ORDER GROWTH FUNCTIONS ON THE POSITIVE SEMI-AXIS

Power functions play an important role in the mathematical analysis. We (naturally and usefully) extend the set of all power functions preserving its some 'good' asymptotic properties in a neighborhood of the infinitely large point.

As is known, it is impossible, generally speaking, to differentiate the main asymptotic relations. Nevertheless, such possibilities are preserved in many important particular cases. But we can not take this occasion when the statement of the problem is to wide. For example, let $g(t) \sim t^k$ for $t \to +\infty$, where k is an negative number. In general, we cannot say anything about the asymptotic relations between derivatives of $g(t)$ and t^k. But if, in addition, it is known that $g(z)$ (as a function of the complex argument z) is holomorphic in an sufficiently small sector S_ε and $g(z) \sim z^k$ for $z \to \infty, z \in S_\varepsilon$, we may assert that

$$g^{(n)}(t) \sim (t^k)^{(n)} = k(k-1)...(k-n+1)t^{k-n}$$

as $t \to +\infty$ for any $n = 1, 2, ...$

Since we chiefly consider problems involving the operation of differentiation, we manly look for so kind of extension when the main asymptotic relations between the considered functions are preserved for their derivatives. In this chapter, we consider power order growth functions of real argument t specified on the positive semi-axis J_+ for $t \gg 1$.

1.   REGULARLY VARYING FUNCTIONS

The well known mathematician J. KARAMATA introduced so called *regularly varying functions*. The extension, considered by Karamata, directly

90

does not connect with the operation of differentiation. The set of all regularly varying functions is sufficiently wide. They are used in many problems of the probability theory, for estimation of series sums, improper integrals etc.

Clearly, for a power function of the form $r(t) = ct^p$ (c and p are real numbers), the relation

$$\frac{r(\lambda t)}{r(t)} = \lambda^p \tag{5.1}$$

is fulfilled for any positive number λ. Karamata restrict the consideration by Lebesgue's *measurable positive functions of a real argument* such that relation (5.1) has to be fulfilled in the limit, that is

$$\lim_{t \to +\infty} \frac{r(\lambda t)}{r(t)} = \lambda^p \tag{5.2}$$

Definition 5.1. A function $r(t)$ is said to be *regularly varying* if it is positive and measurable for $t \gg 1$, and there exists a number $p \in]-\infty, +\infty[$ such that relation (5.2) is valid for any number $\lambda > 0$. The number p is called the *index* of $r(t)$. If $p = 0$, that is,

$$\lim_{t \to +\infty} \frac{r(\lambda t)}{r(t)} = 1, \tag{5.3}$$

then $r(t)$ is said to be a *slowly varying* function. We denote the set of all regularly varying functions by RVF. The designation $R\{r(t)\} = p$ means that $r(t)$ is a regularly varying function of the index p. We denote the set of all slowly varying functions by SVF.

The regular varying is a local property of a function at the point $t_0 = \infty$.

If $R\{r(t)\} = p$, then $R\{r(t)t^{-p}\} \in SVF$. Consequently, the properties of regularly varying functions easily follow from the corresponding properties of slowly varying functions.

Examples 5.2. 1° The most simple example of a regularly varying function is a power function ct^p where c is a positive number and p is a real number. Obviously $R\{ct^p\} = p$. Another simple example gives the function $ct^p(1 + \alpha(t))$ where $\alpha(t)$ is a measurable function for $t \gg 1$ and $\alpha(t) \to 0$ for $t \to +\infty$;

2° $R\{\ln t\} = 0$. Indeed,

$$\frac{\ln \lambda t}{\ln t} = 1 + \frac{\ln \lambda}{\ln t} \to 1 \text{ for } t \to +\infty;$$

$3°$ $1 + (\sin t)/t \in RVF$, but $(\sin t)/t \notin RVF$.

Here we restrict oneself to consideration of the principal properties of slowly varying functions. The following two theorems have fundamental meaning.

Theorem 5.3 (on uniform convergence). *Let $s(t)$ be a slowly varying function. Then $s(\lambda t)/s(t) \to 1$ for $t \to +\infty$ uniformly (in λ) on any fixed bounded segment $[a, b]$.*

Theorem 5.4. *Let $R\{s(t)\} = 0$. Then*

$$s(t) = \exp\left[\eta(t) + \int_T^t \frac{\alpha(t)}{t} dt\right], \qquad (5.4)$$

(for $T \gg 1$) where $\eta(t)$ is a measurable functions and $\alpha(t)$ is a continuous function for $t \gg 1$ such that there exist the limits $\lim_{t\to+\infty} \eta(t) = c \neq \infty$, and $\lim_{t\to+\infty} \alpha(t) = 0$.

Theorem 5.3 is based on the following:

Lemma 5.5. *Let $f(t)$ be a measurable function on any (fixed) interval $[a, b]$ on the real axis for $a \gg 1$, $b > a$, and let for any fixed number γ the following relation hold:*

$$f(t + \gamma) - f(t) \to 0 \ for \ t \to +\infty. \qquad (5.5)$$

Then this relation is fulfilled uniformly in γ belonging to any fixed bounded closed interval.

PROOF. First prove the proposition for $\gamma \in [0, 1]$. Suppose the contrary. Then, there exist a number $\varepsilon > 0$ and sequences $\{t_n\}$, $t_n \to +\infty$ for $n \to \infty$ and $\{\gamma_n\}$, $\gamma_n \in [0, 1]$ such that for any $n = 1, 2, ...,$

$$|f(t_n + \gamma_n) - f(t_n)| \geq \varepsilon. \qquad (5.6)$$

Define the sets U_n and V_n by the relations

$$U_n = \{\gamma : \gamma \in [0, 2], |f(t_m + \gamma) - f(t_m)| < \varepsilon/2, \ \text{for any} \ m \geq n\}, \quad (5.7)$$

$$V_n = \{\lambda : \lambda \in [0, 2], |f(t_m + \gamma_m + \lambda) - f(t_m + \gamma_m)|\} < \varepsilon/2, \ \text{for any} \ m \geq n\}.$$
$$(5.8)$$

Clearly, U_n and V_n are measurable and monotonically extending sets, and $\lim_{n \to \infty} U_n = \lim_{n \to \infty} V_n = [0, 2]$. Consequently, the measures $\mu(U_N) > 3/2$, $\mu(V_N) > 3/2$, for $N \gg 1$. Set $V'_N = V_N + \gamma_N$. Then $\mu(V'_N) = \mu(V_N) > 3/2$. Mark that $U_N \subset [0, 2] \subset [0, 3]$, and $V'_N \subset [0, 3]$. Hence $U_N \cap V'_N \neq \theta$. That is, there is a number $\gamma \in U_N$ such that $\gamma - \gamma_N \in V_N$. Therefore due to (5.7),

$$|f(t_N + \gamma) - f(t_N)| < \varepsilon/2, \qquad (5.9)$$

and from (5.8), $|f(t_N + \gamma_N + (\gamma - \gamma_N)) - f(t_N + \gamma_N)| < \varepsilon/2$, which is equivalent to the inequality

$$|f(t_N + \gamma) - f(t_N + \gamma_N)| < \varepsilon/2, \qquad (5.10)$$

This leads to the inequalities

$$
\begin{aligned}
|f(t_N + \gamma_N) \;-\; & f(t_N)| \\
= \;\; & |[f(t_N + \gamma) - f(t_N)] + [f(t_N + \gamma_N) - f(t_N + \gamma)]| \\
\leq \;\; & |f(t_N + \gamma) - f(t_N)| + |f(t_N + \gamma) - f(t_N + \gamma_N)| \\
< \;\; & \varepsilon.
\end{aligned}
$$

But this contradicts (5.6).

For an arbitrary segment $[\alpha, \beta]$, $(\beta > \alpha)$, put $\tau = (t - \alpha)/(\beta - a)$, $\tilde{f}(\tau) = f(t)$, $\delta = (\gamma - \alpha)/(\beta - \alpha)$. Consequently, $\tau \in [0, 1]$ for $t \in [\alpha, \beta]$; γ. Clearly, if the function $\tilde{f}(\tau)$ is measurable in τ, then $f(t)$ is measurable in t. As it was proved, the relation $\tilde{f}(t + \delta) - \tilde{f}(t) \to 0$ for $t \to +\infty$ uniformly in $\delta \in [a/(\beta - \alpha), b/(\beta - \alpha)]$. This means that relation (5.5) is fulfilled uniformly in $\gamma \in [a.b]$. $\qquad \square$

PROOF *of Theorem* 5.3. Put $f(t) = \ln s(e^t)$. Then $f(t + \gamma) - f(t) = \ln[s(e^{t+\gamma}/s(e^t)] \to 0$ as $t \to +\infty$ for any (fixed) number γ. Clearly, $f(t)$ is a measurable function. Hence we may apply Lemma 5.5. That is, the expression $f(t + \gamma) - f(t)$ is convergent to zero (for $t \to +\infty$) uniformly in γ on any fixed bounded interval of the real axis. Hence $(\lambda = e^\gamma)$ $s(\lambda e^t)/s(e^t)$ is convergent to unity uniformly on $[a, b]$, where $[a, b]$ is an arbitrary (fixed) interval such that $0 < a < b < +\infty$. Clearly, we may replace e^t by t which proves this theorem. $\qquad \square$

Theorem 5.4 is based on the following:

Lemma 5.6. *Let* $f(t)$ *be* n *times differentiable function for* $t \gg 1$, $n \in \{-1, 0, 1, ...\}$ *and*

$$f(t + \gamma) - f(t) \to 0 \quad for \quad t \to +\infty \qquad (5.11)$$

for any real number γ. *Then, for any fixed integer* $m = 0, 1, ...,$ *the function* $f(t)$ *can be represented in the form*

$$f(t) = \eta_m(t) + \int_T^t \alpha_m(t)dt, \qquad (5.12)$$

where $\eta_m(t)$ *is* n *times differentiable function for* $t \gg 1$ *and there exists a finite limit* $\lim_{t \to +\infty} \eta_m(t) = c_m$, $\alpha_m(t)$ *is* $(n+m)$ *times differentiable function for* $t \gg 1$ *and* $\alpha_m(t) \to 0$ *for* $t \to +\infty$.

PROOF. The Proof is made by induction with respect to $m = 0, 1, ...$ Let the Lemma be valid for m. Prove it for $m+1$. For $m = -1$, the prove is the same. We have $f(t) = \eta_m(t) + f_m(t)$, where $f_m(t) = \int_T^t \alpha_m(\tau)d\tau$. Hence

$$|f_m(t + \gamma) - f_m(t)| \leq \int_t^{t+\gamma} |\alpha_m(\tau)|d\tau \leq \sup_{\tau \in [t, t+\gamma]} |\alpha(\tau)| \to 0$$

for $t \to +\infty$. Consequently, the function $f_m(t)$ is $(m+n+1)$ times differentiable and satisfy all the conditions of Lemma 5.5 Consider the identity (which is directly checked)

$$f_m(t) = \int_t^{t+1} [f_m(t) - f_m(\tau)]d\tau + \int_T^{T+1} f_m(\tau)d\tau$$

$$+ \int_T^t [f_m(\tau + 1) - f_m(\tau)]d\tau. \qquad (5.13)$$

Assume

$$\tilde{\eta}_m(t) = \int_t^{t+1} [f_m(t) - f_m(\tau)]d\tau + \int_T^{T+1} f_m(\tau)d\tau,$$

$$\alpha_{m+1}(t) = f_m(t + 1) - f_m(t).$$

Thus, $\alpha_{m+1}(t)$ is $(m + n + 1)$ times differentiable and $\lim \alpha_{m+1}(t) = 0$. Besides (due to Lemma 5.5),

$$\left| \int_t^{t+1} [f_m(t) - f_m(\tau)]d\tau \right| \leq \sup_{\tau \in [t, t+1]} |f_m(t) - f_m(\tau)| \to 0,$$

and hence $\int_t^{t+1} [f_m(t) - f_m(\tau)]d\tau \to 0$ for $t \to +\infty$. Consequently,

$$\tilde{\eta}_m(t) \to \tilde{c}_m = \int_T^{T+1} f_m(\tau)d\tau$$

for any real number γ. Thus, $f(t) = \eta_{m+1}(t) + \alpha_{m+1}(t)$, where $\eta_{m+1}(t) = \eta_m(t) + \tilde{\eta}_m(t)$ which possesses the required property.

PROOF *of Theorem 5.4.* Let $f(t)$ be the function defined in Theorem 5.3. Then $f(t) = c(t) + \varphi(t)$, where $c(t)$ is a measurable function such that $c(t) \to c \neq \infty$ for $t \to +\infty$, $\varphi(t)$ is a continuous function.   □

The inverse theorem also takes place:

Theorem 5.7. *Let $s(t)$ be represented in the form* (5.4) *where $\eta(t)$ and $\alpha(t)$ are measurable bounded functions for $t \gg 1$ such that there are limits $\lim_{t \to +\infty} \eta(t) = c \neq \infty$ and $\lim_{t \to +\infty} \alpha(t) = 0$. Then $R\{s(t)\} = 0$.*

PROOF. Put $f(t) = \ln s(e^t)$. Clearly, $f(t)$ is a measurable function for $t \gg 1$. It follows from (5.3)

$$\ln[s(\lambda t)/s(t)] = f(t + \gamma) - f(t) \to 0 \text{ for } t \to +\infty.$$

Because of Lemma 5.6, $f(t + \gamma) - f(t)$ is convergent uniformly to zero in γ on any bounded closed interval which implies the proved property.   □

Mark that a slowly varying function $s(t)$ can be written in the form $M(t)s_0(t)$ where $M(t)$ is a positive bounded measurable function for $t \gg 1$ and

$$s_0(t) = \exp\left[\int_T^t \frac{\alpha(\tau)}{\tau} d\tau\right] \quad (T \gg 1),$$

where $\alpha(t)$ is continuous and infinitesimal for $t \to +\infty$. It is easy to see that

$$\alpha(t) = ts_0'(t)/s_0(t). \tag{5.14}$$

On the basis of Theorems 5.3 and 5.4, we may formulate the following proposition (which is possible to considered as an equivalent definition of regularly varying function).

Theorem 5.8. *For a function $r(t)$ to be regularly varying of the index p, it is necessary and sufficient to be in the following form*

$$r(t) = M(t)r_0(t), \tag{5.15}$$

where $M(t)$ is a positive measurable bounded function and $s_0(t)$ is a positive differentiable function for $t \gg 1$ and

$$tr_0'(t)/r_0(t) \to p \quad for \ t \to +\infty. \tag{5.16}$$

PROOF. Clearly, $r(t) = t^p s(t)$ where $s(t)$ is a slowly varying function. Because of Theorems 5.6 and 5.7, it is necessary and sufficient for $s(t)$ to be in the form $s(t) = M(t)s_0(t)$, where $s_0(t)$ is a positive differentiable function for $t \gg 1$ and $ts_0'(t)/s_0(t) \to 0$ for $t \to +\infty$ which leads to the required representation. $\square$

2.   CLASS A_T

We disseminate the notion of regularly varying functions to the complex valued functions of the real argument and we consider such functions where (with some exception) the main asymptotic relations for such functions are preserved for their derivatives (respectively). The property of preserving the main asymptotic relations for the derivatives is also a regular property of a function. Classes of functions considered below play analogous role in asymptotic analysis as the power functions in the classical analysis.

Thus, here we consider complex valued functions specified on the positive semi-axis J_+ for sufficiently large argument.

Definition 5.9. We say that a function $f(t)$ possesses the *power order of growth* p for $t \to +\infty$ (p is a number) if $f(t)$ is defined and does not vanish for $t \gg 1$ and

$$\frac{tf'(t)}{f(t)} - p \in C_t. \tag{5.17}$$

We denote the set of all such functions by A_t. The designation

$$P\{f(t)\} = p \tag{5.18}$$

means that $f(t) \in A_t$ and the order of $f(t)$ is equal to p.

Proposition 5.10. *For $P\{f(t)\} = p$ to be valid, it is necessary and sufficient for $f(t)$ to be in the form*

$$f(t) = t^{p+\alpha(t)}, \tag{5.19}$$

where $\alpha(t) \in C_t$ and $\alpha'(t)t\ln t \in C_t$.

To prove this proposition we will require the following:

Lemma 5.11. *Let $\beta(t) \in C_t$. Then*

$$\gamma(t) = \frac{1}{\ln t} \int_{t_0}^{t} \frac{\beta(\tau)}{\tau} d\tau \in C_t \tag{5.20}$$

and $\gamma'(t)t\ln t \in C_t$, where t_0 is a sufficiently large positive number.

PROOF. It is obvious that the function $\gamma(t)$ is defined for $t \gg 1$ and infinitesimal for $t \to +\infty$. We have

$$\gamma'(t)t = \frac{\beta(t) - \gamma(t)}{\ln t}. \tag{5.21}$$

Hence $\gamma'(t)t \to 0$ for $t \to +\infty$. On differentiating relation (5.21) $(m-1)$ times, it is easy to show that $\gamma^{(m)}(t)t^m \to 0$ for $t \to +\infty$. This means that $\gamma(t) \in C_t$. From (5.21) it follows that

$$\gamma'(t)t\ln t = \beta(t) - \gamma(t) \in C_t. \qquad \square$$

PROOF *of Proposition* 5.10. *Necessity.* Let $P\{f(t)\} = p$. Hence $f(t) \neq 0$ for $t \gg 1$ and $f(t)$ is a solution of the equation $ty'/y = p + \beta(t)$, where $\beta(t) \in C_t$. Consequently, there is a number $c \neq 0$ such that $f(t) = ct^{p+\gamma(t)}$, where $\gamma(t)$ is given in (5.20). Hence $f(t)$ is in the form (5.19), where $\alpha(t) = \gamma(t) + c/\ln t$. From Lemma 5.11 it follows that

$$\alpha(t) \in C_t \text{ and } \alpha'(t)t\ln t \in C_t.$$

Sufficiency. From (5.19), it follows that $f(t)$ is defined and does not vanish for $t \gg 1$. Moreover,

$$t\frac{f'(t)}{f(t)} - p = \alpha(t) + \alpha'(t)t\ln t \in C_t. \qquad \square$$

Remark 5.12. Because of Theorem 5.4 and Proposition 5.10 *any positive function $f(t) \in A_t$ is a regularly varying function. Moreover $R\{f(t)\}$ is equal to $P\{f(t)\}$.* Indeed $f(t)$ is measurable for $t \gg 1$ and $f(t) = t^p \exp[c/\ln t + \int_T^t [\beta(\tau)/\tau]d\tau]$, where $\beta(t)$ is continuous for $t \gg 1$ and infinitesimal for $t \to +\infty$.

Examples 5.13. A function of the form

$$\varphi(t) = ct^{p_0}(\ln_1 t)^{p_1}...(\ln_m t)^{p_m}$$

belongs to A_t and $P\{\varphi(t)\} = p_0$. Here c is a complex number, $c \neq 0$, and $p_0, p_1, ..., p_m$ are real numbers; $P\left\{e^{\sqrt{\ln t}}\right\} = 0$; $e^{(\sin t)/t} \notin A_t$ because $e^{(\sin t)/t} = t^{(\sin t)/(t\ln t)}$ and $(\sin t)/(t\ln t) \notin C_t$.

Proposition 5.14. *Let $f(t) = a + \alpha(t)$, where $a \neq 0$ is a number and $\alpha(t) \in C_t$. Then $P\{f(t)\} = 0$ $(f(t) \in A_t)$.*

PROOF. It is obvious that $f(t) \neq 0$ for $t \gg 1$ and

$$f(t) \sim a \neq 0 \text{ for } t \to +\infty.$$

Let us put $\beta(t) = tf'(t)/f(t)$. We have to prove that $\beta(t) \in C_t$. Clearly

$$\beta(t) = \alpha'(t)t/f(t) \to 0 \text{ for } t \to +\infty.$$

On differentiating the last relation $(m-1)$ times and taking into account that $\alpha^{(n)}(t)t^n \to 0$ for any (fixed) natural $n = 1, 2, ..., m$, it is easy to show that $\beta^{(m)}(t)t^m \to 0$ for $t \to +\infty$. That is, $\beta(t) \in C_t$. □

Proposition 5.15. *Let* $P\{f(t)\} = p$; *Then*

(1)
$$p = \lim_{t \to +\infty} \frac{\ln f(t)}{\ln t}; \tag{5.22}$$

(2) $P\{[f(t)]^\sigma\} = p\sigma$, *where* σ *is a number;*

(3) $\Pi\{f(t)\} = \Re p$;

(4) $P\{|f(t)|\} = \Re p$;

(5) *for any* $m = 1, 2, ...$ *there is a function* $\alpha_m(t) \in C_t$ *such that*

$$f^{(m)}(t) = [p(p-1)...(p-m+1) + \alpha_m(t)]f(t)t^{-m}, \tag{5.23}$$

hence

$$\Pi\{f^{(m)}(t)\} \leq \Re p - m, \text{ if } p \neq 0, 1, ..., m-1$$

then

$$P\{f^{(m)}(t)\} = p - m.$$

From (5.23) it follows that if $f(t) \in A_t$ *and* $f(t) \to 0$ *for* $t \to +\infty$ *then* $f(t) \in C_t$.

PROOF. Properties (1)–(4) directly follow from Lemma 5.10. Property (5) is easily proved by induction with respect to m proceeding from (5.17). □

Corollary 5.16. *By Proposition 5.15 (4) if* $P\{f(t)\} = p$ *then* $R\{|f(t)|\}$ *$= \Re p$. This means that for any positive number* λ

$$\lim_{t \to +\infty} \left| \frac{f(\lambda t)}{f(t)} \right| = \lambda^{\Re p}.$$

The following proposition is a direct consequence of Proposition 5.10.

Proposition 5.17. *Let $P\{f(t)\} = p$ and $P\{g(t)\} = q$. Then*

$$P\{f(t)g(t)\} = p + q \text{ and } P\left\{\frac{f(t)}{g(t)}\right\} = p - q.$$

Proposition 5.18 is a simple consequence of Proposition 5.17.

Proposition 5.18. *Let $P\{f(t)\} = p$ and $\Pi\{g(t)\} = q$. Then*

$$\Pi\{f(t)g(t)\} = \Re p + q.$$

PROOF.   We have (see Proposition 5.15 (3)) $\Pi\{f(t)\} = \Re p$ and $\Pi\{[f(t)]^{-1}\} = -\Re p$. Hence $\Pi\{f(t)g(t)\} \le \Re p + q$ and

$$q = \Pi\{f(t)g(t)[f(t)]^{-1}\} \le \Pi\{f(t)g(t)\} - \Re p.$$

Comparing the obtained inequalities, we obtain the required equality.
□

Proposition 5.19. *Let $P\{f(t)\} = p$ and $P\{g(t)\} = q$. Let $f(t)$ and $g(t)$ be comparable for $t \gg 1$, and let $\lim_{t \to +\infty} f(t)/g(t) \ne -1$. Then*

$$P\{f(t) + g(t)\} = \max(p, q).$$

PROOF. Let for definiteness $\lim_{t \to +\infty} f(t)/g(t) = a \ne \infty$. Then

$$f(t) + g(t) = g(t)[1 + a + \alpha(t)],$$

where $\alpha(t) = f(t)/g(t) - a$. We have to prove that $\alpha(t)$ belongs to C_t. If $a = 0$, then this follows from Proposition 5.15 (5). If $a \ne 0$, then $P\{f(t)/g(t)\} = 0$ and hence (see Proposition 5.15 (5))

$$\alpha^{(n)}(t) = \left[\frac{f(t)}{g(t)}\right]^{(n)} = o\left(t^{-n}\right).$$

That is, $\alpha(t) \in C_t$.
□

Proposition 5.20. *Let $P\{f(t)\} = p$, let $g(t)$ be a function such that each derivative $g^{(n)}(t)$ is defined for $t \gg 1$, and*

$$g^{(n)}(t) \sim f^{(n)}(t) \text{ for } t \to +\infty (n = 0, 1, ...).$$

Then $P\{g(t)\} = p$.

PROOF. Since $g(t) \sim f(t)$, we have $g(t) = f(t)(1 + \alpha(t))$ where $\alpha(t) \to 0$ $(t \to \infty)$. Due to Proposition 5.14, it is sufficient to prove that $\alpha(t) \in C_t$. Thus, we have to prove that

$$\alpha^{(n)}(t)t^n \to 0 \ \text{ for } \ t \to +\infty(n = 1, 2, ...).$$

We prove the last relations by induction with respect to $n = 0, 1, ...$ being evidently true for $n = 0$. Let $\alpha^{(m)}(t)t^m = o(1)$ for $m = 0, 1, ..., n - 1$. We have

$$g^{(n)}(t) = f^{(n)}(t)(1 + \alpha(t)) + \sum_{m=1}^{n-1} c_m f^{(n-m)}(t)\alpha^{(m)}(t) + f(t)\alpha^{(n)}(t),$$

where c_m are integers. Hence

$$\alpha^{(n)}(t)t^n = \frac{(g^{(n)}(t) - f^{(n)}(t))t^n}{f(t)} + \sum_{m=1}^{n-1} c_m \frac{f^{(n-m)}(t)t^{n-m}}{f(t)}\alpha^{(m)}(t)t^m.$$

Since $g^{(n)}(t) \sim f^{(n)}(t)$, we have (see Proposition 5.15 (5))

$$\frac{g^{(n)}(t) - f^{(n)}(t)}{f(t)}t^n = o(1).$$

In the same way it easy to show that the expression under the last $\sum$ is $o(1)$, and hence

$$\alpha^{(n)}(t) = o\left(t^{-n}\right) \ \text{ for } \ t \to +\infty. \qquad \square$$

Proposition 5.21. *Let $f(t) \in A_t$. Then $P\{f(\ln t)\} = 0$.*

PROOF. We have to prove (see Definition 5.9) that

$$\gamma(t) \equiv t\frac{df(\ln t)/dt}{f(\ln t)} \in C_t. \qquad (5.24)$$

Let $P\{f(t)\} = p$, hence

$$\frac{\tau f'(\tau)}{f(\tau)} = p + \beta(\tau)$$

where $\beta(\tau) \in C_\tau$. Hence

$$\gamma(t) = \frac{p + \beta(\ln t)}{\ln t}. \qquad (5.25)$$

Clearly $\gamma^{(n)}(t)$ is determined for any (fixed) $n = 0.1, \ldots$ as $t \gg 1$, $\gamma(t) \to 0$, and

$$\gamma'(t)t = -\frac{p + \delta(\ln t)}{\ln^2 t} \tag{5.26}$$

which follows from (5.25) where $\delta(\tau) = \beta(\tau) - \beta'(\tau)\tau \in C_\tau$. Therefore $\gamma'(t)t \to 0$ $(t \to +\infty)$. In the same way $(\gamma'(t)t)'t = \gamma''(t)t^2 + \gamma'(t)t \to 0$, and of cause $\gamma''(t)t^2 \to 0$. By induction with respect to n, on differentiating in succession (5.26) $(n-1)$ times we obtain the relation $\gamma^{(n)}(t)t^n \to 0$ $(t \to +\infty)$. This implies $\gamma(t) \in C_t$. $\qquad\square$

Proposition 5.22. *Let $P\{f(t)\} = p \neq 0$. Then $P\{\ln f(t)\} = 0$.*

PROOF. Because of Definition 5.9 we have to prove that

$$\gamma(t) \equiv t\frac{d(\ln f(t))/dt}{\ln f(t)} = \frac{tf'(t)/f(t)}{\ln f(t)}$$

belongs to C_t. We have $tf'(t)/f(t) = p + \beta(t)$, where $\beta(t) \in C_t$ and (see (5.19)) $\ln f(t) = (p + \alpha(t))\ln t$. Hence

$$\gamma(t) = \frac{p + \beta(t)}{(p + \alpha(t))\ln t}.$$

Taking into account the last expression, it is easy to show that $\gamma(t) \in C_t$. $\qquad\square$

Definition 5.23. A function $f(t)$ is said to be *integrable* in A_t if any antiderivative of the function belongs to A_t.

Proposition 5.24. *Let $P\{f(t)\} = p$ and $\Re p \neq -1$. Then*

$$\gamma(t) = \int_{t_0}^{t} f(t)dt \in A_t \quad and \quad \gamma(t) \sim \frac{f(t)t}{p+1} \quad for \ t \to +\infty.$$

Hence $P\{\gamma(t)\} = p + 1$. Here t_0 is a sufficiently large positive number if $\Re p > -1$ and $t_0 = +\infty$ if $\Re p < -1$.

PROOF. Let $\Re p > -1$. Integrate $\gamma(t)$ by parts. We obtain

$$(p+1)\gamma(t) = f(t)t - f(t_0)t_0 + \int_{t_0}^{t} \alpha(s)f(s)ds,$$

where $\alpha(t) \in C_t$. Denote by

$$\beta(t) = \int_{t_0}^{t} \frac{\alpha(s)f(s)}{f(t)t}ds.$$

Prove that $\beta(t) \to 0$ for $t \to +\infty$. We have

$$|\beta(t)| \leq \delta(t) \equiv \frac{\int_{t_0}^t |\alpha(s)f(s)|ds}{|f(t)t|}.$$

Since $P\{|f(t)t|\} = \Re p + 1 > 0$, clearly, $|f(t)t| \to +\infty$. The numerator of the last fraction does not decrease for $t \gg 1$. If the numerator is bounded for $t \to +\infty$ obviously $\delta(t) \to 0$ for $t \to +\infty$. If the numerator tends to infinity we may apply de l'Hospital's rule to $\delta(t)$. Hence

$$\lim_{t \to +\infty} \delta(t) = \lim_{t \to +\infty} \frac{|\alpha(t)f(t)|}{|f(t)t|'}$$

$$= \lim_{t \to +\infty} \frac{|\alpha(t)|}{\Re p + 1}$$

$$= 0.$$

Consequently, $\lim_{t \to +\infty} \beta(t) = 0$. We have

$$\beta'(t)t = \alpha(t) - t\frac{\beta(t)(f(t)t)'}{f(t)t} = \alpha(t) - (p + 1 + \theta(t))\beta(t) \to 0$$

for $t \to +\infty$. Here $\theta(t) \in C_t$. On differentiating the last relation $(m-1)$ times by induction with respect to m, it is easy to show that

$$\beta^{(m)}(t) = o\left(t^{-m}\right) \quad \text{for } t \to +\infty,$$

which proves the considered case.

Let $\Re p < -1$. We have

$$(p + 1)\gamma(t) = f(t)t + \int_t^{+\infty} \alpha(s)f(s)ds.$$

It is easy to show that

$$\delta(t) \equiv \int_t^{+\infty} \alpha(s)f(s)ds = o(f(t)t) \quad \text{for } t \to +\infty.$$

On differentiating the first relation m times it is easy to show that

$$(p + 1)\gamma^{(m)}(t) \sim (f(t)t)^{(m)}$$

which leads to the required relation.   □

The following proposition is a consequence of Proposition 5.24.

Proposition 5.25. *Let* $P\{f(t)\} = p$, *where* $\Re p \neq -1$. *Then the function* $f(t)$ *is integrable in* A_t.

PROOF. Any antiderivative of $f(t)$ may be represented in the form $J(t) = C + \gamma(t)$, where $\gamma(t)$ is determined in Proposition 5.24 and C is a number. If $C = 0$, the required property is proved in Proposition 5.24. Let $C \neq 0$. Then $P\{C\} = 0$ and taking into account that $\gamma(t) \to 0$ or ∞, C and $\gamma(t)$ comparable for $t \to +\infty$, hence the required property is a simple consequence of Proposition 5.19.

$\square$

3. SUBSETS OF THE CLASS A_T WITH SOME ALGEBRAIC AND ANALYTIC PROPERTIES

To obtain an asymptotic solution to an equation with variable coefficients, it is necessary (as a rule) to reduce the equation to a form acceptable for solution. However the operation of summation of functions of the class A_t may already overstep the limits of the considered class. For example, 1 and $(\sin \ln t)/t - 1$ belong to A_t, but their sum $(\sin \ln t)/t$ does not belong to A_t.

In order to preserve the desired properties of the equation coefficients in process of their transformation, we some contract the class A_t and pick out subsets closed to the used operations. The defined bellow fields of type M and N are differential fields. They consist of the power order functions and zero, they are closed to the main algebraic operations and the operation of differentiation. It is important that such a field can be extended with roots of any algebraic equation with coefficients belonging to the field. Then we consider some more wide sets consisting of asymptotically equal functions to a field of the considered types. In terms of the sets, it may be formulate many important asymptotic problems and their solutions of differential and difference equations.

Definition 5.26. A subset Q of the set $A_t \cup 0$ is said to be a *field of type* N if it is an algebraic field, it contains t and the set $\mathbf{C}$ of all complex numbers, and if $q(t) \in Q$, then $q'(t) \in Q$ and there is a (finite or infinite) limit $\lim_{t \to +\infty} q(t)$.

The simplest example of a field of type N is the set RF of all rational fractions in t with constant complex coefficients. Of course, it is the minimal field of type N.

Proposition 5.27. *Let Q be a field of type N. Let $q(t) \in Q$ and $q(t) \not\equiv 0$. Then there exists a (finite or infinite) limit $\lim_{t \to +\infty} t \ln q(t)$.*

PROOF. If $q(t) \not\to 1$ for $t \to +\infty$, then $\lim_{t \to +\infty} t \ln q(t) = \infty$. Let $\lim_{t \to +\infty} q(t) = 1$, then $q(t) = 1 + s(t)$ where $s(t) \in Q$ and $s(t) \in C_t$. Hence either

$$t \ln q(t) \sim ts(t) \text{ or } t \ln q(t) \equiv 0,$$

which leads to the required relation. $\qquad\square$

Definition 5.28. *We say that G is a field of type M if G is a field of type N and for any function $g(t) \in G$, the function $\bar{g}(t) = \Re g(t) - i \Im g(t)$ belongs to G.*

Proposition 5.29. *Let $g(t) \in G$, where G is a field of type M. Then*

(1) *$\Re g(t)$, $\Im g(t)$, and $|g(t)|^2$ belong to G;*

(2) *if $g(t) \not\equiv 0$, then $P\{g(t)\}$ is a real number;*

(3) *if $|g(t)| \to c \neq \infty$ for $t \to +\infty$ and $g(t) \not\equiv c$, then either $|g(t)| > |c|$ of $|g(t)| < |c|$ for $t \gg 1$.*

PROOF. Prove property (1). We have

$$\Re g(t) = \frac{1}{2}[g(t) + \bar{g}(t)], \quad \Im g(t) = i\frac{1}{2}[\bar{g}(t) - g(t)], \quad \text{and} \quad |g(t)|^2 = g(t)\bar{g}(t)$$

belong to G.

Prove property (2). Let $g(t) \not\equiv 0$, and let (for definiteness) $\Im g(t) \not\equiv 0$. Since $\Im g(t) \in G$, clearly, $\Im g(t) \in A_t$ and $\Im g(t) \neq 0$ for $t \gg 1$. Hence $P\{\Im g(t)\}$ is a real number. Moreover, $\gamma(t) \equiv \Re g(t)/\Im g(t) \in G$. Let $\lim_{t \to +\infty} \gamma(t) = q \neq \infty$. Then $g(t) = \Im g(t)(1 + iq + o(1))$. Hence $P\{g(t)\} = P\{\Im g(t)\}$ (because $1 + iq \neq 0$). Let $\lim_{t \to +\infty} \gamma(t) = \infty$. Then it is easy to see, $P\{g(t)\} = P\{\Re g(t)\}$, which is a real number.

Prove property (3). We have $|g(t)|^2 \in G$ and $|g(t)|^2 \to |c|^2$ for $t \to +\infty$. Consequently, $|g(t)|^2 - |c|^2 = p(t) \in G$. Since $p(t) \not\equiv 0$, we conclude that $p(t) \in A_t$. Moreover, $p(t)$ is a real function. Hence it preserves its sign for $t \gg 1$. If $p(t) > 0$, then $|g(t)|^2 > |c|^2$ and, hence, $|g(t)| > |c|$; if $p(t) < 0$, then, clearly, $|g(t)| < |c|$ $(t \gg 1)$. $\qquad\square$

Definition 5.30. *We say that $Q \subset A_t \cup O_t$ is an asymptotic field of the space Π, if it is an asymptotic space with characteristic function $\Pi\{x(t)\}$, and for any $q_1(t), q_2(t) \in Q$, the following relation holds: $\Pi\{q_1(t)q_2(t)\} = \Pi\{q_1(t)\} + \Pi\{q_2(t)\}$. The last relation is either a number equality or both* sides of them equal to $-\infty$. Besides, if $q(t) \in Q$ and $q(t) \not\asymp 0$, then $1/q(t) \in Q$ and $\Pi\{1/q(t)\} = -\Pi\{q(t)\}$.

Proposition 5.31. *A field Q of type N (M) is an asymptotic field with characteristic function $\Pi\{x(t)\}$.*

PROOF. Let $p(t), q(t) \in Q$. Clearly, $Q \subset \Pi$. Hence $p(t) + q(t) \in \Pi$ and

$$\Pi\{p(t) + q(t)\} \leq \max[\Pi\{p(t)\}, \Pi\{q(t)\}].$$

Consequently Q is an asymptotic space. Moreover, if $p(t)$ and $q(t) \not\equiv 0$ then $p(t)$ and $q(t)$ belong to A_t. Therefore

$$P\{p(t)q(t)\} = P\{p(t)\} + P\{q(t)\} \text{ and } P\{1/q(t)\} = -P\{q(t)\}.$$

Hence (for any $p(t), q(t) \in Q$ and $q(t) \neq 0$) we have (see Proposition 5.18)

$$\Pi\{p(t)q(t)\} = \Pi\{p(t)\} + \Pi\{q(t)\} \text{ and } \Pi\{1/q(t) = -\Pi\{q(t)\}. \qquad \square$$

Definition 5.32. Let Q be a field of type N. By $\{Q\}$ we denote a set of all functions such that if $q(t) \in \{Q\}$ then for each natural m there exists a function $q_m(t) \in Q$ such that $\Pi\{q(t) - q_m(t)\} < -m$. The set $\{Q\}$ is said to be a *resulting space* of the field Q. Every function $q(t) \in \{Q\}$ is said to be *asymptotically close* to the set Q.

Proposition 5.33. *Let Q be a field of type N, then*

(1) *if $q(t) \in \{Q\}$, then $q(t) + \theta(t) \in \{Q\}$ for any $\theta(t) \asymp 0$. In particular $\theta(t) \in \{Q\}$; there is a finite or infinite limit $\lim_{t \to +\infty} q(t)$.*

(2) *the resulting space $\{Q\}$ is an asymptotic field.*
If (in addition) Q is a field of type M then for $q(t) \in \{Q\}$.

(3) *$\bar{q}(t), \Re q(t), \Im q(t),$ and $|q(t)|^2$ belong to $\{Q\}$. If $\lim_{t \to +\infty} q(t) = c \neq \infty$ and $q(t) \not\equiv c$ for $t \gg 1$, then either $|q(t)| > |c|$ or $|q(t)| < |c|$ for $t \gg 1$.*

PROOF. If $q(t) = q_m(t) + \alpha_m(t)$ where $q_m(t) \in Q$ and $\Pi\{\alpha_m(t)\} < -m$, then $q(t) + \theta(t) = q_m(t) + \tilde{\alpha}_m(t)$, where $\tilde{\alpha}_m(t) = \alpha_m(t) + \theta(t)$. Clearly, for any $\theta(t) \asymp 0$ $\Pi\{\tilde{\alpha}_m(t)\} \leq \max[\Pi\{\alpha_m(t)\}, \Pi\{\theta(t)\}] < -m$. Hence if $q(t) \in \{Q\}$ then $q(t) + \theta(t) \in \{Q\}$. Since $0 \in Q$, we conclude that any $\theta(t) \asymp 0$ belongs to $\{Q\}$.

If $\Pi\{q(t)\} < 0$ then $\lim_{t \to +\infty} q(t) = 0$. If $\Pi\{q(t)\} \geq 0$ then $q(t) = q_0(t) + \alpha_0(t)$ where $q_0(t) \in Q$, and $\Pi\{\alpha_0(t) < 0$. Clearly, $\Pi\{q_0(t)\} = \Pi\{q_0(t)\} \geq 0$. Hence $q_0(t) \in A_t$ and $q(t) = q_0(t)(1 + \beta(t))$ where $\beta(t) = \alpha(t)/q_0(t)$. Consequently, (see Proposition 5.18) $\Pi\{\beta(t)\} < 0$. That is, $\beta(t) = o(1)$

for $t \to +\infty$. We have $\lim_{t \to +\infty} q(t) = \lim_{t \to +\infty} q_0(t)$. Thus, point (1) is proved.

Prove point (2). Let $q(t) \in \{Q\}$ and $\Pi\{q(t)\} = q > -\infty$. then $q(t) = \tilde{q}(t) + \{\tilde{\alpha}(t)\}$, where $q(t) \to Q$ and $\Pi\{\tilde{\alpha}(t)\} < \Pi\{q(t)\}$. Hence $q(t) = \tilde{q}(t)(1 + \beta(t))$ where $\beta(t) \in C_t$. Consequently (see Proposition 5.14) $P\{q(t)\} = P\{\tilde{q}(t)\}$. That is, $p(t) \in A_t$. Thus, we have proved that if $p(t) \in \{Q\}$ then $p(t) \in A_t \cup O_t$. It is easy to prove that if $p(t)$ and $q(t)$ belong to $\{Q\}$, then $p(t) + q(t)$ and $p(t)q(t)$ belong to $\{Q\}$. Moreover, if in addition $q(t) \in A_t$ then $1/q(t) \in \{Q\}$. For instance, prove that $p(t) + q(t) \in \{Q\}$. Indeed, if $p(t) + q(t) \asymp 0$, then (see point (1) of this proposition) $p(t) + q(t) \in \{Q\}$. If $p(t) + q(t) \not\asymp 0$ then for any (fixed) $m \gg 1$ $p_m(t) + q_m(t) + \alpha_m(t)$, where $p_m(t), q_m(t) \in Q$, and $\Pi\{\alpha_m(t)\} < -m$ ($m > \Pi\{p(t) + q(t)\}$). Since $p_m(t) + q_m(t) \in Q$ we conclude that $p(t) + q(t) \in \{Q\}$. For $p(t), q(t) \in A_t$ the relation $P\{p(t)q(t)\} = P\{p(t)\} + P\{q(t)\}$ is obvious. If at least one of the functions belong to O_t, then both the sides of last equality is $-\infty$. Thus, $\{Q\} = \{q(t)\}$ is an asymptotic field with the characteristic function $\Pi\{q(t)\}$. Property (3) is proved in the same way as in Proposition 5.29.

$\square$

Remark 5.34. Clearly, Q is isomorphic to the factor field with respect to $\{Q\}$. Consequently, instead of the factor space we may consider the field Q.

4.   ALGEBRAIC EXTENSIONS OF FIELDS OF TYPE N_S

It is possible to extend fields of type N by roots of algebraic equations with coefficients belonging to fields of type N. Beforehand it will be necessary to consider some properties of algebraic equations with variable coefficients. Let

$$H(t, y) \equiv y^n + a_1(t)y^{n-1} + \ldots + a_n(t) = 0 \qquad (5.27)$$

be an algebraic equation with continuous coefficients for $t \gg 1$. We say that $\lambda(t)$ is a *root* of this equation (or of the polynomial $H(t, y)$) in a domain $D \subset J_+$ if $\lambda(t)$ is a continuous solution of equation (5.27) in the domain. Let

$$\Lambda = \{\lambda_1(t), \lambda_2(t), \ldots, \lambda_n(t)\} \qquad (5.28)$$

be a set of roots of equation (5.27) (some of them may be equal in pairs) in the considered domain.

Definition 5.35. Let the identity

$$H(t, y) = (y - \lambda_1(t))(y - \lambda_2(t))\ldots(y - \lambda_n(t)) \qquad (5.29)$$

hold in D. We say that Λ is a *complete set* (*complete system*) of *roots* of equation (5.27) and of the polynomial $H(t, y)$ (in D).

Definition 5.36. Let (5.28) be a complete set of roots of equation (5.27) (of the polynomial $H(t, y)$) in a domain $[T, +\infty[$ where T is a sufficiently large positive number. A root $\lambda(t) \in \Lambda$ is said to be *asymptotically $k-$multiple* (or, which is the same, *asymptotically multiple of order k*) for $t \to +\infty$ if there are exactly k roots of the set Λ (including the considered one) equivalent to $\lambda(t)$ for $t \to +\infty$. If $k = 1$, then we say that $\lambda(t)$ is an *asymptotically simple root* of equation (5.27) (polynomial $H(t, y)$) for $t \to +\infty$.

Example 5.37. Equation

$$y^3 - (2t + 1)y^2 + (t^2 + t)y - t^2 + t = 0$$

has a complete set of roots $\{1, t + \sqrt{t}, t - \sqrt{t}\}$, where 1 is an asymptotically simple root; $t + \sqrt{t}$ and $t - \sqrt{t}$ are asymptotically double roots for $t \to +\infty$.

The next Proposition directly follows from the continuous dependence of a root on the coefficients of the polynomial. Consider an equation of the form

$$P(x, t) \equiv (q_0 + \alpha_0(t))x^n + (q_1 + \alpha_1(t))x^{n-1} + ... + (q_n + \alpha_n(t)) = 0. \quad (5.30)$$

Proposition 5.38. *Let q_j be numbers such that there is at least one number $q_r \neq 0$ for $r < n$, $\alpha_j(t) \in C_t$ $(j = 0, 1, ..., n)$. Let us designate the limiting equation by*

$$P(\mu, \infty) \equiv q_0\mu^n + q_1\mu^{n-1} + ... + q_n\mu^n = 0. \quad (5.31)$$

Then if the limiting equation has a k-multiple zero root, then equation (5.30) has exactly k infinitesimal roots for $t \to +\infty$ of its complete set of roots. If the limiting equation has k-multiple root $\mu \neq 0$, then equation (5.30) has asymptotically k-multiple roots equivalent to μ for $t \to +\infty$.

Lemma 5.39. *Consider an equation of the form*

$$u(t) = \alpha(t) + A(t, u), \quad (5.32)$$

where

$$A(u, t) = (q_2 + \alpha_2(t))u^2 + ... + (q_n + \alpha_n(t))u^n.$$

Here q_i are numbers, $\alpha(t)$ and $\alpha_i(t)$ are functions belonging to C_t. Then there exists a unique infinitesimal solution $u^*(t)$ of equation (5.32), $u^*(t) \sim \alpha(t)$ for $t \to +\infty$, and $u^*(t) \in C_t$. Moreover, if $\alpha(t) \in A_t$, then $P\{u^*(t)\} = P\{\alpha(t)\}$.

PROOF. The existence and uniqueness of the infinitesimal solution $u^*(t)$ is a simple consequence of Proposition 5.38. Substitute the solution in (5.32). We obtain the identity

$$u^*(t) = \alpha(t) + A(t, u^*(t)).$$

Hence

$$u^*(t) = \alpha(t)(1 + o(1)) \to 0 \text{ for } t \to +\infty.$$

Consequently,

$$u^*(t) \sim \alpha(t) \text{ for } t \to +\infty.$$

On differentiating the identity m times (by induction with respect to $m = 1, 2, ...$) it is easy to show that

$$u^{(m)}(t) = o\left(t^{-n}\right) \text{ for } t \to +\infty.$$

Hence $u^*(t) \in C_t$. Let $\alpha(t) \in A_t$. As it follows from the considered identity (taking into account that $u^*(t) \in C_t$)

$$u^*(t) = \alpha(t)(1 + \beta(t)),$$

where $\beta(t) \in C_t$. On the basis of Propositions 5.14 and 5.17 we conclude that $P\{u^*(t)\} = P\{\alpha(t)\}$. $\square$

Lemma 5.40. *Let $a_i(t) \in Q$ $(i = 1, 2, ..., n)$, where Q is a field of type N. Let $\lambda(t) \not\equiv 0$ be a root of equation (5.27). Then there exist a function $q(t) \in Q$ and a rational number r such that*

$$\lambda(t) \sim q^r(t) \quad for \ t \to +\infty.$$

PROOF. Since $\lambda(t) \not\equiv 0$, then at least one $a_i(t) \not\equiv 0$. Hence there exists a function $g(t)$ of the greatest growth of the set

$$\{[a_j(t)]^{1/j}\} \ (j = 1, 2, ..., n)$$

such that $g(t) \neq 0$ for $t \gg 1$. Put $y = g(t)x$. This leads to equation (5.30) where $q_0 + \alpha_0(t) = 1$. We distinguish the two cases:

(1) $\lambda(t) \sim \mu g(t)$ for $t \to +\infty$ where μ is a non-zero solution of equation (5.31),

(2) $\lambda(t) = o(g(t))$ for $t \to +\infty$.

In the first case the function $\mu g(t)$ possesses the required property. In the second case, there exists a zero root of equation (5.31). Hence $q_n = 0$. Suppose $q_{n-s} \neq 0$ and $q_{n-s+1} = \ldots = q_n = 0$. Since $q_j = 1$ for at least one $j \in \{1, 2, \ldots, n\}$, clearly, $s < n$. Consider the set

$$\{[\alpha_{n-s+j}(t)]^{1/j}\} \quad (j = 1, 2, \ldots, s).$$

It is obvious that at least one of the functions does not vanish. Let $g_1(t)$ be a function of the greatest growth of the considered set which is a rational power of a function of field Q. Make the substitution $x = g_1(t)u$ in (5.27). This leads to the equation

$$\beta_0(t)u^n + \ldots + \beta_{n-s-1}(t)u^{s+1} + u^s$$
$$+(b_1 + \beta_{n-s+1}(t))u^{s-1} + \ldots + (b_s + \beta_n(t)) = 0, \qquad (5.33)$$

where $\alpha_i(t) \in C_t$ $(i = 0, 1, \ldots, n)$, b_j $(j = 1, 2, \ldots, s)$ are numbers such that not all b_j are zero. We distinguish two cases. The first case: there exists a number $\mu \neq 0$ such that

$$\lambda(t) \sim \mu g(t)g_1(t) \text{ for } t \to +\infty,$$

and hence the function $\mu g(t)g_1(t)$ possesses the required property. In the second case

$$\lambda(t) = o\left(g(t)g_1(t)\right) \text{ for } t \to +\infty.$$

But as before, by means of finite steps (at most $n-1$) we may show that there exist functions $g(t), g_1(t), \ldots, g_{n-2}(t)$ and a number $\mu \neq 0$ such that

$$\lambda(t) \sim \mu g_0(t)g_1(t)\ldots g_{n-2}(t),$$

where each $g_i(t)$ is equal to a rational power of a function belonging to Q. The last product can be rewritten as $q^r(t)$ where $q(t) \in Q$ and r is a rational number. Hence

$$\lambda(t) \sim q^r(t) \text{ for } t \to +\infty. \qquad \square$$

Lemma 5.41. *Assume the hypothesis of Lemma 5.40 and in addition let $\lambda(t)$ be an asymptotically simple root of equation (5.27) for $t \to +\infty$. Then $\lambda(t) \in A_t$.*

PROOF. In view of Lemma 5.40 the root $\lambda(t)$ is equivalent to a function $q^r(t)$ for $t \to +\infty$. Consequently, on substituting $y = q^r(t)(1 + u)$ in (5.27) we get

$$(q_0 + \alpha_0(t))u^n + (q_1 + \alpha_1(t))u^{n-1} + \ldots + (q_s + \alpha_s(t))u^{n-s}$$

$$+\alpha_{s+1}(t)u^{n-s-1} + ... + \alpha_n(t) = 0$$

where $q_i = \text{const}$ and

$$\alpha_j(t) \in C_t \quad (i = 0, 1, ..., s, j = 0, 1, ..., n)$$

and $q_s \neq 0$. This means that there exist s asymptotically multiple roots of equation (5.27) (equivalent to $q^r(t)$). Since $\lambda(t)$ is asymptotically simple, clearly, $s = 1$ and the last equation can be represented in the form (5.32). That is, $u(t)$ is a unique infinitesimal solution and $u(t) \in C_t$. Hence $P\{\lambda(t)\} = rP\{q(t)\}$, which means that $\lambda(t) \in A_t$.   □

Theorem 5.42. *Let $a_i(t) \in Q$ $(i = 1, 2, ..., n)$ where Q is a field of type N, and let $\lambda(t) \not\equiv 0$ be a root of equation (5.27). Then there exists a field P of type N such that $Q \subset P$ and $\lambda(t) \in P$.*

PROOF. The proof is made by induction on n being trivial for $n = 1$. Assume the statement for any natural $p < n$:

1. Here we prove that $\lambda(t) \in A_t$. In the process of the proof we extend the field Q by finitely many roots of polynomials of degrees no more than $n-1$. For the proof simplification, taking into account the supposition of induction, we may assume that Q already contains all such functions. We prove this assertion by induction with respect to the asymptotic multiplicity m of the root. The case $m = 1$ is proved in Lemma 5.41. Suppose $m > 1$. Let us retrace the formation of the asymptotic approximation of the root $\lambda(t)$ in the proof of Lemma 5.41. Consider the set

$$\{[a_j(t)]^{1/j}\} \quad (j = 1, 2, ..., n).$$

We may suppose that the function $g(t)$ of the greatest growth of this set belongs to Q. Indeed, otherwise $g(t)$ is a unique function of the greatest growth and $g(t) = [a_n(t)]^{1/n}$. Then the substitution $y = [a_n(t)]^{1/n}x$ in (5.27) leads to equation (5.30) where $q_0 = q_n = 1$ and $q_1 = q_2 = ... = q_{n-1} = 0$. Hence its limiting equation turns into equation $\mu^n + 1 = 0$. This means that all the roots of equation (5.27) are asymptotically simple, but it is impossible. Let us substitute $y = q^r(t)x$ in (5.27). It leads to equation (5.30) where the limiting equation possesses an m-multiple root $\mu = 1$. Hence the substitution $x = 1 + v$ leads to the equation

$$\begin{aligned}(b_0 + \beta_0(t))v^n \quad &+ \quad ... + (b_{n-m-1}\beta_{n-m-1}(t))v^{m+1} + v^m \\ &+ \quad (b_{n-m+1} + \beta_{n-m+1}(t))v^{m-1} + ... \qquad (5.34) \\ &+ \quad (b_n + \beta_n(t)) = 0,\end{aligned}$$

where b_i are numbers, $\beta_i(t) \in C_t$ and $\beta_i(t) \in Q$ $(i = 0, 1, ..., n)$. For any root $v(t) \to 0$ for $t \to +\infty$ there exists a root $y(t) = q^r(t)(1 + v(t))$ of equation (5.27). Now we make a substitution of the form $v = w + h(t)$, where $h(t) \in C_t$ and $h(t) \in Q$ so that the equation (in w) may have all infinitesimal roots with asymptotic multiplicity $m_1 < m$. To this and, it suffices to obtain the equation where the term of the power $m-1$ vanishes. On substituting and making the corresponding simplification in the obtained equation, we get the equation

$$(b_0 + \gamma_0(t))v^n + ... + (b_{n-m-1}\gamma_{n-m-1}(t))v^{m+1} + v^m$$
$$+(b_{n-m+1} + \gamma_{n-m+2}(t))v^{m-2} + ... + (b_n + \gamma_n(t)) = 0, \qquad (5.35)$$

where $\gamma_i(t)$ are infinitesimal functions $(i = 0, 1, ..., n)$ and $h(t)$ is an infinitesimal solution to the equation

$$(b_0 + \beta_0(t))\binom{n}{m-1}u^{n-m+1} + ... \quad + \quad (b_{n-m-1} + \beta_1(t))\binom{n-1}{m-1}u^{m+1} + ...$$
$$+ \quad \mu + \beta_{n-m+1}(t) = 0. \qquad (5.36)$$

Since $m > 1$, the power of the last equation is no more than $n - 1$. All its coefficients belong to Q. Hence each its solution belongs to Q. But equation (5.36) possesses a unique infinitesimal solution $h(t)$ which is the required function.

There exist exactly m infinitesimal solutions of equation (5.36). If $\gamma_{n-m+j}(t) = 0$ $(j = 2, ..., m)$, then

$$\lambda(t) = q^r(t)(1 + h(t)) \in Q.$$

Otherwise equation (5.36) possesses infinitesimal roots whose asymptotic multiplicity is no more than $m - 1$. Indeed, if the asymptotic multiplicity is m, then all the infinitesimal solutions $w_i(t)$ are equivalent to the same function (say $\alpha(t)$) for $t \to +\infty$. Hence the coefficient $\beta_{n-m-1}(t)$ of the equation is equivalent to the function $-m\alpha(t)$ for $t \to +\infty$, but it is impossible because $-m\alpha(t) \not\equiv 0$ for $t \gg 1$. Thus,

$$\lambda(t) = q^r(t)(1 + h(t) + w(t)),$$

where $h(t) + w(t) \in Q$, and since the last function is infinitesimal, the function belongs to C_t. Consequently (see Propositions 5.14, 5.15, and 5.17), $\lambda(t) \in A_t$.

2. Prove the existence of the required algebraic extension of the field Q. Let P be the minimal algebraic extension of the field Q containing the root $\lambda(t)$. Check the fulfillment of hypothesis of a field of type N. Since $Q \subset P$, then $\mathbb{C} \cup t \subset P$. We may suppose (without loss of generality) that (5.27)

is a minimal polynomial of the root $\lambda(t)$. Then P is an n-dimensional space over the field Q possessing a basis

$$\{1, \lambda(t), \lambda^2(t), ..., \lambda^{n-1}(t)\}.$$

If $p(t) \in P$, then there are functions $q_0(t), q_1(t), ..., q_{n-1}(t)$, belonging to Q such that

$$p(t) = q_0(t) + q_1(t)\lambda(t) + ... + q_{n-1}(t)\lambda^{n-1}(t). \tag{5.37}$$

The function $p(t)$ is a root of a (not identical zero) polynomial with coefficients belonging to Q. In view of point (1), $p(t) \in A_t \cup 0$ and

$$p(t) \sim q^r(t) \text{ for } t \to +\infty,$$

and hence there exists a (finite or infinite) limit

$$\lim_{t \to +\infty} p(t) = \lim_{t \to +\infty} q^r(t).$$

Now it is enough to check that $p'(t) \in P$. Clearly, it suffices to show that $\gamma(t) \in P$. Since $H(t, y)$ is a minimal polynomial, we have $\partial H(\lambda(t), t)/\partial y \not\equiv 0$, and

$$\lambda'(t) = -\frac{\partial H(\lambda(t), t)/\partial t}{\partial H(\lambda(t), t)/\partial y} \in P. \qquad \square$$

Remark 5.43. As it was shown in the Proof of the Theorem (see point 2), *the minimal field of type* N, *which contains* Q *and the root* $\lambda(t)$, *is a minimal algebraic extension of the field* Q *by the root* $\lambda(t)$.

Theorem 5.44. *Let* $a_i(t) \in G$ *(*$i = 1, 2, ..., m$*), where* G *is a field of type* M, *and let* $\lambda(t)$ *be a root of equation (5.27). Then there exists a field* P *of type* M *such that it contains the field* G *and the root* $\lambda(t)$.

PROOF. The function $\overline{\lambda}(t)$ is a root of the polynomial

$$H^*(t, y) \equiv y^n + \overline{a}_1(t)y^{n-1} + ... + \overline{a}_n(t). \tag{5.38}$$

Since $\overline{a}_i(t) \in G$, clearly, $\overline{\lambda}(t)$ is an algebraic element over the field Q. Consider a minimal algebraic extension P of the field Q by the functions $\lambda(t)$ and $\overline{\lambda}(t)$ which is a field of type N. Any function belonging to P is represented as a polynomial in $\lambda(t)$ and $\overline{\lambda}(t)$ and with coefficients belonging to Q. Consequently, if $p(t) \in P$, the conjugate function $\overline{p}(t)$ is a polynomial in $\lambda(t)$ and $\overline{\lambda}(t)$ with coefficients belonging to Q. Hence $\overline{p}(t) \in P$, which proves this Theorem.

In algebra there exists a theorem about the algebraic closure existence and uniqueness (with accuracy of the algebraic equivalence) of an algebraic field. Like this there exists a theorem about fields of type N and M. A field P is an *algebraic closure* of algebraic field Q if it is a minimal algebraic field containing Q and any root of every polynomial (which does not vanish identically) with coefficients belonging to P. Indeed, the field P consists of all such finite algebraic extensions of the field Q, and hence every element of P is a root of a polynomial (which does not vanish identically) with coefficients belonging to Q. This leads to the following proposition.

Theorem 5.45. *An algebraic closure P of the field Q of type N (M) is a field of type N (M).*

PROOF. Let us check the hypothesis of a field of type N (M). Clearly, $C \cup t \subset P$. If $p(t) \in P$, then $p(t)$ belongs to a finite algebraic extension of Q which is a field of type N (M). Therefore $p'(t) \in P$ and there exists a finite or infinite limit $\lim_{t \to +\infty} p(t)$ (and $\bar{p}(t) \in P$ if Q is a field of type M). $\qquad\square$

5.   ALGEBRAIC EXTENSIONS BY ROOTS OF AN ALGEBRAIC SYSTEM

If it is given a system of algebraic equations with coefficients belonging to a field Q of type N (M), we may extend the field by each component of its solution almost without restrictions.

For convenience of the reader we give a short outline of the theory of algebraic systems.

Let us set

$$P_1(x) = a_0 x^n + a_1 x^{n-1} + ... + a_n = 0, \quad P_2(x) = b_0 x^m + b_1 x^{m-1} + ... + b_m = 0,$$

where all the coefficients are numbers.

Definition 5.46. A determinant of $n + m$ order of the form

$$R(P_1, P_2) = \left. \begin{vmatrix} a_0 & a_1 & ... & a_n & & & \\ & a_0 & a_1 & ... & a_n & & \\ & & & & & & \\ & & a_0 & a_1 & ... & a_n \\ b_0 & b_1 & ... & b_m & & & \\ & b_0 & b_1 & ... & b_m & & \\ & & & & & & \\ & & b_0 & b_1 & ... & b_m \end{vmatrix} \right\} \begin{matrix} m \\ \\ n \end{matrix} \qquad (5.39)$$

is called the *resultant* of the polynomials P_1 and P_2 (everywhere at the places where it is nothing written, we imply zeros).

We give the following assertion without proof. *If $R(P_1, P_2)$ equals to zero, then either the polynomials have a common different of a constant multiplier or $a_0 = b_0 = 0$. The inverse assertion is also fulfilled.* Consider a system of n algebraic equations with n unknowns with coefficients belonging to Q of the form

$$
\begin{cases}
P_1(X) \equiv \varphi_{10}(\tilde{X})x_n^{n_1} + \varphi_{11}(\tilde{X})x_n^{n_1-1} + ... + \varphi_{1n_1}(\tilde{X}) = 0, \\
P_2(X) \equiv \varphi_{20}(\tilde{X})x_n^{n_2} + \varphi_{21}(\tilde{X})x_n^{n_2-1} + ... + \varphi_{2n_2}(\tilde{X}) = 0, \\
\quad\quad \cdots\cdots\cdots\cdots\cdots\cdots\cdots\cdots \\
P_n(X) \equiv \varphi_{n0}(\tilde{X})x_n^{n_n} + \varphi_{n1}(\tilde{X})x_n^{n_n-1} + ... + \varphi_{nn_n}(\tilde{X}) = 0,
\end{cases}
$$
$$(5.40)$$

where $\tilde{X} = (x_1, ..., x_{n-1})$, and $\varphi_{ij}(\tilde{X})$ are polynomials in $\tilde{X}$. We will prove the following

Theorem 5.47. *Let all the coefficients of system* (5.40) *belong to* Q. *Let* $X_0 = (x_{10}, ..., x_{n0})$ *be an isolate solution of system* (5.40). *Then, for each* $i = 1, 2, ..., n$, x_{i0} *is an algebraic element of* Q.

Remark 5.48. X_0 is called an *isolated solution* if there exists a number $\varepsilon > 0$ such that there are no solutions in the domain $U_\varepsilon = \{X : 0 < ||X - X_0|| < \varepsilon\}$.

Beforehand, we discuss the problem how it is possible to solve system (5.40). We consider a routine procedure which may lead to the desired result. Let the system have an isolate solution $X_0 = (x_{10}, x_{20}, ..., x_{n0})$. For definiteness, we are going to look for the coordinate x_{10}. Consider the first two equations of the system and eliminate the unknown x_n. To this end, compose their resultant $R(P_1, P_2)$ which is a polynomial in the variable $\tilde{X}$. Since both the considered equations have the solution X_0, clearly, the equation $R(P_1, P_2) = 0$ possesses the solution $\tilde{X}_0 = (x_{10}, x_{20}, ..., x_{n-10})$. As it follows from (5.40), all the coefficients of the last polynomial belong to Q. In the same way, we compose the equations $R(P_1, P_3) = 0, ..., R(P_1, P_n) = 0$. This leads to a system of $n-1$ algebraic equations with $n-1$ unknowns. After that we eliminate the unknown x_{n-1}, x_{n-2}, and so on. Finally, we arrive to one algebraic equation in the unknown x_1 possessing the solution x_{10}.

Example 5.49. Given the system

$$
\begin{cases}
y^2 + x^2 - 3x - 2 = 0 \\
2y \quad\quad - x - 3 = 0
\end{cases}
$$

To eliminate the unknown y, we compose the resultant of the equations and equate the obtained determinant to zero

$$\begin{vmatrix} 1 & 0 & x^2 - 3x - 2 \\ 2 & -x - 3 & 0 \\ 0 & 2 & -x - 3 \end{vmatrix} = 0.$$

We have $5x^2 - 6x + 1 = 0$. The last equation has two roots $x_1 = 1$ (correspondingly, y=2) and $x = 1/5$ (y=8/5). Thus, we solve the given system and obtained all its solutions (1,2) and (1/5, 8/5).

PROOF *of Theorem* 5.47. Prove the Theorem for the first coordinate x_{10} of the solution X_0 (for the other coordinates the prove is made in the same way). We have to prove that there exists a polynomial (say $p(x)$) with coefficients belonging to Q such that $p(x) \not\equiv const$ and $p(x_{10}) = 0$. If the system permits to apply the routine procedure, then, clearly, x_{10} is an algebraic element. However, the routine procedure may lead to the case when the final polynomial vanishes identically. We shall not discuss the cause how this case may arose. We will give a rule to avoid this difficulty. Compose a perturbed system of the form

$$\begin{cases} P_{1\varepsilon}(X) \equiv P_1(X) + \varepsilon P_{11}(X) = 0, \\ P_{2\varepsilon}(X) \equiv P_2(X) + \varepsilon P_{21}(X) = 0, \\ \cdots\cdots\cdots\cdots\cdots\cdots \\ P_{n\varepsilon}(X) \equiv P_n(X) + \varepsilon P_{n1}(X) = 0. \end{cases}$$

Here $P_{j1}(X)$ are polynomials in X with coefficients belonging to Q $(j = 1, 2, ..., n)$. Clearly, it is possible to choose the polynomials with the following restrictions: the system must have finitely many solutions for $\varepsilon = 1$ and all the solutions must be isolated and can be obtained by means of the routine procedure. In all other respects the polynomials are arbitrarily chosen. Consequently, the final polynomial has a perturbed solution $x_{10\varepsilon} = x_{10} + f(x_{10}, \varepsilon))$, and $p_\varepsilon(x)$ cannot be constant identically because it is impossible at the point $\varepsilon = 1$. Owing to the resultant definition, the last polynomial (in x) is simultaneously a polynomial in ε. Hence we may it write down in the following form

$$p_\varepsilon(x) = p_0(x) + \varepsilon p_1(x) + ... + \varepsilon^s p_s(x),$$

where $p_j(x)$ are polynomials in x with coefficients belonging to Q $(j = 0, 1, ..., s)$. Of cause, $p_0(x) \equiv 0$ and $s > 0$. Moreover, at least one of the polynomials is not a constant. Let k be the least number such that $p_k(x) \not\equiv$

0. Then the equation $p_\varepsilon(x) = 0$ can be rewritten in the form

$$p_k(x) + \varepsilon p_{k+1}(x) + \ldots + \varepsilon^{s-k} p_s(x) = 0.$$

Clearly, x_{10} is a root of the polynomial $p_k(x) = 0$.   □

Corollary 5.50. *If all the coefficients of system* (5.40) *belong to a field* Q *of type* N (M) *and the system possesses an isolate solution* $X_0(t) = (x_{10}(t), x_{20}(t), \ldots, x_{n0}(t))$, *then there exists a field* $\tilde{Q}$ *of type* N (M) *such that* $\tilde{Q}$ *contains* Q *and all the functions* $x_{10}(t), x_{20}(t), \ldots, x_{n0}(t)$.

6.   NON-ALGEBRAIC EXTENSIONS

Consider some possibilities to extend a field of type N by non-algebraic functions. Consider a differential equation of the form

$$F(t, y, y') \equiv y' + \frac{P_1(t, y)}{P_2(t, y)} = 0, \tag{5.41}$$

where $P_1(t, y), P_2(t, y)$ are polynomials in y with variable coefficients.

Proposition 5.51. *Let all coefficients of equation* (5.41) *belong to a field* Q *of type* N (M). *Let equation* (5.41) *have a solution* $g(t) \in \{Q\}$ *where* $\Pi\{g(t)\} > -\infty$. *Then there exist a field* G *of type* N (M) *and a solution* $y(t)$ *of equation* (5.41) *such that* $y(t) \asymp g(t)$, $y(t) \in G$, *and* $Q \subset G$.

PROOF. Let Q be a field of type N. Consider a polynomial of the form

$$H(t, y) = y^n + a_1(t) y^{n-1} + \ldots + a_n(t) \tag{5.27}$$

with coefficients $a_i(t)$ ($i = 1, 2, \ldots, n$) belonging to Q. Since $g(t) \in \{Q\}$, only the two following cases are possible:

(1) there exists a polynomial of the form (5.27) such that $H(t, g(t)) \asymp 0$,

(2) for any polynomial of the form (5.27), $\Pi\{H(t, g(t))\} > -\infty$.

Case (1). Let

$$\Lambda \equiv \{\lambda_1(t), \lambda_2(t), \ldots, \lambda_n(t)\} \tag{5.28}$$

be a complete set of roots of polynomial (5.27), and let G be the algebraic extension of the field Q by the roots Λ. Clearly, G is also a field of type N. Because of the relation $H(t, g(t)) \asymp 0$, there exists at least one root $\lambda(t) \in G$ such that $\lambda(t) \asymp g(t)$. Show that the function $y(t) = \lambda(t)$ is the required solution. Indeed, it is obvious that $F(t, y(t), y'(t)) \in G$ and $F(t, y(t), y'(t)) \in O_t$. But any function of G belongs to $A_t \cup 0$. Hence $F(t, y(t), y'(t)) = 0$.

Case (2). Let G be a set of all rational functions in $x = g(t)$ with coefficients belonging to Q. Let us check that G is a field of type N. Since $g(t) \in \{Q\}$, and any polynomial (which does not vanish identically) does not belong to O_t, we conclude that any such rational function either belongs to A_t or is an identical zero. Clearly, $Q \subset G$. Hence $\mathbb{C} \cup t \subset G$. Moreover, $G \subset \{Q\}$. Consequently, there exists a (finite or infinite) limit $\lim_{t \to +\infty} f(t)$ for any $f(t) \in G$. By the identity $F(t, g(t), g'(t)) = 0$ we have $g'(t) \in G$. Since $f'(t)$ is a linear function in $g'(t)$ with coefficients belonging to G, we have $f'(t) \in G$. Thus, G satisfies all the axioms of a field of type N.

Let Q be a field of type M.

Case (1). Clearly $y(t)$ is an algebraic element over the field Q. That is, G is a field of type M which contains Q and $y(t)$.

Case (2). The set G_1 of all rational functions in $x = g(t)$ is a field of type N. Consider the equation

$$F^*(t, y, y') \equiv y' + \frac{\overline{P}_1(t, \overline{y})}{\overline{P}_2(t, \overline{y})} = 0.$$

The function $\overline{g}(t)$ is a solution to the last equation. If the function $\overline{g}(t)$ is an algebraic element of the field G_1 then the minimal algebraic extension G of the field G_1 is an n-dimensional linear space over the field G_1. Hence if $f(t) \in G$ then $\overline{f}(t) \in G$. That is, G is a field of type M. Otherwise, the required field G is a set of all rational fractions of two variables in $x = g(t)$ and $y = \overline{g}(t)$. We have if $f(t) \in G$ then $\overline{f}(t) \in G$. Thus, G is a field of type M. □

Example 5.52. Consider the function

$$e(t) = \int_t^{+\infty} \frac{e^{t-\tau}}{\tau} d\tau$$

(see (1.12)). It is obvious that $e(t) \in \{RF\}$ where RF is a field of all rational fractions in t with constant complex coefficients (RF is a minimal field of type M). The function $e(t)$ is a solution of the equation

$$y' - y + \frac{1}{t} = 0,$$

$e(t) \sim 1/t$ for $t \to +\infty$, consequently, $P\{e(t)\} = -1$. Hence all the conditions of Proposition 5.51 are satisfied. Due to this Proposition there is a field $G \supset RF$ of type M and a solution $y^*(t) \in G$ of the obtained equation such that $y^*(t) \asymp e(t)$. The considered equation is linear and all its solutions can be given in the form $y(t) = Ce^t + y^*(t)$, where C is an arbitrary constant. If $C \neq 0$ then, clearly, $h\{y(t)\} = h\{e^t\} = +\infty$. This means that $y^*(t)$ is its unique solution with the estimate $h\{y^*(t)\} < +\infty$. Thus, $e(t) = y^*(t) \in G$.

Proposition 5.53. *Let all the coefficients of the polynomials $P_1(t)$ and $P_2(t)$ belong to a field Q of type N. Let $y(t)$ be a solution of equation (5.41) belonging to A_t. Let, for any (fixed) function $q(t) \in Q$, only the following two cases be possible: $y(t) = o(q(t))$ or $q(t) = o(y(t))$ for $t \to +\infty$. Then the field R of all rational fractions in $y(t)$ with coefficients belonging to Q is a field of type N.*

PROOF. Consider a polynomial of the form (5.27) with coefficients belonging to Q. Show that the set

$$\{p_j(t) = a_j(t)y^{n-j}(t)\} \ (j = 0, 1, ...n, \ a_0(t) = 1)$$

has one and only one function of the greatest growth. Indeed, if all $a_i(t) \equiv 0$ $(i = 1, 2, ..., n)$ then $y^n(t)$ is the required function. Let $a_j(t) \not\equiv 0$ for at least one number $j \in \{1, 2, ..., n\}$ and let $p_s(t)$ and $p_r(t)$ be two non-zero functions of the considered set. Let for definiteness $s < r$. Then

$$\frac{p_s(t)}{p_r(t)} = \frac{a_s(t)}{a_r(t)} y^{s-r}(t)$$

tends to 0 or ∞, which proves our assertion. As a consequences there are only the two following possibilities:

(1)
$$H(t, y(t)) = q(t)y^m(t)(1 + \alpha(t)),$$

where $q(t) \in Q$ and m is a natural number, $\alpha(t)$ is a sum of infinitesimal functions belonging to A_t. Hence $\alpha(t) \in C_t$. Consequently, $H(t, y(t)) \in A_t$.

(2) $H(t, y(t)) = 0$, then $H(t, y) \equiv 0$.

Let us prove that R is a field of type N. Indeed, $C \cup t \subset R$; if $r(t) \in R$, then $r(t)$ is a ratio of two functions belonging to A_t or $r(t) \equiv 0$, hence $R \subset A_t \cup 0$. Moreover $r(t) \sim q(t)y^s(t)$ for $t \to +\infty$ where $q(t) \in Q$ and s is an integer. Hence there is a (finite or infinite) limit $\lim_{t \to +\infty} r(t)$. To prove that $r'(t)$ belongs to R it is enough to prove that $y'(t)$ belongs to R. The last follows from the identity $F(t, y(t), y'(t)) = 0$. □

Proposition 5.54. *Let all the conditions of Proposition 5.53 be fulfilled, and let, in addition, Q be a field of type M, and $y(t)$ is a real function. Then the field R obtained in Proposition 5.53 is a field of type M.*

PROOF. We have to prove that for any $r(t) \in R$, the function $\bar{r}(t)$ belongs to R. But, clearly, $\bar{r}(t)$ is a rational fraction in $y(t)$ with coefficients belonging to Q, and hence $\bar{r}(t) \in R$. □

Example 5.55. $1°$ Owing to Proposition 5.54 the minimal extension of the field RF by the function $\ln t$ is a field of type M. Indeed, it is easy to check all the conditions of this Proposition. In particular, the function $\ln t$ is a solution of the equation $y' - 1/t = 0$. In the same way we can extend successively the field RF by the functions $\ln_m t$ $(m = 1, 2, ...)$.

$2°$ Prove that the minimal extension of the field L of all power-logarithmic functions by the function $y(t) = e^{\sqrt{\ln t}}$ is a field of type M. Indeed, check the fulfillment of the conditions of Proposition 5.54: (1) we have

$$t\frac{y'(t)}{y(t)} = \frac{1}{2\sqrt{\ln t}} \in C_t.$$

Hence $P\{y(t)\} = 0$ $(y(t) \in A_t)$; (2) $y(t)$ is a solution to the equation

$$y' - \frac{y}{2t\sqrt{\ln t}} = 0$$

with coefficients belonging to L; (3) let $l(t) \in L$. Then

$$l(t) \sim ct^{k_0}(\ln_1 t)^{k_1}...(\ln_m t)^{k_m} \text{ for } t \to +\infty$$

where c is a complex number, $k_0, k_1, ..., k_m$ are real numbers. Hence $l(t) = o\left(e^{\sqrt{\ln t}}\right)$ if $k \le 0$, and $e^{\sqrt{\ln t}} = o(l(t))$ for $t \to +\infty$ if $k > 0$ which leads to the required property.

7. ESTIMATES OF ROOTS OF ALGEBRAIC EQUATIONS WITH COMPARABLE COEFFICIENTS

In this paragraph we give some simple and useful estimates of the roots of an algebraic equation with variable coefficients. The main results follow from Lemma 5.58. They are used for estimation the proximity of the logarithmic derivatives of solutions of a linear differential equation to the corresponding root of its characteristic equation. They also have independent meaning.

Definition 5.56. We say that a set of functions U is a *field of comparable functions* or (which is the same) a *field of type CF* if it is a field containing the set c of all complex numbers, and if $u(t) \in U$ then there exists a (finite or infinite) limit $\lim_{t \to +\infty} u(t)$.

Remark 5.57. Because of the Definition for any two functions $u_1(t)$, $u_2(t) \in U$ (where at least one of them is not an identical zero) there exists at least one finite limit

$$\lim_{t \to +\infty} \frac{u_1(t)}{u_2(t)} \quad \text{or} \quad \lim_{t \to +\infty} \frac{u_2(t)}{u_1(t)}.$$

Indeed, let (for definiteness) $u_2(t) \not\equiv 0$. Since U is a field, $u_1(t)/u_2(t) \in U$ which implies the required property.

Given the equation

$$H(t, y) \equiv (y - \lambda_1(t))...(y - \lambda_n(t)) = \alpha(t). \tag{5.42}$$

Lemma 5.58. *Let all the functions $\lambda_i(t)$ and $\alpha(t)$ belong to an algebraic closed field U of type CF. Then there exists the identity*

$$\tilde{H}(t, y) \equiv (y - \lambda_1(t))...(y - \lambda_n(t)) - \alpha(t) = (y - \tilde{\lambda}_1(t))...(y - \tilde{\lambda}_n(t)), \tag{5.43}$$

where

$$\tilde{\lambda}_i(t) = \lambda_i(t) + \varphi_i(t).$$

Here

$$\varphi_i(t) = O\left(\alpha^{1/n}(t)\right) \quad for \quad t \to +\infty \quad (i = 1, 2, ..., n).$$

PROOF. Let $\lambda_1(t)$ be an asymptotically $s-$multiple root of the equation $H(t, y) = 0$ (for $t \to +\infty$), and let there be the finite limits $\lim_{t \to +\infty} [\lambda_i(t)/\lambda_1(t)]$ for any root $\lambda_i(t)$ of the equation. Substitution $y = \lambda_1(t)u$ in (5.42) leads to the equation

$$(u - 1 - \mu_1(t))...(u - 1 - \mu_s(t))(u - \nu_1(t))...(u - \nu_{n-s}(t)) = \frac{\alpha(t)}{\lambda_1^n(t)}, \tag{5.44}$$

where all $\mu_i(t) = o(1)$, all $\nu_j(t) = O(1)$ and $\nu_j(t) \not\to 1$ ($i = 1, 2, ..., s$ and $j = 1, 2, ..., n - s$). Being trivial for $n = 1$ we suppose (by induction) that this Lemma is true for any polynomial of the considered type of any power $m < n$. Consider all the possible cases:

(1) Let

$$\lim_{t \to +\infty} \frac{\alpha(t)}{\lambda_1^n(t)} = 0, \quad \text{and} \quad s < n.$$

Let $\tilde{u}_i(t) = \tilde{\lambda}_i(t)/\lambda_1(t)$. Since the roots of equation (5.44) continuously depend on $\alpha(t)/\lambda^n(t)$, all the functions $\tilde{u}_i(t)$ are bounded for $t \geq 1$. Moreover, $\tilde{u}_j(t) \to 1$ and $\tilde{u}_k(t) \not\to 1$ for $j = 1, 2, ..., s$, $k = s + 1, ..., n$ ($t \to +\infty$). Let

$$\tilde{u}(t) \in \{\tilde{u}_1(t), ..., \tilde{u}_s(t)\}.$$

Then $\tilde{u}(t)$ is a root of the equation

$$(u - 1 - \mu_1(t))(u - 1 - \mu_2(t))...(u - 1 - \mu_s(t)) = \psi(t),$$

where

$$\psi(t) = \frac{\alpha(t)/\lambda^n(t)}{(\tilde{u}_k(t) - \nu_1(t))...(\tilde{u}_k(t) - \nu_{n-s}(t))} = O\left(\alpha(t)/\lambda_1^n(t)\right)$$

for $t \to +\infty$. By induction all the roots $u_m^*(t)$ $(m = 1, 2, ..., s)$ of the last equation have the estimates

$$u_m^*(t) = 1 + \mu_m(t) + O\left(\alpha(t)/\lambda_1^n(t)\right) \quad \text{for } t \to +\infty.$$

Since $\tilde{u}(t)$ is one of the roots, clearly, the corresponding root of equation (5.43) has the estimate

$$\tilde{\lambda}_j(t) = \lambda_j(t) + O\left(\alpha(t)\lambda_1^{s-n}(t)\right).$$

Taking into account that $s - n < 0$ and $\alpha(t) = o\left(\lambda^n(t)\right)$, we get $[y(t)]^{m-n} = o\left([\alpha(t)]^{(m-n)/n}\right)$ which implies the required relation

$$y_j(t) = y_j(t) + O\left(\alpha^{1/n}(t)\right) \quad \text{for } t \to +\infty.$$

The function $\tilde{u}_k(t)$ $(k = s + 1, ..., n)$ is a root of the equation

$$(u - \nu_1(t))(u - \nu_2(t))...(u - \nu_{n-s}(t)) = \psi^*(t),$$

where

$$\psi^*(t) = \frac{\alpha(t)/\lambda_1^n(t)}{(\tilde{u}_k(t) - 1 - \mu_1(t))...(\tilde{u}_k(t) - 1 - \mu_s(t))} = O\left(\alpha(t)/\lambda_1^n(t)\right)$$

for $t \to +\infty$. It easy to show (in the same way as in the last investigation)

$$\tilde{\lambda}_k(t) = \lambda_k(t) + O\left(\alpha^{1/n}(t)\right).$$

Thus, the considered case is proved.

Let

$$\lim_{t \to +\infty} [\alpha(t)/\lambda^n(t)] = 0 \quad \text{and} \quad s = n.$$

Substitution $y = \lambda_1(t)u$ in (5.42) leads to the equation

$$(u - 1 - \mu_1(t))...(u - 1 - \mu_n(t)) = \psi(t),$$

where all $\mu_i(t) = o(1)$, and $\psi(t) = \alpha(t)/\lambda_1^n(t) = o(1)$ $(i = 1, 2, ..., n; t \to +\infty)$. Substitute $u = 1 + v + \delta(t)$ in the last equation where

$$\delta(t) = (\mu_1(t) + ... + \mu_n(t))/n.$$

We have
$$(v - \delta_1(t))(v - \delta_2(t))...(v - \delta_n(t)) = \psi(t).$$

Here $\delta_i(t) \to 0$ for $t \to +\infty$ and
$$\delta_1(t) + \delta_2(t) + ... + \delta_n(t) = 0.$$

We suppose (without loss of generality) that
$$\lim_{t \to +\infty} \delta_i(t)/\delta_1(t) \neq \infty \text{ for any } i = 1, 2, ..., n$$

(otherwise we may change the numeration of the roots). Notice that if at least one of the functions $\delta_i(t)$ does not vanish then $\delta_1(t)$ can not be n-asymptotically multiple root of the equation
$$(v - \delta_1(t))(v - \delta_2(t))...(v - \delta_n(t)) = 0.$$

Indeed, if $\delta_1(t)$ is n-asymptotically multiple then the sum
$$\delta_1(t) + \delta_2(t) + ... + \delta_n(t)$$

must not be equal to zero. If all the functions $\delta_i(t)$ are zeros then equation (5.42) may be written in the form
$$(y - \lambda(t))^n = \alpha(t)$$

which leads to the required relations. In the other case (as it was proved above), the corresponding roots of equation (5.42) can be written in the form
$$\tilde{u}_i(t) = 1 + \mu_i(t) + O\left(\alpha^{1/n}(t)/\lambda_1(t)\right)$$

which lead to the required estimates. Thus, case (1) is proved.

(2) Let
$$\lim_{t \to +\infty} \alpha(t)/\lambda_1^n(t) = \infty.$$

It is easy to show that equation (5.42) has a complete set of roots $\{\tilde{u}_j(t)\}$ $(j = 1, 2, ..., n)$ such that,
$$u_j(t) \sim \varepsilon_j \alpha^{1/n}(t)/\lambda_1(t) \text{ for } t \to +\infty,$$

where ε_j are different roots of nth degree from 1, which implies the required estimates for the roots $\tilde{\lambda}_j(t)$ $(j = 1, 2, ..., n)$.

(3) Let
$$\lim_{t \to +\infty} \alpha(t)/\lambda_1^n(t) \neq 0, \infty.$$

Then all the roots of equation (5.42) have finite limits. The last assertion can be written in the form

$$\tilde{u}_j(t) = \frac{\lambda_j(t)}{\lambda_1(t)} + O(1)$$

or (which means the same)

$$\tilde{u}_j(t) = \frac{\lambda_j(t)}{\lambda_1(t)} + O\left(\frac{\alpha^{1/n}(t)}{\lambda_1(t)}\right).$$

Hence

$$\tilde{\lambda}_j(t) = \lambda_j(t) + O\left(\alpha^{1/n}(t)\right) \quad \text{for } t \to +\infty \text{ and any } j = 1, 2, ..., n. \quad \square$$

Consider an equation of the form

$$(y - \lambda_1(t))...(y - \lambda_n(t)) + \alpha_0(t) + \alpha_1(t)y + ... + \alpha_{n-1}(t)y^{n-1} = 0. \quad (5.45)$$

Proposition 5.59. *Let $\lambda_i(t)$ and $\alpha_j(t)$ belong to U where U is an algebraic closed field of type CF $(i = 1, 2, ..., n, \; j = 1, 2, ..., n-1)$. Let (for definiteness) $\lambda_{k+1}(t) = O(\lambda_k(t))$ and any $k = 1, 2, ..., n - 1$; let $\alpha_r(t) = o(\lambda_1(t)...\lambda_{n-r}(t))$ and any $r = 0, 1, ..., n - 1$ $(t \to +\infty)$. Then equation (5.45) can be rewritten in the following equivalent form:*

$$\tilde{H}(t, y) \equiv (y - \tilde{\lambda}_1(t))...(y - \tilde{\lambda}_n(t)) = 0, \qquad (5.46)$$

where all $\tilde{\lambda}_i(t)$ belong to U and $\tilde{\lambda}_i(t) \sim \lambda_i(t)$ as $t \to +\infty$ for any $i = 1, 2, ..., n$.

PROOF. The case $\lambda_1(t) = 0$ is trivial. Let $\lambda_1(t) \neq 0$, then the limits

$$\lim_{t \to +\infty} [\lambda_i(t)/\lambda_1(t)] = c_i \quad \text{for any } i = 1, 2, ..., n$$

exist and finite (that is $c_i \neq \infty$; clearly, $c_1 = 1$). Substitute $y = \lambda(t)u$ in equation (5.45). As the result we obtain the equation

$$(u - c_1 - \mu_1(t))...(u - c_n - \mu_n(t)) + \beta_0(t) + \beta_1(t)u + ... + \beta_{n-1}u^{n-1} = 0,$$

$$(5.47)$$

where all $\mu_i(t)$ and $\beta_j(t)$ are infinitesimal functions for $t \to +\infty$. On the basis of the theorem on continuous depends of the roots on an algebraic equation, we may assert that the last equation has a complete set of roots $\{u_i(t) = c_i + o(1)\}$. Thus, for any $c_i \neq 0$ the corresponding root

$$\tilde{\lambda}_i(t) = \lambda_1(t)(c_i + \mu_i(t)) \sim \lambda_i(t)$$

and for any $c_j = 0$ $\tilde{\lambda}_j(t) = o\,(\lambda_1(t))$ for $t \to +\infty$. Let $c_m \neq 0$ and $c_{m+1} = 0$ (it means that $c_{m+1} = c_{m+2} = ... = c_n = 0$). Let us divide equation (5.45) by $(y - \tilde{\lambda}_1(t))...(y - \tilde{\lambda}_m(t))$. As the result we obtain an algebraic equation of the degree $m - n$ which can be written in the following form:

$$F(t,y) \equiv (y - \lambda_{m+1}(t))...(y - \lambda_n(t)) + \beta_0(t) + \beta_1(t)y + ... + \beta_{n-m}(t)y^{n-m-1}$$

$$= \frac{(y - \lambda_1(t))...(y - \lambda_n(t)) + \alpha_0(t) + \alpha_1(t)y + ... + \alpha_{n-1}(t)y^{n-1}}{(y - \tilde{\lambda}_1(t))...(y - \tilde{\lambda}_m(t))}.$$

$$(5.48)$$

Let us estimate all the functions $\beta_j(t)$. Since the last expression is an identity, we may put $y = 0$ in it. Hence

$$\beta_0(t) =$$

$$(-1)^{n-m+1}\left\{\lambda_{m+1}(t)...\lambda_n(t)\left[1 - \frac{\lambda_1(t)...\lambda_m(t)}{\tilde{\lambda}_1(t)...\tilde{\lambda}_m(t)}\right] - \frac{\alpha(t)}{\tilde{\lambda}_1(t)...\tilde{\lambda}_m(t)}\right\},$$

that leads to the estimate

$$\beta_0(t) = o\,(\lambda_{m+1}(t)...\lambda_n(t)) \quad \text{for } t \to +\infty.$$

Differentiating identity (5.48) j times with respect to y and putting $y = 0$, we may estimate the function $\beta_j(t)$. We have $\beta_j(t) = o\,(\lambda_{m+1}(t)... \lambda_{m+n-j}(t))$ for $t \to +\infty$. Thus, the proposition (being trivial for $n = 1$) is proved by induction with respect to n.

Proposition 5.60. *In the polynomial $H(t,y)$ (see(5.27)), let all the coefficients $a_i(t)$ belong to $\{Q\}$ where Q is an algebraic closed field of type N (M) $(i = 1, 2, ..., n)$. Let $\lambda(t)$ be a root of this polynomial. Then*

(1) *there exists a function $\lambda^*(t) \in \{Q\}$ such that $\mathrm{h}\{\lambda(t) - \lambda^*(t)\} = -\infty$;*
(2) *if, in addition,*

$$\mathrm{h}\{\lambda(t) - \tilde{\lambda}(t)\} > -\infty$$

for each other root $\tilde{\lambda}(t)$ of the polynomial, then $\lambda(t) \in \{Q\}$.

PROOF. Choose the functions $a_{im}(t) \in Q$ so that

$$\Pi\{a_i(t) - a_{im}(t)\} < -m, \quad \text{where } m = 1, 2, ...$$

Clearly, the polynomial

$$H_m(t,y) = y^n + a_{1m}(t)y^{n-1} + ... + a_{nm}(t)$$

has a root

$$\lambda_m^*(t) \in Q \text{ and } h\{\lambda(t) - \lambda_m^*(t)\} \to -\infty \text{ for } m \to \infty.$$

On the basis of Theorem 4.17 there exists an asymptotic limit $\lambda^*(t) \in \{Q\}$ of the sequence $\{\lambda_m^*(t)\}$ (that is, $\lambda^*(t)$ satisfies the following relation $\Pi\{\lambda^*(t) - \lambda_m^*(t)\} \to -\infty$ for $m \to \infty$). Clearly,

$$h\{\lambda(t) - \lambda^*(t)\} = h\{(\lambda(t) - \lambda_m^*(t)) + (\lambda_m^*(t) - \lambda^*(t))\}$$

$$\leq h\{\lambda(t) - \lambda_m^*(t))\} + h\{\lambda^*(t) - \lambda_m^*(t))\}$$

$$\to -\infty$$

for $m \to \infty$. Since the difference $\lambda(t) - \lambda^*(t))$ is independent of m, we conclude that

$$h\{\lambda(t) - \lambda^*(t))\} = -\infty$$

and case (1) is proved.

Prove property (2). First, let $h\{\lambda(t)\} = -\infty$. Since

$$h\{\lambda(t) - \tilde{\lambda}(t))\} > -\infty,$$

we have $h\{a_{n-1}(t)\} > -\infty$ and hence $a_{n-1}(t) \in A_t$. Moreover, $a_n(t) \in O_t$ (differently, the estimate of any root is more than $-\infty$). The estimate $h\{\lambda'(t)\} = -\infty$ follows from the relation

$$\lambda'(t) = -\frac{a_n'(t) + a_{n-2}'(t)\lambda^2(t) + \dots + a_1'(t)\lambda^{n-1}(t)}{a_{n-1}(t) + 2a_{n-2}(t)\lambda(t) + \dots + (n-1)a_1(t)\lambda^{n-2}(t) + n\lambda^{n-1}(t)}.$$

On differentiating the last relation $(m-1)$ times, we obtain (by induction with respect to $m = 1, 2, \dots$) the estimates $h\{\lambda^{(m)}(t)\} = -\infty$. The last implies the estimate $\lambda(t) \asymp 0$. Let $h\{\lambda(t)\} > -\infty$. Substitute $y = \lambda^*(t) + u$ in the equation $H(t, y) = 0$ where u is a new unknown. The obtained equation (in u) has the root

$$\delta(t) \equiv \lambda(t) - \lambda^*(t) = O\left(t^{-\infty}\right).$$

And as it was proved above $\delta(t) \asymp 0$, that is, $\lambda(t) \in \{Q\}$. $\qquad\square$

In the Proposition 5.60 the condition (2) is essential as the following example shows: Given the equation $(y - 1)^2 - e^{-2t}\sin t = 0$. The equation coefficient belong to the space $\{RF\}$ because $e^{-2t}\sin t \in O_t$. But the roots $\lambda_{1,2}(t) = 1 \pm e^{-t}\sqrt{\sin t}$ does not belong to Π (and hence they does not belong to any space $\{Q\}$ where Q is a field of type N). Here only $h\{\lambda_1(t) - \lambda_2(t)\} = h\{e^{-2t}\sin t\} = -\infty$.

8.   OPERATORS IN A FIELD OF TYPE N

Here we will consider some simplifications which we can get if the equation $y = \lambda(t) + R(y)$ is considered in a space $\{Q\}$ where Q is a field of type N (see Definition 5.32). Clearly, the resulting space $\{Q\}$ is a complete asymptotic space. Let $\lambda(t) \in \{Q\}$. We shall consider the equation in the region

$$W_\lambda = \{y(t) : y(t) \in \{Q\}, y(t) \sim \lambda(t) \text{ for } t \to +\infty\}$$

and in the region

$$W_{\lambda h} = \{y(t) : y(t) \in \{Q\}, \Pi\{y(t) - \lambda(t)\} \le h\},$$

where h is a real number, $h < \Pi\{\lambda(t)\}$. Lemma 4.24 on the solution existence of the equation may be applied for the equation considered in W_λ or $W_{\lambda h}$ and the terms of the sequence $\{s_m(t)\}$ and the formal solution $s(t)$ belong to W_λ ($W_{\lambda h}$).

Since Q is a field, we may give some other examples of operators $R(y)$ of power type.

Proposition 5.61. *Let* $\lambda(t) + R(W_\lambda) \subset W_\lambda$ *($W_{\lambda h}$), let* $R(y)$ *be continuous in* W_λ *($W_{\lambda h}$) and of power type with a majorant* $d > -\infty$. *Let* $\Pi\{R(\lambda(t))\} = d$. *Then* $1/R(y)$ *is of the power type at the point* $\lambda(t)$ *(in* $W_{\lambda h}$*) with a majorant* $-d$ *at the point* $\lambda(t)$ *(in* $U_\lambda h$*).*

PROOF. We prove this proposition in the case when $R(y)$ is defined on W_λ (the other case is proved in the same way). We have

$$\frac{1}{R(y_1(t))} - \frac{1}{R(y_2(t))} = \frac{R(y_1(t)) - R(y_2(t))}{R(y_1(t))R(y_2(t))}$$

for any $y_1(t), y_2(t) \in W_l a$. Let $y(t) \in W_\lambda$. Since

$$\Pi\{R(y_1(t))\} = \Pi\{R(y_2(t))\} = \Pi\{R(y(t))\} = \Pi\{R(\lambda(t))\}$$

we have $\Pi\{1/R(y(t))\} = -d$ and

$$\Pi'\{1/R(\lambda(t))\} = \Pi\{R(\lambda(t))\} - 2d = -d - \Pi\{\lambda(t)\}. \qquad \square$$

Example 5.62. Consider a polynomial of the form

$$H(y) = y^n + a_1(t)y^{n-1} + \ldots + a_n(t) \equiv (y - \lambda_1(t))\ldots(y - \lambda_n(t)),$$

where all the functions $\lambda_j(t)$ belong to $\{Q\}$. Let $\lambda(t) \in \{Q\}$ and $\lambda(t) \not\sim \lambda_j(t)$ as $t \to +\infty$ for $j = 1, 2, ..., n$. Clearly, $\Pi\{y(t) - \lambda_j(t)\} = p_j \equiv \max[\Pi\{\lambda(t)\}, \Pi\{\lambda_j(t)\}]$. Hence $H(y)$ is continuous on W_λ and $\Pi\{H(\lambda(t))\} = r \equiv p_1 + ... + p_n$ and it is of power type with a majorant $r = p_1 + ... + p_n$ at $\lambda(t)$. Consequently, $1/H(y)$ is of power type with majorant $\tilde{r} = -p_1 - ... - p_n$ at $\lambda(t)$. For instance, the polynomial $H(y) = y^2 + yt + t^2$ has the roots

$$\lambda_{1,2}(t) = \frac{(-1 \pm i\sqrt{3})t}{2} \not\sim t \text{ for } t \to +\infty,$$

$\Pi\{H(t)\} = \Pi\{3t^2\} = 2$. Hence it is of power type with a majorant $r = 2$ at the point t. Clearly, $H(y)$ is continuous at this point. Thus, $R(y) = -y'/(y^2 + yt + t^2)$ (as a product of two operators of power type) is of power type with a majorant $r = -2$ at the point t.

Chapter 6

POWER ORDER GROWTH FUNCTIONS OF THE COMPLEX ARGUMENT

In this chapter we consider functions of the power order of growth in a central sector S or in a sufficiently small sector S_ε on the complex plane.

1. CLASSES A_S AND A_ε

Definition 6.1. We say that a function $f(z)$ possesses the *power order of growth* p in S (S_ε) (p is a number) if $f(z)$ is holomorphic in $[S]$ (S_ε) for $|z| \gg 1$ does not vanish in this region and

$$\frac{zf'(z)}{f(z)} - p \in C_S \ (C_\varepsilon). \tag{6.1}$$

The set of all such functions we denote by A_S (A_ε). The designation $P_S\{f(z)\} = p$ $(P_\varepsilon\{f(z)\} = p)$ means that $f(z) \in A_S$ $(f(z) \in A_\varepsilon)$ and its order of growth is equal to p.

Remark 6.2. By Proposition 4.2 if $f(z) \in A_\varepsilon$ then $f(t) \in A_t$ on the *positive semi-axis*.

Proposition 6.3. *For $P_S\{f(z)\} = p$ to be valid, it is necessary and sufficient for $f(z)$ to be in the form*

$$f(z) = z^{p+\alpha(z)}, \tag{6.2}$$

where $\alpha(z) \in C_S$ and $\alpha'(z)z \ln z \in C_S$.

PROOF. *Sufficiency.* Let $f(z)$ be in the form (6.2). Then $f(z)$ is holomorphic in $[S]$ for $|z| \gg 1$, $f(z) \neq 0$ in this region and

$$z\frac{f'(z)}{f(z)} = p + \beta(z) \text{ where } \beta(z) = \alpha(z) + \alpha'(z)z \ln z \in C_S.$$

128

Hence $P\{f(z)\} = p$.

Necessity. Let $f(z) \in A_S$. Hence $f(z)$ is a solution of the equation

$$zy' = y(p + \delta(z)),$$

where $\delta(z) \in C_S$. It is a linear equation. Its general solution may be represented in the form

$$y = Cz^{p+\beta(z)}, \quad \text{where } \beta(z) = \frac{1}{\ln z} \int_{z_0}^z \frac{\delta(\tau)}{\tau} d\tau,$$

z_0 is a point belonging to a permissible domain of the function $\delta(z)$ in S, C is an arbitrary constant.

Let us choose (by arbitrariness) a subsector S^* (S^* is a closed central sector such that its boundary rays are interior rays of the sector S). It is sufficient to prove that any (fixed) solution of the family possesses the required property in S^*, that is, we have to prove that

$$\alpha(z) = \frac{\ln C}{\ln z} + \frac{1}{\ln z} \int_{z_0}^z \frac{\delta(\tau)}{\tau} d\tau \to 0 \text{ and } \alpha'(z)z\ln z \to 0$$

for $z \to 0, z \in S^*$. Since any function of the form $\varphi(z) = A/\ln z$ belongs to C_S, and $\varphi'(z)z\ln z \in C_S$ (A =const), it is sufficient to prove that there exist at least one number $z_1 \in S^*$ such that

$$\beta_1(z) = \frac{1}{\ln z} \int_{z_1}^z \frac{\delta(\tau)}{\tau} d\tau \to 0 \text{ for } z \to \infty, z \in S^*.$$

Indeed then

$$\beta_1'(z)z\ln z = \delta(z) - \beta_1(z) \to 0$$

and

$$\beta(z) = C^*/\ln z + \beta_1(z),$$

where

$$C^* = \int_{z_0}^{z_1} \frac{\delta(z)}{z} dz.$$

Take (by arbitrariness) a number $\varepsilon > 0$ and choose a number $z_2 \in S^*$ such that $|z_2| \geq |z_1|$ and $|\delta(z)| < \varepsilon$ for $|z| \geq |z_2|, z \in S^*$. Hence

$$\beta_1(z)\ln z = B + \int_{z_2}^z \frac{\delta(\tau)}{\tau} d\tau, \quad \text{where } B = \int_{z_1}^{z_2} \frac{\delta(\tau)}{\tau} d\tau = \text{const.}$$

Hence $|\beta_1(z)\ln z - B| \leq \varepsilon|\ln z - \ln z_2|$ for $|z| \geq |z_1|$. The last inequality may be rewritten in the form $|\beta_1(z)| \leq 2\varepsilon$ for $|z| \gg 1$. Taking into account the arbitrariness of ε we conclude that $\beta_1(z) \to 0$ ($z \to \infty, z \in S^*$). $\square$

List several properties of the functions of A_S whose proof is almost no differ from the corresponding cases of the real argument.

Proposition 6.4.
$$P_S\{a + \alpha(z)\} = 0, \qquad (6.3)$$
where a is a non-zero number and $\alpha(z) \in C_S$.

Proposition 6.5. Let $P_S\{f(z)\} = p$. Then

(1)
$$p = \lim_{z\to\infty, z\in[S]} \frac{\ln f(z)}{\ln z}; \qquad (6.4)$$

(2) $P_S\{[f(z)]^\sigma\} = p\sigma$, where σ is a number;

(3) $\Pi_S\{f(z)\} = \Re p$;

(4) there exist functions $\alpha_m(z) \in C_S$ for each $m = 1, 2, ...$ such that

$$f^{(m)}(z) = [p(p-1)...(p-m+1) + \alpha_m(z)]f(z)z^{-m}, \qquad (6.5)$$

hence $\Pi_S\{f^{(m)}(z)\} \leq \Re p - m$. If $p \neq 0, 1, ..., m-1$ then $P_S\{f^{(m)}(z)\} = p - m$.

Remark 6.6. If $f(z) \in A_S$ and $f(z) \to 0$ for $z \to \infty, z \in [S]$ then (see Definition 4.40) $f(z) \in C_S$.

Proposition 6.7. Let $P_S\{f(z)\} = p$ and $P_S\{g(z)\} = q$. Then

$$P_S\{f(z)g(t)\} = p+q \text{ and } P_S\left\{\frac{f(z)}{g(z)}\right\} = p - q.$$

Proposition 6.8. Let $P_S\{f(z)\} = p$ and $\Pi_S\{g(z)\} = q$. Then

$$\Pi_S\{f(z)g(z)\} = \Re p + q.$$

Proposition 6.9. Let $P_S\{f(z)\} = p$ and $P_S\{g(z)\} = q$. Let $f(z)$ and $g(z)$ be comparable for $z \in [S], |z| \gg 1$, and let

$$\lim_{z\to\infty, z\in[S]} \frac{f(z)}{g(z)} \neq -1.$$

Then $P_S\{f(z) + g(z)\} = \max(p, q)$.

Proposition 6.10. Let $P_S\{f(z)\} = p$, let $g(z)$ be a holomorphic function for $|z| \gg 1, z \in [S]$ and $g(z) \sim f(z)$ for $z \to \infty, z \in [S]$. Then $P_S\{g(t)\} = p$.

Proposition 6.11. *Let $f(z) \in A_\varepsilon$. Then $P_S\{f(\ln z)\} = 0$ in any sector S containing the positive semi-axis as its interior ray.*

Proposition 6.12. *Let $P_S\{f(z)\} = p \neq 0$. Then $P_S\{\ln f(z)\} = 0$.*

Definition 6.13. *A function $f(z)$ is said to be* integrable *in A_S if any its antiderivative belongs to A_S.*

Proposition 6.14. *Let $P_S\{f(z)\} = p$, where $\Re p \neq -1$. Then*

$$\gamma(z) \equiv \int_{z_0}^{z} f(s)ds \in A_S \ and \ \gamma(z) \sim \frac{f(z)z}{p+1} \ for \ z \to \infty, z \in [S]. \quad (6.6)$$

Hence $P_S\{\gamma(z)\} = p + 1$. Here z_0 is a point belonging to a permissible domain of the function $f(z)$ in S if $\Re p > -1$, and $z_0 = \infty$ if $\Re p < -1$.

PROOF. Taking the integral $\gamma(z)$ by parts, we obtain

$$\gamma(z) = f(z)z + C - \int_{z_0}^{z} sf'(s)ds. \quad (6.7)$$

Here $C = -f(z_0)z_0$ if $\Re p > -1$, and $C = -\lim_{z \to \infty, z \in [S]} f(z)z = 0$ if $\Re p < -1$.

Let $\Re p > -1$. Hence $z_0 \neq \infty$. Taking into account that $f(z) \in A_S$, we have $zf'(z) = pf(z) + \alpha(z)f(z)$, where $\alpha(z) \in C_S$. Consequently, we may rewrite expression (6.7) in the following form

$$\int_{z_0}^{z} f(s)ds = \frac{f(z)z}{p+1}(1 + \alpha_1(z)) + \int_{z_0}^{z} \alpha_2(s)f(s)ds.$$

Here

$$\alpha_1(z) = -\frac{f(z_0)z_0}{f(z)z} \in C_S, \ and \ \alpha_2(z) = -\frac{\alpha(z)}{p+1} \in C_S.$$

To prove the required property of the function $\gamma(z)$, it is enough to prove that

$$\int_{z_0}^{z} \alpha_2(s)f(s)ds = o(f(z)z) \ for \ z \to \infty, z \in [S].$$

Let us chose a sector S^*, and let φ_1^*, φ_2^* be its boundary angels. Put $z = re^{i\varphi}$, $z_0 = r_0 e^{i\varphi_0}$, and $z^* = re^{i\varphi^*}$, where $\varphi \in [\varphi_1^*, \varphi_2^*]$ ($\varphi^* = $ const, $i = \sqrt{-1}$). Without loss of generality, we may suppose that $\varphi^* \in [\varphi_1^*, \varphi_2^*]$. We represent the last integral as a sum of two integrals:

$$\int_{z_0}^{z} \alpha_2(s)f(s)ds = I_1(z) + I_2(z),$$

where

$$I_1(z) = \int_{z_0}^{z^*} \alpha_2(s)f(s)ds \text{ and } I_2(z) = \int_{z^*}^{z} \alpha_2(s)f(s)ds.$$

The first integral is taken along the segment connecting the points z_0 and z. The second one is taken along the circumference connecting the points z^* and z. We have

$$I_1(z) = \int_{r_0}^{r} \alpha_2(te^{i\varphi_0})f(te^{i\varphi_0})e^{i\varphi_0}dt$$

$$= o(f(z)z) \text{ for } r \to \infty$$

which is proved in the same way as in Proposition 6.3.

$$I_2(z) = \int_{\varphi_0}^{\varphi} ire^{it}\alpha_2(re^{it})f(re^{it})dt.$$

Let us estimate the difference $\alpha_2(re^{it}) - \alpha_2(z)$. Since $\alpha(z) \in C_S$ and $\alpha'(z)z \ln z \in C_S$ we have

$$\alpha_2(re^{it}) - \alpha_2(z) = \int_{z}^{re^{it}} a'(s)ds$$

$$= \int_{\varphi}^{t} ire^{is}\alpha'(re^{is})ds$$

$$= o\left(\frac{1}{\ln z}\right) \text{ for } z \to \infty, z \in S^*.$$

Let us estimate the expression $f(re^{it})$. Since $f(z) \in A_S$ we have (see (6.2))

$$f(re^{it}) = [re^{it}]^{p+\alpha(re^{it})}$$
$$= [re^{i\varphi}e^{i(t-\varphi)}]^{p+\alpha(re^{i\varphi})+(\alpha(re^{it})-\alpha(re^{i\varphi}))}$$
$$= f(z)[re^{i\varphi}]^{\alpha(re^{it})-\alpha(re^{i\varphi})}[e^{i(t-\varphi)}]^{p+\alpha(re^{it})}.$$

Since $\alpha(re^{it}) - \alpha(re^{i\varphi}) = O(1/\ln r)$ for $r \to \infty$, clearly, $f(re^{it}) = f(z)O(1)$ for $r \to \infty$ and $t \in [\varphi_1^*, \varphi_2^*]$. Consequently, $a_2(s)f(s) = f(z)o(1)$ for $r \to \infty$ uniformly in $t \in [\varphi_1^*, \varphi_2^*]$. Hence

$$I_2(z) = zf(z)\int_{\varphi_0}^{\varphi} o(1)dt$$
$$= o(zf(z)) \text{ for } z \to \infty, z \in S^*,$$

which proves the considered case.

Let $\Re p < -1$. Then $z_0 = \infty$ and $C = 0$ in (6.6). Hence

$$\gamma(z) = \frac{f(z)z}{p+1} + \int_{z_0}^{z} \beta(s)f(s)ds,$$

where $\beta(z) = -\alpha(z)/(p+1) \in C_S$. To prove the required property of the function $\gamma(z)$ it is enough to prove that

$$\int_{z_0}^{z} \beta(s)f(s)ds = o(f(z)z) \text{ for } z \to \infty, z \in [S].$$

Let φ^* be an interior angle of the sector S. We represent the last integral as a sum of two integrals:

$$\int_{z_0}^{z} \beta(s)f(s)ds = J_1(z) + J_2(z),$$

where

$$J_1(z) = \int_{\infty}^{z^*} \beta(s)f(s)ds \text{ and } J_2(z) = \int_{z^*}^{z} \beta(s)f(s)ds,$$

where $z^* = re^{i\varphi^*}$. The first integral is taken along the central ray passing through the point z (connecting the points ∞ and z^*). The second one is taken along the circumference connecting the points z^* and z. We have

$$J_1(z) = \int_{+\infty}^{r} \beta(re^{i\varphi^*})f(re^{i\varphi^*})e^{i\varphi^*} dr = o(f(z)z) \text{ for } r \to \infty,$$

which is proved in the same way as in Proposition 6.3. $J_2(z) = o(zf(z))$ for $z \to \infty, z \in [S]$, which is proved in the same way as in the case $\Re p > -1$. The last remark proves this Proposition. □

Proposition 6.15. *Let $P_S\{f(z)\} = p$, where $\Re p \neq -1$. Then the function $f(z)$ is integrable in A_S.*

2. SUBSETS OF CLASS A_S WITH SOME ALGEBRAIC AND ANALYTIC PROPERTIES

Definition 6.16. A subset Q of the set $A_S \cup 0$ $(A_\varepsilon \cup 0)$ is said to be a *field of type* N_S (N_ε) if Q is an algebraic field, $\mathbb{C} \cup z \subset Q$ ($\mathbb{C}$ is the set of all complex numbers), and if $q(z) \in Q$ then $q'(z) \in Q$ and there exists a (finite or infinite) limit

$$\lim_{z\to\infty, z\in[S]} q(z) \; (\lim_{z\to\infty, z\in S_\varepsilon} q(z)).$$

Definition 6.17. A field Q of type N_S is said to be *normal* if, for any $q(z) \in Q$ such that $q(z) \neq 0$, $P_S\{q(z)\}$ is a real number and there exists a finite limit

$$\lim_{z \to \infty, z \in [S]} \Im[\ln q(z) - P_S\{q(z)\} \ln z]. \qquad (6.8)$$

Proposition 6.18. *A field Q of type N_S is normal if and only if for any function $q(z) \in Q$ such that $q(z) \not\equiv 0$, there exists a finite limit*

$$\lim_{z \to \infty, z \in \Gamma} \Im \ln q(z) \qquad (6.9)$$

for at least one interior ray $\Gamma = \{z : \arg z = \varphi, z \neq \infty\}$ in the sector S. Here φ is a real number (the angle of the ray Γ).

PROOF. *Necessity.* Let Q be a normal field of type N_S, $q(z) \in Q$, and $q(z) \not\equiv 0$. Then taking into account that $P_S\{q(z)\}$ is a real number we obtain

$$\lim_{z \to \infty, z \in [S]} \Im[\ln q(z) - P_S\{q(z)\} \ln z] = \lim_{z \to \infty, z \in \Gamma} \Im[\ln q(z) - i\varphi P_S\{q(z)\}]$$

$$= \lim_{z \to \infty, z \in \Gamma} \Im \ln q(z) - \varphi P_S\{q(z)\}.$$

Since the considered limit is finite (see (6.8)) the limit $\lim_{z \to \infty, z \in \Gamma} \Im \ln q(z)$ is finite ($i = \sqrt{-1}$).

To prove the sufficiency of the considered property we will require:

Lemma 6.19. *Let $\alpha(z) \in C_S$ and $\alpha'(z) z \ln z \in C_S$. Then*

$$\alpha(z) \ln z = \alpha(|z|e^{i\varphi}) \ln(|z|e^{i\varphi}) + o(1) \ for \ z \to \infty, z \in [S].$$

PROOF. Put $z = |z|e^{it}$, where $\varphi_1 < t < \varphi_2$ (φ_1 and φ_2 are the boundary angles of the sector S). Let us put

$$g(t, |z|) = \alpha(z) \ln z - \alpha(|z|e^{i\varphi}) \ln(|z|e^{i\varphi}).$$

Hence

$$g(t, |z|) = \int_{|z|e^{i\varphi}}^{z} [\alpha(s) \ln s]' ds.$$

Consequently,

$$|g(t, |z|)| \leq \sup_{z \in [S]} [|\alpha'(z) z \ln z| + |\alpha(z)|] |\varphi_2 - \varphi_1| \to 0$$

for $z \to \infty, z \in [S]$, which leads to the required relation. □

PROOF *of Proposition 6.18 sufficiency.* Since $q(z) \in A_S$ and because of the limit (6.9) is finite, we have (see (6.4))

$$
\begin{aligned}
p &= \lim_{z \to \infty, z \in [S]} \frac{\ln f(z)}{\ln z} \\
&= \lim_{z \to \infty, z \in [S]} \frac{\Re \ln f(z)}{\ln z} + \lim_{z \to \infty, \arg z = \varphi} \frac{i \Im \ln f(z)}{\ln z} \\
&= \lim_{z \to \infty, z \in [S]} \frac{\Re \ln f(z)}{\ln |z|}.
\end{aligned}
$$

The last limit, clearly, is a real number. That is, p is a real number. Moreover,

$$
\lim_{z \to \infty, z \in \Gamma} \Im[(p + \alpha(z)) \ln z] = p\varphi + \lim_{z \to \infty, z \in \Gamma} \Im[\alpha(z) \ln z].
$$

By Lemma 6.19

$$
\lim_{z \to \infty, z \in \Gamma} \Im \ln q(z) = p\varphi + \lim_{z \to \infty, z \in [S]} \Im[\ln q(z) - p \ln z],
$$

which leads to the required property. □

Definition 6.20. Let Q be a subset of a field of type N_S and let S' be a subsector of S (which in particular may coincide with S). We denote by $\{Q\}_{S'}$ the set of all function such that if $q(z) \in \{Q\}_{S'}$, then there exists a function $q_m(z) \in Q$ such that

$$
\Pi_{S'}\{q(z) - q_m(z)\} < -m
$$

for any natural m. Any function $q(z) \in \{Q\}_{S'}$ is said to be *asymptotically close* to the set Q in S'.

Consider an equation of the form

$$
H(y, z) \equiv y^n + a_1(z)y^{n-1} + \dots + a_n(z) = 0. \tag{6.10}
$$

Here $a_i(z)$ are functions of z defined in a region D. By a *root* of this equation (or of the polynomial $H(y, z)$) we set to be any continuous solution to the equation for $z \in D$.

In the same way as in the real case it is possible to prove:

Theorem 6.21. *Let* $a_i(z) \in Q$ $(i = 1, 2, ..., m)$, *where* Q *is a field of type* N_S, *and let* $\lambda(z) \not\equiv 0$ *be a root of equation* (6.10). *Then the minimal algebraic extension* $\widetilde{Q}$ *of the field* Q *by the root* $\lambda(z)$ *is a field of type* N_S.

Theorem 6.22. *Let the hypothesis of Theorem* 6.21 *be fulfilled. Let, in addition,* Q *be a normal field of type* N_S. *Then the field* $\widetilde{Q}$ *obtained in Theorem* 6.21 *is normal.*

PROOF. Let $p(z) \in \widetilde{Q}$ and $p(z) \not\equiv 0$. Then (see Lemma 5.40) $p(z) \sim q^r(z)$ for $z \to \infty, z \in [S]$, where $q(z) \in Q$ and r is a rational number. Hence

$$\lim_{z \to \infty, z \in [S]} \Im[\ln p(z) - P_S\{p(z)\} \ln z] = \lim_{z \to \infty, z \in [S]} r\Im[\ln p(z) - P_S\{p(z)\} \ln z].$$

This proves the Theorem.   □

The following theorem is proved in the same way as Theorem 5.45.

Theorem 6.23. *The algebraic closure* P *of the field* (*normal field*) Q *of type* N_S *is a field of type* N_S (*normal field of type* N_S).

Examples 6.24.

$1°$ The simplest example of a field (normal field) of type N_S is the set of all rational functions in z with complex coefficients (which is denoted by RF). Indeed, RF is an algebraic field $\mathbb{C} \cup z \subset RF$. If $r(z) \in RF$ and $r(z) \not\equiv 0$ then $r'(z) \in RF$, and there exists a monomial $p(z) = cz^n$, where n is an integer and c is a (generally speaking, complex) non-zero number such that $r(z) \sim cz^\sigma$ for $|z| \to \infty$. Consequently, if $\sigma > 0$ ($\sigma = 0, \sigma < 0$) then $\lim_{|z| \to \infty} r(z) = \infty$ (respectively, c and 0). Moreover, $\lim_{t \to +\infty} \lim \Im c$. The last relation proves our assertion.

$2°$ *Logarithmic field of rank zero.*

Consider a monomial of the form $M(z) = cz^k$, where k is a real number, c is a non-zero (complex) number. The set of all such functions we denote by Φ_0. We say that a set of functions L_S^0 is a *logarithmic field of rank zero* in the sector S if (i) L_S^0 is an algebraic field; (ii) $\Phi_0 \subset L_S^0$; (iii) if $f(z) \in L_S^0$ and $f(z) \not\equiv 0$ then $f(z)$ is a holomorphic function in $[S]$ for $|z| \gg 1$, and there is a monomial $M(z) \in \Phi_0$ such that $f(z) \sim M(z)$ for $z \to \infty, z \in [S]$; (iv) $f'(z) \in L_S^0$.

The field L_S^0 is a normal field of type N_S. In fact $\mathbb{C} \cup z \subset L_S^0$; if $f(z) \in L_S^0$ and $f(z) \not\equiv 0$ for $z \in [S], |z| \gg 1$, we have $f'(z) \in L_S^0$ and there exists a monomial $M(z) = cz^k \in \Phi_0$ such that $f(z) \sim M(z)$ for $z \in$

$[S]$, $z \to \infty$. Hence if $k > 0$ then $\lim_{z\to\infty, z\in[S]} f(z) = \infty$ if $k < 0$ then $\lim_{z\to\infty, z\in[S]} f(z) = 0$ and if $k = 0$ then the limit is equal to c. Thus L_S^0 is a field of type N_S. Moreover for an arbitrary (fixed) ray $z = te^{i\varphi}$ we have $\lim_{z\to\infty, z=te^{i\varphi}, t\to+\infty} \Im \ln f(z) = \arg c + k\varphi \neq \infty$. On the basis of Proposition 6.18 we conclude that L_S^0 is normal.

3. POWER-LOGARITHMIC FUNCTIONS

Fields of type N and M play an important role in our consideration. The set of all power-logarithmic functions (which we define below) is the most important example of a field of type M. It is convenient to consider the power logarithmic functions in the complex plane.

Definition 6.25. We say that a (generally speaking, multivalued) function $l(z)$ belongs to the *class* L^* if it can be obtain from $\mathbb{C} \cup z$ by means of a finite number of the operations of summation, multiplication, raising to a real power, and taking the logarithm (here $\mathbb{C}$ is the set of all complex numbers). Each single-valued branch of any function $l(z) \in L^*$ in the complex plane is said to be a *power-logarithmic function*. The set of all such functions we denote by L_0.

Later on we somewhat extend the notion of power-logarithmic functions.

A simple example of a power-logarithmic function gives a monomial of the form

$$M(z) = cz^{k_0}(\ln_1 z)^{k_1}...(\ln_m z)^{k_m}, \tag{6.11}$$

where c is a complex number, $k_0, k_1, ..., k_m$ are real numbers.

Let us mark some simple properties of the power-logarithmic functions:

(1) every power-logarithmic function is a single-valued analytic function in any sector of the complex plain without removable singularities (see *Conventions and Notation* in the beginning of this book);

(2) if $l(z) \in L_0$ then $l'(z) \in L_0$ and $\bar{l}(\bar{z}) \in L_0$.

Property (2) is easily proved by induction with respect to the number of actions by means of the function $l(z)$ can be obtained from the set $\mathbb{C} \cup z$.

We will prove that any function $l(z) \in L_0$ belongs to $A_S \cup 0$ in any sector S and (if $l(z) \not\equiv 0$) has a real order. This follows from the following assertion (which is almost obvious) that $l(z)$ is equivalent to a function of the form (6.11). Then $l(z) = M(z)(1 + \alpha(z))$, where $P_S\{M(z)\} = k_0$, and $\alpha(z)$ is a power-logarithmic function which is infinitesimal in any sector S of the complex plane, hence $\alpha(z) \in C_S$. And $\alpha(t) \in C_t$ on the positive semi-axis. Consequently, $P_S\{l(z)\} = P_S\{M(z)\} = k_0$, i.e. $l(z) \in A_S$ and $l(t) \in A_t$ on the positive semi-axis.

The complete proof of this assertion is given below.

Any function $l(z) \in L_0$ can be obtained from $\mathbb{C} \cup z$ (by the definition) using different methods. Let us consider all methods of obtaining $l(z)$ by the minimal number of such operations and form pairs $[l(z), n]$ where $n = 1, 2, \ldots$ We associate a number $\dim[l(z), n]$ with each pair. The number is defined by induction. If $l(z) \in \mathbb{C}$ then $\dim[l(z), n] = -1$ (in this case we suppose that the number of the methods to be equal to 1). If $l(z) \neq$ constant and the method of formation of the pair includes only the operations of summation, multiplication, and raising to a power, then $\dim[l(z), n] = 0$. Let $l(z) \neq$ constant and let the last operation to obtain the pair $[l(z), n]$ be the operation on the functions $p(z)$ and $q(z)$ (or only on the function $p(z)$) obtained from the methods defined by pairs $[p(z), n_1]$ and $[q(z), n_2]$. Let $\dim[p(z), n_1] = p$ and $\dim[q(z), n_2] = q$. Then $\dim[p(z) + q(z), n] = \dim[p(z)q(z), n] = \max(p, q)$ and $\dim[p^d(z), n] = p$, where d is a real number. If $l(z) = \ln p(z)$ then $\dim[\ln p(z), n] = p + 1$.

Definition 6.26. Let $l(z) \in L_0$. We say that the number

$$\dim_0 l(z) = \min_n \dim[l(z), n]$$

is the *length of the logarithmic chain* of the function $l(z)$.

Definition 6.27. We say that a function $l(z)$ *possesses the property H_k* in S if it is defined for $|z| \gg 1, z \in S$ and in the considered domain it is a sum of absolutely convergent series of the form

$$z^{r_0} l_0(\ln z) + z^{r_1} l_1(\ln z) + \ldots + z^{r_m} l_m(\ln z) + \ldots \qquad (6.12)$$

and

$$h_S\{l(z) - s_m(z)\} \to -\infty \text{ for } m \to \infty.$$

Here r_i are real numbers such that

$$r_0 > r_1 > \ldots > r_m > \ldots, \quad r_m \to -\infty \text{ for } m \to \infty;$$

$l_i(z)$ are power-logarithmic functions such that

$$\dim_0 l_i(z) \leq k \text{ and } h_S\{l_i(z)\} < +\infty;$$

$s_m(z)$ is the m-partial sum of the series (6.12) $(m = 1, 2, \ldots)$.

Proposition 6.28. *Let $l(z)$ be a power-logarithmic function and let $\dim_0 l(z) = k \geq 0$. Then $l(z)$ is a holomorphic function for $|z| \geq 1, z \in S$ in any (fixed) sector S and $l(z)$ possesses the property H_{k-1}.*

PROOF. We prove this Proposition by induction with respect to $k = 0, 1, \ldots$ Let $k = 0$. This case is proved by induction with respect to the number of the operations by means of the function obtained from $\mathbb{C} \cup z$. Let the last operation to obtain the function $l(z)$ be the operation on the functions $p(z)$ and $q(z)$ (or only on the function $p(z)$). Hence $p(z)$ and $q(z)$ are sums of absolutely convergent generalized power series. Let

$$p(z) = a_0 z^{p_0} + a_1 z^{p_1} + \ldots + a_m z^{p_m} + \ldots$$

and

$$q(z) = b_0 z^{q_0} + b_1 z^{q_1} + \ldots + b_m z^{q_m} + \ldots .$$

We can suppose without loss of generality that all a_m and b_m are non-zero numbers ($m = 0, 1, \ldots$). The function $p(z) + q(z)$ possesses the property H_0. Indeed, the sum is a holomorphic function in the considered domain. Let us chose a natural N and collect the sum $s_N(z)$ which consists of all terms belonging to the sums $p_1(z)$ and $p_2(z)$ with powers at most N. On tending N to infinity we obtain a generalized series $s(z)$ (or $s(z) \equiv 0$) which is the sum of $p(z)$ and $q(z)$, and $s(z)$ possesses the property H_0. In the same way we prove that $p(z)q(z)$ is a generalized power series possessing the property H_0. Let $l(z) = p^d(z)$, where $d \neq 0$ is a real number. We have $l(z) = a_0^{d_0} z^{p_0 d}(1 + r(z))^d$, where $r(z) \sim (a_1/a_0)z^{p_1 - p_0}$, and hence $r(z) \to 0$ for $z \to \infty, z \in S$. Let us expend the function $(1 + r(z))^d$ into a series in $r(z)$. We have

$$l(z) = 1 + \sum_{p=1}^{m} c_p r^p(z) + O(r^{m+1}(z)) \text{ for } z \to \infty, z \in S.$$

where $c_p = d(d-1)\ldots(d-p+1)/p!$. The function

$$l_m(z) = \frac{a_1}{a_0} z^{p_1 - p_0} \left(1 + \sum_{p=1}^{m} c_p r^p(z)\right)$$

(which consists of a finite sum and of a product of functions possessing the property H_0) possesses the property H_0. Let us collect all the terms in the generalized series $l_m(z)$ such that their powers are at most $T = h_S\{r^m(z)\} + p_0 d$. Denote the sum by $s_T(z)$. Clearly, $T \to -\infty$ for $m \to \infty$. On tending m to infinity we obtain a generalized power series which absolutely convergent in the considered domain. We have $l(z) - s_T(z) = O(z^{-T})$ for $z \to \infty, z \in S$. Hence $l(z)$ possesses the property H_0. Thus, the case $k = 0$ is proved. Let us note that the function

$$l(z) = \ln p(z) = p_0 \ln z + \ln(1 + r(z))$$

(where $\dim_0 l(z) = 1$) also possesses the required properties. In the same way passing from $k - 1$ to k it is easy to show that the function $l(z)$ possesses all the required properties. □

Proposition 6.29. *Let $l(z) \in L_0$, $l(z) \not\equiv 0$ and $\dim_0 l(z) = m$. Then in any sector S:*

(1) *there exist functions $f(z)$, $\alpha(z)$ belonging to L_0, and a real number k such that $\dim_0 f(z) \le m - 1$, $\Pi_S\{\alpha(z)\} < k$, and*

$$l(z) = z^k f(\ln z) + \alpha(z); \qquad (6.13)$$

(2) *there exists a monomial of type (6.11), where $k = k_0, k_1, ..., k_m$ are real numbers and c is a non-zero (complex) number such that*

$$l(z) \sim M(z) \ \ for \ |z| \to \infty \qquad (6.14)$$

and hence $\Pi_S(l(z)) = k$ in any sector S ($l(z) \in A_S$).

PROOF. In view of Proposition 6.28 the function $l(z)$ is an asymptotic sum of the series (6.12) for $z \to \infty, z \in S$, hence

$$l(z) = z^{r_0} l_0(\ln z) + O(z^{r_1} l(\ln z)) \ \ for \ z \to \infty, z \in S,$$

where $r_0 > r_1$, $l_0(z) \in L_0$, $\dim_0 l_1(z) \le m - 1$ and $l_0(z) \not\equiv 0$. Re-designate by $k = r_0$, $l_0(z) \equiv f(z)$ and $z^{r_1} l_1(\ln z) \equiv \alpha(z)$ we obtain the required relation. Relation (6.14) is easily proved by induction with respect to m. □

Proposition 6.30. *The set L_0 is a normal field of type N_S in any sector S of the complex plane*

PROOF. Obviously $\mathbb{C} \cup z \in L_0$. If $l(z) \in L_0$ then in view of Proposition 6.29 $l(z) \in A \cup 0$; $l'(z) \in L_0$; $\lim_{z \to \infty} l(z) = 0$ if $(k_0, k_1, ..., k_m) \prec (0, 0, ..., 0)$; $\lim_{z \to \infty} l(z) = \infty$ if $(0, 0, ..., 0) \prec (k_0, k_1, ..., k_m)$ and $l(z) \to$ const if $(k_0, k_1, ..., k_m) = (0, 0, ..., 0)$. Hence there exists a (finite or infinite) limit $\lim_{z \to \infty} l(z)$ for any function $l(z) \in L_0$. Moreover if $l(z) \not\equiv 0$ then there exists a finite limit $\lim_{t \to +\infty} \Im \ln l(t)$. Indeed, since the function $l(t)$ is equivalent to a monomial of type (6.11) we have $\Im \ln l(t) = \Im c + \ln(1 + o(1))$ for $t \to +\infty$. Hence $\lim_{t \to +\infty} \Im \ln l(t) = \Im c$. Consequently, (see Proposition 6.28) is a normal field of type N_S. □

Proposition 6.31. *The set L_0 is a field of type M (on the positive semi-axis).*

PROOF. In Proposition 6.30 it is proved that L_0 is a field of type N. We must only prove that the conjugate function $\overline{l(t)}$ belong to L_0 (for any $l(t) \in L_0$). This property is easily proved by induction on the minimal operations n by means of the function $l(t)$ can be obtained from the set $c \cup z$ according to Definition 6.25. □

Proposition 6.32. *If* $l_1(z), l_2(z) \in L_0$, $l_2(z) \to 0$ *or* ∞ *and* $l_1(z) = o(l_2(z))$ *for* $|z| \to \infty$ *then* $l_1'(z) = o(l_2'(z))$ *for* $|z| \to \infty$.

PROOF. We prove this Proposition with respect to $m = -1, 0, 1, \ldots$ where $m = \max[\dim_0 l_1(z), \dim_0 l_2(z)]$. Being trivial for $m = -1$ let the proposition be valid for $m-1$, then

$$f(\ln z) \sim c(\ln_1 z)^{k_1} \ldots (\ln_m z)^{k_m} \text{ for } z \to \infty, z \in S,$$

which leads to the required relation. □

Proposition 6.33. *Let* $l(z) \in L_0$ *and* $l(z) \to 0$ *for* $|z| \to \infty$. *Then*

$$l'(z)z \ln_1 z \ldots \ln_s z \to 0 \text{ as } |z| \to \infty \tag{6.15}$$

for any natural s.

PROOF. Let $l(z) \not\equiv 0$ (the case $l(z) \equiv 0$ is trivial). The prove is made by induction on $m \equiv \dim_0 l(z) = 0, 1, \ldots$ (the case $m = -1$ is impossible). For $m = 0$, clearly, $P_S\{f(z)\} < 0$, hence $P_S\{f'(z)\} < -1$ in any sector S which proves the considered case. Assume the statement for $m - 1$. Obviously we have only to consider the case $P_S\{l(z)\} = 0$. Hence in view of Proposition 6.29 (1) $l(z) = f(\ln z) + \alpha(z)$, where $f(z) \in L_0$, $f(z) \to 0$ for $|z| \to \infty$, $\dim_0\{f(z)\} \le m - 1$, and $P_S\{\alpha(z)\} < 0$. Thus,

$$l'(z) = O\left(\frac{df(\ln z)}{d \ln z} \ln_1 z \ldots \ln_s z\right) \to 0$$

for $|z| \to \infty$, which leads to the required property. □

Definition 6.34. The algebraic closure of the set L_0 (which is a normal field of type N_S in any sector S and a field of type M on the positive semi-axis) we denote by L. Any function of the set L is said to be a *power-logarithmic function*. We say that L is the *set of all power-logarithmic functions*.

Remark 6.35. Since any function $l(z) \in L$ belongs to $A_S \cup 0$ in any sector S and its power is independent of S (for $l(z) \ne 0$), we will omit the designation of the sector S and write $P\{l(z)\}$, $\Pi\{l(z)\}$, and $l(z) \in A$ instead of $P_S\{l(z)\}$, $\Pi_S\{l(z)\}$, and $l(z) \in A_S$, respectively.

Definition 6.36. Let $l(z) \in L$ and let

$$H(y, z) = y^n + a_1(z)y^{n-1} + \ldots + a_n(z) \tag{6.16}$$

be the minimal polynomial over the field L_0 containing the root $l(z)$. Then the number

$$\dim l(z) = \max_{i=1,2,\ldots,n} \dim_0 a_i(z) \tag{6.17}$$

is said to be the *length of the logarithmic chain* of the function $l(z)$.

Functions of the class L possess the main properties of the class L_0. We will consider only the following two propositions.

Proposition 6.37. *Let* $l(z) \in L$, $P\{l(z)\} = k$ *and* $\dim l(z) = m$. *Then there exist functions* $f(z)$ *and* $\alpha(z) \in L$ *such that* $P\{\alpha(z)\} < k$, $\dim f(z) \leq m - 1$ *and*

$$l(z) = z^k f(\ln z) + \alpha(z). \tag{6.18}$$

Beforehand we have to prove the following:

Lemma 6.38. *Consider an equation of the form*

$$R(y, \ln z) \equiv y^m + b_1(\ln z)y^{m-1} + \ldots + b_m(\ln z) = \beta(z). \tag{6.19}$$

Let $b_j(z) \in L_0$, $\dim_0\{b_j(z)\} \leq m - 1$ $(j = 1, 2, \ldots, m,)$ $\beta(z) \in L$ *and* $P\{\beta(z)\} < 0$. *Let equation* (6.19) *have a root* $l(z)$ *with the estimate* $P\{l(z)\} = 0$. *Then there exist a root* $f(\ln z)$ *of the equation* $R(y, \ln z) = 0$, $\dim f(z) \leq m - 1$ *and a function* $\alpha(z) \in L$ *with the estimate* $P\{\alpha(z)\} < 0$, *such that* $l(z) = f(\ln z) + \alpha(z)$.

PROOF. This assertion is a simple consequence of Proposition 6.29. Let the equation $R(y, \ln z) = 0$ have a complete set of solutions which we denote by $\{f_i(\ln z)\}$ $(i = 1, 2, \ldots, m)$. Then equation (6.19) has a complete set of roots $\{x_i(z) = f_i(\ln z) + \beta_i(z)\}$, where $f_i(z) \in L$, $\dim f_i(z) \leq m - 1$, $\beta_i(z) \in L$ and $\beta_i(z) = O\left([\beta(z)]^{1/n}\right)$ for $z \to \infty$ in any (fixed) sector S. Hence $P\{\beta_i(z)\} = h\{\beta_i(z)\} \leq (1/n)P\{\beta(z)\} < 0$. □

Proof of Proposition 6.37. Without loss of generality we may suppose that $k = 0$, because if $k \neq 0$ we may pass to the case $k = 0$ by means of the substitution $y = z^k u$. Thus, we have $P\{l(z)\} = 0$. Let (6.16) be the minimal polynomial for the function $l(z)$ over the field L_0. It means that all the coefficients of the polynomial belong to L_0 and $\dim_0 a_i(z) \leq m$ $(i = 1, 2, ..., n)$. Since $P\{l_i(z)\} = 0$ there exists at least one function $a_s(z)$ with the estimate $P\{a_s(z)\} \geq 0$. Let us put $a_0(t) = 1$ and $P\{a_i(t)\} = k_i$ for any $i = 0, 1, ..., n$. Let the number $j \in \{0, 1, ..., n\}$ be introduced as follows: for any $i < j$, $k_i < k_j$ and for any $i > j$, $k_i \leq k_j$. Clearly, $k_j \geq 0$ and if $k_j = 0$ then $j = 0$. On the basis of Proposition 6.29 there are functions $p_i(z) \in L_0$ and $\delta_i(t) \in L_0$ where $\dim p_i(z) \leq m - 1$, $P\{\delta_i(z)\} < 0$, such that $a_i(z) = z^{k_i}[p_i(\ln z) + \delta_i(z)]$. Besides we introduce the functions $b_i(\ln z)$ and $\alpha_i(t)$ in the following way. If $k_i < k_j$ then $b_i(\ln z) \equiv 0$ and $\alpha_i(z) = z^{-k_j} a_i(z)$; if $k_i = k_j$ then $b_i(\ln z) = p_i(\ln z)$ and $\alpha_i(z) = \delta_i(z)$. Clearly, $b_i(\ln z) = 0$ for $i = 1, 2, ..., j - 1$, $b_j(\ln z) = p_j(\ln z)$ and $P\{\alpha_i(z)\} < 0$ for any $i = 0, 1, ..., n$. Equation $H(y, z) = 0$ may be rewritten in the following form

$$R(y, z) \equiv b_j(\ln z)y^{n-j} + b_{j+1}(\ln z)y^{n-j-1} + ... + b_n(\ln z) = \sum_{i=0}^{n} \alpha_i(z)y^{n-i}.$$

$$(6.20)$$

Substitute $y = l(z)$ in the sum of equation (6.20). Hence we conclude that $l(z)$ is a root of the equation (6.19) where

$$\beta(z) = \sum_{i=0}^{n} \alpha_i(z)l^{n-1}(z) \in L.$$

Obviously $P\{\beta(z)\} < 0$.   □

The following Proposition is a simple consequence of Proposition 6.37.

Proposition 6.39. *Assume the hypothesis an notation of Proposition 6.37. Let $l(z) \neq 0$. Then there exist a complex numbers $c \neq 0$ and real numbers $k, k_1, ..., k_m$ such that*

$$l(z) \sim cz^k (\ln_1 z)^{k_1} ...(\ln_m z)^{k_m} \quad for \ |z| \to +\infty.$$

Chapter 7

INTEGRALS

1. IMPROPER INTEGRALS

In this section we consider integrals of the form

$$\int_{-\infty}^{+\infty} f(t)dt \text{ or } \int_0^{+\infty} f(t)dt, \tag{7.1}$$

which are possible to solve using asymptotic methods. In some rare cases integrals considered may be find exactly. For example

$$\int_0^{+\infty} \frac{\sin t}{t} dt = \frac{\pi}{2} \text{ and } \int_{-\infty}^{+\infty} e^{-t^2} dt = \sqrt{\pi}.$$

Certain integrals may be computed by means of analytic methods. Show how to compute the integral

$$I = \int_{-\infty}^{+\infty} e^{-t^2} dt \tag{7.2}$$

because it is of importance for the further consideration. We find (7.2) using double integrals.

$$I^2 = \int_{-\infty}^{+\infty} e^{-x^2} dx \int_{-\infty}^{+\infty} e^{-y^2} dy$$

$$= \lim_{R \to \infty} \iint_{C_R} e^{x^2 + y^2} dx dy, \tag{7.3}$$

where C_R is a central circle of radius R. The last integral may be determined in the polar coordinate. We have

$$I^2 = \int_0^{2\pi} d\varphi \int_0^{+\infty} e^{-r^2} r dr = \pi,$$

144

which leads to the required relation.

An approximate result easily may be obtained by the numerical methods. They involve (as a part) the asymptotic methods. We will show the application of asymptotic methods by two examples. Consider (7.2)

$$\frac{I}{2} = \int_0^{+\infty} e^{-t^2} dt$$

$$= \int_0^T e^{-t^2} dt + \int_T^{+\infty} e^{-t^2} dt,$$

where T is a positive number. The last integral can be easily estimate:

$$\int_T^{+\infty} e^{-t^2} dt \leq \frac{1}{2T} \int_T^{+\infty} 2t e^{-t^2} dt$$

$$= \frac{1}{2T} e^{-T^2}.$$

This expression decreases very quickly (when T increases) and becomes negligible small. Hence I is approximately equal (up to the desired precision) to the definite integral $2 \int_0^T e^{-t^2} dt$. We may compute the last integral applying any appropriate calculus method. For example we may use the Simpson formula.

In the general case if we have to find an improper integral of the form

$$I = \int_0^{+\infty} f(t) dt \tag{7.4}$$

we may operate in the same way. We rewrite (7.4) in the form of a sum of two integrals

$$I = \int_0^T f(t) dt + \int_T^{+\infty} f(t) dt. \tag{7.5}$$

The estimate of the last integral may be not so easy as in the previous case. For example compute the integral

$$J = \int_0^\infty \frac{\sin^2 t}{t^2} dt \tag{7.6}$$

with accuracy to within 10^{-7}. Clearly,

$$\int_T^{+\infty} \frac{\sin^2 t}{t^2} dt < \int_T^{+\infty} \frac{1}{t^2} dt$$

$$= \frac{1}{T}.$$

Consequently we have to take $T \geq 10^7$. In reality the estimate $1/T$ may be reduced to $1/(2T)$. But it can not be improved. The number T is too large that creates difficulties in the integral $\int_0^T (\sin^2 t / t^2) dt$ calculation. Perhaps $T \approx 10$ will be acceptable. We have

$$J = J_1(T) + J_2(T),$$

where

$$J_1(T) = \int_0^T \frac{\sin^2 t}{t^2} dt \text{ and } J_2(T) = \int_T^{+\infty} \frac{\sin^2 t}{t^2} dt. \tag{7.7}$$

Clearly,

$$J_2(T) = \int_T^{+\infty} \frac{1 - \cos 2t}{2t^2} dt$$

$$= \frac{1}{2T} - \int_T^{+\infty} \frac{\cos 2t}{2t^2} dt.$$

Let us introduce the following notation

$$I_n(T) = \int_T^{+\infty} \frac{\cos 2t}{t^{2n}} dt \ (n = 1, 2, ...). \tag{7.8}$$

Taking the last integral by parts we obtain

$$\begin{aligned} I_n(T) &= \frac{1}{2} \int_T^{+\infty} \frac{1}{t^{2n}} d \sin 2t \\ &= -\frac{\sin 2T}{2T^{2n}} + n \int_T^{+\infty} \frac{\sin 2t}{t^{2n+1}} dt \\ &= -\frac{\sin 2T}{2T^n} - \frac{n}{2} \int_T^{+\infty} \frac{1}{t^{2n+1}} d \cos 2t \\ &= -\frac{\sin 2T}{2T^n} + \frac{n \cos 2T}{2T^{2n+1}} - \frac{n(2n+1)}{2} \int_T^{+\infty} \frac{\cos 2t}{t^{2n+2}} dt. \end{aligned}$$

That is,

$$I_n(T) = -\frac{\sin 2T}{2T^{2n}} + \frac{n \cos 2T}{2T^{2n+1}} - \frac{n(2n+1)}{2} I_{n+1}(T). \tag{7.9}$$

The last integral has the following estimate:

$$\begin{aligned} |I_{n+1}(T)| &< \int_T^{+\infty} \frac{1}{t^{2n+2}} dt \\ &= \frac{1}{(2n+1)T^{2n+1}}. \end{aligned} \tag{7.10}$$

To obtain the desired accuracy (and simplify the calculation) put $T = 7\pi/2$. Due to relation (7.9)

$$
\begin{aligned}
I_1(T) &\approx -\frac{\sin 2T}{2T^2} + \frac{\cos 2T}{2T^3} + \frac{3\sin 2T}{4T^4} - \frac{3\cos 2T}{2T^5} \\
&\quad -\frac{15\sin 2T}{4T^6} + \frac{45\cos 2T}{4T^7} + \frac{315\sin 2T}{8T^8} - \frac{315\cos 2T}{4T^9} \\
&= -\frac{1}{2(7\pi/2)^3} + \frac{3}{2(7\pi/2)^5} - \frac{45}{4(7\pi/2)^7} + \frac{315}{4(7\pi/2)^9} \\
&\approx -0.00036732
\end{aligned}
$$

and $J_2(7\pi/2) \approx 0.04565649$. We obtain (using the Simpson formula) $J_1(7\pi/2) \approx 1.52513984$. Consequently

$$
J = I_1(T) + I_2(T) \approx 1.57079633,
$$

where all the first 7 decimal digits are true. The exact magnitude is $J = \pi/2$.

2.   INTEGRALS WITH A VARIABLE LIMIT

Here we consider integrals of the form

$$
I(t) = \int_t^{+\infty} \exp\left[\int_t^\tau \varphi(\xi)d\xi\right]\alpha(\tau)d\tau, \tag{7.11}
$$

and

$$
J(t) = \int_T^t \exp\left[\int_\tau^t \varphi(\xi)d\xi\right]\alpha(\tau)d\tau, \tag{7.12}
$$

where $\varphi(t)$ and $\alpha(t)$ are complex-valued functions, T is a sufficiently large positive number. We look for asymptotic approximation to the integrals for $t \to +\infty$. They are used for numerical calculation of some improper integrals and to estimate some particular solutions of differential and difference equations.

Lemma 7.1. *Let* $\varphi(t) \in A_t$, $\Pi\{\varphi(t)\} > -1$, $\alpha(t) \in \Pi$ *and*

$$
\lim_{\tau \to +\infty} \beta(\tau)\exp\left[\int_t^\tau \Re\varphi(\xi)d\xi\right] = 0 \tag{7.13}
$$

for any $\beta(t) \in H$ *with the estimate* $h\{\beta(t)\} \le \Pi\{\alpha(t)\} - \Pi\{\varphi(t)\}$. *Then integral* (7.11) *exists for* $t \gg 1$ *and*

$$
I(t) \asymp \delta_0(t) + \delta_1(t) + ... + \delta_m(t) + ..., \tag{7.14}
$$

where $\delta_0(t) = -\alpha(t)/\varphi(t)$, *and* $\delta_j(t) = -\delta'_{j-1}(t)/\varphi(t)$ *for* $j = 1, 2, ...$

PROOF. First prove the $I(t)$ existence. Integrate (7.11) by parts ($t \gg 1$):

$$I(t) = \int_t^{+\infty} \frac{\alpha(\tau)}{\varphi(\tau)} d\exp\left[\int_t^\tau \varphi(\xi)d\xi\right]$$

$$= \left[\frac{\alpha(\tau)}{\varphi(\tau)} \exp\left[\int_t^\tau \varphi(\xi)d\xi\right]\right]_t^{+\infty}$$

$$- \int_t^{+\infty} \left[\frac{\alpha(\tau)}{\varphi(\tau)}\right]' \exp\left[\int_t^\tau \varphi(\xi)d\xi\right] d\tau$$

$$= \delta_0(t) + R_1(t). \tag{7.15}$$

Here

$$R_1(t) = \int_t^{+\infty} \alpha_1(\tau) \exp\left[\int_t^\tau \varphi(\xi)d\xi\right] d\tau,$$

where $\alpha_1(t) = -[\alpha(t)/\varphi(t)]'$. Relation (7.15) is valid and $I(t)$ exists if $R_1(t)$ exists.

In the same way on taking (7.11) by parts in succession m times we obtain the relation

$$I(t) = s_m(t) + R_{m+1}(t). \tag{7.16}$$

Here

$$s_m(t) = \delta_0(t) + \delta_1(t) + ... + \delta_m(t) \tag{7.17}$$

and

$$R_{m+1}(t) = \int_t^{+\infty} \alpha_{m+1}(\tau) \exp\left[\int_t^\tau \varphi(\xi)d\xi\right] d\tau,$$

where $\alpha_j(t) = -[\alpha_{j-1}(t)/\varphi(t)]'$ ($j = 1, 2, ...$). Here we use the relation

$$\lim_{\tau \to +\infty} \frac{\alpha_m(\tau)}{\varphi(\tau)} \exp\left[\int_a^t \Re\varphi(\xi)d\xi\right] = 0.$$

It is correct because (see (7.13)) $\Pi\{\alpha_m(t)/\varphi(t)\} \le \Pi\{\alpha_{m-1}(t)\} - \sigma$, where $\sigma = 1 + \Pi\{\varphi(t)\} > 0$. Clearly, relation (7.16) is valid and $I(t)$ exists if $R_{m+1}(t)$ exists. Indeed, because of the relation $R_m(t) = \delta_{m-1} + R_{m+1}(t)$ it follows that $R_m(t)$ exists. Hence (in the same way) $R_{m-1}(t)$ exists and so on. Finally we conclude that $I(t)$ exists. Let us estimate $R_m(t)$ for $m \gg 1$. By induction on m we easily obtain the relation $\Pi\{\alpha_m(t)\} \le \Pi\{a(t)\} - m\sigma$. Hence $\Pi\{\alpha_m(t)\} \ll -1$ and $\Pi\{\alpha_m(t)\} \to -\infty$ for $m \to \infty$. We have

$$|R_m(t)| \le \sup_{\tau \ge t} |\alpha_m(\tau)\tau^{-b+2}| \int_t^{+\infty} \tau^{-2}\tau^b \exp\left[\int_t^\tau \Re\varphi(\xi)d\xi\right] d\tau,$$

$$\le t^{-1} \sup_{\tau \ge t} |\alpha_m(\tau)\tau^{-b+2}| \ll 1 \text{ for } t \gg 1, \tag{7.18}$$

where $b = \Pi\{\alpha(t)\} - \Pi\varphi(t)$. Due to (7.13) we conclude that $R_m(t)$ exists, $h\{R_m(t)(t)\} \ll -1$ and

$$h\{R_m(t)(t)\} \to -\infty \text{ for } n \to +\infty.$$

Thus, we have proved that $I(t)$ exists.

Prove (7.14). Taking into account the last relation we obtain $h\{I(t) - s_m(t)\} \to -\infty$ for $m \to \infty$, where

$$s_m(t) = \delta_0(t) + \delta_1(t) + ... + \delta_m(t).$$

On differentiating (7.11) we obtain $I'(t) + \varphi(t)I(t) + \alpha(t) = 0$. It means that the function (7.11) is a solution to the equation

$$y' + \varphi(t)y + \alpha(t) = 0. \tag{7.19}$$

Rewrite the equation in the form $y = \delta_0(t) - y'/\varphi(t)$. It has a formal solution $y(t) \sim \varphi_0(t)$ for $t \to +\infty$, that may be proved using Lemma 4.24. In our simple case, it can be also obtained directly from Theorem 4.17. Consider the iteration sequence:

$$s_m(t) = \varphi_0(t) - \frac{s'_{m-1}(t)}{\varphi(t)}, \quad s_0(t) = \delta_0(t), \quad m = 1, 2, ...$$

Clearly, $\Pi\{\delta_m(t)\} \le \Pi\{\delta_{m-1}(t)\} - \sigma$ and $\Pi\{\delta_m(t)\} \to -\infty$ $(m \to \infty)$. On the basis of Theorem 4.17 there is a function $\hat{I}(t) \in \Pi$ such that

$$\Pi\{\hat{I}(t) - s_m(t)\} \to -\infty \text{ for } m \to +\infty$$

and $\theta(t) \equiv \hat{I}'(t) + \varphi(t)\hat{I}(t) + \alpha(t) \asymp 0$. Because of the estimate of the functions $R_m(t)$ we have $h\{I(t) - \hat{I}(t)\} = -\infty$. Put $y = \hat{I}(t) + u$ in (7.19). We obtain the equation

$$u' + \varphi(t)u + \theta(t) = 0. \tag{7.20}$$

As it was proved $u(t) = I(t) - \hat{I}(t) = O(t^{-\infty})$ is a solution to the equation. Since this equation is linear there exists a number C such that

$$u(t) = C \exp\left[-\int_T^t \varphi(t)dt\right] + R(t).$$

Here T is a sufficiently large number (and if it is possible $T = +\infty$).

$$R(t) = \int_t^{+\infty} \tilde{\theta}(t) \exp\left[\int_t^\tau \varphi(\xi)d\xi\right] d\tau,$$

where $\tilde{\theta} = -\theta/\varphi(t) \asymp 0$. In the same way as for $R_m(t)$ it is easy to obtain the estimate $h\{R(t)\} = -\infty$. Since $h\{u(t)\} = -\infty$ we conclude that

$$\psi(t) \equiv C \exp\left[-\int_T^t \varphi(t)dt\right] = O(t^{-\infty}).$$

Taking into account that $\varphi(t) \in \Pi$, clearly, $\psi^{(n)}(t) = f_n(t)\psi(t)$ for any $n = 1, 2, ...$, where $f_n(t) \in \Pi$. Hence

$$h\{\psi(t)\} \leq h\{f_n(t)\} + h\{\psi(t)\} = -\infty$$

which implies $\Pi\{\psi(t)\} = -\infty$. In the same way, we may also obtain the estimate $\Pi\{R(t)\} = -\infty$. Thus, $\theta(t) \asymp 0$. $\square$

Examples 7.2. 1° Consider an integral of the form

$$I(t) = \int_t^{+\infty} e^{t-\tau}\alpha(\tau)d\tau,$$

where $\alpha(t) \in \Pi$. Here $\varphi(t) = -1 \in A_t$, hence $\Pi\{\varphi(t)\} = 0$, and $\lim_{\tau \to +\infty} p(\tau)e^{t-\tau} = 0$ for any $p(t) \in \Pi$. Consequently all the conditions of Lemma 7.1 are satisfied. We have $\delta_m(t) = \alpha^{(m)}(t)$ $(m = 0, 1, ...)$. Therefore

$$I(t) \asymp \alpha(t) + \alpha'(t) + ... + \alpha^{(m)}(t) + ...$$

2° Consider an integral of the form

$$I(t) = \int_t^{+\infty} e^{i(\tau-t)}\alpha(\tau)d\tau,$$

where $\Pi\{\alpha(t)\} \leq 0$. Here $\varphi(t) = i \in A_t$ $(i = \sqrt{-1})$. Hence $\Pi\{\varphi(t)\} = 0$. We cannot directly apply Lemma 7.1 because the function $\beta(t)$ can be equal to $\ln t$ $(\Pi\{\ln t\} = 0)$. Then $\lim_{\tau \to +\infty} \beta(\tau)e^{i(\tau-t)}$ may be equal to ∞. But

$$I(t) = i\alpha(t) + i\int_t^{+\infty} \alpha'(\tau)e^{i(\tau-t)}d\tau.$$

Clearly, $\Pi\{\alpha'(t)\} \leq -1$, and Lemma 7.1 is applicable to the integral

$$I_1(t) = \int_t^{+\infty} \alpha'(t)e^{i(\tau-t)}d\tau.$$

Finally we have

$$I(t) \asymp i\alpha(t) - \alpha'(t) + ... + i^{m+1}\alpha^{(m)}(t) + ...$$

Lemma 7.3. *Let $\varphi(t) \in A_t$, $\Pi\{\varphi(t)\} > -1$. Let either $\Re\varphi(t)t$ be bounded above or $\Re\varphi(t)t \to -\infty$ for $t \to +\infty$. Let $\alpha(t) \in \Pi$. Then there is a number C (generally speaking, C depends on T) such that (see (7.13))*

$$J(t) = C \exp\left[\int_T^t \varphi(\tau)d\tau\right] + G(t), \qquad (7.21)$$

where $T = \text{const} \gg 1$, and

$$G(t) \asymp \delta_0(t) + \delta_1(t) + ... + \delta_m(t) + ... \qquad (7.22)$$

Here $\delta_0(t) = \alpha(t)/\varphi(t)$ and $\delta_j(t) = \delta'_{j-1}(t)/\varphi(t)$ for $j = 1,2,...$ In particular if

$$C \exp\left[\int_T^t \Re\varphi(\tau)d\tau\right] \asymp 0 \text{ then } J(t) \asymp G(t). \qquad (7.23)$$

PROOF. On differentiating $J(t)$ we obtain $J'(t) = \alpha(t) + \varphi(t)J(t)$. Hence $J(t)$ is a solution of the linear differential equation

$$y' - \varphi(t)y - \alpha(t) = 0. \qquad (7.24)$$

Consider the iteration sequence $\{s_m(t)\}$, where $s_0(t) = \delta_0(t)$ and for $m = 1,2,...$

$$s_m(t) = \delta_0(t) + \frac{s'_{m-1}(t)}{\varphi(t)}.$$

Clearly,

$$s_m(t) = \delta_0(t) + \delta_1(t) + ... + \delta_m(t).$$

It is easy to check (see Proof of Lemma 7.1) that $\Pi\{\delta_m(t)\} \to -\infty$ for $m \to +\infty$. On the basis of Theorem 4.17 there is a function $\hat{J}(t) \in \Pi$ such that

$$\Pi\{\hat{J}(t) - s_m(t)\} \to -\infty \text{ for } m \to +\infty$$

and $\theta(t) \equiv \hat{I}'(t) - \varphi(t)\hat{I}(t) - \alpha(t) \asymp 0$. Put $y = u + \hat{J}(t)$. We obtain the equation

$$u' - \varphi(t)u - \theta(t) = 0. \qquad (7.25)$$

Since it is linear, its general solution may be presented in the form $u = C\eta(t) + Q(t)$, where $\eta(t) = \exp\left[\int_T^t \varphi(\xi)d\xi\right]$, and

$$Q(t) = \int_T^t \exp\left[\int_\tau^t \varphi(\xi)d\xi\right]\theta(\tau)d\tau.$$

To prove this Lemma it is enough to represent the function $Q(t)$ in the form $Q(t) = A\eta(t) + R(t)$, where A is a number and $R(t) \asymp 0$.

Let $\Re\varphi(t)t$ be bounded below. Then

$$A = \int_T^{+\infty} \exp\left[-\int_T^{\tau} \varphi(\xi)d\xi\right] \theta(\tau)d\tau$$

exists, and by Lemma 7.1,

$$R(t) = \int_t^{+\infty} \exp\left[-\int_t^{\tau} \varphi(\xi)d\xi\right] \theta(\tau)d\tau \asymp 0.$$

That is $Q(t)$ is represented in the required form.

Let $\Re\varphi(t)t \to -\infty$ for $t \to +\infty$. Then $t^{-n} \exp\left[-\int_T^t \Re\varphi(\xi)d\xi\right] \to \infty$ for any n. Estimate $Q(t)t^n$. We have

$$|Q(t)| \leq \int_T^t \exp\left[\int_\tau^t \Re\varphi(\xi)d\xi\right] |\theta(\tau)|d\tau.$$

$$|Q(t)t^n| \leq f(t,n) \equiv \frac{\int_T^t \exp\left[-\int_T^\tau \Re\varphi(\xi)d\xi\right] |\theta(\tau)|d\tau}{t^{-n} \exp\left[-\int_T^t \Re\varphi(\xi)d\xi\right]}.$$

The denominator of the last fraction tends to infinity for $t \to +\infty$. Hence if the numerator of the fraction is bounded, then, clearly, $f(t,n) \to 0$ for $t \to +\infty$. If the numerator tends to infinity we can apply de l'Hospital rule.

$$\lim_{t \to +\infty} f(t,n) = \lim_{t \to +\infty} -\frac{|\theta(t)|t^{n-1}}{\Re\varphi(t)t + n} = 0.$$

Consequently $\lim_{t \to +\infty} Q(t)t^n = 0$ for any n. This means that $h\{Q(t)\} = -\infty$. On differentiating $Q(t)$ in succession m we easily obtain that $h\{Q^{(m)}(t)\} = -\infty$ for any $m = 1, 2, \ldots$ Thus, $\Pi\{Q(t)\} = -\infty$. $\square$

Consider an equation of the form

$$y' + g(t)y + a(t) = 0. \tag{7.26}$$

The following Theorem is a simple consequence of Lemmata 7.1 and 7.3.

Theorem 7.4. *Let $g(t) \in A_t$ and $\Pi\{g(t)\} > -1$. Let $\alpha(t) \in \Pi$. Let one of two following conditions be fulfilled:*

(1)
$$\lim_{\tau \to +\infty} \beta(\tau) \exp\left[\int_t^\tau \Re g(\xi)d\xi\right] = 0 \ for \ t \gg 1,$$

and for any $\beta(t) \in \Pi$ such that $\Pi\{\beta(t)\} \leq \Pi\{\alpha(t)\} - \Pi\{g(t)\}$;

(2) *either* $\lim_{t \to +\infty} \Re g(t)t = +\infty$ *or the function* $\Re g(t)t$ *is bounded above.*
Then equation (7.26) *possesses at least one solution*

$$y(t) \asymp \delta_0(t) + \delta_1(t) + \ldots + \delta_n(t) + \ldots \qquad (7.27)$$

where $\delta_0(t) = -a(t)/\varphi(t)$ *and* $\delta_m(t) = -\delta'_{m-1}(t)/\varphi(t)$ *for* $m = 1, 2, \ldots$

PROOF. Let condition (1) be fulfilled. On the basis of Lemma 7.1 (where $\varphi(t) = g(t)$) the function $I(t)$ (see (7.12)) is a solution to the equation. And it satisfies relation (7.27).

Let condition (2) be fulfilled. Set $\varphi(t) = -g(t)$. Consequently either $\lim_{t \to +\infty} \Re \varphi(t)t = -\infty$ or the function $\Re \varphi(t)$ is bounded from below. Thus, all the conditions of Lemma 7.3 are satisfied. Hence the function $J(t)$ (see (7.13)) is represented in the form (7.21) and is a solution to equation (7.26). Clearly, the function $G(t)$ is also a solution to the equation and it satisfies (7.27). $\qquad \square$

In the subsequent sections we consider integrals of the form

$$I(\lambda) = \int_a^b k(\lambda, x) f(x) dx, \qquad (7.28)$$

where λ is a numerical parameter. The function $k(\lambda, x)$ is said to be a *kernel* of integral (7.28). Such integrals play an important role in many divisions of mathematical analysis. In particular they are used for investigation some differential and difference equations. We look for the asymptotic behavior of $I(\lambda)$ for $\lambda \to \infty$. First, consider so called Laplace integrals.

3.   THE LAPLACE TRANSFORMATION

Integrals of the form

$$F(p) = \int_0^{+\infty} e^{-pt} f(t) dt \qquad (7.29)$$

are called the *Laplace integrals*. They form a special but important part of integrals of type (7.28). Here $f(t)$ is a complex-valued function (of the real variable t), p is a complex parameter. Integrals of this type are widely used in many problems of the mathematical analysis. The operational calculation permits to fined integrals (7.29) without the process of integration for many (sufficiently complicated and widely met functions). Here we consider only its main properties. We limit our consideration only to so called *original functions* $f(t)$.

Definition 7.5. $f(t)$ is said to be an *original function*, and $F(p)$ (obtained from (7.29)) is called its *Laplace transform* if $f(t)$ satisfies the following conditions:

(1) $f(t) = 0$ for $t < 0$;

(2) $f(t)$ is (Lebesgue) integrable on any interval $[0, T] \subset J_+$ $(T > 0)$;

(3) there exist positive numbers k and A such that $|f(t)| < Ae^{kt}$ for all $t > 0$.

The relation

$$f(t) \risingdotseq F(p) \text{ or } F(p) \fallingdotseq f(t) \tag{7.30}$$

means that $f(t)$ is an original function, $F(p)$ exists and determined by (7.29).

Thus, integral (7.29) associates with an original function $f(t)$ a function $F(p)$ of the complex variable p.

The first condition is convenient for the further consideration and, clearly, does not influence on the consideration generality.

Owing to the second condition $e^{-pt} f(t)$ is an absolutely integrable function on any finite interval of the positive semi-axis J_+. But (7.29) is an improper integral dependent on the complex variable p. Integral (7.29) may not be convergent in the entire complex plane. Owing to property (3) there is at least one number s such that (7.29) is absolutely convergent in the right half plane $\Re p > s$.

Definition 7.6. The value

$$I\{f(t)\} = \varlimsup_{t \to +\infty} \frac{\ln|f(t)|}{t} \tag{7.31}$$

is said to be the *index of the order of growth* of the function $f(t)$ (or, simply, the *index*).

Proposition 7.7. Let $f(t)$ be an original function. Then $I\{f(t)\} < +\infty$ and for any real number $s > I\{f(t)\}$ there exists a number $M > 0$ (which, generally speaking, depends on s) such that the inequality $|f(t)| < Me^{st}$ is valid for any $t \geq 0$ and not valid for any $s < I\{f(t)\}$.

PROOF. Fix a number $s > I\{f(t)\}$. Due to Definition 7.6 there exists a number $T > 0$ such that $|f(t)| < e^{st}$ for $t \geq T$. Hence $I\{f(t)\} < s < +\infty$. Moreover, $|f(t)|$ is bounded on the interval $[0, T]$. That is, there is a number $M \geq 1$ such that $|f(t)| < M \max[1, e^{sT}]$. Consequently $|f(t)| < Me^{st}$ on the interval $[0, T]$ and, clearly, on the entire positive semi-axis. Suppose that the required inequality is fulfilled for $s < I\{f(t)\}$. Then (from (7.31)) it follows that $I\{f(t)\} \leq s$. The obtained contradiction proves this Proposition. $\square$

On the basis of properties (2), (3) the integral (7.29) is absolutely convergent for any p such that $\Re p > I\{f(t)\}$ (see (7.31)) and it is an analytic function in the considered domain. Indeed, take a number s such that $\Re p > s > I\{f(t)\} \equiv a$. There is a number $M > 0$ such that $|f(t)| < Me^{st}$ for $t \geq 0$. Integral (7.29) is absolutely convergent because majorized by a convergent integral:

$$
\begin{aligned}
|F(p)| &= \left| \int_0^{+\infty} e^{-pt} f(t)\,dt \right| \\
&\leq M \int_0^{+\infty} e^{-at} e^{st}\,dt \\
&= \frac{M}{s-a},
\end{aligned}
$$

moreover, $F'(p) = -\int_0^{+\infty} e^{-pt} t f(t)\,dt$. Clearly, the index of the order of growth of the function $tf(t)$ is equal to the same value as for $f(t)$. Hence $F'(p)$ exists for $\Re p > I\{f(t)\}$ which implies the desired property. $\qquad\square$

As a consequence if $I\{f(t)\} = -\infty$ then $F(p)$ is a holomorphic function in the entire (finite) complex plane $\mathbb{C}$. Besides, we may formulate the following assertion.

Proposition 7.8. *Let $f(t)$ be an original function and let $f(t)$ be bounded on the positive semi-axis. Then $I\{f(t)\} \leq 0$. Moreover, if (in addition)*

(i) there is a number $N > 0$ such that $\sup_{t \gg 1} |f(t)| > N$ (that is, there are numbers $t_1, t_2, \ldots, t_n, \ldots$, where $t_n \to +\infty$ for $n \to \infty$, and $f(t_n) > N$ for any n), then $I\{f(t)\} = 0$ and (as a consequence) $F(p)$ is a holomorphic function for $\Re p > 0$.

(ii) $f(t) = 0$ for $t \gg 1$, then $I\{f(t)\} = -\infty$ and (as a consequence) $F(p)$ is a holomorphic function in the entire (finite) complex plane.

Example 7.9. Consider the function

$$
q(t) = \begin{cases} 0 & \text{for } t < 0 \text{ and } t > 1, \\ 1 & \text{for } \quad 0 \leq t \leq 1. \end{cases}
$$

Clearly, $I\{q(t)\} = -\infty$. Let $Q(p)$ be the Laplace transform of the function $q(t)$. We have

$$
\begin{aligned}
Q(p) &= \int_0^1 e^{-pt}\,dt \\
&= \frac{1}{p}\left(1 - e^{-p}\right) \quad \text{for } p \neq 0.
\end{aligned}
$$

Besides $Q(0) = 1$ and $Q'(0) = -\int_0^1 te^{-pt}dt\big|_{p=0} = -1/2$. We can directly verify that $Q(p)$ is a holomorphic function in the entire complex plane. Indeed, $Q(p)$ has only one point $p = 0$, where we may suppose that it is not holomorphic. Since

$$\lim_{p\to 0}\frac{1}{p}\left(1 - e^{-p}\right) = 1 \quad \text{and} \quad \lim_{p\to 0}\left[\frac{1}{p}\left(1 - e^{-p}\right)\right]' = -1/2$$

we conclude that $Q(p)$ is holomorphic at the point $p = 0$, which leads to the required property.

The next proposition follows from 7.5 (1)–(3).

Proposition 7.10. *Let* $f(t) \risingdotseq F(p)$. *Then*

$$F(p) = O(1/p) \tag{7.32}$$

for $p \to \infty$ *in any central closed sector* S^* *which its any point is an interior point of the right half-plan* $S^+ = \{p : -\frac{\pi}{2} < \arg p < \frac{\pi}{2}\}$. *That is,* $|F(p)| \leq (C/|p|)$ *for* $p \to \infty, p \in [S^+]$, *where* C *is a positive number. Besides,* $\Pi_{S^+}\{F(p)\} \leq -1$. *As a simple consequence* $F(p) \to 0$ *for* $p \to \infty, p \in [S^+]$.

PROOF. Let $|f(t)| < Me^{st}$. We have

$$|F(p)| \leq \left|\int_0^{+\infty} Me^{-pt}e^{st}dt\right|$$

$$\leq \frac{M}{\Re p - s} \quad \text{for } \Re p \gg 1$$

in any (fixed) central closed sector S^* with its boundary angeles $\varphi_1^* = \frac{\pi}{2} - \varepsilon_1$ and $\varphi_2^* = -\frac{\pi}{2} + \varepsilon_2$ ($\varepsilon_{1,2}$ are positive numbers, $\varepsilon_{1,2} < \pi/2$). Hence

$$F(p) = \frac{M}{\Re p(1 - s/\Re p)} \sim \frac{M}{\Re p} \quad \text{for } \Re p \to +\infty.$$

Put $\alpha = \min[\sin\varphi_1^*, \sin\varphi_2^*]$. Then $|p/\Re p| \leq C = M/\sin\alpha < +\infty$. Hence $|F(p)| \leq C/|p|$. That is $F(p) = O(1/p)$ for $p \to \infty, p \in S^*$. Taking into account the arbitrariness of S^* we conclude that $F(p) = O(1/p)$ for $p \to \infty, p \in [S^+]$. Moreover, $h_{S^+}\{F(p)\} \leq -1$. Because of analysis of $F(p)$ we have $\Pi_{S^+}\{F(p)\} \leq -1$.

Example 7.11. 1° Prove that

$$t^k \risingdotseq \Gamma(k + 1)p^{-k-1} \quad \text{and} \quad I\{t^k\} = 0, \tag{7.33}$$

where k is a non-negative real number and $\Gamma(k) = \int_0^{+\infty} e^{-t} t^{k-1} dt$ is the so called Euler's Gamma function. In particular if k is a non-negative integer then

$$\int_0^{+\infty} e^{-pt} t^k \, dx = \frac{k!}{p^{k+1}}. \qquad (7.34)$$

Indeed, $I\{t^k\} = \overline{\lim}_{t \to +\infty} (\ln t^k)/t = 0$. We have

$$t^k \doteqdot F(p) \equiv \int_0^{+\infty} e^{-pt} t^\sigma \, dt.$$

Let p be a positive number. Substitute $pt = \tau$. Hence

$$F(p) = p^{-(k+1)} \int_0^{+\infty} e^{-\tau} \tau^k \, d\tau = \Gamma(k+1) p^{-(k+1)}.$$

Clearly, the last formula holds true for any p belonging to the half plane $\Re p > 0$.

In fact we may suppose that $k > -1$ in (7.33). Indeed, we have

$$F(p) = \frac{1}{\sigma + 1} \int_0^{+\infty} e^{-pt} \, dt^{k+1} = \frac{p}{k+1} \int_0^{+\infty} e^{-pt} t^{k+1} \, dt.$$

Hence $F(p)$ exists for any $k > -1$ and $\Re p > 0$. Clearly, formula (7.33) holds true. But it must be noted that the function t^k for $-1 < k < 0$ does not satisfy definition 7.5 because it is unbounded for $t \ll 1$.

Example 7.12. Let $f(t)$ be an original function. Let there exist a series

$$a_0 t^{k_0} + a_1 t^{k_1} + \dots + a_m t^{k_m} + \dots, \qquad (7.35)$$

where k_j are real numbers, $0 \le k_0 < k_1 < \dots < k_m < \dots$, a_m are (complex) numbers, such that $f(t) - s_m(t) = o(t^{k_m})$ for $t \to 0$, where

$$s_m(t) = a_0 t^{k_0} + a_1 t^{k_1} + \dots + a_m t^{k_m}$$

$(m = 0, 1, \dots)$. Then $f(t) \doteqdot F(p)$, where

$$
\begin{aligned}
F(p) =\ & a_0 \Gamma(k_0 + 1) p^{-(k_0+1)} + a_1 \Gamma(k_1 + 1) p^{-(k_1+1)} + \dots \\
& + a_m \Gamma(k_m + 1) p^{-(k_m+1)} + o\left(p^{-(k_m+1)}\right) \qquad (7.36)
\end{aligned}
$$

as $\Re p \to +\infty$ for any $m = 0, 1, \dots$ In particular if the original function $f(t)$ is a sum of its Taylor's series:

$$f(t) = f(0) + \frac{f'(0)}{1!} t + \dots + \frac{f^{(m)}(0)}{m!} t^m + \dots \quad (t \ll 1) \qquad (7.37)$$

then

$$F(p) \asymp \frac{f(0)}{p} + \frac{f'(0)}{p^2} + \ldots + \frac{f^{(n)}(0)}{p^{m+1}} + \ldots \text{ for } p \to \infty, p \in [S^+]. \quad (7.38)$$

Indeed, owing to (7.36)

$$s_m(t) \doteq F_m(p) = a_0 \Gamma(k_0 + 1) p^{-(k_0+1)} + a_1 \Gamma(k_1 + 1) p^{-(k_1+1)}.$$

We have $f(t) = f_m(t) + o(t^{k_m})$ for $t \to 0$ and for any positive number δ

$$\int_\delta^{+\infty} e^{-pt} f(t) dt = O(\Re p^{-\infty})$$

(such relations are examined in detail in the next section). So that $|F(p) - F_m(p)| = \int_0^\delta e^{-pt} o(t^{km}) dt + \theta(p)$, where $\theta(p) = O(\Re p^{-\infty})$. Hence for any $\varepsilon > 0$ it is possible to take a number δ such that

$$|F(p) - F_m(p)| \leq \varepsilon \int_0^\delta e^{-\Re p t} t^{km} dt$$

$$\leq \varepsilon \int_0^\infty e^{-\Re p t} t^{km} dt$$

$$\leq \varepsilon \Gamma(k_m + 1) \Re p^{-(k_m-1)}$$

which leads to the required relation.

It should be noted that it is possible to extend the class of functions $f(t)$ which admit the Laplace transformation, that the following proposition shows.

Proposition 7.13. *Let $f(t)$ be defined for $t \geq 0$ and let there be a complex number p_0 such that the integral $\int_0^{+\infty} e^{-p_0 t} f(t) dt$ converges. Then the integral $F(p) = \int_0^{+\infty} e^{-pt} f(t) dt$ exists for $Re(p - p_0) > 0$.*

PROOF. Put $G(t) = -\int_t^{+\infty} e^{-p_0 \tau} f(\tau) d\tau$. Let $T_2 > T_1 > 0$ and $\Re(p - p_0) > 0$.

$$I(T_1, T_2) \equiv \int_{T_1}^{T_2} e^{-pt} f(t) dt$$

$$= \int_{T_1}^{T_2} e^{-(p-p_0)t} dG(t)$$

$$= e^{-(p-p_0)T_2} G(T_2) - e^{-(p-p_0)T_1} G(T_1)$$

$$+ (p - p_0) \int_{T_1}^{T_2} e^{-(p-p_0)t} G(t) dt.$$

Since $\Re(p - p_0) > 0$ and (because of the existence of $G(t)$) $G(t) \to 0$ for $t \to +\infty$, each term of the last expression tends to zero for $T_1 \to \infty$. Consequently $I(T_1, T_2) \to 0$ for $T_1 \to +\infty$. By the Cauchy test the considered integral converges. $\square$

It is easy to see that in the last case the integral $F(p)$ converges absolutely and $F(p)$ is analytic for $\Re p > \Re p_0$.

In many problems the Laplace transform of an unknown function is easily found. Thus, we may use the relation (7.29) in order to investigate the original function $f(t)$. The function $f(t)$ exists and may be calculated by means of an improper integral for certain sufficient conditions imposed on $F(p)$. For the further consideration we will require several proposition which directly follows from the Cauchy integral properties. Consider an integral of the form

$$I = \frac{1}{2\pi i} \int_{x-i\infty}^{x+i\infty} \frac{F(z)}{z - p} dz \equiv \frac{1}{2\pi i} \lim_{A \to +\infty} \int_{x-iA}^{x+iA} \frac{F(z)}{z - p} dz. \qquad (7.39)$$

Lemma 7.14. *Let $F(z)$ be a holomorphic function for $\Re z > a$ (a is a real number) and $\lim_{z \to \infty, \Re z > a} F(z) = 0$ uniformly in the half-plane $\Re z > a$. Then $I = -F(p)$ independently of x satisfying the inequality $a < x < \Re p$.*

PROOF.   Consider a closed contour γ (see Fig. 3) consisting of the segment $l_R = [x - iR, x + iR]$ and the arc $C_R = \{z : \Re z \geq x, |z - x| = R\}$, connecting the endpoints of the segment. The function $F(z)/(z - p)$ has only one singularity $z = p$ which is a pole of the first order located inside the contour γ for $R \gg 1$.

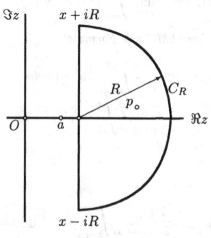

Fig. 3

The integral $\tilde{I} = \frac{1}{2\pi i} \int_{\gamma} [F(z)/(z-p)]dp$, taken over the contour traversed in the positive direction, is a Cauchy integral and it is equal to $F(p)$. Clearly, $\tilde{I} = -I_{l_R} + I_{C_R}$, where

$$I_{l_R} = \int_{x-iR}^{x+iR} \frac{F(z)dz}{z-p} \quad \text{and} \quad I_{C_R} = \int_{C_R} \frac{F(z)dz}{z-p}.$$

The last integral is taken over the arc C_R in the positive direction. I_{C_R} can be easily estimate.

$$|I_{C_R}| \leq \frac{1}{2\pi} M_R \int_{-R}^{R} \frac{2\pi dR}{R - |\Re p|}$$

$$\leq 2M_R \frac{R}{R - |\Re p|}$$

$$\to 0 \text{ for } R \to \infty.$$

Here $M_R = \sup_{|z|=R, \Re z>0} |F(z)| \to 0$ for $R \to \infty$. Since I is independent of R, we conclude that $I = -\tilde{I}$. That is $I = -F(p)$.   □

Theorem 7.15. *Let the function $F(p)$ satisfy the following conditions:*

(a) $F(p)$ is an analytic function in a half plane $\Re p > a$ where a is a real number;

(b) In the half plane $\Re p > a$, $F(p)$ tends to zero for $p \to +\infty$ uniformly in $\arg p$;

(c) the following limit is finite for any $x > a$:

$$M(x) = \lim_{A \to +\infty} \int_{-A}^{A} |F(x+iy)|dy. \qquad (7.40)$$

Then $f(t) \doteqdot F(p)$, where (Mellin's Formula)

$$f(t) = \frac{1}{2\pi i} \int_{x-i\infty}^{x+i\infty} e^{pt} F(p)dp$$

$$\equiv \lim_{A \to +\infty} \int_{x-iA}^{x+iA} e^{pt} f(p)dp \qquad (7.41)$$

and $I\{f(t)\} \leq a$.

PROOF. Clearly,

$$|f(t)| \leq \frac{1}{2\pi} \lim_{A \to +\infty} \int_{x-iA}^{x+iA} |e^{pt} F(p)||dp|$$

$$= \frac{e^{xt}}{2\pi} \lim_{A\to+\infty} \int_{-A}^{A} |F(x+iy)| dy$$

$$= \frac{M(x)}{2\pi} e^{xt} \tag{7.42}$$

which proves the existence of $f(t)$. As a consequence of the last inequality it follows that integral (7.41) uniformly converges with respect to the parameter t on any finite interval $[0, T] \in J_+$.

(1) Prove that $f(t)$ (defined by means of relation (7.41)) is independent of x. Consider a closed contour Γ (see fig. 4) consisting of the segments $[x_1 - iA, x_2 - iA]$,

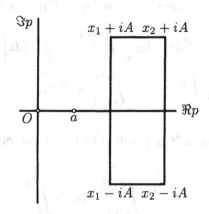

$$\Im p$$

$$x_1 + iA \quad x_2 + iA$$

$$O \quad a \quad \Re p$$

$$x_1 - iA \quad x_2 - iA$$

Fig. 4.

$[x_1 + iA, x_2 + iA]$ (parallel to the real axis), and $[x_1 - iA, x_1 + iA]$, $[x_2 - iA, x_2 + iA]$ (parallel to the imaginary axis). Here $a < x_1 < x_2$ and $A > 0$. Since $e^{pt} F(p)$ is analytic for $\Re p > a$ then by the Cauchy theorem the integral of the function around the contour Γ is equal to zero. Tend A to infinity and let x_1, x_2 be constants. Since $F(p) \to 0$ for $p \to +\infty$ uniformly in $\arg p$, the integral over the horizontal segments tends to zero. This leads to the equality

$$\int_{x_1-i\infty}^{x_1+i\infty} e^{pt} F(p) dp = \int_{x_2-i\infty}^{x_2+i\infty} e^{pt} F(p) dp$$

which means the required property.

(2) Prove that $f(t) = 0$ for $t < 0$. Let γ be the closed contour considered in Lemma 7.14 (see Fig. 3) On the basis of the Cauchy theorem the integral of $e^{zt} F(z)$ over the contour γ is equal to zero. The relation $\int_{C_R} e^z t F(z) dz \to 0$ for $R \to \infty$ follows from the conditions $t < 0$ and

$F(z) \to 0$ for $|p| \to \infty, \Re z > a$. Hence $\int_{x-iR}^{x+iR} e^{zt} F(z) dz \to 0$ for $R \to \infty$. This implies the relation $\lim_{R \to \infty} f(t) = 0$. Since $f(t)$ does not depend on R we conclude that $f(t) = 0$ $(t < 0)$.

(3) Prove that $I\{f(t)\} \leq a$. Indeed, the relation

$$\frac{\ln|f(t)|}{t} \leq \frac{M(x)}{2\pi t} + x.$$

Hence $\overline{\lim}_{t \to +\infty}(\ln f(t))/t \leq x$. This follows from condition (c) of the Theorem. Taking into account the arbitrariness of $x > a$ we conclude that $I\{f(t)\} \leq a$.

(4) It remains only to show the fulfillment of relation (7.29). We have

$$\int_0^{+\infty} e^{-pt} f(t) dt = \frac{1}{2\pi i} \int_0^{+\infty} e^{-pt} dt \left[\lim_{A \to +\infty} \int_{x-iA}^{x+iA} e^{-zt} F(z) dz \right]$$

$$= \lim_{A \to +\infty} \frac{1}{2\pi i} \int_0^{+\infty} e^{-pt} dt \int_{x-iA}^{x+iA} e^{-zt} F(z) dz.$$

Here $a < x < \Re p$. In the last integral it is possible to invert the order of integration. We obtain

$$\frac{1}{2\pi i} \int_0^{+\infty} e^{-pt} dt \int_{x-iA}^{x+iA} e^{-zt} F(z) dz = \frac{1}{2\pi i} \int_{x-iA}^{x+iA} F(z) dz \int_0^{+\infty} e^{(z-p)t} dt$$

$$\to -\frac{1}{2\pi i} \int_{x-i\infty}^{x+i\infty} \frac{F(z)}{z-p} dz$$

$$= F(p) \quad (A \to +\infty).$$

That is $f(t) \risingdotseq F(p)$.   □

The following assertion also holds.

Theorem 7.16. *Let $F(p)$ be the transform of a piecewise smooth function $f(t)$ such that $I\{f(t)\} = a < +\infty$. Then Mellin's formula is valid:*

$$f(t) = \frac{1}{2\pi i} \int_{x-i\infty}^{x+i\infty} e^{pt} F(p) dp$$

for $a < x < \Re p$.

Let us enumerate (without proof) the main elementary properties of the Laplace transformation of original functions.

Proposition 7.17 (Linearity). *Let* $\varphi(t) \risingdotseq \Phi(p)$ *and* $\psi(t) \risingdotseq \Psi(p)$. *Then*

$$f(t) \equiv \alpha\varphi(t) + \beta\psi(t) \risingdotseq \alpha\Phi(p) + \beta\Psi(p) \qquad (7.43)$$

for any numbers α *and* β. *Here* $I\{f(t)\} \leq \max[I\{\varphi(t)\}, I\{\psi(t)\}]$.

Let $f(t) \risingdotseq F(p)$.

Proposition 7.18 (Similarity). *For any number* $\alpha > 0$

$$f(\alpha t) \risingdotseq \frac{1}{\alpha} F\left(\frac{p}{\alpha}\right), \quad I\{f(\alpha t)\} = \alpha I\{f(t)\}. \qquad (7.44)$$

Proposition 7.19 (differentiation of the original function). *If in addition* $f(t)$ *is continuous and* $f'(t)$ *is an original function then*

$$f'(t) \risingdotseq pF(p) - f(0) \ \ for \ \Re p > I\{f(t)\}. \qquad (7.45)$$

The index of the derivative is connected with the index of the function $f(t)$ *by the following relation* $I\{f(t)\} \leq \max[0, I\{'(t)\}]$.

In general if $f^{(n-1)}(t)$ is a continuous function for $t \gg 1$ there are the finite limits $\lim_{t \to +0} f^{(k)}(t) = f^k(0)$ $(k = 1, 2, ..., n-1)$, and $f^{(n)}(t)$ is an original function of the Laplace transform, then

$$f^{(n)}(t) \risingdotseq p^n F(p) - p^{n-1}f(0) - p^{n-2}f'(0) - \cdots - f^{(n-1)}(0) \qquad (7.46)$$

$$for \ \Re p > I\{f^{(n-1)}(t)\},$$

$n = 1, 2, ...$ *In particular if* $f(0) = f'(0) = ... = f^{(n-1)}(0) = 0$ *then*

$$f^{(n)}(t) \risingdotseq p^n F(p). \qquad (7.47)$$

Proposition 7.20 (Differentiation of a transform).

$$F'(p) \risingdotseq (-1)tf(t). \qquad (7.48)$$

In general

$$F^{(n)}(p) \risingdotseq (-1)^n t^n f(t) \ (n = 1, 2, ...). \qquad (7.49)$$

$I\{t^n f(t)\} = I\{f(t)\}$.

Proposition 7.21

$$g(t) \equiv \int_0^t f(\tau)d\tau$$

$$\doteqdot \frac{F(p)}{p} \quad \text{for } \Re p > I\{f(t)\}. \tag{7.50}$$

Proposition 7.22 (integration of a transform). *If the integral* $\int_p^\infty F(z)dz$ *is convergent for* $\Re p > a$ *then*

$$\frac{f(t)}{t} \doteqdot \int_p^\infty F(z)dz \quad for \; \Re p > a. \tag{7.51}$$

Proposition 7.23 (Time delay theorem). *Consider a function of the form*

$$f_\sigma(t) = \begin{cases} 0 & \text{if } t < \sigma, \\ f(t - \sigma) & \text{if } t \geq \sigma. \end{cases} \tag{7.52}$$

Then

$$f_\sigma(t) \doteqdot F_\sigma(p) = e^{-p\sigma} F(p). \tag{7.53}$$

Proposition 7.24 (theorems of multiplication) *Let* $g(t) \doteqdot G(p)$ *($f(t) \doteqdot F(p)$). Then*

(1)
$$F(p)G(p) \doteqdot \gamma(t) \equiv \int_0^t f(\tau)g(t - \tau)d\tau \tag{7.54}$$

and $I\{\gamma(t)\} \leq \max[I\{f(t)\}, I\{g(t)\}].$

(2)
$$f(t)g(t) = \frac{1}{2\pi i} \int_{\alpha-i\infty}^{\alpha+i\infty} f(q)G(p - q)dq, \tag{7.55}$$

where $\alpha > I\{f(t)\}$ *and* $\Re p > \alpha + I\{g(t)\}.$

Example 7.25. Find the transform $F(p)$ of the function $t \sin t$. First, we find the transform $\widetilde{F}(p) \doteqdot \sin t$. We have

$$
\begin{aligned}
\widetilde{F}(p) &\doteqdot \int_0^{+\infty} e^{-pt} \sin t\, dt \\
&= \int_0^{+\infty} e^{-pt} \frac{e^{it} - e^{-it}}{2i} dt \\
&= \frac{1}{2i}\left(\frac{1}{p - i} - \frac{1}{p + i} \right) \\
&= \frac{1}{p^2 + 1}. \\
F(p) &= \frac{2p}{(p^2 + 1)^2}.
\end{aligned}
$$

As a more simple and natural application of the Laplace transformation, we consider linear non-homogeneous differential equations with constant coefficients given in the form

$$a_0 x^{(n)} + a_1 x^{(n-1)} + \dots + a_n x = f(t), \qquad (7.56)$$

where $a_0, a_1, \dots, a_n$ are complex numbers, $f(t)$ is a given function and there are given the following initial conditions: $x(0) = x_0, x'(0) = x_1, \dots, x^{(n-1)}(0) = x_{n-1}$. Let $f(t) \risingdotseq F(p)$ and let there exist a transform $X(p)$ for the considered solution to the equation. Hence taking into account formula (7.46) we obtain

$$a_0 p^n \left[X(p) - \frac{x_0}{p} - \dots - \frac{x_{n-1}}{p^{n-1}} \right] + a_1 p^{n-1} \left[X(p) - \frac{x_0}{p} - \dots - \frac{x_{n-2}}{p^{n-2}} \right] + \dots$$

$$+ a_n X(p) = F(p).$$

The obtained equation can be represented in the form

$$H(p) X(p) = Q(p) + F(p), \qquad (7.57)$$

where

$$H(p) = a_0 p^n + a_1 p^{n-1} + \dots + a_n$$

is said to be the *characteristic polynomial* of the equation. $Q(p)$ is a polynomial with constant coefficients. Its power is $\le n - 1$. We have

$$X(p) = \frac{Q(p)}{H(p)} + \frac{F(p)}{H(p)}.$$

Finally $x(t) \risingdotseq X(p)$.

Notice that *the transform of the required solution is obtained from the equation only by means of algebraic operations.*

The function $g(t) \risingdotseq F(p)/H(p)$ is a partial solution to equation (7.57) with zero initial conditions $g(0) = g'(0) = \dots = g^{(n-1)}(0) = 0$. The function $u(t) \risingdotseq Q(p)/H(p)$ is a solution to the homogeneous equation

$$a_0 x^{(n)} + a_1 x^{(n-1)} + \dots + a_n x = 0$$

with the given initial conditions. It depends on all the initial values $x_0, x_1, \dots, x_{n-1}$ which may be considered as a numerical parameters of the solution.

Example 7.26. Consider the equation

$$x'' + x = \cos t.$$

Here $x(t) \risingdotseq X(p)$. Clearly $\cos t = (\sin t)' \risingdotseq p/(p^2 + 1)$ (see Proposition 7.19 and Example 7.25). We have $x''(t) \risingdotseq p^2 X(p) - x_0 p - x_1$. Hence

$$(p^2 + 1) X(p) = x_0 p + x_1 + \frac{p}{p^2 + 1}.$$

So that

$$X(p) = \frac{x_0 p + x_1}{p^2 + 1} + \frac{p}{(p^2 + 1)^2}.$$

$$\frac{x_0 p + x_1}{p^2 + 1} \risingdotseq x_0 \cos t + x_1 \sin t = u(t) \text{ and } \frac{p}{(p^2 + 1)^2} \risingdotseq \frac{1}{2} t \sin t = g(t).$$

Finally

$$\begin{aligned} x(t) &= u(t) + g(t) \\ &= x_0 \cos t + x_1 \sin t + \frac{1}{2} t \sin t. \end{aligned}$$

4.   THE LAPLACE METHOD

The Laplace method is applied to integrals of the form

$$J(\lambda) = \int_a^b e^{\lambda h(x)} g(x) dx, \tag{7.58}$$

where $h(x)$ is a real function. The function $g(x)$ can be real or complex as well (x is a real variable), λ is a sufficiently large positive parameter, a and b are points of the real axis and each of the points may be infinitely large so that $-\infty \le a < b \le +\infty$.

In the simplest cases integral (7.58) may be reduced to the Laplace integral (7.29). Let $a = 0, b = +\infty$ and the derivative $h'(x)$ exists for $x \ge 0$. Let there be an inverse function $x = \psi(t)$ of the original function $t = -h(x)$ such that $dx = -dt/h'(\psi(t))$. Let $f(t) = g(\psi(t))/h'(\psi(t))$ and let $f(t) \risingdotseq F(\lambda)$. Then $J(\lambda) = -F(\lambda)$.

Example 7.27.   1° Consider an integral of the form

$$I(\lambda) = \int_0^{+\infty} e^{-\lambda x^2} x^\mu dt, \tag{7.59}$$

where μ is a real number, $\mu > -1$. Substitute $x = \sqrt{t}$. Hence

$$I(\lambda) = \frac{1}{2} \int_0^{+\infty} e^{-\lambda t} t^{(\mu-1)/2} dt = \frac{1}{2} \Gamma\left(\frac{\mu+1}{2}\right) \lambda^{-(\mu+1)/2}, \tag{7.60}$$

where $\Gamma(t)$ is Euler's Gamma function. In particular

$$\int_0^{+\infty} e^{\lambda x^2} dx = \frac{1}{2} \sqrt{\frac{\pi}{\lambda}}. \tag{7.61}$$

Definition 7.28. Let c be an interior point of the segment $[a, b]$ and δ be a sufficiently small positive number. The integral

$$J_\delta(\lambda) = \int_{c-\delta}^{c+\delta} e^{\lambda h(x)} g(x) dx \qquad (7.62)$$

is called the *contribution* to the integral (7.58) at the point c. If $c = a \neq -\infty$ then the contribution is equal to the integral $\int_a^{a+\delta} e^{\lambda h(x)} g(x) dx$. In the same way if $c = b \neq +\infty$ then the contribution is equal to $\int_{b-\delta}^b e^{\lambda h(x)} g(x) dx$.

The contribution depends on δ. Evidently $J_0(\lambda) = 0$ for any λ. But in some points the limit $\lim_{\delta \to 0} J_\delta(\lambda) = 0$ is not uniform in λ for $\lambda \gg 1$ and there exists a function $\varphi(\lambda)$ which is independent of δ and

$$J_\delta(\lambda) \sim \varphi(\lambda) \text{ as } \lambda \to +\infty$$

(for any $\delta \neq 0$ on the considered interval). A similar situation takes place in many asymptotic problems. For example for the function $y(\lambda, \delta) = 1 - e^{-\lambda \delta}$, clearly, $y(\lambda, 0) = 0$. But $y(\lambda, \delta) = 1 + o(1)$ as $\lambda \to +\infty$ for any $\delta > 0$.

Laplace's method is applied in the cases when a critical point of $J(\lambda)$ exists and the integral is equivalent to the main part of the contribution at the point.

The following proposition gives us a typical example of the Laplace method application.

Example 7.29. Prove that

$$I(\lambda) = \int_{-\infty}^{+\infty} e^{-\lambda x^2 (1+|x|)} dx \sim \sqrt{\frac{\pi}{\lambda}} \text{ for } \lambda \to +\infty.$$

Indeed, introduce the notation

$$I(a, b) = \int_a^b e^{-\lambda x^2 (1+|x|)} dx.$$

Let $\delta \ll 1$. We have $I(\lambda) = I(-\infty, -\delta) + I(-\delta, \delta) + I(\delta, +\infty)$. Clearly, for $\lambda \gg 1$

$$I(-\infty, -\delta) = I(\delta, \infty) \leq e^{-\lambda \delta^2} \int_\delta^{+\infty} e^{-\lambda \delta^2 x} dx = O(\lambda^{-\infty}).$$

Besides

$$2 \int_0^{+\infty} e^{-\lambda x^2 (1+\delta)} dx < I(-\delta, \delta) < 2 \int_0^{+\infty} e^{-\lambda x^2 (1-\delta)} dx.$$

As it follows from the result obtained in Example 7.27

$$\sqrt{\frac{\pi}{\lambda(1+\delta)}} < I(-\delta, \delta) < \sqrt{\frac{\pi}{\lambda(1-\delta)}}.$$

Taking into account the arbitrariness of δ and the obtained estimates for $I(-\infty, -\delta)$ and $I(\delta, \infty)$ we obtain $I(\lambda) \sim \sqrt{\pi/\lambda}$ for $\lambda \to +\infty$.

The idea of the Laplace method is simple: In many cases the contribution to the integral (7.58) at the supremum point of the function $h(x)$ gives the principal term of the integral asymptotics. Let, for example, the supremum be attained at an interior point c of the integration path, and let there be a finite derivative on a neighborhood of this point. Then c is a point of maximum of the function $h(x)$. Without loss of generality we may suppose that $a < 0$, $b > 0$, $c = 0$ and $h(0) = h'(0) = 0$. In other cases when $h(0) \neq 0$, we may represent integral (7.58) in the form

$$I(\lambda) = e^{\lambda h(x_0)} \int_{a_1}^{b_1} e^{\lambda h_1(x)} dx,$$

where $a_1 = a - x_0$, $b_1 = b - x_0$ and $h_1(x) = h(x + x_0) - h(x_0)$, thus $a_1 < 0$, $b_1 > 0$ and $h_1(0) = 0$.

First, we examine some simple integrals. Consider an integral of the form

$$I(\lambda) = \int_0^{+\infty} e^{\lambda h(x)} dx. \tag{7.63}$$

Proposition 7.30. *Let*

$$h(x) = -x^2(1 + \alpha(x)),$$

where $\alpha(x)$ is a continuous function specified on $[0, +\infty[$ and $\alpha(0) = 0$. Let $h(x) < 0$ for $x > 0$ and let there be numbers $p > 0$ and $E > 0$ such that $h(x) < -p$ for $|x| > E$, and let integral (7.63) be convergent for $\lambda = 1$. Then

$$I(\lambda) \sim \sqrt{\frac{\pi}{\lambda}} \text{ for } \lambda \to +\infty. \tag{7.64}$$

PROOF. Choose by arbitrariness a sufficiently small number $\delta > 0$. $I(\lambda) = I(0, \delta) + I(\delta, +\infty)$, where

$$I(0, \delta) = \int_0^\delta e^{\lambda h(x)} dx \text{ and } I(\delta, +\infty) = \int_\delta^{+\infty} e^{\lambda h(x)} dx.$$

To estimate the integral $I(\delta, +\infty)$ put

$$q(\delta) = \max[-p, \sup_{\delta \leq x \leq E} h(x)].$$

Since the function $h(x)$ is negative and continuous on the interval $[\delta, E]$, clearly, $q(\delta) < 0$ for any $\delta > 0$. Hence

$$I(\delta, +\infty) \leq e^{(\lambda-1)q(\delta)} \int_{\delta}^{+\infty} e^{h(x)} dx = O(\lambda^{-\infty}) \text{ for } \lambda \to +\infty.$$

Estimate the integral $I(0, \delta)$ which is the contribution of $I(\lambda)$ at the point $c = 0$. Let $|\alpha(x)| < \varepsilon$ for $x \leq \delta$ ($\varepsilon \ll 1$ if $\delta \ll 1$). Hence

$$\int_0^{\delta} e^{-x\lambda x^2(1+\varepsilon)} dx \leq I(0, \delta) \leq \int_0^{\delta} e^{-\lambda x^2(1-\varepsilon)} dx.$$

That is (see example 7.29)

$$\frac{1}{2}\sqrt{\frac{\pi}{\lambda(1+\varepsilon)}} \leq I(0, \delta) \leq \frac{1}{2}\sqrt{\frac{\pi}{\lambda(1-\varepsilon)}}$$

which leads (see Example 7.29) to (7.64). □

The next proposition is a simple consequence of Proposition 7.30, but its hypothesis some easier.

Proposition 7.31. *Let $h(x)$ be a continuous function and $h(x) < 0$ for $|x| > 0$. Let $h(0) = h'(0) = 0$ and $h''(0) < 0$. Let there be numbers $p > 0$ and $E > 0$ such that $h(x) < -p$ for $|x| > E$. Then (see (7.63))*

$$I(\lambda) \sim \sqrt{-\frac{2\pi}{h''(0)\lambda}} \text{ for } \lambda \to +\infty. \tag{7.65}$$

PROOF. Put $f(t) = h(x) - \frac{h''(0)}{2}x^2$. Hence $f(0) = f'(0) = f''(0) = 0$. Mark, the existence of $f''(0)$ implies the existence of $f'(x)$ for $|x| \ll 1$. We have $f'(x)/x \to 0$ for $x \to 0$. Hence $f'(x) = o(x)$ ($x \to 0$). We may apply Lagrange's formula of finite increments. We have $f(x) - f(0) = f'(\theta x)x$, where $0 < |\theta| < 1$. Thus, $f(x) = o(x^2)$ ($x \to 0$). That is

$$h(x) - \frac{h''(0)}{2}x^2 = o(x^2).$$

Thus,

$$h(x) = -\frac{h''(0)}{2}x^2[1 + \alpha(x)],$$

where $\alpha(x)$ is continuous and $\alpha(0) = 0$. Moreover since $h(x)$ is continuous and negative on a closed region $[-\delta, -E] \cup [\delta, E]$, where δ is a positive number, and taking into account the inequality $h(x) < -p$ for $|x| > E$, we conclude that there is a number $q > 0$ such that $h(x) \leq -q$ for $|x| \geq \delta$. Consequently $I(\lambda)$ exists for $\lambda = 1$. So that we may apply Proposition 7.30 where $-\lambda h''(0)/2$ is rewritten as λ which leads the required relation.

$\square$

Proposition 7.32. *Let the function $h(x)$ satisfy all the conditions of Proposition 7.31. Let $g(x)$ be a continuous function for $x > 0$. Let the integral $I(1) = \int_0^{+\infty} e^{h(x)} g(x) dx$ be convergent. Let for $x \ll 0$ $g(x) = x^{\mu+\beta(x)}$ where $\mu \leq -1$, $\beta(x)$ is continuous for $x \ll 1$ and $\beta(0) = 0$. Then $\Pi\{I(\lambda)\} = -(\mu+1)/2$.*

PROOF. First, prove that $h\{I(\lambda)\} = -(\mu+1)/2$. Just as in the Proof of Example 7.29 it is easy to show that

$$I(\lambda) \sim \int_0^{\delta} e^{\lambda h(x)} g(x) dx \text{ for } \lambda \to +\infty \ (\delta \ll 1). \qquad (7.66)$$

Moreover for an arbitrary (fixed) sufficiently small positive numbers ε_1, ε_2 it is possible to choose a number $\delta > 0$, such that

$$\int_0^{\delta} e^{-\lambda x^2 (1+\varepsilon_1)} x^{\mu+\varepsilon_2} dx \leq \int_0^{\delta} e^{\lambda h(x)} g(x) dx \leq \int_0^{\delta} e^{-\lambda x^2 (1-\varepsilon_1)} x^{\mu-\varepsilon_2} dx.$$

Clearly,

$$\begin{aligned} I_1(\lambda) &= \int_0^{\delta} e^{-\lambda x^2 (1+\varepsilon_1)} x^{\mu+\varepsilon_2} dx \\ &\sim \frac{1}{2} \Gamma\left(\frac{\mu+1+\varepsilon_2}{2}\right) [\lambda(1+\varepsilon_1)]^{-(\mu+1+\varepsilon_2)/2} \end{aligned}$$

for $\lambda \to +\infty$. Consequently $h\{I_1(\lambda)\} = -(\mu+1+\varepsilon_2)/2$. Hence $h\{I(\lambda)\} \geq -(\mu+1)/2 - \varepsilon_2/2$. In the same way we obtain the inequality $h\{I(\lambda)\} \leq -(\mu+1)/2 + \varepsilon_2/2$. Since $h\{I(\lambda)\}$ is independent of ε_2 we conclude that $h\{I(\lambda)\} = -(\mu+1)/2$.

To prove this Proposition it remains to show that $I(\lambda)$ is an analytic function in λ in a sufficiently small sector $S_\varepsilon = \{\lambda : |\arg \lambda| < \varepsilon \ll 1\}$. In fact, since formally $I'(\lambda) = \int_0^{+\infty} e^{\lambda h(x)} g(x) h(x) dx$, we have to prove that the integral

$$\tilde{J}(\lambda) = \int_0^{\infty} e^{\lambda h(x)} g(x) h(x) dx$$

uniformly converges in S_ε for $|\lambda| \gg 1$. Because of convergence of the integral $I(1)$ we have $e^{h(x)}g(x) \to 0$ for $x \to +\infty$. Then there is a number M such that $|e^{h(x)}g(x)| < M$ for $x > 0$. So that

$$\tilde{J}(\lambda) \le M \int_0^\infty e^{(\Re\lambda-2)h(x)} e^{h(x)} |h(x)| dx.$$

The function $e^{-t}t$ has its maximum for $t \ge 1$ at the point $t = 1$. Consequently (taking into account that $h(x) < 0$ for $x > 0$) we have $e^{h(x)}|h(x)| \le 1/e$ which leads to the required estimate. $\qquad\square$

Example 7.33. 1° Derive the Stirling formula for $n!$.
$n!$ may be given in the form

$$n! \equiv I(n) = \int_0^{+\infty} e^{-t} t^n dt. \qquad (7.67)$$

We may suppose that n is an arbitrary positive sufficiently large number. We are going to reduce integral (7.67) to the form considered in Proposition 7.32. Substitution $t = n(1 + x)$ leads to the relation

$$I(n) = n^{n+1}e^{-n} \int_{-1}^{+\infty} e^{-nx}(1+x)^n dx \sim 2n^{n+1}e^{-n} \int_0^{+\infty} e^{n[-x+\ln(1+x)]} dx.$$

The function

$$h(x) = -x + \ln(1 + x)$$

has a continuous derivative for $x > -1$, the point 0 is a point of $h(x)$ maximum, $h(x) < 0$ in the region $]-1,0[$, $]0,\infty[$, $h(0) = h'(0) = 0$, and $h''(0) = -1 < 0$. Thus, we have $q = 2$ and $\sigma = 0$. Let us deduce the inverse function to $-h(x)$ for $|x| \ll 1$. We have

$$x = \tau - \ln(1 + \tau) \sim \tau^2/2 \quad (\tau \to 0).$$

Consequently $\tau \sim \sqrt{2x}$ for $x \to 0$ and finally we obtain

$$n! \sim n^n e^{-n} \sqrt{2\pi n} \text{ for } n \to +\infty.$$

5.   THE SADDLE POINT METHOD

In this section we consider the asymptotic behavior of integrals of the form

$$J(\lambda) = \int_\gamma e^{\lambda h(z)} g(z) dz \text{ for } \lambda \to +\infty, \qquad (7.68)$$

where γ is an integration contour in the complex plane z connecting points a and b. Each of the points may be infinitely large. The saddle point method is introduced by B. Riemann and P. Debye and it is a very strong analytic method of investigation of the asymptotic behavior of integrals (7.68) with a large positive parameter λ. It can be applied for investigation of some linear differential equations with variable coefficients. The saddle point method may be considered as a spreading of the Laplace method for integrals in the complex plane.

We suppose that $h(z)$ and $g(z)$ are analytic functions in a single-connected domain D containing the contour γ. Thus, the integral may be rewritten in the form

$$J(\lambda) = \int_a^b e^{\lambda h(z)} g(z)\,dz \text{ in } D, \qquad (7.69)$$

where the path of integration may be anyone satisfying the conditions of the Cauchy theorem for integration of analytic functions.

Later on throughout the section we suppose that any curve under consideration satisfies such conditions.

The saddle point method consists of two parts. In the first one we choose a contour of integration C (connecting the points a and b in D instead of the contour γ) such that it has to be convenient for investigation of integral (7.69). The second part is the technique of obtaining the exact asymptotic formula which (as a rule) is like the Laplace technique of investigation of integrals on the real axis.

The first step is more difficult because we have a lot of integration contours. And the saddle point method in essence consists of some reasons or rules how to obtain the contour C and of a collection of the well chosen examples. But it is possible to formulate some precise propositions to solve such a problem.

We begin with a simple but important example. Examine the integral

$$J(\lambda) = \int_{-1}^1 e^{i\lambda z^2}\,dz \qquad (7.70)$$

$(i = \sqrt{-1})$. If z is real we have

$$|J(\lambda)| \leq \int_{-1}^1 |e^{i\lambda z^2}|\,dz = \int_{-1}^1 dz = 2.$$

But the obtained estimate is sufficiently rough. To obtain the estimate more accurate let us choose the contour of $J(\lambda)$ integration consisting of a polygonal line connecting the points $-1, -1-i, 1+i, 1$. The line passes through the origin O and the integral is equal to a sum of three integrals

$$\int_{-1}^{-1-i} + \int_{-1-i}^{1+i} + \int_{1+i}^1 e^{i\lambda z^2}\,dz.$$

In the first integral we substitute $z = -1 - it$, hence

$$J_1(\lambda) \equiv \int_{-1}^{-1-i} e^{i\lambda z^2} dz = -i \int_0^1 e^{i\lambda(1-t^2)-2\lambda t} dt.$$

Consequently

$$|J_1(\lambda)| \leq \int_0^1 e^{-2\lambda t} dt = \frac{1}{2\lambda}(1 - e^{-2\lambda t}) = O\left(\frac{1}{\lambda}\right) \text{ for } \lambda \to +\infty.$$

In the same way we obtain

$$J_3(\lambda) \equiv \int_{1+i}^1 e^{i\lambda z^2} dz = O\left(\frac{1}{\lambda}\right) \quad (\lambda \to +\infty).$$

In the integral

$$J_2(\lambda) = \int_{-1-i}^{1+i} e^{i\lambda z^2} dz$$

substitute $z = (1+i)t$. Hence

$$
\begin{aligned}
J_2(\lambda) &= (1+i) \int_{-1}^1 e^{-2\lambda t^2} dt \\
&= (1+i) \int_{-\infty}^{+\infty} e^{-2\lambda t^2} dt + O(\lambda^{-\infty}) \\
&= (1+i)\sqrt{\frac{\pi}{2\lambda}} + O(\lambda^{-\infty}).
\end{aligned}
$$

Consequently

$$J(\lambda) = (1+i)\sqrt{\frac{\pi}{2\lambda}} + O\left(\frac{1}{\lambda}\right) \text{ for } \lambda \to +\infty. \tag{7.71}$$

Let us give some simple reasons how to choose the desired contour C. One is obvious, C must not contain double points because the contribution to integral (7.68) on any loop is equal to zero. It is the best (but may be not so simple) that $h(z)$ remains a real function on the path C. Thus, we choose the curve $\Im h(z) = 0$.

Let (for simplicity) the curve $\Im h(z) = 0$ connect the points a and b in the domain D and let the function $\Re h(z)$ attain its supremum at an interior point z_0 of the curve. Let us put $z = x + iy$, $u(x,y) = \Re h(z)$, $v(x,y) = \Im h(z)$ and $z_0 = x_0 + iy_0$. Since z_0 is an extremal point we have

$$\frac{\partial u(x_0, y_0)}{\partial x} = \frac{\partial u(x_0, y_0)}{\partial y} = 0.$$

Since the function $h(z)$ is analytic in D, taking into account the Cauchy–Riemann conditions, we obtain

$$\frac{\partial v(x_0, y_0)}{\partial x} = \frac{\partial v(x_0, y_0)}{\partial y} = 0$$

which implies the relation $h'(z_0) = 0$. In a neighborhood of any point of an analytic function $h(z)$, the maximum of its modulus (if $h(z)$ is, indeed, not a constant) is attained only on the boundary. Hence there are points in any sufficiently small neighborhood which are found up and down of the point $|h(z_0)|$. But the tangent plane at the point z_0 is horizontal. A point such z_0 is named a *saddle point* of the surface $|h(z)|$. Thus, the contour C passes through the saddle point. In the last example the function $h(z) = iz^2$ has only one saddle point $z_0 = 0$ and so we choose the integration path containing this saddle point.

The direction of the tangent line of the curve $\Im h(z) = 0$ is determined by the vector

$$\vec{s} = -\frac{\partial v(x, y)}{\partial y}\vec{e}_1 + \frac{\partial v(x, y)}{\partial x}\vec{e}_2,$$

where $\vec{e}_1$ and $\vec{e}_2$ are the unit vectors of the axes x and y, respectively. On the basis of the Cauchy–Riemann conditions $\vec{s}$ is equal to

$$\frac{\partial u(x, y)}{\partial x}\vec{e}_1 + \frac{\partial u(x, y)}{\partial y}\vec{e}_2 = \mathrm{grad}\, u(x, y).$$

The vector $\mathrm{grad}\, u(x, y)$ determines the greatest variation of the level line $\Re h(z) = \mathrm{const}$ at the point z. Hence $\Im h(z) = 0$ is the line of the greatest variation of the functions $\Re h(z)$ and $|e^{h(z)}|$. It is referred to as a *steepest line*. The maximum of the function is strongly expressed on this way, and the line $\Im h(z) = 0$ is convenient for the function $e^{h(z)}$ and for integral (7.69) estimate. Let the contribution at the saddle point be equivalent to the function $\varphi(\lambda)$ for $\lambda \to +\infty$. We may hope that $J(\lambda) \sim \varphi(\lambda)$ for $\lambda \to +\infty$. Otherwise the Laplace method indeed can not be applied to the considered problem.

Some additional remarks: (1) if $h(z_0) \neq 0$ we represent integral (7.68) in the form $J(\lambda) = e^{\lambda h(z_0)}J_1(\lambda)$ where $J_1(\lambda) = \int_a^b e^{\lambda h^*(z)}g(z)dz$ and $h^*(z) = h(z) - h(z_0)$. Thus, without loss of generality we may suppose that $h(z_0) = 0$;

(2) We have deform the contour C such that it should pass through the chosen saddle point. Thus, the line $\Im h(z) = 0$ convenient for this purpose. If it does not pass through the points a, b we have to connect the points with the line $\Im h(z) = 0$ by means of suitable lines such that the contribution to integral (7.68) on these lines must be $(o(\varphi(\lambda))$ for $\lambda \to +\infty$;

(3) consider an infinitesimal neighborhood of the saddle point z_0. Let $h''(z_0) = q \neq 0$ and $g(z_0) \neq 0$. Then the main term of the function $h(z)$ is equal to the function $q(z - z_0)^2/2$, and the curve $\Im h(z) = \text{const}$ may be interchanged by a segment of a strait line connecting the points $z_0 - \delta/\sqrt{-q}$ and $z_0 + \delta/\sqrt{-q}$, which is a tangent line to the curve $\Im h(z) = 0$ at the point z_0. Here $\delta = \text{const} > 0$, $\delta \ll 1$. Contribution to the integral (7.68) is equivalent to the principal term of the integral

$$J_\delta = \int_{z_0 - \delta/\sqrt{-q}}^{z_0 + \delta/\sqrt{-q}} e^{\lambda q z^2/2} g(z_0) dz. \tag{7.72}$$

Substitute $z = z_0 + t/\sqrt{-q}$. We obtain

$$J_\delta = \frac{g(z_0)}{\sqrt{-q}} \int_{-\delta}^{\delta} e^{-\lambda t^2/2} dt$$

$$\sim \frac{g(z_0)}{\sqrt{-q}} \int_{-\infty}^{+\infty} e^{-\lambda t^2/2} dt$$

$$= g(z_0) \sqrt{-\frac{2\pi}{q\lambda}} \quad \text{for } \lambda \to +\infty.$$

We result the above mentioned reason with a simple proposition.

Lemma 7.34. (1) *Let D be a simple connected (open) domain containing the points a and b;*

(2) *let the functions $h(z)$ and $g(z)$ be analytic in D and there be a point $z_0 \in D$ such that $h(z_0) = h'(z_0) = 0$ and $h''(z_0) \neq 0$. Let $g(z_0) \neq 0$;*

(3) *consider the line $\eta(z) = \{z : \Im h(z) = 0\}$, and let there be two domains D_a and D_b which possess the following properties:*

(i) *$D_a \cup D_b \subset D$, $a \in D_a$ and $a \notin D_b$, $b \in D_b$ and $b \notin D_a$, $z_0 \in \overline{D_a} \cap \overline{D_b}$;*

(ii) *the function $\Re h(z)$ is negative in $D_a \cup D_b$;*

(iii) *the line $\eta(z)$ passes from the domain D_a to the domain D_b trough the point z_0;*

(iv) *there is at least one contour C such that it is a continuous line consisting of three segments: the first one connects the point a with a point $z_1 \in \gamma_1$ in D_a, the third one connects the point b with a point $z_2 \in \gamma_1$ in D_b, and the second one is the segment of γ_1 connecting the points z_1 and z_2; let there exist a number $\lambda_0 > 0$ such that the integral $\int_C |e^{\lambda_0 h(z)} g(z) dz|$ is convergent. Then*

$$J(\lambda) = g(z_0) \sqrt{-\frac{2\pi}{\lambda h''(z_0)}} + O(\lambda^{-3/2}) \quad for \ \lambda \to +\infty. \tag{7.73}$$

PROOF. We evaluate the integral $J(\lambda)$ along the contour C. It is obvious that the contribution to the integral along C except a neighborhood of the point z_0 is $O(\lambda^{-\infty})$. The contribution at the point z_0 to equals to

$$\int_{z_0-\delta/\sqrt{-h''(z_0)}}^{z_0+\delta/\sqrt{-h''(z_0)}} e^{\lambda[h''(z_0)\frac{z^2}{2}+O((z-z_0)^3)]}[g(z_0) + O(z - z_0)]dz$$

for $z \to z_0$ which implies (7.63). Here δ is a sufficiently small positive number.

$\square$

Some remarks to Lemma 7.34. (1) When we use Lemma 7.34, surely, we may not find exactly the domains D, D_a, D_b and the lines. It remains to show that they exist and satisfy the required properties;

(2) the cases when at least one of the points a or b lies on the curve $\Re h(z) = 0$ or they are singular points of the function $e^{\lambda h(z)}g(z)$ are possible. Then indeed formula (7.63) may not be correct. To solve this problem we have to check that integral (7.68) does not depend on any contour C in the region $D \cup a$ or $D \cup b$, respectively, and we have to add the contribution at the corresponding point to the considered integral.

Examples 7.35. $1°$ Here we continue the examination of the integral $\int_{-1}^{1} e^{iz^2} dz$ (see (7.70)). In this example $h(z) = iz^2$; $h(0) = h'(0) = 0$ and $h''(0) = 2i \neq 0$; $z_0 = 0$ is a saddle point. Let us put $z = x + iy$, then $\Re h(z) \equiv -2xy = 0$ which implies the equation $xy = 0$. The curve consists of two branches: $x = 0$ and $y = 0$. There exist two domains $D_1 = \{(x,y) : x < 0, < 0\}$ and $D_2 = \{(x,y) : x > 0, y > 0\}$ where the function $\Re h(z) < 0$. Besides, the equation $\Im h(z) = 0$ implies the curve $x^2 - y^2 = 0$ which consists of two strait lines $y = -x$ and $y = x$. The chosen contour C connects the points $z = -1$ and $z = 1$ such that it lies in the domains D_1 and D_2 and contains a segment of the line $y = x$. The contribution at the point $z = 0$ is equal to

$$J_\delta = (1 + i)\sqrt{\frac{\pi}{2\lambda}} + O(\lambda^{-\infty})$$

for $\lambda \to +\infty$. But the points $z = -1$ and $z = +1$ lie on the line $\Re h(z) = 0$. Hence we have to add the contributions at these points. As it was beforehand shown the contribution at each point is $O\left(\frac{1}{\lambda}\right)$. Let us find it more accurate. The contribution at the point $z = -1$ is equal to

$$\int_{-1}^{-1-i\delta} e^{i\lambda z^2} dz = -\frac{i}{\lambda} \int_0^{\lambda\delta} e^{i\lambda-2u-iu^2/\lambda} du$$

$$= -\frac{ie^{i\lambda}}{2\lambda} + O\left(\frac{1}{\lambda^2}\right) \quad \text{for } \lambda \to +\infty.$$

Here $z = -1 - iu/\lambda$. We have the same contribution at the point $z = 1$. Consequently

$$\int_{-1}^{1} e^{i\lambda z^2} dz = (1+i)\sqrt{\frac{\pi}{2\lambda}} - \frac{ie^{i\lambda}}{\lambda} + o\left(\frac{1}{\lambda}\right) \qquad (7.74)$$

for $\lambda \to +\infty$. We can obtain the subsequent terms of the decomposition of integral (7.74) considering the decomposition of the function iz^2 in the neighborhood of the points $z = -1$ and $z = 1$. But we do it by using a method (which is perhaps not typical for the considered problems but it is found in many problems in calculus). The function $J(\lambda)$ is analytic for $|\lambda| \gg 1$ and we may differentiate it with respect to λ under the sign of the integral. Hence

$$J'(\lambda) = \int_{-1}^{1} iz^2 e^{i\lambda z^2} dz$$

$$= \frac{1}{2\lambda} \int_{-1}^{1} z de^{i\lambda z^2}$$

$$= \frac{1}{2\lambda}\left[ze^{i\lambda z^2}\Big|_{-1}^{1} - \int_{-1}^{1} e^{i\lambda z^2} dz \right].$$

Therefore

$$J'(\lambda) = \frac{1}{2\lambda}\left(2e^{i\lambda} - J(\lambda) \right).$$

Thus, the function $J(\lambda)$ is a solution to the differential equation (in λ)

$$y' + \frac{1}{2\lambda} y = \frac{1}{\lambda} e^{i\lambda}$$

with the asymptotic estimate

$$J(\lambda) \sim (1+i)\sqrt{\frac{\pi}{2\lambda}} \quad \text{for } \lambda \to +\infty.$$

Let us substitute $y = ue^{i\lambda}$ in the last equation. We obtain

$$u' + \left(i + \frac{1}{2\lambda}\right) u = \frac{1}{\lambda}, \qquad (7.75)$$

where the solution satisfies the following relation

$$u(\lambda) \sim \frac{(1+i)\sqrt{\pi}}{\sqrt{2\lambda}} e^{-i\lambda} \quad \text{for } \lambda \to +\infty.$$

It is easy to see that equation (7.75) possesses a formal solution $s(t)$ which is an asymptotic sum of a formal power series of the form

$$\frac{c_1}{\lambda} + \frac{c_2}{\lambda^2} + ... + \frac{c_n}{\lambda^n} + ...$$

In order to calculate the coefficients c_n substitute the series in (7.75). We have $c_1 = -i$, $c_2 = -1/2, ...$ It is easy to see that

$$c_n = \frac{(2n-3)c_{n-1}}{2i} \quad \text{for } n = 2, ...$$

which leads to the relations

$$c_1 = -i \text{ and } c_n = \frac{(2n-3)!!}{i^n 2^n} \quad (n = 2, ...).$$

Let us prove that there exists an asymptotic solution $g(t) \asymp s(t)$. To this end substitute $u = v + s(t)$ in (7.75). We obtain the equation

$$v' + \left(i + \frac{1}{2\lambda}\right) v = \alpha(t),$$

where $\alpha(t) \asymp 0$. Its general solution is represented in the form

$$v(t) = C \frac{e^{-i\lambda}}{\sqrt{\lambda}} + \beta(t),$$

where

$$\beta(t) = -\int_t^{+\infty} e^{\int_t^\tau (i+1/(2s))ds} \alpha(\tau)d\tau \asymp 0$$

and C is an arbitrary constant. Clearly, the equation has a unique solution $v^*(t) \asymp 0$. This leads to the following estimate

$$\int_{-1}^1 e^{i\lambda x^2} dx \asymp (1+i)\sqrt{\frac{\pi}{2\lambda}} - \frac{1}{\lambda}\left(i + \frac{q(t)}{\lambda}\right) e^{i\lambda}, \qquad (7.76)$$

where

$$q(t) \asymp \frac{1}{2} + ... + \frac{(2n-1)!!}{i^{n-1}2^n \lambda^{n-1}} + ... \qquad (7.77)$$

2° Let us investigate the behavior of the Euler gamma function $\Gamma(z)$ in the complex plane when $y \equiv \Im z \to \infty$. Euler's gamma function is defined by the following relation

$$\Gamma(1+z) = \int_0^{+\infty} e^{-t} t^z dt, \qquad (7.78)$$

for $x \equiv \Re z \geq 0$, where the integration contour coincides with the positive semi-axis. First, we consider integral (7.78) for $y \to +\infty$ and for any x belonging to the interval $[0.1]$. (Thus, we suppose for y to be a large positive parameter $z = x + iy$, $i = \sqrt{-1}$). A rough estimate is obtained from the relation

$$|\Gamma(1+z)| \leq \int_0^{+\infty} e^{-t}t^x \leq \int_0^1 e^{-t}dt + \int_1^{+\infty} e^{-t}tdt < \infty.$$

Hence $\Gamma(1+z)$ is a bounded function for $y \to +\infty$ uniformly in $x \in [0,1]$. To obtain the estimate more precisely let us substitute $t = y\tau$ in (7.78). We have

$$\Gamma(1+z) = y^{x+1+iy} \int_0^{+\infty} e^{-y\tau}\tau^{x+iy}d\tau. \qquad (7.79)$$

The function $F(\tau) = e^{-y\tau}\tau^{x+iy}$ is analytic in the domain $D = \{\tau : |\arg\tau| < \pi, \tau \neq 0, \tau \neq \infty\}$ and $\tau = 0$ is a singular point. $\Gamma(1+z)$ is continuous at the point $z = 0$ (it is easy to check that $\lim_{z\to 0}\Gamma(1+z) = 1$). Hence the integral $\int e^{-y\tau}\tau^{x+iy}d\tau$ is independent of any contour connecting the point $\tau = 0$ with any (fixed) point $\tau \in D$. Rewrite the last integral in the form $\int_0^{+\infty} e^{yf(\tau)}\tau^x d\tau$, where $f(\tau) = -\tau + i\ln\tau$ ($\ln\tau = \ln|\tau| + i\arg\tau$ in D). We have $f'(\tau) = 1 - i/\tau$. Consequently there is only one point $\tau = i$ where the derivative vanishes and it is a saddle point of the function $f(\tau)$. We have $f(i) = -i + i\ln i = -i - \pi/2$. Therefore we rewrite integral (7.78) in the form

$$\Gamma(1+y) = y^{x+1+iy}e^{-y(\pi/2+i)} \int_0^{+\infty} e^{yh(\tau)}\tau^x d\tau, \qquad (7.80)$$

where $h(\tau) = -\tau + i\ln\tau + i + \pi/2$, $h(i) = h'(i) = 0$ and $h''(i) = -i/\tau^2|_{\tau=i} = i$. We take the last integral

$$J_1(y) = \int_0^{+\infty} e^{h(\tau)}\tau^x d\tau$$

in D in the following way. Its contour of integration C begins from the point $\tau = 0$ at an angle $3\pi/4$ with the axis $\Re\tau$ to the intersection point with the curve $\Im h(\tau) = 0$, and then the contour coincides with the line $\Im h(\tau) = 0$ which passes trough the saddle point at an angle of $\pi/4$ and then asymptotically approaches to the positive semi-axis. The contribution at the saddle point is equal to the integral

$$\int_{i-\delta\exp(\frac{i\pi}{4})}^{i+\delta\exp(\frac{i\pi}{4})} e^{yh(\tau)}\tau^x d\tau \sim e^{i(\frac{\pi}{4}(x+1))} \int_{-\infty}^{+\infty} e^{-yu^2/2}du$$

$$= e^{i(\frac{\pi}{2}x+\frac{\pi}{4})}\sqrt{\frac{2\pi}{y}}$$

for $y \to +\infty$.

Let us find the contribution at the singular point $\tau = 0$. It is equal to the integral

$$\int_0^{\delta e^{3i\pi/4}} e^{yh(\tau)} \tau^x d\tau = e^{-3y\pi/4} O(1)$$

$$= e^{-y\frac{\pi}{2}} O(y^{-\infty})$$

for $y \to +\infty$. Finally we obtain

$$\Gamma(1+z) \sim \sqrt{2\pi} e^{i\frac{\pi}{4}(2x+1) - y(\frac{\pi}{2}+i)} y^{z+\frac{1}{2}} \quad \text{for } y \to +\infty. \tag{7.81}$$

It is easy to check that the given estimate is uniform in $x \in [0, 1]$.

In the same way it is easy to obtain the following estimate

$$\Gamma(1+\bar{z}) \sim \sqrt{2\pi} e^{-i(\frac{\pi}{4}(2x+1) - y(\frac{\pi}{2}-i)}(y)^{\bar{z}+\frac{1}{2}} \quad \text{for } y \to +\infty \tag{7.82}$$

uniformly in $x \in [0, 1]$. It easily follows from the last relation that

$$\Gamma(1+z) \sim \sqrt{2\pi} e^{-i\frac{\pi}{4}(2x+1) + y(\frac{\pi}{2}-i)}(-y)^{z+\frac{1}{2}} \quad \text{for } y \to -\infty \tag{7.83}$$

uniformly in $x \in [0, 1]$.

Formulae (7.81) and (7.83) result in the following relation.

$$|\Gamma(1+z)| \sim \sqrt{2\pi} e^{-|y|\frac{\pi}{2}} |y|^{x+1/2} \quad \text{for } |y| \to \infty \tag{7.84}$$

uniformly in $x \in [0, 1]$.

Examples 7.36. Here we find the asymptotics (for $t \to +\infty$) of the first Airy function to the Airy equation

$$x'' - tx = 0. \tag{7.85}$$

For the first time this equation was considered by G.B. AIRY in his works in optics. The Airy function is determined by means of the improper integral

$$\text{Ai}(t) = \frac{1}{2\pi i} \int_\gamma \exp\left(t\tau - \frac{\tau^3}{3}\right) d\tau, \tag{7.86}$$

where γ is a contour of the τ plane consisting of two strait lines connecting the points $\tau = \infty e^{-2\pi i/3}$, $\tau = 0$, and $\tau = 0$, $\tau = \infty e^{2\pi i/3}$. The curve may be deform according to the Caushy theorem. The second function

$$\text{Bi}(t) = i e^{4\pi i/3} \text{Ai}(e^{4\pi i/3} t) - i e^{2\pi i/3} \text{Ai}(e^{2\pi i/3} t). \tag{7.87}$$

Here t is supposed to be real. Solutions of linear differential equations in the complex plane will be given in Chapter 12.

We can directly verify that Ai(t) is a solution to (7.85). However the method of reducing to the function has more general character. The Airy functions may be arise when we try to obtain a solution to equation (7.87) using contour improper integrals. We look for a solution to equation (7.85) in the form:

$$x(t) = \int_C u(\tau)e^{t\tau}d\tau. \tag{7.88}$$

We have to choose the contour C and the function $u(\tau)$. To this end let us substitute (7.88) in (7.85), and suppose that all the used transformations are true. Hence

$$\int_C \tau^2 u(\tau)e^{t\tau}d\tau - t\int_C u(\tau)e^{t\tau}d\tau = 0.$$

On integrating the last integral by parts we obtain

$$-u(\tau)e^{t\tau}|_C + [u'(\tau) + \tau^2 u(\tau)]\int_C u(\tau)e^{t\tau}d\tau. \tag{7.89}$$

If C is chosen such that the first term of (7.88) vanish and $u(t)$ satisfies the differential equation

$$u'_\tau + \tau^2 u = 0 \tag{7.90}$$

then $x(t)$ is a solution to the Airy equation. We may choose $u(\tau) = e^{-\tau^3/3}$ and $C = \gamma$ (the last was given above). Hence

$$x(t) = \int_C e^{t\tau - \tau^3/3}d\tau.$$

The integral is absolutely convergent and it has derivatives (of any order) with respect to t. If we shall consider $x(t)$ in a sufficiently small sector S_ε containing the positive semi-axis J_+, we may be convinced that one is analytic on J_+. We have

$$\text{Ai}(t) = \frac{1}{2\pi i}x(t).$$

We can obtain the asymptotic behavior of the function $x(t)$ applying the saddle point method. First, by the substitution $\tau = \sqrt{t}z$ ($\sqrt{t} > 0$ for $t > 0$), we obtain $x(t)$ in the form $x(t) = \int_\gamma e^{\lambda h(z)}dz$, where $\lambda = t\sqrt{t}$ and $h(z) = z - z^3/3$, $h'(z) = 1 - z^2$. Hence the function has two saddle points $z = \pm 1$. Clearly, for our purpose we have to choose the point $z = -1$. Since $h(-1) = -2/3$, we represent $x(t) = e^{-2/3\lambda}\tilde{x}(t)$, where $\tilde{x}(t) = \int_\gamma e^{\lambda \tilde{h}(z)}dz$, and $\tilde{h}(z) = z - z^3/3 + 2/3$. Thus $\tilde{h}(-1) = \tilde{h}'(-1) = 0$ and $\tilde{h}''(-1) = 2$. Moreover put $z = u + iv$. Hence $\Im\tilde{h}(z) = v(1 - u^2 + v^2/3)$. Consequently

$\Im\tilde{h}(z) = 0$, on the curve $1 - uv + v^2 = 0$ is the steepest line of $h(z)$. It is a hyperbola. The straight lines $arg\,z = \pm(2/3)\pi$ are its asymptotes. Hence we may deform the contour γ to the left part of the hyperbola which indeed pass through the saddle point $z = -1$. The contribution at the saddle point is equal to the integral $x(t)$ equals to

$$e^{-2/3\lambda} \int_{-1-i\delta}^{-1+i\delta} e^{(z+1)^2/2} dz \quad \sim \quad ie^{-2/3\lambda} \int_{-\infty}^{+\infty} e^{-\lambda s^2/2} ds$$

$$= \quad \frac{\sqrt{\pi}}{2\lambda} \quad \text{for } t \to +\infty.$$

Finally

$$\text{Ai}(t) \sim \frac{t^{-1/4}}{2\sqrt{\pi}} e^{-\frac{2}{3}t^{-3/2}} \quad \text{for } t \to +\infty. \tag{7.91}$$

Chapter 8

LINEAR DIFFERENTIAL EQUATIONS

1. SYSTEMS OF LINEAR DIFFERENTIAL EQUATIONS

Here we consider a so called (scalar) *normal system* of n *ordinary linear differential equations* which is a system of the form

$$\begin{cases} x'_1 = a_{11}(t)x_1 + a_{12}(t)x_2 + ... + a_{1n}(t)x_n + f_1(t), \\ x'_2 = a_{21}(t)x_1 + a_{22}(t)x_2 + ... + a_{2n}(t)x_n + f_2(t), \qquad (8.1) \\ \quad \cdot \quad \cdot \quad \cdot \quad \cdot \quad \cdot \quad \cdot \quad \cdot \quad \cdot \quad \cdot \quad \cdot \quad \cdot \quad \cdot \\ x'_n = a_{n1}(t)x_1 + a_{n2}(t)x_2 + ... + a_{nn}(t)x_n + f_n(t), \end{cases}$$

where $a_{ij}(t)$ and $f_i(t)$ are complex-valued functions defined on an interval $]a, b[$ of the real axis (one or both of the points a and b can be infinitely large, $i, j = 1, 2, ..., n$). System (8.1) is equivalent to a single matrix equation of the form

$$X' = A(t)X + F(t). \qquad (8.2)$$

Here

$$A(t) = (a_{ij}(t))_n, \quad F(t) = (f_1(t), f_2(t), ..., f_n(t))^T,$$
$$X = (x_1, x_2, ..., x_n)^T, \quad \text{and} \quad X' = (x'_1, x'_2, ..., x'_n)^T.$$

We also say that (8.2) is a *system of n (scalar) linear differential equations*.

Definition 8.1. A point $t^* \in]a, b[$ is said to be *regular* of equation (8.2) if the matrix-functions $A(t)$ and $F(t)$ are continuous at this point. Any other point is said to be *singular* of the equation.

Let us mark that (owing expression (8.2)) *any solution to (8.2) at a regular point t^* must be differentiable.*

Throughout this paragraph (if it is not stipulated apart) $A(t)$ and $F(t)$ *are supposed to be matrices continuous on the interval* $]a, b[$.

Theorem 8.2. *Let* $X_0 = (x_{01}, x_{02}, ..., x_{0n})^T$ *be a (constant) numerical column matrix and let* $t_0 \in]a, b[$. *Then there exists a unique solution* $X(t)$ *to system (8.2) on the interval* $]a, b[$ *satisfying the condition* $X(t_0) = X_0$.

Remark. 8.3. The matrix X_0 is named an *initial condition* of matrix equation (8.2) and $x_{01}, x_{02}, ..., x_{0n}$ are called initial conditions of system (8.1).

PROOF. Let us prove the existence and uniqueness of the required solution on $[t_0, b[$ (the case $t \in]a, t_0]$ is proved in the same way). Consider the integral equation

$$X(t) = X_0 + \int_{t_0}^{t} F(s)ds + \int_{t_0}^{t} A(s)X(s)ds. \qquad (8.3)$$

Clearly, any continuous solution to this equation is also a solution to equation (8.2) with the initial condition $X(t_0) = X_0$. And vice versa, any solution $X(t)$ to equation (8.2) satisfying the condition $X(t_0) = X_0$ is also a solution to equation (8.3). Thus, it is sufficient to prove that equation (8.3) has a unique continuous solution. Put

$$L[X(t)] = X_0 + \int_{t_0}^{t} [F(s) + A(s)X(s)]ds. \qquad (8.4)$$

We may consider $L[X(t)]$ as an integral operator in the Banach space $C_n[t_0, \tau]$ of continuous matrix functions on the interval $[t_0, \tau]$ where $\tau - t_0 \ll 1$. The norm of a matrix $X(t)$ is determined by the relation

$$||X(\tau)||^* = \sup_{t \in [t_0, \tau]} |x_1(t)| + ... + \sup_{t \in [t_0, \tau]} |x_n(t)|. \qquad (8.5)$$

We have $||F(\tau)||^* = \sup_{t \in [t_0, \tau]} |f_1(t)| + ... + \sup_{t \in [t_0, \tau]} |f_n(t)|$ and

$$||A(\tau)||^* = \sum_{i,j=1}^{n} \sup_{t \in [t_0, \tau]} |a_{i,j}(t)|. \qquad (8.6)$$

We prove the case when $t_0 < t < b$ (the case $a < t < t_0$ is proved in the same way). First we prove that any continuous solution $X(t)$ to integral equation (8.3) satisfies the inequality

$$||X(t)|| \leq \left(||X_0|| + \int_{t_0}^t ||F(s)||ds \right) \exp \left[\int_{t_0}^t ||A(s)||ds \right], \qquad (8.7)$$

where (as usual) $||X(t)|| = |x_1(t)| + ... + |x_n(t)|$, $F||(t)|| = |f_1(t) + ... + |f_n(t)|$, and $||A(t)|| = \sum_{i,j=1}^n |a_{ij}(t)|$. To this end prove the following lemma:

Lemma 8.4. *Let the functions $c(t)$, $q(t)$ and $u(t)$ be continuous on an interval $[t_0, \tau]$; $q(t) \geq 0$, $u(t) \geq 0$ for $t_0 < t \leq \tau$. Let $c(t)$ be a non-decreasing function on this interval and $c(t) > 0$ ($\tau < +\infty$). Let*

$$u(t) \leq c(t) + \int_{t_0}^t q(s)u(s)ds \ \ for \ any \ t \in [t_0, \tau]. \qquad (8.8)$$

Then

$$u(t) \leq c(t) \exp \left[\int_{t_0}^t q(s)ds \right] \ \ for \ any \ t \in [t_0, \tau]. \qquad (8.9)$$

PROOF. Inequality (8.9) is trivial for $t = t_0$. For $t > t_0$ put $y(t) = \int_{t_0}^t q(s)u(s)ds$ (clearly, $y(t)$ is differentiable, $y(t_0) = 0$ and $y(t) \geq 0$). We have

$$
\begin{aligned}
y'(t) &= q(t)u(t) \\
&\leq (c(t) + y(t))q(t)
\end{aligned}
$$

and $c(t) + y(t) > 0$ for any $t \in]t_0, \tau]$. Whence $y'(t)/(c(t) + y(t)) \leq q(t)$ and (taking into account that $c(s) \leq c(t)$ for $t_0 \leq s \leq t$) we obtain

$$\int_{t_0}^t \frac{y'(s)}{c(t) + y(s)} ds \leq \int_{t_0}^t q(s)ds.$$

Thus (because of $y(t_0) = 0$),

$$\ln(c(t) + y(s))|_{t_0}^t = \ln(c(t) + y(t)) - \ln c(t)$$

$$\leq \int_{t_0}^t q(s)ds.$$

Finally,

$$
\begin{aligned}
u(t) &\leq c(t) + y(t) \\
&\leq c(t) \exp \left[\int_{t_0}^t q(s)ds \right].
\end{aligned}
$$

$\square$

Continuation of the proof of Theorem 8.2. If $\|X_0\| + \int_{t_0}^{t_1} \|F(s)\| ds = 0$ at a point $t_1 > t_0$, then, clearly, $X_0 = 0$ and $F(t) = 0$, and $X(t) = 0$ is a unique solution to equation (8.3) on the interval $[t_0, t_1]$. Consequently we may suppose that $\|X_0\| + \int_{t_0}^{t} \|F(s)\| ds > 0$ for any $t > t_0$. Prove inequality (8.7). Let there exist a solution $X(t)$ on an interval $[t_0, \tau] \in [t_0, b[$. Then (see (8.3)) we obtain inequality (8.8) where $u(t) = \|X(t)\|$, $c(t) = \|X_0\| + \int_{t_0}^{t} \|F(s)\| ds > 0$, and $q(t) = \|A(t)\|$. On the basis of Lemma 8.4 the solution satisfies (8.7).

Prove that integral equation (8.3) has a unique continuous solution on $[t_0, \tau]$. In the space $\mathbb{C}_n[t_0, \tau]$ consider a sphere of the form

$$U_\tau = \left\{ X(t) : \|X(\tau)\|^* \le 2 \left(\|X_0\| + \int_{t_0}^{\tau} \|F(s)\|^* ds \right) \right\}.$$

We have (for any $X(t), X_1(t),$ and $X_2(t) \in U_\tau$)

$$
\begin{aligned}
\|L[X(t)]\|^* &\le \left(\|X_0\| + \int_{t_0}^{t} \|F(s)\|^* ds \right) + \int_{t_0}^{t} \|A(s)\|^* \|X(s)\|^* ds \\
&\le \left(\|X_0\| + \int_{t_0}^{t} \|F(s)\|^* ds \right) + \|X(t)\|^* \int_{t_0}^{t} \|A(s)\|^* ds \\
&\le \left(\|X_0\| + \int_{t_0}^{t} \|F(s)\|^* ds \right) \left(1 + 2 \int_{t_0}^{t} \|A(s)\|^* ds \right) \\
&\le 2 \left(\|X_0\| + \int_{t_0}^{t} \|F(s)\|^* ds \right)
\end{aligned}
$$

because $A(t)$ is bounded on $[t_0, \tau]$ and $\tau - t_0 \ll 1$. Clearly, here $\|X(s)\|^* = \sup_{t \in [t_0, s]} |x_1(t)| + \ldots + \sup_{t \in [t_0, s]} |x_n(t)|$, and similarly for $\|F(s)\|^*$ and $\|A(s)\|^*$. Consequently, $L[U_\tau] \subset U_\tau$. Moreover

$$
\begin{aligned}
\|L[X_1(\tau) - X_2(\tau)]\|^* &\le \int_{t_0}^{\tau} \|A(s)\|^* \|X_1(s) - X_2(s)\|^* ds \\
&\le \|X_1(\tau) - X_2(\tau)\|^* \int_{t_0}^{\tau} \|A(s)\|^* ds \\
&\le \frac{1}{2} \|X_1(\tau) - X_2(\tau)\|^*.
\end{aligned}
$$

Consequently we can apply the principle of the contractive mappings to integral equation (8.3). Thus, the equation has a unique continuous solution $X(t) = X_1(t)$ on the interval $[t_0, \tau]$. If $\tau \ge b$ then the assertion is proved. Let $t_1 \equiv \tau < b$. Set

$$X_1 = X_0 + \int_{t_0}^{t_1} [F(s) + A(s)X(s)] ds.$$

Clearly, there exists a unique continuous solution $X_2(t)$ on the interval $[t_1, t_2]$. satisfying the initial condition $X_2(t_1) = X_1$. Thus, there exists a matrix function (say $X(t)$). $X(t)$ coincides with $X_1(t)$ on the interval $[t_0, t_1]$ and with $X_2(t)$ on $]t_1, t_2]$. It is obvious that $X(t)$ is continuous at the point t_1, and hence, it is a solution to (8.2) on the entire interval $[t_0, t_2]$. In the same way proceeding with this process we extend $X(t)$ on an interval $[t_0, t_3]$ and so on. We obtained an increasing numerical sequence $\{t_m\}$. If $t_m \geq b$ for a number m then the desired solution exists. If $t_m < b$ for any m then there exists a limit $\lim_{m \to \infty} t_m = t^* \leq b$ and $X(t)$ is continuous on the interval $[t_0, t^*[$. Suppose $t^* < b$. By identity (8.6) and inequality (8.7), there exists a finite limit $X^* = \lim_{t \to t^*-0} X(t)$. And hence $X(t)$ can (putting $X(t^*) = X^*$) be extended (as a solution to (8.2)) on some more wide interval. Thus, we conclude that $t^* = b$. $\qquad\square$

Remark 8.5. *The solution $X(t)$ obtained in Theorem 8.2 (as it follows from the principle of contractive mappings) can be represented as a limit of the iterate sequence*

$$X_m(t) = X_0 + \int_{t_0}^t [F(s) + A(s)X_{m-1}(s)]ds, \quad X_0(t) = \theta, \quad m = 1, 2, ...,$$

(8.10)

That is, $X(t) = \lim_{m \to \infty} X_m(t)$.

Corollary 8.6. *Let $A(t)$ and $F(t)$ possess continuous derivatives of mth order for any $t \in]a, b[$. Then any solution $X(t)$ of equation (8.2) possesses a continuous derivative $X^{(m+1)}(t)$ on $]a, b[$. Indeed, the solution $X(t)$ satisfies the identity*

$$X'(t) = A(t)X(t) + F(t).$$

Hence $X'(t)$ is a continuous matrix on $]a, b[$. On differentiating equation (8.2) in succession m times and substituting the continuous derivatives of the solution $X(t)$ in the right side, we conclude that the matrices $X^{(2)}(t), ...,$ $X^{(m+1)}(t)$ are continuous on $]a, b[$.

Remark 8.7. *Let the hypothesis of Theorem 8.2 be fulfilled. Let (in addition) $A(t)$, $F(t)$ and X_0 be real matrices. Then the solution $X(t)$ obtained in Theorem 8.2, is a real matrix. Indeed, this property immediately follows from Remark 8.5 because (in the considered case) all the matrices $X_m(t)$ are real and the limiting matrix $X(t)$ must be real.*

If $F(t) \equiv 0$ in (8.2), then we have an equation of the form

$$X' = A(t)X$$

(8.11)

which is a particular case of (8.2) called a *homogeneous linear differential matrix equation* (or *scalar system*). Let $X(t)$ be a continuous solution of equation (8.11) on the interval $]a, b[$ which is a column matrix. The matrix $X(t) = 0$ (clearly, is a solution of the homogeneous equation) is called a *trivial solution*. If $X(t_0)$ is equal to 0 at a point $t_0 \in]a, b[$ then (due to Theorem 8.2) $X(t) = 0$ for any $t \in]a, b[$.

Theorem 8.8. *The set of all solutions to equation (8.11) forms n-dimensional linear space over the field of all complex numbers.*

PROOF. Let $X_1(t)$ and $X_2(t)$ be solutions to (8.11), and let c_1 and c_2 be complex numbers. Then obviously the matrix function $c_1 X_1(t) + c_2 X_2(t)$ is a solution to the equation. Thus, the considered space is linear. To prove that the set is n-dimensional we have to show that there exists a system

$$\{X_1(t), X_2(t), ..., X_n(t)\}$$

of linear independent solutions to the equation such that any solution $X(t)$ of the equation is a linear combination of the system. Consider the system of column matrices (vectors)

$$\{E_1 = (1, 0, ..., 0)^T, E_2 = (0, 1, 0, ..., 0)^T, ..., E_n = (0, ..., 0, 1)^T\}$$

(which is linearly independent in the space $\mathbb{C}_n$). Then on the basis of Theorem 8.2 there exist unique solutions $X_i(t)$ such that $X_i(t_0) = E_i$ $(i = 1, 2, ..., n)$. They form a linearly independent system. Indeed, let us consider a linear combination

$$X(t) = c_1 X_1(t) + c_2 X_2(t) + ... + c_n X_n(t)$$

where $c_1, c_2, ..., c_n$ are complex numbers. Then $X(t)$ is a solution to equation (8.11) with the initial condition

$$X(t_0) = (c_1, c_2, ..., c_n)^T.$$

If $X(t)$ is a zero matrix then $X(t_0)$ is a zero matrix. Hence $c_1 = c_2 = ... = c_n = 0$. Thus, the solutions $X_1(t), X_2(t), ..., X_n(t)$ form a linear independent system. If $X(t)$ is an arbitrary (fixed) solution to the equation and

$$X(t_0) = c_1 E_1 + c_2 E_2 + ... + c_n E_n,$$

where c_i are complex numbers $(i = 1, 2, ..., n)$, then we have

$$X(t) = c_1 X_1(t) + c_2 X_2(t) + ... + c_n X_n(t). \qquad \square$$

Definition 8.9. Any linearly independent system of solutions

$$\{X_1(t), X_2(t), ..., X_n(t)\}$$

to equation (8.11) is said to be a *basis* (or a *fundamental system, fundamental set*) of solutions (briefly *FSS*) of system (8.11). A square matrix $\Phi(t)$, where all its columns form a basis of system (8.11), is said to be a *fundamental matrix* (*FM*) of the system.

Remark 8.10. It is possible somewhat to strengthen the assertion of Theorem 8.8. *The set of all solutions to equation* (8.11) *forms n-dimensional set at any* (*fixed*) *point* $t^* \in]a, b[$ (*in the space* C_n). And as a simple consequence if $\Phi(t)$ is a fundamental matrix then

$$\det \Phi(t^*) \neq 0 \text{ for any } t^* \in]a, b[.$$

Indeed, suppose the contrary. Let $\{X_1(t), X_2(t), ..., X_n(t)\}$ be an independent set of solutions to (8.1) and let there be a vector $\bar{e} = (c_1, ..., c_n) \neq 0$ such that $c_1 X_1(t^*) + c_2 X_2(t^*) + ... + c_n X_n(t^*)$ is a zero matrix. Consequently (by Theorem 8.2) the solution $X(t) = c_1 X_1(t) + ... + c_n X_n(t)$ (because of it is a zero matrix of the point t^*) is trivial. The last means that the set $\{X_1(t), X_2(t), ..., X_n(t)\}$ is dependent. The contradiction obtained proves our assertion.

Remark 8.10 is not evident because a linearly independent system of variable vectors, generally speaking, can be linearly dependent for some values of t or, moreover, for any $t \in]a, b[$. For example the vectors (1,0) and (t,0) are linearly independent, but they linearly dependent at any (fixed) point t.

A square matrix $\Phi(t)$, where each its column is a solution to equation (8.11), is said to be a *matrix solution* of equation (8.11). Clearly,

$$\Phi'(t) = A(t)\Phi(t) \ (t \in]a, b[). \tag{8.12}$$

In particular the relation is valid for any fundamental matrix.

Proposition 8.11. *Let* $X(t)$ *be a solution and* $\Phi(t)$ *be a fundamental matrix of system* (8.11). *Then there exists a constant column matrix* C *such that* $X(t) = \Phi(t)C$.

PROOF. Let $\varphi_i(t)$ be the ith column of $\Phi(t)$ $(i = 1, 2, ..., n)$. Then there exists a column matrix $C = (c_1, c_2, ..., c_n)^T$ such that

$$X(t) = c_1 \varphi_1(t) + c_2 \varphi_2(t) + ... + c_n \varphi_n(t)$$

which is equivalent to the relation $X(t) = \Phi(t)C$.   □

Theorem 8.12. *For a square matrix $\Phi(t)$ to be fundamental it is neces-sary and sufficient that $\Phi'(t) = A(t)\Phi(t)$ at any point $t \in]a, b[$ and $\det \Phi(t^*) \neq 0$ for at least one point $t^* \in]a, b[$.*

PROOF. *Necessity.* If $\Phi(t)$ is fundamental then clearly identity (8.12) holds and (see Remark 8.10)) $\det \Phi(t^*) \neq 0$ for any point $t^* \in]a, b[$.

Sufficiency. Let $\Phi'(t) = A(t)\Phi(t)$ then $\Phi(t)$ consists of columns solutions to equation (8.11). Since $\det \Phi(t^*) \neq 0$, the columns of the matrix $\Phi(t^*)$ form a linearly independent system. Thanks to the uniqueness of the so-lutions the columns of the matrix $\Phi(t)$ is also linearly independent. Thus, $\Phi(t)$ is fundamental.   □

Theorem 8.13. *Let $\Phi(t)$ be a matrix solution to equation (8.11) and $t_0 \in]a, b[$. Then*

$$\det \Phi(t) = \det \Phi(t_0) \exp \left[\int_{t_0}^{t} \mathrm{Sp}A(s)ds \right], \qquad (8.13)$$

where

$$\mathrm{Sp}A(t) = a_{11}(t) + a_{22}(t) + \ldots + a_{nn}(t).$$

PROOF. By $\Phi_i(t)$ we denote a matrix obtained from $\Phi(t)$ replacing its column $\varphi_i(t)$ by $\varphi_i'(t)$. Since $\Phi(t)$ is a matrix solution

$$\varphi_i'(t) = a_{i1}(t)\varphi_1(t) + a_{i2}(t)\varphi_2(t) + \ldots + a_{in}\varphi_n(t).$$

Taking into account that a determinant with two equal columns vanishes we conclude that $\det \Phi_i'(t) = a_{ii}(t) \det \Phi(t)$. This leads to the relation

$$[\det \Phi(t)]' = \mathrm{Sp}A(t) \det \Phi(t).$$

Thus, the function $\det \Phi(t)$ is a partial solution of the equation $y' = \mathrm{Sp}A(t)y$. The last equation has a general solution of the form

$$y = C \exp \left[\int_{t_0}^{t} \mathrm{Sp}A(s)ds \right],$$

where C is an arbitrary constant. Hence there is a number C_0 such that

$$\det \Phi(t) = C_0 \exp \left[\int_{t_0}^{t} \mathrm{Sp}A(s)ds \right].$$

On substituting $t = t_0$ we obtain $C_0 = \det \Phi(t_0)$ which proves this Theorem.

$\square$

Remark 8.14. Let $\Phi(t)$ be an FM to matrix-equation (8.11) on the interval $]a, b[$, and let $\varphi_i(t)$ be its column matrices $(i = 1, 2, ..., n)$. We say that the expression

$$C_1\varphi_1(t) + C_2\varphi_2(t) + ... + C_n\varphi_n(t) \qquad (8.14)$$

is a *general solution* to equation (8.11) on the interval $]a, b[$, where C_1, $C_2, ..., C_n$ are numerical parameters called *arbitrary constants*. On substituting (fixed) numbers c_i instead of C_i in (8.14) $(i = 1, 2, ..., n)$, we obtain a matrix-function $\varphi^*(t)$ (clearly, it is a solution to (8.11)) which is called a *partial solution* to the equation. Any solution to (8.11) may be obtained from (8.14) picking out a suitable vector $(c_1, c_2, ..., c_n)$, so that we may consider the general solution as a totality of all solutions to equation (8.11) on the interval $]a, b[$. The expression

$$C_1\varphi_1(t) + C_2\varphi_2(t) + ... + C_n\varphi_n(t) + \varphi(t), \qquad (8.15)$$

where $\varphi(t)$ is a (partial) solution to equation (8.2), is called a *general solution* to equation (8.2).

Theorem 8.15. *Let $\Phi(t)$ be a fundamental matrix of equation (8.11) and let $t_0 \in]a, b[$. Then the column matrix*

$$\varphi(t) = \Phi(t) \int_{t_0}^{t} \Phi^{-1}(s) F(s) ds \qquad (8.16)$$

is a solution to equation (8.2) on the interval $]a, b[$ with the initial condition $\varphi(t_0) = 0$.

Remark 8.15a. It is possible to take a or b as the lower limit in (8.16) instead of $t_0 \in]a, b[$ if the corresponding integral is convergent. Then $\lim_{t \to a+0} \varphi(t) = 0$, $\lim_{t \to b-0} \varphi(t) = 0$ respectively.

PROOF. On substituting (8.16) in (8.2), we are directly convinced that the matrix-function (8.16) is a solution of equation (8.2). $\square$

Remark 8.15b. Formula (8.16) can be deduced by the method which is called the *Lagrange method of variation of constant parameters*. The method consists of the following procedure. Let $\Phi(t)$ be an FM of (8.11). Then any solution to the equation can be written in the form $X = \Phi(t)C$, where C is a constant column matrix. Let the required solution be in the form $X = \Phi(t)U$, where U is a new unknown matrix (instead of $C = $ const). We have

$$\Phi'(t)U + \Phi(t)U' = A(t)\Phi(t)U + F(t).$$

Taking into account that $\Phi'(t) = A(t)\Phi(t)$ we obtain the relation $\Phi(t)U' = F(t)$ which implies $U' = \Phi^{-1}(t)F(t)$. This leads to (8.16).

Let us mark the following proposition.

Corollary 8.16. *The general solution to equation* (8.2) *can be written in the form*
$$X(t) = \Phi(t)C + \varphi(t), \qquad (8.17)$$
where C is an arbitrary constant column-matrix, $\Phi(t)$ is a fundamental matrix of equation (8.11), *and $\varphi(t)$ is a (partial) solution of equation* (8.2) *(which can be taken in the form* (8.16)).

Proposition 8.17. *Let*
$$X' = A(t)X \text{ and } X' = A^*(t)X \qquad (8.18)$$
be two systems such that $A(t)$ and $A^*(t)$ are continuous matrices and let the systems have the same set of solutions on the interval $]a, b[$. Then $A(t) = A^*(t)$ on $]a, b[$.

PROOF of this proposition is based on the following theorem of linear algebra: Consider a homogeneous linear system of the form $PX = \theta$, where P is a (constant) square matrix, and X is an unknown column matrix. Then the set E_X of all solutions of the equation forms a linear space and the relation $r + e = n$ holds where r is the rank of the matrix P, e is the dimension of the space E_X and n is the order of the considered system. In particular if $e = n$ then $P = \Theta$.

Let $X(t)$ be a solution of systems (8.18). Then $(A(t) - A^*(t))X(t) = \theta$. Since the set of solutions $X(t)$ forms an n-dimensional linear space at any point $t \in]a, b[$, $A(t) - A^*(t) = \Theta$. □

Given a substitution of the form
$$Y = Q(t)X \qquad (8.19)$$

in (8.11) where $Q(t)$ is a square matrix, $Y = (y_1, y_2, ..., y_n)^T$. Let $Q(t)$ and $Q'(t)$ be continuous and let there exist an inverse matrix $Q^{-1}(t)$ continuous on an interval $]a_1, b_1[\in]a, b[$. Let us substitute (8.19) in (8.11). We obtain the equation

$$Y' = B(t)Y, \qquad (8.20)$$

where

$$B(t) = Q^{-1}(t)Q'(t) + Q^{-1}(t)A(t)Q(t). \qquad (8.21)$$

Equations (8.11) and (8.20) are equivalent on the interval $]a_1, b_1[$ in the following sense: there exists an one-to-one correspondence between the solutions to equations (8.11) and (8.20) which is given by the relation $Y = Q(t)X$ or (which is the same correspondence) $X = Q^{-1}(t)Y$.

Proposition 8.18. *Let* $\det Q(t_0) = 0$ *where* $t_0 \in]a, b[$ ($Q(t)$ *is given in* (8.19)). *Then* t_0 *is a singular point of the equation* (8.20).

PROOF. Suppose the contrary, i.e. the matrix $B(t)$ (see (8.21)) is continuous at t_0. This is possible if $B(t)$ is determined at the points $\{t : |t - t_0| < \varepsilon, t \neq t_0, \varepsilon = \text{const} \ll 1\}$ and there exist the finite limits

$$\lim_{t \to t_0 - 0} B(t) = \lim_{t \to t_0 + 0} B(t).$$

Consequently the set $\{Y(t)\}$ of all the solutions of equation (8.11) at the point t_0 form a linear space of n dimension. Because of (8.19) the dimension is equal to the rank of the matrix $Q(t_0)$, hence it is less than n because $\det Q(t_0) = 0$. The contradiction obtained proves this Proposition.

$\square$

2.   SINGLE LINEAR DIFFERENTIAL EQUATIONS OF NTH ORDER

Here we consider a linear differential equation of nth order of the form

$$F(t, x) \equiv x^{(n)} + a_1(t)x^{(n-1)} + \ldots + a_n(t)x = f(t). \qquad (8.22)$$

Throughout this paragraph *we suppose for* $a_i(t)$ *and* $f(t)$ *to be continuous on the interval* $]a, b[$ ($i = 1, 2, \ldots, n$).

Let us put $x_1 = x, x_2 = x', \ldots, x_n = x^{(n-1)}$ and $X = (x_1, x_2, \ldots, x_n)^T$. This leads to the system

$$X' = A(t)X + F(t), \qquad (8.23)$$

where

$$A(t) =$$

$$
\begin{pmatrix}
0 & 1 & 0 & 0 & \cdots & 0 & 0 \\
0 & 0 & 1 & 0 & \cdots & 0 & 0 \\
\multicolumn{7}{c}{\cdot \quad \cdot \quad \cdot \quad \cdot} \\
0 & 0 & 0 & 0 & \cdots & 0 & 1 \\
-a_n(t) & -a_{n-1}(t) & -a_{n-2}(t) & -a_{n-3}(t) & \cdots & -a_2(t) & -a_1(t)
\end{pmatrix}
$$

$$(8.24)$$

and $F(t) = (0, ..., 0, f(t))^T$. Equation (8.22) and system (8.23) are equivalent (on $]a, b[$) in the following sense: for each solution $x(t)$ to equation (8.22) there exists a (unique) solution

$$X(t) = (x(t), x'(t), ..., x^{(n-1)})^T$$

of system (8.23) and for each solution

$$X(t) = (x_1(t), x_2(t), ..., x_n(t))^T$$

of system (8.23) there exists a (unique) solution $x(t)$ of equation (8.19) such that

$$x(t) = x_1(t), x'(t) = x_2(t), ..., x^{(n-1)}(t) = x_n(t).$$

On the basis of Theorem 8.8 it is possible to formulate the following:

Theorem 8.19.   *Let $x_0, x_1, ..., x_{n-1}$ be (constant) complex numbers. Then there exists a unique solution $x(t)$ to equation (8.22) such that the function $x^{(n)}(t)$ is continuous on the interval $]a, b[$,*

$$x(t_0) = x_0, x'(t_0) = x_1, ..., x^{(n-1)}(t_0) = x_{n-1}$$

and the inequality

$$\|x^{(m)}(t)\|^* \le \left(c + \left|\int_{t_0}^t |f(s)|^* ds\right|\right) \exp \left|\int_{t_0}^t \|q(s)\|^* ds\right| \qquad (8.25)$$

holds where

$$c = |x_0| + |x_1| + ... + |x_{n-1}|, \quad |f(t)|^* = \sup_{\tau \in [t_0, t]} |f(\tau)|$$

and

$$q(t) = \sup_{\tau \in [t_0, t]} |a_1(\tau)| + ... + \sup_{\tau \in [t_0, t]} |a_n(\tau)|.$$

The set of the numbers $x_0, x_1, ..., x_n$ is called the *initial conditions* for equation (8.22) *and $t_0 \in]a, b[$.*

If $f(t) = 0$ then (8.22) is called a homogeneous linear differential equation of nth order. Thus, it is an equation of the form

$$F(t, x) \equiv x^{(n)} + a_1(t)x^{(n-1)} + ... + a_n(t)x = 0. \qquad (8.26)$$

The corresponding homogeneous system of linear differential equations is a system of the form

$$X' = A(t)X, \qquad (8.27)$$

where the matrix $A(t)$ is of the form (8.24).

The following theorem is proved in the same way just as Theorem 8.8.

Theorem 8.20.   *The set of all solutions to equation (8.27) forms n-dimensional linear space over the field of all complex numbers.*

Definition 8.21. A set of n linearly independent solutions to equation (8.27) are called a *fundamental system of solutions (FSS)* or a *basis* of the equation.

Let

$$\varphi_1(t), \varphi_2(t), ..., \varphi_n(t) \qquad (8.28)$$

be solutions to equation (8.27). Consider the matrix

$$\Phi(t) = \begin{pmatrix} \varphi_1(t) & \varphi_2(t) & \cdots & \varphi_n(t) \\ \varphi_1'(t) & \varphi_2'(t) & \cdots & \varphi_n'(t) \\ \vdots & & & \\ \varphi_1^{(n-1)}(t) & \varphi_2^{(n-1)}(t) & \cdots & \varphi_n^{(n-1)}(t) \end{pmatrix} \qquad (8.29)$$

which is a matrix-solution to system (8.27). Its determinant $\det \Phi(t)$ is called the *wronskian* of functions (8.28) and denoted by $W(\varphi_1(t), ..., \varphi_n(t))$. Thus

$$W(\varphi_1(t), ..., \varphi_n(t)) = \begin{vmatrix} \varphi_1(t) & \varphi_2(t) & \cdots & \varphi_n(t) \\ \varphi_1'(t) & \varphi_2'(t) & \cdots & \varphi_n'(t) \\ \vdots & & & \\ \varphi_1^{(n-1)}(t) & \varphi_2^{(n-1)}(t) & \cdots & \varphi_n^{(n-1)}(t) \end{vmatrix}. \qquad (8.30)$$

In this case relation (8.13) (taking into account that $\mathrm{Sp}A(t) = -a_1(t)$) turns into the relation

$$W(\varphi_1(t), ..., \varphi_n(t)) = W(\varphi_1(t_0), ..., \varphi_n(t_0)) \exp\left[-\int_{t_0}^{t} a_1(s)ds \right]. \qquad (8.31)$$

The following proposition is a simple consequence of Theorem 8.8.

Theorem 8.22. *Collection (8.28) of n solutions to equation (8.26) forms an FSS if and only if*

$$W(\varphi_1(t_0), ..., \varphi_n(t_0)) \neq 0$$

for at least one point $t_0 \in]a, b[$.

Definition 8.23. Let (8.28) form an *FSS* of equation (8.27) in $]a, b[$. Let $C_1, C_2, ..., C_n$ be an arbitrary constants. The expression

$$C_1\varphi_1(t) + C_2\varphi_2(t) + ... + C_n\varphi_n(t) \tag{8.32}$$

is said to be a *general solution* to equation (8.26) (on the interval $]a, b[$).

Theorem 8.24. *Any solution $x(t)$ to equation (8.26) is a linear combination of functions belonging to FSS.*

PROOF. Any linear combination of the form (8.32) is obviously a solution to equation (8.26). Let $x(t)$ be a solution to equation (8.26) with the initial conditions

$$x(t_0) = x_0, x'(t_0) = x_1, ..., x^{(n-1)}(t_0) = x_{n-1}.$$

Consider the system

$$c_1\varphi_1^{(j)}(t_0) + c_2\varphi_1^{(j)}(t_0) + ... + c_n\varphi_n^{(j)}(t_0) = x_j \ (j = 0, 1, ..., n-1),$$

where the set (8.28) forms the considered *FSS* of the equation. Since $W(\varphi_1(t_0), ..., \varphi_n(t_0)) \neq 0$ the system has a unique solution $(c_1, c_2, ..., c_n)$ and hence there exists a unique linear combination

$$x(t) = c_1\varphi_1(t) + c_2\varphi_2(t) + ... + c_n\varphi_n(t). \qquad \square$$

Remark 8.25. In view of Theorem 8.24 expression (8.31) (with n numerical parameters $c_1, c_2, ..., c_n$) can be considered as a set of all solutions to equation (8.26).

The following proposition is a simple consequence of Proposition 8.17.

Proposition 8.26. *Let*

$$F_1(t, x) \equiv x^{(n)} + a_{11}(t)x^{(n-1)} + ... + a_{n1}(t)x = 0$$

and

$$F_2(t, x) \equiv x^{(n)} + a_{12}(t)x^{(n-1)} + \dots + a_{n2}(t)x = 0$$

be two equations such that the functions $a_{i1}(t)$ and $a_{i2}(t)$ are continuous and the equations have the same set of solutions on the interval $]a, b[$. Then $F_1(t, x) \equiv F_2(t, x)$.

The following proposition is a simple consequence of Propositions 8.26.

Proposition 8.27. *Let each function of (8.28) have continuous derivative of nth order on an interval $]a, b[$ and*

$$W(\varphi_1(t), \dots, \varphi_n(t)) \neq 0$$

for any $t \in]a, b[$. Then there exists a unique homogeneous differential equation of nth order with continuous coefficients of the form (8.26) such that each function $\varphi_i(t)$ $(i = 1, 2, \dots, n)$ is a solution to the equation. This equation may be written in the form:

$$F(t, x) \equiv (-1)^n \frac{W(x, \varphi_1(t), \dots, \varphi_n(t))}{W(\varphi_1(t), \dots, \varphi_n(t))} = 0. \qquad (8.33)$$

Here $W(x, \varphi_1(t), \dots, \varphi_n(t))$ is a wronskian of $n + 1$ order of the functions $x, \varphi_1(t), \dots, \varphi_n(t)$, i.e. it is a determinant where its first row consists of the elements $x, \varphi_1(t), \dots, \varphi_n(t)$ and the subsequent rows are consecutive derivatives of the first row to the nth order inclusive.

PROOF. It is obviously that (8.33) is a linear differential equation of the form (8.26). On the basis of Proposition 8.26 it is sufficient to check that all the functions $\varphi_1(t), \dots, \varphi_n(t)$ are solutions to equation (8.33). But indeed, $W(\varphi_i(t), \varphi_1(t), \dots, \varphi_n(t)) \neq 0$ for any $i = 1, 2, \dots, n$ because the last determinant have two equal rows which leads to the required. $\qquad \square$

Theorem 8.28. *Let the set of functions (8.28) form an FSS to equation (8.26) and let $t_0 \in]a, b[$. Then*

$$\varphi(t) = \sum_{j=1}^{n} \varphi_j(t) \int_{t_0}^{t} \frac{W_j(\varphi_1(s), \dots, \varphi_n(s))}{W(\varphi_1(s), \dots, \varphi_n(s))} ds \qquad (8.34)$$

is a solution to equation (8.22) on the interval $]a, b[$ with the initial conditions $\varphi(t_0) = 0, \varphi'(t_0) = 0, \dots, \varphi^{(n-1)}(t_0) = 0$. Here

$$W_j(\varphi_1(t), \dots, \varphi_n(t))$$

is a determinant which is obtained from $W(\varphi_1(t), \dots, \varphi_n(t))$ replacing its jth column by the column $(0, \dots, 0, 1)^T$.

PROOF. Relation (8.34) can be checked directly. But we shall deduce (8.34) by so called *Lagrange's method of variation of constant parameters.* The method consists of the following. Any solution to equation (8.26) can be written in the form

$$x(t) = c_1\varphi_1(t) + c_2\varphi_2(t) + \ldots + c_n\varphi_n(t),$$

where $c_1, c_2, \ldots, c_n$ are constants (complex numbers). Let the required solution be in the form

$$\varphi(t) = u_1\varphi_1(t) + u_2\varphi_2(t) + \ldots + u_n\varphi_n(t),$$

where u_i are new unknown functions of t (instead of $c_i = \text{const}$). Let us put

$$u_1'\varphi_1^{(j)}(t) + u_2'\varphi_2^{(j)}(t) + \ldots + u_n'\varphi_n^{(j)}(t) = 0 \qquad (8.35)$$

for $j = 0, 1, \ldots, n - 2$. Hence

$$\varphi^{(k)}(t) = u_1\varphi_1^{(k)}(t) + u_2\varphi_2^{(k)}(t) + \ldots + u_n\varphi_n^{(k)}(t) \qquad (8.36)$$

for $k = 0, 1, \ldots, n - 1$ and

$$\varphi^{(n)}(t) = u_1\varphi_1^{(n)}(t) + u_2\varphi_2^{(n)}(t) + \ldots + u_n\varphi_n^{(n)}(t)$$

$$+u_1'\varphi_1^{(n-1)}(t) + u_2'\varphi_2^{(n-1)}(t) + \ldots + u_n'\varphi_n^{(n-1)}(t).$$

Let us substitute the last relation and (8.36) in (8.22). Taking into account that $\varphi_j(t)$ is a solution to equation (8.26) we obtain

$$u_1'\varphi_1^{(n-1))}(t) + u_2'\varphi_2^{(n-1)}(t) + \ldots + u_n'\varphi_n^{(n-1)}(t) = f(t). \qquad (8.37)$$

Equations (8.35) and (8.37) form an algebraic system of linear equations with the unknowns $u_1', u_2', \ldots, u_n'$ at any point $t \in]a, b[$. Determinant of the system $W(\varphi_1(t), \ldots, \varphi_n(t))$ does not equal to zero, hence the system has a unique solution $(u_1'(t), u_2'(t), \ldots, u_n'(t))$, where

$$u_j'(t) = \frac{W_j(\varphi_1(t), \ldots, \varphi_n(t))}{W(\varphi_1(t), \ldots, \varphi_n(t))}$$

and

$$u_j(t) = \int_{t_0}^t \frac{W_j(\varphi_1(s), \ldots, \varphi_n(s))}{W(\varphi_1(s), \ldots, \varphi_n(s))} ds$$

which leads to the required.   □

Remark 8.29. It is possible to take a or b in the capacity of the lower limit in (8.34) instead of $t_0 \in]a, b[$, if the corresponding integral is convergent. Then

$$\lim_{t \to a+0} \varphi^{(j)}(t) = 0 \quad \text{or} \quad \lim_{t \to b-0} \varphi^{(j)}(t) = 0 \quad \text{for } j = 0, 1, \ldots, n - 1.$$

In some sense scalar linear differential equations are some more simple than linear systems of differential equations. Therefore we consider a linear transformation $Y = Q(t)X$ which reduced a linear system of the form (8.11) to a linear system $Y' = B(t)Y$, where $B(t)$ have to be a Frobenius matrix (of the form (8.24)). We suppose that the matrix $A(t)$ possesses a continuous derivative of $n-2$ order on the interval $]a.b[$.

In order to obtain the required transformation choose a linear form

$$y = e(t)X \tag{8.38}$$

where $e(t) = (e_1(t), e_2(t), ..., e_n(t))$ is (generally speaking) a variable vector so that

$$y = e_1(t)x_1 + e_2(t)x_2 + ... + e_n(t)x_n.$$

We also suppose that $e(t)$ has a continuous derivative of $n-1$ order on the interval $]a, b[$. Differentiate (8.38) in succession $(n-1)$ times according to system (8.11). Clearly, $y'(t) = e'(t)X + e(t)X'$. Hence

$$y' = [e'(t) + e(t)A(t)]X \equiv e^{[2]}(t)X,$$

$e^{[2]}(t)$ is a vector with corresponding coordinate

$$e_1^{[2]} = e_1'(t) + e_1(t)a_{11}(t) + ... + e_n(t)a_{n1}(t),$$

$$e_2^{[2]} = e_2'(t) + e_1(t)a_{12}(t) + ... + e_n(t)a_{n2}(t),$$

$$\cdots \cdots \cdots \cdots \cdots \cdots \cdots \cdots \cdots \cdots$$

$$e_n^{[2]} = e_n'(t) + e_1(t)a_{1n}(t) + ... + e_n(t)a_{nn}(t)$$

and so on. Thus, $y^{(m)} = e^{[m+1]}(t)X$, where $e^{[1]}(t) = e(t)$. For $m = 2, ..., n-1$, clearly,

$$\left[e^{[m-1]}(t)X\right]' = \left(\left[e^{[m-1])}(t)\right]' + e^{[m-1]}(t)A(t)\right)X.$$

That is,

$$e^{[m]}(t) = \left[e^{[m-1]}(t)\right]' + e^{[m-1]}(t)A(t).$$

Thus we may compose a square matrix $Q(t)$ consisting of the rows $e^{[1]}(t)$, $e^{[2]}(t), ..., e^{[n]}(t)$. Let $\det Q(t) \neq 0$ for any $t \in]a, b[$. On substituting $Y = Q(t)X$ in (8.11) we obtain the system $Y' = B(t)Y$ where $B(t)$ must be in the form (8.24). Indeed, any solution $Y(t) = (y_1(t), y_2(t), ..., y_n(t))$ to the system can be obtained from a corresponding solution $X(t)$ to system (8.11). The matrix $Q(t)$ is composed such that

$$y_2(t) = y_1(t)', ..., y_n(t) = y_{n-1}'$$

for any solution. On the basis of Proposition 8.17 we are convinced that
the matrix $B(t)$ of the required form.

If we substitute a solution $X(t)$ of equation (8.11) instead of X in (8.38)
we obtain a function $y(t) = e(t)X(t)$. The set of all such functions forms
a linear space E. The space E (matrix $Q(t)$, substitution $Y = Q(t)X$)
is said to be the *resulting space (matrix and substitution) of the system*
$X' = A(t)X$ *and of the linear form* $y = e(t)X$.

Example 8.30. Consider the system $X' = \left(\begin{smallmatrix} 1 & 2 \\ 2 & 3t \end{smallmatrix}\right) X$. Put $y = x_1$ and
differentiate the linear form according to the considered system. We have
$y' = x_1 + 2x_2$. Hence

$$Y = \left(\begin{smallmatrix} 1 & 0 \\ 1 & 2 \end{smallmatrix}\right) X \quad \left(\det Y = \det \left(\begin{smallmatrix} 1 & 0 \\ 1 & 2 \end{smallmatrix}\right) = 2 \neq 0\right).$$

Then

$$Y' = \left(\begin{smallmatrix} 1 & 0 \\ 1 & 2 \end{smallmatrix}\right) X' = \left(\begin{smallmatrix} 1 & 0 \\ 1 & 2 \end{smallmatrix}\right) \left(\begin{smallmatrix} 1 & 2 \\ 2 & 3t \end{smallmatrix}\right) \left(\begin{smallmatrix} 1 & 0 \\ 1 & 2 \end{smallmatrix}\right)^{-1} Y.$$

Consequently

$$Y' = \left(\begin{smallmatrix} 0 & 1 \\ 2-3t & 1+3t \end{smallmatrix}\right) Y$$

and

$$y'' - (3t + 1)y' + (3t - 2)y = 0.$$

3.   SUBSTITUTION $X' = YX$

Solution of linear equation (8.26) can be reduced to solution of a non-
linear equation of $n-1$ order where we have to find only n partial solutions.
Consider the relation

$$x' = yx \quad \left(' = \frac{d}{dt}\right), \tag{8.39}$$

where t is a real or complex argument, x and y are differentiable functions (so many times as required). We have $x'' = yx' + y'x$, hence, taking
into account (8.39) we obtain $x'' = (y' + y^2)x$. On differentiating the latter
expression we obtain $x''' = (y^3 + 3yy' + y')x$ and so on. We put

$$x^{(m)} = \Phi_m(y)x \quad (m = 1, 2, ..., n) \tag{8.40}$$

and $\Phi_0(y) = 1$. The expression $\Phi_m(y)$ can be obtained by differentiating
(8.39) $(m-1)$ times. Thus,

$$\Phi_1(y) = y, \Phi_2(y) = y' + y^2, \quad \Phi_3(y) = y'' + 3yy' + y^3, ...$$

Let us substitute (8.40) in (8.26). This leads to the relation $F(t, x) = \Phi(t, y)x$, where

$$\Phi(t, y) \equiv \Phi_n(y) + a_1(t)\Phi_{n-1}(y) + ... + a_n(t)\Phi_0(y). \tag{8.41}$$

First we investigate the differential expressions $\Phi_m(y)$. On differentiating the relation $x^{(m-1)} = \Phi_{m-1}(y)x$ we obtain

$$x^{(m)} = \left(\Phi_{m-1}(y)y + \frac{d\Phi_{m-1}(y)}{dt} \right) x.$$

Hence

$$\Phi_m(y) = \Phi_{m-1}(y)y + \frac{d\Phi_{m-1}(y)}{dt}, \quad m = 1, 2, \dots \qquad (8.42)$$

Expression $\Phi_m(y)$ can be represented as a sum of terms of the following form

$$c y^{k_0} (y')^{k_1} \dots (y^{(q)})^{k_q}, \qquad (8.43)$$

where $c, k_0, \dots, k_q$ are non-negative integers and q is an arbitrary integer such that $q \geq m - 1$.

We say that the number

$$k_0 + k_1 + \dots + k_q$$

is the *power* of expression (8.43) and

$$k_1 + 2k_2 + \dots + qk_q$$

is the *index* of (8.43).

By (8.42) the decomposition of $\Phi_m(y)$ can be represented as a sum of different members of form (8.31) such that *the sum of the power and the index of each member is equal to m*, that is

$$k_0 + 2k_1 + \dots + (q+1)k_q = m. \qquad (8.44)$$

Indeed, for $m = 0$ decomposition of $\Phi_0(y)$ consists of only one number 1 and the required sum of its power and index is equal to zero. Due to (8.42) upon proceeding from $\Phi_{m-1}(y)$ to $\Phi_m(y)$, each term of $\Phi_{m-1}(y)$ is either multiplied by y which increases the power of this term by 1, or differentiated which increases the index by 1.

Theorem 8.31. *The decomposition*

$$\Phi_m(y) = \sum_{k_1 \geq 0, \dots, k_q \geq 0} c(k_1, \dots, k_q)(y')^{k_1} \dots (y^{(q)})^{k_q} \frac{d^r y^m}{dy^r} \qquad (8.45)$$

takes place for each $m = 0, 1, \dots$, where

$$c(k_1, \dots, k_q) = \frac{1}{(2!)^{k_1} k_1!} \dots \frac{1}{[(q+1)!]^{k_q} k_q!}, \qquad (8.46)$$

$$r = 2k_1 + 3k_2 + ... + (q+1)k_q. \qquad (8.47)$$

Here q is an arbitrary number such that $q \geq m - 1$.

PROOF. Let us show that relation (8.44) takes place for the general term in (8.45). Indeed, its power is equal to $k_1 + k_2 + ... + k_q + m - r$ and its index is equal to $k_1 + 2k_2 + ... + qk_q$. Their sum is equal to $2k_1 + 3k_2 + ... + (q+1)k_q + m - r$, and (see (8.47)) equal to m.

Relation (8.45) is proved by induction with respect to m being trivial for $m = 0$. Let it be true for $m-1$. Suppose that the coefficient $c(k_1 ..., k_q)$ depends on m (we have to prove that this coefficient has form (8.46) and thus does not depend on m). Let us supply the coefficient with the index m then it will be $c_m(k_1, ..., k_q)$ or simply c_m. On the basis of formula (8.42) the term under the sum in (8.45) is obtained from the corresponding terms of the decomposition of the expression $\Phi_{m-1}(y)$ by multiplication by y or differentiating with respect to t. Thus, we have (taking into account the supposition of induction)

$$c_m(y')^{k_1}(y'')^{k_2}...(y^{(q)})^{k_q}\frac{d^r y^m}{dy^r}$$

$$= \quad c(k_1, ..., k_q)(y')^{k_1}(y'')^{k_2}...(y^{(q)})^{k_q}y\frac{d^r y^{m-1}}{dy^r}$$

$$+ \quad c(k_1 - 1, k_2, ..., k_q)(y')^{k_1-1}(y'')^{k_2}...(y^{(q)})^{k_q}d\left[\frac{d^{r-2}y^{m-1}}{dy^{r-2}}\right]/dt$$

$$+ \quad \sum_{p=1}^{q-1}c(k_1, ..., k_{p-1}, k_p + 1, k_{p+1} - 1, k_{p+2}, ..., k_q)(y')^{k_1}$$

$$... \quad (y^{(p-1)})^{k_{p-1}}\frac{d(y^{(p)})^{k_p+1}}{dt}(y^{(p+1)})^{k_{p+1}-1}(y^{(p+2)})^{k_{p+2}}...(y^{(q)})^{k_q}\frac{d^{r-1}y^{m-1}}{dy^{r-1}}.$$

Hence

$$mc_m = \quad c(k_1, ..., k_q)\left[m - r + \frac{c(k_1 - 1, k_2, ..., k_q)}{c(k_1, ..., k_q)}\right.$$

$$\left. + \sum_{p=1}^{q-1}\frac{(k_p + 1)c(k_1, ..., k_{p-1}, k_p + 1, k_{p+1} - 1, k_{p+2}, ..., k_q)}{c(k_1, ..., k_q)}\right].$$

We have from (8.46)

$$\frac{c(k_1 - 1, k_2, ..., k_q)}{c(k_1, ..., k_q)} = 2k_1$$

and

$$\frac{c(k_1, ..., k_{p-1}, k_{p+1}, k_{p+1} - 1, k_{p+2}, ..., k_q)}{c(k_1, ..., k_q)} = \frac{(j+2)k_{j+1}}{k_j + 1}.$$

Hence we obtain

$$mc = c(k_1, ..., k_q)[m - r + \sum_{p=1}^{q-1}(p+2)k_{q+1}]$$

which leads to the required.

Remark 8.32. Because of (8.46), *the coefficient* $c(k_1, ..., k_q)$ *may be rewritten in the form*

$$c(k_1, ..., k_q) = c_1(k_1)...c_q(k_q), \tag{8.48}$$

where

$$c_j(k_j) = \frac{1}{[(j+1)!]^{k_j}k_j!} \qquad (j = 1, 2, ..., q). \tag{8.49}$$

Formula (8.45) *may be rewritten in the form*

$$\Phi_m(y) = \sum \frac{m!}{r!}c_1(k_1)...c_q(k_q)y^{m-r}(y')^{k_1}(y'')^{k_2}...(y^{(q)})^{k_q}, \tag{8.50}$$

where the sum is extended over all terms such that the sum of its power and index is equal to m.

Definition 8.33. The expression

$$H(t, y) \equiv y^n + a_1(t)y^{n-1} + ... + a_n(t) = 0 \tag{8.51}$$

(polynomial $H(t, y)$) is said to be the *characteristic equation* (*polynomial*) of equation (8.26).

Theorem 8.34. *The decomposition*

$$\Phi(t, y) = \sum_{k_1 \geq 0, ..., k_q \geq 0} c_1(k_1)...c_q(k_q)(y')^{k_1}(y'')^{k_2}...(y^{(q)})^{k_q}\frac{\partial^r H(y, t)}{\partial y^r}$$

$$\tag{8.52}$$

takes place. Here q *is an arbitrary number such that* $q \geq n - 1$.

PROOF. Relation (8.52) is a simple consequence of (8.45). Indeed, let us substitute the relations

$$x^{(m)} = \Phi_m(y)x \quad (m = 1, 2, ..., n)$$

in (8.26) where $\Phi_m(y)$ are taken from (8.42). Hence

$$\Phi(t, y) = \sum_{k_1 \geq 0, ..., k_q \geq 0} c_1(k_1)...$$

$$c_q(k_q)(y')^{k_1}(y'')^{k_2}...(y^{(q)})^{k_q}\frac{\partial^r(y^n + a_1(t)y^{n-1} + ... + a_n(t))}{\partial y^r}$$

which means (8.52).   □

4.   SYSTEM OF LINEAR DIFFERENTIAL EQUATIONS IN THE COMPLEX PLANE

Here we consider a normal system of linear differential equations or (which is the same) a single matrix equation of the form

$$X' = A(z)X + F(z),\qquad\qquad (8.53)$$

where $A(z) = (a_{ij}(z))_n$ is a square matrix of nth order and

$$F(z) = (f_1(z), f_2(z), ..., f_n(z))^T$$

is a column matrix. We suppose that *all the matrices are holomorphic in a domain D of the complex plain such that its boundary is a single piecewise smooth curve.*

Theorem 8.35. *Let $z_0 \in D$. Let*

$$X_0 = (x_{01}, x_{02}, ..., x_{0n})^T$$

be a column matrix consisting of the components x_{0i} ($i = 1, 2, ..., n$) which are (constant) complex numbers [X_0 is called an initial condition *of matrix equation (8.53) or (which is the same)* initial conditions *for system (8.53)]. Then there exists a unique solution $X(z)$ in D of the equation such that it is holomorphic in D and $X(z_0) = X_0$.*

PROOF of this Theorem is made in the same way as the proof of Theorem 8.2. The solution $X(z)$ can be represented as a limit (for $m \to \infty$) of the iteration sequence of matrix-functions $\{X_m(z)\}$ where $X_0(z) = X_0$ and

$$X_m(z) = X_0 + \int_{z_0}^{z} [A(s)X_{m-1}(s) + F(s)]ds \ \ for \ m = 1, 2, ... \quad (8.54)$$

Here the integral is taken along an arbitrary piecewise smooth curve $\Gamma[z_0, z]$ in D. It is easy to show that the series

$$X_0 + (X_1(z) - X_0(z)) + ... + (X_m(z) - X_{m-1}(z)) + ...$$

uniformly converges in a sufficiently small neighborhood of the point z_0. Its sum $X(z)$ is obviously a solution to equation (8.53). Clearly, all $X_m(z)$ are holomorphic in D and $X(z)$ is a holomorphic matrix-function (as a sum of uniformly convergent series consisting of holomorphic matrix-functions) in a neighborhood at the point z_0. In the same way as was shown in Proof of Theorem 8.2 we can show that the solution can be analytically continue into the entire domain D. A theorem on existence and uniqueness (like the theorem on a real domain) takes place for a single nth order linear homogeneous differential equation of the form

$$x^{(n)} + a_1(z)x^{(n-1)}(z) + \ldots + a_n(z)x = 0 \qquad (8.55)$$

with holomorphic coefficients $a_i(z)$ $(i = 1, 2, ..., n)$ in the domain D. Namely *there exists a unique solution $x(z)$ of the equation in D such that*

$$x(z_0) = x_0, x'(z_0) = x_1, ..., x^{(n-1)}(z_0) = x_{n-1},$$

*where $z_0 \in D$. Besides the solution is holomorphic in D. Here $x_0, x_1, ..., x_n$ are given complex numbers.*Moreover for equations (8.53) and (8.55) all the corresponding theorems of chapter 6(with some evident alterations) hold.

Chapter 9

GENERAL ASYMPTOTIC PROPERTIES OF LINEAR DIFFERENTIAL EQUATIONS

1.   GENERAL ESTIMATES

Here we obtain (may be sufficiently rough but important) estimates of solutions to a linear differential equation of the form

$$F(t, x) \equiv x^{(n)} + a_1(t)x^{(n-1)} + \ldots + a_n(t)x = 0 \qquad (9.1)$$

for $t \to +\infty$. Here $a_i(t)$ are functions specified on an interval $]a, b[$, where a, b are points of the real axis (one or both of the points a and b can be infinitely large, $i = 1, 2, \ldots, n$). First, the equation is reduced to an equivalent (in some sense) system written in the form

$$X' = A(t)X, \qquad (9.2)$$

where $X = (x_1, x_2, \ldots, x_n)^T$, $X' = (x'_1, x'_2, \ldots, x'_n)^T$, and $A(t) = (a_{ij}(t))_n$ is a square matrix specified on $]a, b[$.

The most simple estimates gives the following inequality.

Proposition 9.1. *Let $A(t)$ be a matrix continuous on the interval $]a, b[$ and let $t_0 \in]a, b[$. Then for the solution $X(t)$ satisfying the initial condition $X(t_0) = X_0$ the following inequality is fulfilled:*

$$\|X(t)\| \leq \|X_0)\| \exp\left[\left|\int_{t_0}^{t} \|A(s)\| ds\right|\right] \quad for \ any \ t \in]a, b[. \qquad (9.3)$$

Here $\|X(t)\| = |x_1(t)| + \ldots + |x_n(t)|$ and $\|A(t)\| = \sum_{i,j=1}^{n} |a_{ij}(t)|$.

PROOF. This inequality was proved in Theorem 8.2 for the case $t \geq t_0$ and based on Lemma 8.4. The case $t < t_0$ is proved in the same way and based on the following assertion: *Let the functions $c(t)$, $q(t)$ and $u(t)$ be continuous on an interval $[\tau, t_0]$; $q(t) \geq 0$, $u(t) \geq 0$ for $\tau \leq t \leq t_0$. Let $c(t)$ be a non-increasing function on this interval and $c(t) > 0$ $(-\infty > \tau)$. Let*

$$u(t) \leq c(t) + \int_t^{t_0} q(s)u(s)ds$$

for any $t \in [\tau, t_0]$. Then

$$u(t) \leq c(t) \exp\left[\int_t^{t_0} q(s)ds\right]$$

for any $t \in [\tau, t_0]$ which is proved in the same way as Lemma 8.4. $\square$

Below we consider some different transitions from equation (9.1) to system (9.2).

(1) The substitutions $x_1 = x, x_2 = x_1', ..., x_n = x_{n-1}'$ in (9.1) lead to system (9.2) where $A(t)$ is a (Frobenius) matrix of the form

$$A(t) =$$
$$\begin{pmatrix}
0 & 1 & 0 & 0 & \cdots & 0 & 0 \\
0 & 0 & 1 & 0 & \cdots & 0 & 0 \\
\cdot & \cdot & \cdot & \cdot & \cdots & \cdot & \cdot \\
0 & 0 & 0 & 0 & \cdots & 0 & 1 \\
-a_n(t) & -a_{n-1}(t) & -a_{n-2}(t) & -a_{n-3}(t) & \cdots & -a_2(t) & -a_1(t)
\end{pmatrix}$$
$$(9.4)$$

(2) Let all $a_i(t)$ $(i = 1, 2, ..., n)$ belong to a space $\{Q\}$ where Q is a field of type N. Let $\Pi\{a_j(t)\} > -j$ for at least one $j \in \{1, 2, ..., n\}$. Let $g(t)$ be a function of the greatest growth of the set $\{[a_s(t)]^{1/s}\}$ $(s = 1, 2, ..., n)$. Then $\Pi\{g(t)\} > -1$. The substitution $\tau = \int g(t)dt$ reduces equation (9.1) to the equation

$$x_\tau^{(n)} + q_1(t)x_\tau^{(n-1)} + ... + q_n(t)x = 0. \qquad (9.5)$$

Here $q_j(t) = q_j^* + \alpha_j(t)$, $\alpha_j(t) \in C_t$, $j = 1, 2, ..., n$, where

$$q_j^* = \lim_{t \to +\infty} \frac{a_j(t)}{g^j(t)}. \qquad (9.6)$$

Indeed, it is easy to show (by induction with respect to $s = 1, 2, ...$) that

$$x_t^{(s)} = g^s(t)[x_\tau^{(s)} + \tau^{-1}(c_{s1} + \alpha_{s1}(t))x_\tau^{(s-1)} + ... + \tau^{1-s}(c_{ss-1} + \alpha_{ss-1}(t))x_\tau'], \qquad (9.7)$$

where c_{sr} are numbers, $\alpha_{sr}(t) \in C_t$ ($r = 1, 2, ..., s - 1$). On substituting (9.7) in (9.1) we obtain (9.5).

Let us put $x_1 = x$, $x_2 = dx_1/d\tau, ..., x_n = dx_{n-1}/d\tau$ in (9.1). This leads to the system $X'_\tau = Q(t)X$ where $Q(t)$ is a Frobenius matrix where its last row is in the form

$$-q_n(t) - q_{n-1}(t)... - q_1(t).$$

As the result we may rewrite down the system in the following form

$$X' = g(t)Q(t)X. \tag{9.8}$$

(3) Let us reduce equation (9.1) to a system by means of the substitutions $x_1 = x, x_2 = tx'_1, ..., x_n = tx'_{n-1}$. As the result, we obtain a system of the form

$$tX' = Q(t)X \tag{9.9}$$

for the unknown $X(x_1, x_2, ..., x_n)$. Here $Q(t)$ is a Frobenius matrix where each element of the last row is a linear combination of functions of the form $a_j(t)t^j$ and constants.

Here we consider only several systems obtained from (9.1). If there is some additional information about the coefficients of the equation, then maybe another equivalent system to equation (9.1) is useful.

Lemma 9.2. *Let all $a_i(t)$ be continuous on the interval $[t_0, +\infty[$ ($i = 1, 2, ..., n$). Put*

$$p(t) = |a_1(t)| + ... + |a_n(t)|. \tag{9.10}$$

Let $x(t)$ be a solution to equation (9.1) on this interval. Then

$$|x^{(j)}(t)| \leq \left[|x(t_0)| + ... + |x^{(n-1)}(t_0)|\right] e^{(n-1)t} \exp\left[\int_{t_0}^t p(s)ds\right] \tag{9.11}$$

for any $j = 0, 1, ..., n - 1$. As a simple consequence

$$\|X(t)\| = O\left(e^{(n-1)t} \exp\left[\int_{t_0}^t p(\tau)d\tau\right]\right) \quad for \ t \to +\infty. \tag{9.12}$$

Here $X(t) = (x(t), x'(t), ..., x^{(n-1)}(t))^T$.

PROOF. Let $x(t)$ be a solution to equation (9.1) on $[t_0, +\infty[$. Let $X(t)$ be the corresponding solution of system (9.2) with matrix (9.4). Hence $X(t)$ is also a solution of the integral equation

$$X(t) = X(t_0) + \int_{t_0}^t A(s)X(s)ds. \tag{9.13}$$

Clearly, $\|A(s)\| = n - 1 + p(s)$. On the basis of inequality (9.3) we obtain the inequality

$$\|X(t)\| \leq \|X(t_0)\| \exp\left[\int_{t_0}^{t} (n - 1 + p(s))ds\right]$$

which implies (9.12). $\qquad\qquad\qquad\qquad\qquad\qquad\qquad\qquad\qquad\quad$ □

Example 9.3. *Consider equation (9.1) where all the properties of Lemma 9.2 hold, and (in addition) $a_i(t) \to a_i \neq \infty$ for $t \to +\infty$ ($i = 1, 2, ..., n$). Then there exists a number p such that the following estimates hold:*

$$x^{(j)}(t) = O(e^{pt}) \text{ as } t \to +\infty \qquad\qquad (9.14)$$

for any (fixed) solution $x(t)$ and $j = 0, 1, ..., n$. Indeed, if $j = 0, .., n - 1$ then relation (9.14) is a simple consequence of Lemma 9.2. If $j = n$ then the required relation follows from the identity $x^{(n)}(t) = -a_1(t)x^{(n-1)}(t) - ... - a_1(t)x(t)$.

Proposition 9.4. *Let all $a_j(t)$ belong to a space $\{Q\}$ where Q is a field of type N. Let $\{\lambda_1(t), ..., \lambda_n(t)\}$ be a complete set of roots of the characteristic equation*

$$H(t, y) \equiv y^n + a_1(t)y^{n-1} + ... + a_n(t) = 0 \qquad\qquad (9.15)$$

and $\lambda(t)$ be its function of the greatest growth (for $t \to +\infty$). Let $\Re P\{\lambda(t)\} > -1$. Then there exists a number q such that the following estimates hold:

$$x^{(j)}(t) = O\left(\exp\left[q|\lambda(t)|t\right]\right) \text{ as } t \to +\infty \qquad\qquad (9.16)$$

for any (fixed) solution $x(t)$ to equation (9.1) and for any $j = 0, 1, ..., n$.

PROOF. Let $g(t)$ be a function of the greatest growth (as $t \to +\infty$) for the set

$$[a_j(t)]^{1/j} \quad (j = 1, 2, ..., n).$$

Since $\Pi\{\lambda(t)\} > -\infty$ at least one $a_j(t)$ belongs to A_t. Consequently $g(t) \neq 0$ for $t \gg 1$. Substitute $y = g(t)u$ in (9.15). This leads to the equation

$$\tilde{H}(t, u) \equiv u^n + \frac{a_1(t)}{g(t)}u^{n-1} + ... + \frac{a_n(t)}{g^n(t)} = 0.$$

We have $\tilde{\lambda}_i(t) = \lambda_i(t)/g(t)$, where $\tilde{\lambda}_i(t)$ is the corresponding root of the last equation to the root $\lambda_i(t)$, and let $\tilde{\lambda}(t) = \lambda(t)/g(t)$. By the definition of $g(t)$ all the coefficients of the equation $\tilde{H}(u, t) = 0$ have finite limits and at least one of the coefficients is equal to 1, which is, indeed, their function of the greatest growth. Hence, there exists a number $\mu \neq 0$ such that $\mu \tilde{\lambda}(t) \sim 1$, which implies the relation $g(t) \sim \mu \lambda(t)$ $(t \to +\infty)$. As a consequence $P\{g(t)\} = P\{\lambda(t)\}$. Thus, $P\{g(t)\} > -1$. Substitute $\tau = \int g(t)dt$ in (9.1). Whence we obtain equation (9.5) and equivalent system (9.8). In the same way as in the proof of Lemma 9.2 we obtain the estimates

$$x^{(j)}(t) = O\left(\exp\left[O\left(\int g(t)dt\right)\right]\right)$$

for $t \to +\infty$ and $j = 0, 1, ..., n-1$. This implies relation (9.16) for $j = 0, 1, ..., n-1$. For $j = n$ the required relation follows from identity $x^{(n)}(t) = -a_1(t)x^{(n-1)}(t) - ... - a_n(t)x(t)$. Consequently (see (9.10)), there is a number $q \neq 0$ such that

$$\begin{aligned} x^{(n)}(t) &= O\left(p(t)\exp[O(\lambda(t)t)]\right) \\ &= O\left(\exp[q|\lambda(t)|t + \ln|p(t)|]\right) \text{ for } t \to +\infty. \end{aligned}$$

Clearly, $p(t) \in A_t$, hence $\Pi\{\ln|p(t)|\} = 0$ and since $\Re P\{\lambda(t)t\} > 0$, we have $\ln p(t) = o(\lambda(t)t)$ which implies the relation

$$x^{(n)}(t) = O\left(\exp[O(\lambda(t)t)]\right) \ (t \to +\infty). \qquad \square$$

Proposition 9.5. *Let $a_i(t)$ be continuous functions on the interval* $]t_0, +\infty[$ *(t_0 is a number). Put*

$$q(t) = 1 + |a_1(t)|t + |a_2(t)|t^2 + ... + |a_n(t)|t^n. \qquad (9.17)$$

Then

$$x^{(j)}(t) = O\left(\exp\left[\int_{t_0}^t O\left(\frac{q(s)}{s}ds\right)\right]\right) \text{ as } t \to +\infty \qquad (9.18)$$

for any (fixed) solution $x(t)$ to equation (9.1) and for any $j = 1, 2, ..., n-1$.

PROOF. The required relation follows from representation (9.9) in the same way as in the proof of Proposition 9.2. $\qquad \square$

Examples 9.6. $1°$ Let us estimate the solutions of equation (9.1) in the case when all the conditions of Proposition 9.4 hold and, in addition, the characteristic equation have an asymptotic n−multiple root $\lambda(t)$ for $t \to +\infty$.

It is easy to see that $\lambda(t) \sim -a_1(t)/n$ for $t \to +\infty$ and $\Pi\{a_1(t)\} > -1$. Substitute

$$x = u \exp\left[-\int \frac{a_1(t)}{n} dt\right]$$

in (9.1). We obtain the equation

$$u^{(n)} + b_2(t)u^{(n-2)} + \ldots + b_n(t)u = 0,$$

where $b_j(t) \in \{Q\}$ $(j = 2, \ldots, n)$, and any root of the corresponding characteristic equation is $o(\lambda(t))$ for $t \to +\infty$. By Proposition 9.4 the following estimate holds:

$$u(t) = O(\exp[o(\lambda(t)t)]) \text{ as } t \to +\infty$$

for any solution $u(t)$ of the last equation. Thus, finally we obtain the estimate

$$x(t) = O\left(\exp\left[-\int \left(\frac{a_1(t)}{n} + o(a_1(t))\right) dt\right]\right) \text{ as } t \to +\infty \qquad (9.19)$$

for any (fixed) solution $x(t)$ to equation (9.1);

$2°$ *Let all the conditions of Proposition 9.5 hold. And let (in addition) the functions $a_i(t)t^s$ be bounded for $t \gg 1$ $(i = 1, 2, \ldots, n)$. Then there is a number p such that any solution $x(t)$ to equation (9.1) has the estimate*

$$x^{(j)}(t) = O(t^p) \text{ as } t \to +\infty \qquad (9.20)$$

for any (fixed) solution and $j = 1, 2, \ldots, n-1$.

2. THE CASE WHEN THE COEFFICIENTS RAPIDLY TEND TO ZERO

Theorem 9.7. *Let $\Pi\{a_j(t)\} < -j$ for all $j = 1, 2, \ldots, n$. Let us denote by*

$$\sigma = -\max_{j=1,2,\ldots,n}[j + \Pi\{a_j(t)\}] > 0.$$

Then equation (9.1) has an FSS in the form

$$\{x_k(t) = t^{k-1}(1 + \alpha_k(t))\} \ (k = 1, 2, \ldots, n), \qquad (9.21)$$

where $\Pi\{\alpha_k(t)\} < -\sigma < 0$. Moreover,

$$\gamma_k(t) \equiv \frac{x_k'(t)}{x_k(t)} = \frac{k-1}{t} + \beta_k(t), \qquad (9.22)$$

where

$$\Pi\{\beta_k(t)\} \leq -\sigma - 1 < -1 \ (k = 1, 2, \ldots, n).$$

PROOF. We prove the theorem by induction with respect to $n = 1, 2, \dots$ For $n = 1$, (9.1) is written in the form

$$x' + a(t)x = 0. \tag{9.23}$$

Here $a(t) = a_1(t)$. Clearly, its *FSS* consists of one function

$$x(t) = e^{\int_t^{+\infty} a(s)ds}.$$

This is correct because $\Pi\{a(t)\} < -1$. Hence the last integral is convergent for $t \gg 1$. Clearly, $x(t) \to 1$ for $t \to +\infty$. Let us estimate the difference $\alpha(t) = x(t) - 1$. We have $\alpha(t) \sim \int_t^{+\infty} a(s)ds$. Hence $h\{\alpha(t)\} \le h\{a(t)\} + 1$. Moreover $\alpha'(t) = -a(t)x(t)$. Consequently $h\{\alpha'(t)\} \le h\{a(t)\}$. We have $\alpha''(t) = -a'(t)x(t) + a(t)\alpha(t)$. Consequently $h\{\alpha''(t)\} \le h\{a(t)\} - 1$ and so on. On differentiating the identity $\alpha'(t) = -a(t)x(t)$ in succession $(m - 1)$ times, we obtain (by induction with respect to m) that $h\{\alpha^{(m)}(t)\} \le h\{a(t)\} - m + 1$. This means that $\Pi\{\alpha(t)\} \le \Pi\{a(t)\} + 1 = -\sigma < 0$. Besides,

$$\gamma(t) \equiv \frac{x'(t)}{x(t)} = -a(t).$$

Hence $\Pi\{\gamma(t)\} = \Pi\{a(t)\} = -\sigma - 1$. Thus, the case $n = 1$ is proved.

Remark. Equation (9.23) may be solved as follows. We pass from (9.23) to the integral equation $x(t) = 1 + \int_t^{+\infty} a(s)x(s)ds$. Since $\Pi\{a(t)\} < -1$, the operator $l(x(t)) = 1 + \int_t^{+\infty} a(s)x(s)ds$ operates in the space of all bounded continuous functions on $[t, +\infty[$ and transforms any considered function to a function $1 + o(1)$ $(t \to +\infty)$. Consequently we may apply the principle of the contractive mappings to the operator equation $x(t) = l(x(t))$. Thus, it has a unique continuous solution $x(t)$, and (as it is described above) $\Pi\{x(t) - 1\} \le -\sigma < 0$. In the general case (for an arbitrary n) we also will use the method to reduce the considered equation to an appropriate linear integral equation.

Let the theorem hold for $n-1$. Introduce the following notation:

$$\int_t^{+\infty} f(s)ds^{[m]} = \int_t^{+\infty} ds_1 \int_{s_1}^{+\infty} ds_2 \dots \int_{s_{m-1}}^{+\infty} f(s_m)ds_m \tag{9.24}$$

for any function $f(t)$. The inequality $\Pi\{f(t)\} < -m$ is enough for the integral convergence.

Bellow by formal integration by parts we obtain several identities (that is, we suppose for the functions $f(t)$ and $x(t)$ to possess all the properties such that all the considered transformations have to be hold). Suppose that

$$\lim_{t \to +\infty} f(t)x(t) = 0.$$

We have

$$\int_t^{+\infty} f(s)x'(s)ds^{[m]} = -\int_t^{+\infty} f(s)x(s)ds^{[m-1]} - \int_t^{+\infty} f'(s)x(s)ds^{[m]}.$$

The last expression is generalized by the following formula.

$$\int_t^{+\infty} f(s)x^{(m)}(s)ds^{[m+1]} =$$

$$(-1)^m \left\{ \int_t^{+\infty} f(s)x(s)ds^{[1]} + m\int_t^{+\infty} f'(s)x(s)ds^{[2]} + \dots \right.$$

$$\left. + \binom{m}{p}\int_t^{+\infty} f^{(p)}(s)x(s)ds^{[p+1]} + \dots + \int_t^{+\infty} f^{(m)}(s)x(s)ds^{[m+1]} \right\},$$

$$(9.25)$$

where

$$\lim_{t \to +\infty} f^{(r)}(t)x^{(k)}(t) = 0 \text{ for } r, k = 0, 1, \dots, m.$$

Prove that equation (9.1) has a continuous bounded solution $x(t)$ for $t \gg 1$. Indeed, integrate the equation n times and apply formula (9.25) to any of the obtained integrals. This leads to the equation

$$x(t) = 1 + L(x(t)), \tag{9.26}$$

where $L(x)$ is a linear combination of integrals of the form

$$\int_t^{+\infty} \delta_m(\tau)x(\tau)d\tau^{[m]}, \quad m = 1, 2, \dots, n,$$

where the functions $\delta_m(t)$ have the estimates $\Pi\{\delta_m(t)\} \leq -m - \sigma$. Clearly, the solution $x(t)$ is also a solution to equation (9.1). Apply the principle of the contractive mappings to equation (9.26). It is easy to show that the equation has a continuous solution $x(t)$ and $|x(t)| < 2$ for $t \gg 1$. On substituting this solution in (9.26) we conclude that $\Pi\{\gamma(t)\} = \Pi\{a(t)\} = -\sigma - 1$. Hence $x(t) = 1 + \alpha_1(t)$, where $\Pi\{\alpha_1(t)\} \leq -\sigma$. Let us make the substitutions $x = u(1 + \alpha_1(t))$ and $u' = v$ in (9.1). Since $1 + \alpha_1(t)$ is a solution to equation (9.1) we obtain an equation of $n - 1$ order which is written in the form

$$v^{(n-1)} + b_1(t)v^{(n-2)} + \dots + b_{n-1}(t)v = 0, \tag{9.27}$$

where $b_j(t) \in \Pi$ and $\max_{j=1,2,...,n-1}[j + \Pi\{b_j(t)\}] \leq -\sigma < 0$. The last equation has (by induction) an FSS of the form

$$\{v_k = t^k(1 + \beta_k(t))\} \quad k = 0, 1, ..., n - 2,$$

where $\Pi\{b_k(t)\} \leq -\sigma$. Consequently, equation (9.1) has an FSS $\{x_k(t)\}$ which consists of the functions $x(t) \equiv x_1(t) = 1 + \alpha_1(t)$, and for $k = 2, ..., n$

$$x_k(t) = -(1 + \alpha_1(t)) \int_t^{+\infty} s^{k-2}(1 + \beta_{k-2}(s))ds.$$

Hence $x_k(t) = t^{k-1}(1 + \alpha_k(t))$ where, as it is easy to see, $\Pi\{\alpha_k(t)\} \leq -\sigma$. Thus, relations (9.21) are proved. Formulae (9.22) are easily follow from (9.21). $\qquad\square$

Example 9.8. The equation

$$x'' + \frac{\sin \ln t}{t^3} x = 0$$

satisfies all the conditions of Proposition 9.7 because $\Pi\{(\sin \ln t)/t^3\} = -3$. Hence the equation has an FSS

$$\{x_1(t) = 1 + \alpha_1(t), \quad x_2(t) = t(1 + \alpha_2(t))\},$$

where $\Pi\{\alpha_{1,2}(t)\} \leq -1$.

3.   ALMOST DIAGONAL AND L-DIAGONAL SYSTEMS

Here we consider the asymptotic behavior of solutions to homogeneous system

$$X' = A(t)X \tag{9.2}$$

for $t \to +\infty$, $A(t)$ is a matrix-function of nth order, $X = (x_1, ..., x_n)^T$ and $X' = (x_1', ..., x_n')^T$. We begin with a system which is close to a diagonal system, that is, with a system of the form

$$X' = (\Gamma(t) + B(t))X, \tag{9.28}$$

where

$$\Gamma(t) = \text{diag}(\gamma_1(t), \gamma_2(t), ..., \gamma_n(t))$$

is a diagonal matrix-function and $B(t) = (b_{ij}(t))_n$ is a square matrix-function.

Throughout this subsection *we suppose for* $\Gamma(t)$ *and* $B(t)$ *to be continuous for* $t \gg 1$.

Designate by

$$d_{ij}(t) = \gamma_i(t) - \gamma_j(t) \quad (i, j = 1, 2, ..., n).$$ (9.29)

Definition 9.9. System (9.28) is said to be *almost diagonal* if $\|B(t)\| \to 0$ for $t \to +\infty$.

Definition 9.10. System (9.28) is said to be *L-diagonal* if $\|B(t)\| \in L_1[T, +\infty[$, i.e.

$$\int_T^{+\infty} \|B(t)\| dt < \infty,$$

where T is a positive number.

Asymptotic behavior of solutions of an almost diagonal system is close to the corresponding solutions of diagonal system $Y' = \Gamma(t)Y$ for certain general additional conditions. The last system disintegrates into n separate scalar differential equations $x_i' = \gamma_i(t)x_i$ and its *FM* may be represented in the form

$$\Phi(t) = \operatorname{diag}\left(\int \gamma_1(t)dt, \int \gamma_2(t)dt, ..., \int \gamma_n(t)dt\right).$$

In order to investigate almost diagonal and $L-$diagonal systems the following Lemmata can be useful.

Lemma 9.11. *Consider an expression of the form*

$$I(t) = \sum_{i=2}^{n} \left| \int_{t_i}^{t} \exp\left[\int_{\tau}^{t} \operatorname{Re} d_{i1}(s) ds \right] \|B(\tau)\| d\tau \right|,$$ (9.30)

where t_i is either a sufficiently large number or $t_i = +\infty$ $(i = 2, ..., n)$. Let it be possible to choose the values t_i such that $I(t) \to 0$ for $t \to +\infty$. Then for $t \gg 1$ system (9.28) has a continuous solution

$$X(t) = (x_1(t), x_2(t), ..., x_n(t))^T$$

such that

$$x_i(t) = u_i(t) \exp\left[\int (\gamma_1(t) + \delta(t))dt \right] \quad (i = 1, 2, ..., n),$$

where

$$u_1(t) = 1, \delta(t) = b_{11}(t) + O(\|B(t)\| I(t))$$

and $u_j(t) = O(I(t))$ *for* $j = 2, ..., n$ $(t \to +\infty)$. *Hence*

$$\frac{x_1'(t)}{x_1(t)} - \gamma_1(t) \to 0 \text{ and } \frac{x_i(t)}{x_1(t)} \to 0 \text{ for } t \to +\infty \ (i = 2, ..., n).$$

PROOF. Make the following substitutions: $x_1' = yx_1$ and $x_i = u_i x_1$ for $i = 2, ..., n$. Hence

$$y = \gamma_1(t) + \sum_{j=1}^{n} b_{1j}(t)u_j. \tag{9.31}$$

Clearly, $u_1 = 1$ and for $i = 2, ..., n$

$$u_i' = (\gamma_i(t) - \gamma_1(t))u_i - u_i \sum_{j=1}^{n} b_{1j}(t)u_j + \sum_{j=1}^{n} b_{ij}(t)u_j.$$

The last relation implies $u_i(t) = F_i(t, U)$, where

$$F_i(t, U) = \int_{t_i}^{t} \exp\left[\int_{\tau}^{t} d_{i1}(s)ds\right] f_i(\tau, U(\tau))d\tau.$$

Here $U = (u_2, ..., u_n)$ and

$$f_i(t, U) = \sum_{j=1}^{n} b_{ij}(t)u_j - u_i \sum_{j=1}^{n} b_{1j}(t)u_j.$$

Consider the integral expression

$$F(t, Y) = (F_2(t, Y), ..., F_n(t, Y))$$

in a region

$$V = \{Y : |y_i| \le 1, t \ge \tilde{T}, i = 2, ..., n\},$$

where $Y = (y_2, ..., y_n)^T$ and $\tilde{T}$ is a sufficiently large positive number. We have $|f_i(t, Y(t))| \le 2I(t)$ for $Y(t) \in V$ and $i = 2, ..., n$. Hence the operator $F(t, Y)$ transforms any continuous vector-function $Y(t) \in V$ to a continuous vector-function in V on $[\tilde{T}, +\infty[$.

We have

$$|f_i(t, Y_1(t)) - f_i(t, Y_2(t))| \le 3\|B(t)\|\|Y_1(t) - Y_2(t)\|$$

for $Y_1(t), Y_2(t) \in V$. Consequently

$$\|F(t, Y_1(t)) - F(t, Y_2(t))\|^* \le \frac{1}{2}\|Y_1(t) - Y_2(t)\|^*.$$

Here

$$\|F(t, Y_1(t)) - F(t, Y_2(t))\|^* = \sup_{t \geq \tilde{T}} \|F(t, Y_1(t)) - F(t, Y_2(t))\|$$

and

$$\|Y_1(t) - Y_2(t)\|^* = \sup_{t \geq \tilde{T}} \|Y_1(t) - Y_2(t)\|.$$

On the basis of the principle of the contractive mappings, there is a unique continuous solution

$$U(t) = (u_2(t), ..., u_n(t))^T \in V$$

of the equation $Y(t) = F(t, Y)$. Since $\|u_i(t)\| \leq 1$ on substituting $U(t)$ to the last equation we obtain

$$\|u_i(t)\| \leq 2I(t) \text{ for } t \geq \tilde{T}.$$

Besides, due to (9.31)

$$y(t) = \gamma_1(t) + b_{11}(t) + O(\|B(t)\|I(t)). \qquad \square$$

Let us give some sufficient conditions of fulfillment of the property $I(t) \to 0$ for $t \to +\infty$.

Lemma 9.12. *Let system (9.28) be almost diagonal. Let there exist a number $c > 0$ such that either*

$$\Re(\gamma_i(t) - \gamma_1(t)) \leq -c, \text{ or } \Re(\gamma_i(t) - \gamma_1(t)) \geq c \text{ as } t \gg 1$$

for each $i \in \{2, ..., n\}$. Hence we can break up the set $H = \{2, ..., n\}$ into two subsets H_1 and H_2 $(H_1 \cup H_2 = H)$ such that $i \in H_1$ if $\Re(\gamma_i(t) - \gamma_1(t)) \leq -c$, and $i \in H_2$ if $\Re(\gamma_i(t) - \gamma_1(t)) \geq c$ $(t \gg 1)$. Then

$$I(t) = \sum_{i=2}^{n} \left| \int_{t_i}^{t} \exp\left[\int_{\tau}^{t} \Re d_{i1}(s) ds \right] \|B(\tau)\| d\tau \right| \to 0 \qquad (9.32)$$

$(t \to +\infty)$. Here $t_i = T$ if $i \in H_1$ (T is a sufficiently large positive number) and $t_i = +\infty$ if $i \in H_2$.

PROOF. Let $i \in H_1$ and

$$I_i(t) = \int_{T}^{t} \exp\left[\int_{t}^{\tau} \Re d_{1i}(s) ds \right] \|B(\tau)\| d\tau.$$

We have

$$0 \le I(t) \le \int_T^t e^{c(\tau-t)} \|B(\tau)\| d\tau.$$

It is obvious that there is a (finite or infinite) limit $\lim_{t \to +\infty} I_i(t)$. Let us choose positive numbers ε and T_1 such that $\|B(t)\| < \varepsilon$ for $t > T_1$. Hence for $t \ge T_1$

$$I_i(t) \le e^{-ct} K + \varepsilon \int_{T_1}^t e^{c(\tau-t)} d\tau \le K e^{-ct} + \frac{\varepsilon}{c}(1 - e^{c(T_1-t)}) \le K e^{-ct} + \frac{\varepsilon}{c},$$

where

$$K = \int_T^{T_1} e^{c(\tau-t)} \|B(\tau)\| d\tau.$$

Consequently $\lim_{t \to +\infty} I_i(t) \le \varepsilon/c$. Taking into account the arbitrariness of ε we conclude that $\lim_{t \to +\infty} I_i(t) = 0$.

Let $i \in H_2$. Consider the integral

$$I_i(t) = \int_t^{+\infty} \exp\left[\int_\tau^t \Re d_{i1}(s) ds\right] \|B(\tau)\| d\tau.$$

We have

$$0 \le I_i(t) \le \sup_{\tau \ge t} \|B(\tau)\| \int_t^{+\infty} e^{c(t-\tau)} \|d\tau = \sup_{\tau \ge t} \frac{1}{c} \|B(\tau)\|.$$

Hence $\lim_{t \to +\infty} I_i(t) = 0$. Thus, it is proved that

$$\lim_{t \to +\infty} I_i(t) = 0 \text{ for any } i = 2, ..., n$$

and hence

$$I(t) = I_2(t) + ... + I_n(t) \to 0 \text{ for } t \to +\infty. \qquad \square$$

We have the following propositions for the L-diagonal system.

Lemma 9.13. *Let*

(1) *the difference $\Re d_{i1}(t)$ preserve its sign for $t \gg 1$ for any $i = 2, ..., n$;*

(2) *$\|B(t)\| \in L_1[T, +\infty[$, where T is a sufficiently large positive number, i.e. $\int_T^{+\infty} \|B(t)\| dt$ is convergent.*

Then (in (9.32)) it is possible to choose the values t_i such that $I(t) \to 0$ for $t \to +\infty$.

PROOF. Since $\Re d_{i1}(t)$ preserves its sign for $t > T$, there is a finite or infinite limit

$$p_i = \lim_{t \to +\infty} \int_T^t \Re d_{i1}(s)ds.$$

We put $t_i = +\infty$ if $p_i > -\infty$, and $t_i = T$ if $p_i = -\infty$. Let $p_i > -\infty$. Then there is a real number E such that

$$\int_T^t \Re d_{i1}(s)ds > E \text{ for any } t > T.$$

We have

$$I_i(t) = \int_t^{+\infty} e^{\int_\tau^t \Re d_{i1}(s)ds} \|B(\tau)\|d\tau \le e^{-E} \int_t^{+\infty} \|B(\tau)\|d\tau \to 0$$

and hence $I_i(t) \to 0$ for $t \to +\infty$. Let $p_i = -\infty$. Let us choose positive numbers ε and $T_1 > T$ such that

$$\int_t^{+\infty} \|B(\tau)\|d\tau < \varepsilon \text{ for } t > T_1.$$

We have

$$
\begin{aligned}
I_i(t) &\equiv \int_T^t \exp\left[\int_\tau^t \Re d_{i1}(s)ds\right] \|B(\tau)\|d\tau \\
&= K \exp\left[\int_T^t \Re d_{i1}(s)ds\right] + \int_{T_1}^T \exp\left[\int_\tau^t \Re d_{i1}(s)ds\right] \|B(\tau)\|d\tau \\
&\le K \exp\left[\int_T^t \Re d_{i1}(s)ds\right] + \varepsilon,
\end{aligned}
$$

where

$$K = \int_T^{T_1} \exp\left[\int_\tau^T \Re d_{i1}(s)ds\right] \|B(\tau)\|d\tau.$$

Hence $0 \le \lim_{t \to +\infty} I_i(t) \le \varepsilon$. Taking into account the arbitrariness of $\varepsilon > 0$ we conclude that $\lim_{t \to +\infty} I_i(t) = 0$.

Thus, it is proved that $\lim_{t \to +\infty} I_i(t) = 0$ for any $i = 2, ..., n$ and hence $I(t) = I_2(t) + ... + I_n(t) \to 0$ for $t \to +\infty$. $\square$

Remark. The Proof shows that condition (2) of Lemma 9.13 can be somewhat extended: *let the set* $H = \{2, ..., n\}$ *can be broken up into two subsets* H_1 *and* H_2, $H_1 \cup H_2 = H$, *such that* $i \in H_1$ *if there is a number* E *such that*

$$\int_{t_1}^{t_2} \Re d_{i1}(s)ds > E$$

for any $T \le t_1 \le t_2 < +\infty$, and $i \in H_2$ if

$$\lim_{t \to +\infty} \int_T^t \Re d_{i1}(s)ds = -\infty.$$

Lemma 9.14. Condition (1) of Lemma 9.13 can be replaced by the condition that the matrix $\Gamma(t)$ belongs to $\{G\}$ where G is a field of type M.

Indeed, since G is an algebraic field, then $d_{i1}(t) \in \{G\}$. Hence $\Re d_{i1}(t) \in \{G\}$. Consequently $\Re d_{i1}(t) \in A_t \cup O_t$. If $\Re d_{i1}(t) \in A_t$ then

$$\Re d_{i1}(t) \ne 0 \text{ for } t \gg 1$$

and hence $\Re d_{i1}(t)$ preserve its sign. If $\Re d_{i1}(t) \in O_t$ then

$$\int_t^{+\infty} \Re d_{i1}(s)ds \to 0 \text{ for } t \to +\infty$$

and therefore

$$\int_t^{+\infty} \exp\left[\int_\tau^t \Re d_{i1}(s)ds\right] \|B(\tau)\|d\tau = O\left(\int_t^{+\infty} \|B(\tau)\|d\tau\right)$$
$$= o(1) \text{ for } t \to +\infty.$$

Lemma 9.13 implies the following:

Proposition 9.15. *Let*

(1) the set $H = \{1, 2, ..., n\}$ for each number $k \in \{1, 2, ..., n\}$ can be broken up into two subsets H_{k1} and H_{k2} $(H_{k1} \cup H_2 = H)$ such that $i \in H_{k1}$ if there is a number E such that

$$\int_{t_1}^{t_2} \Re d_{ik}(s)ds > E$$

for any $T \le t_1 \le t_2 < +\infty$, and $i \in H_{k2}$ if

$$\lim_{t \to +\infty} \int_T^t \Re d_{ik}(s)ds = -\infty.$$

Here T is a sufficiently large positive number.

(2) $\|B(t)\| \in L_2[T, +\infty[$.

Then FM of system (9.28) can be written in the following form $\Phi(t) = (x_{ik}(t))_n$, where

$$x_{ik}(t) = u_{ik}(t) \exp\left[\int (\gamma_k(t) + \delta_k(t))dt\right], \quad k = 1, 2, ..., n.$$

Here

$$\delta_k(t) = \beta_{kk}(t) + O(\|B(t)\|I_k(t)), \ u_{kk}(t) = 1,$$

and

$$u_{ik}(t) = O(I_k(t)) \text{ for } i = 1, 2, ..., k-1, k+1, ..., n;$$

$I_k(t)$ *is given in* (9.32) *where* $t_{ik} = +\infty$ *if* $i \in H_{k1}$, *and* $t_{ik} = T$ *if* $i \in H_{k2}$.
Hence

$$x_{kk}(t) = (1 + \alpha_{kk}(t)) \exp\left[\int \gamma_k(t)dt\right]$$

and

$$x_{ik}(t) = \alpha_{ik} \exp\left[\int \gamma_k(t)dt)\right]$$

for $i = 1, 2, ..., k-1, k+1, ..., n$, *where the functions* $\alpha_{ij}(t)$ *are continuous for* $t \gg 1$ *and infinitesimal for* $t \to +\infty$.

Remark 9.16. Condition (1) in this Proposition can be replaced by each of the following conditions: (1) *any difference* $\Re d_{ij}(t)$ *preserves its sign for* $t \gg 1$ $(i, j = 1, 2, ..., n)$; or (2) *the matrix* $\Gamma(t)$ *belongs to* $\{G\}$ *where* G *is a field of type* M.

We give (without proof) several important theorems related to the almost diagonal and $L-$ diagonal systems.

Theorem 9.17 (PERRON). *Let system* (9.28) *be almost diagonal. Let there be a number* $c > 0$ *such that*

$$\Re(\gamma_{j+1}(t) - \gamma_j(t)) \ge c \text{ for any } j = 1, 2, ..., n-1.$$

Then system (9.28) *has a fundamental matrix* $\Phi(t)$ *of the form*

$$\Phi(t) = [E + \Lambda(t)] \exp\left[\int_T^t (\Gamma(\tau) + \Delta(\tau))d\tau\right], \tag{9.33}$$

where $\Delta(t)$ *is a diagonal matrix and*

$$\lim_{t \to +\infty} \|\Lambda(t)\| = \lim_{t \to +\infty} \|\Delta(t)\| = 0. \tag{9.34}$$

The asymptotic formula (9.33) is rather rough and it cannot be define more precisely without additional hypothesis about the behavior of the matrix-function $B(t)$ (even for $n = 1$).

Theorem 9.18 (HARTMAN–WINTNER). *Let*

$$\|B(t)\| \in L_2[T, +\infty[,$$

where T is a sufficiently large positive number, i.e.

$$\int_T^{+\infty} \|(B(t)\|^2 dt < \infty.$$

Let there exist a positive number c such that

$$|\Re d_{jk}(t)| > c \ \ for \ \ t \ge T$$

for all $j, k, \ j \ne k$. Then system (9.28) has an FM of the form

$$\Phi(t) = (E + \Lambda(t)) \exp\left[\int_T^t [\Gamma(\tau) + \mathrm{diag}\, B(\tau)] d\tau\right], \qquad (9.35)$$

where $\lim_{t\to+\infty} \|\Lambda(t)\| = 0$.

Theorem 9.19 (LEVINSON). *Consider a system of the form*

$$X' = (A + V(t) + R(t))X, \qquad (9.36)$$

where A is a constant matrix with different characteristic roots $\mu_1, \mu_2,$ ..., μ_n. Let Q be a non-singular matrix which reduces the matrix A to a diagonal form, i.e.

$$Q^{-1}AQ = \mathrm{diag}(\mu_1, \mu_2, ..., \mu_n).$$

Let the matrix $V(t)$ be differentiable

$$\int_0^{+\infty} \|V'(t)\| dt < \infty \qquad (9.37)$$

and let $\lim_{t\to+\infty} \|V(t)\| = 0$. Let $R(t)$ be integrable and

$$\int_0^{+\infty} \|R(t)\| dt < \infty.$$

Let us denote the roots of the equation $\det(A + V(t) - \lambda E) = 0$ by $\lambda_j(t)$ $(j = 1, 2, ..., n)$. Obviously it is possible (if it is necessary) renumber ρ_j so that $\lim_{t\to+\infty} \lambda_j(t) = \rho_j$. Put

$$D_{kj}(t) = \Re(\lambda_k(t) - \lambda_j(t))$$

for a given (fixed) k. Suppose that any $j \in \{1, 2, ..., n\}$ belongs to one of the two classes I_1 and I_2 where

$$j \in I_1 \ \ if \ \ \int_0^t D_{kj}(\tau) d\tau \to +\infty \ \ for \ \ t \to +\infty$$

and

$$\int_{t_1}^{t_2} D_{kj}(\tau)d\tau < K \quad (K = rmconst, \ t_2 \geq t_1 \geq 0).$$

Let p_k be a characteristic vector corresponding to the root μ_k so that $Ap_k = \mu_k p_k$. Then there exists a solution $\varphi_k(t)$ of system (9.36) and a number t_0 $(0 \leq t_0 < +\infty)$ such that

$$\lim_{t \to +\infty} \varphi_k(t) \exp\left[-\int_{t_0}^{t} \lambda_k(\tau)d\tau\right] = p_k. \tag{9.38}$$

Example 9.20. Consider system (9.28) where

$$A = \begin{pmatrix} 0 & 0 \\ 0 & 2 \end{pmatrix}, \ V(t) = t^{-\alpha} \begin{pmatrix} 0 & 1 \\ 1 & 0 \end{pmatrix}, \ \|R(t)\| = 0, \ 0 < \alpha \leq 1.$$

The characteristic roots of the matrix $A + V(t)$ equal to

$$\lambda_1(t) = 1 - \sqrt{1 + t^{-2\alpha}} \sim -\frac{1}{2}t^{-2\alpha}$$

and

$$\lambda_2(t) = 1 + \sqrt{1 + t^{-2\alpha}} = 2 + O(t^{-2\alpha})$$

for $t \to +\infty$, so that $\Re d_{21}(t) = 2 + O(t^{-2\alpha})$. The Hartman–Wintner Theorem is applicable for $\alpha > \frac{1}{2}$. In this case a fundamental matrix of the system can be written as follows

$$\Phi(t) = (E + o(1)) \begin{pmatrix} 1 & 0 \\ 0 & e^{2t} \end{pmatrix} \quad \text{for } t \to +\infty.$$

Since $\|V'(t)\| = O\left(t^{-\alpha-1}\right)$ for $t \to +\infty$ the conditions of Levinson theorem is satisfied and we have

$$\Phi(t) = (E + o(1)) \begin{pmatrix} \exp\left[\int^t \lambda_1(s)ds\right] & 0 \\ 0 & \exp\left[\int^t \lambda_2(s)ds\right] \end{pmatrix}.$$

In particular for $\alpha = 1/2$

$$\Phi(t) = (E + o(1)) \begin{pmatrix} t^{1/2} & 0 \\ 0 & t^{1/2}e^{2t} \end{pmatrix}.$$

4.   SINGLE LINEAR EQUATION WITH ASYMPTOTICALLY SIMPLE ROOTS OF THE CHARACTERISTIC POLYNOMIAL

We considered equation (9.1) in two cases: (1) the roots $\lambda_i(t)$ of the characteristic equation have the same growth for $t \to +\infty$; (2) all the roots $\lambda_i(t)$ have the same growth with except of one of the roots whose growth is less than the rest for $t \to +\infty$. Consider the first case.

Theorem 9.21. (FEDORJUK). *Let all the coefficients of equation*

$$F(t, x) \equiv x^{(n)} + a_1(t)x^{(n-1)} + \ldots + a_n(t)x = 0 \qquad (9.1)$$

have continuous derivatives of the second order and $a_n(t) \neq 0$ for $t \gg 1$. Besides

(1) *all the roots $\lambda_k(t)$ $(k - 1, 2, \ldots, n)$ of the characteristic equation*

$$H(t, y) \equiv y^n + a_1(t)y^{n-1} + \ldots + a_n(t) = 0 \qquad (9.16)$$

have the same growth for $t \to +\infty$. This means that all the limits

$$\lim_{t \to +\infty} a_k(t)a_n^{-k/n}(t) = q_k$$

are finite $(k = 1, 2, \ldots, n)$. The asymptotics of the roots can be written in the form

$$\lambda_k(t) = (\rho_k + o(1))a_n^{1/n}(t) \;\; for \; t \to +\infty,$$

where ρ_k are constants $(k = 1, 2, \ldots, n)$;

(2) *the substitution $y = a_n^{1/n}(t)u$ in (9.16) leads to the equation*

$$u^n + (q_1 + o(1))u^{n-1} + \ldots + (q_{n-1} + o(1))u + 1 = 0.$$

On tending t to infinity we obtain the limiting equation

$$\rho^n + q_1\rho^{n-1} + \ldots + q_{n-1}\rho + 1 = 0. \qquad (9.39)$$

It has a complete set of roots $\rho_1, \rho_2, \ldots, \rho_n$. We assume that all the roots are different. Thus, $\rho_i \neq \rho_k$ for all $i \neq k$.

Let us mark that all $\rho_i \neq 0$. This follows from condition (1);

(3) *Let us denote by*

$$p_k(t) = -\frac{1}{2}\lambda_k'(t)\frac{\partial^2 H(t, \lambda_k(t))/\partial y^2}{\partial H(t, \lambda_k(t))/\partial y}. \qquad (9.40)$$

Let the difference

$$\Re(\lambda_j(t) - \lambda_k(t) + p_j(t) - p_k(t)) \tag{9.41}$$

preserve its sign for a fixed number j, k ≠ j, and t ≫ 1;

(4) $\int_T^{+\infty} \alpha(t)dt < \infty$, *where T is a sufficiently large positive number*

$$\alpha(t) = \sum_{k=1}^{n} |a_k'(t)|^2 |a_n(t)|^{-(2k+1)/n} + |a_k''(t)| \|a_n(t)|^{-(k+1)/n}. \tag{9.42}$$

Let us denote by

$$\tilde{x}_j(t) = \exp\left[\int_T^t (\lambda_j(s) + p_j(s))ds\right].$$

Then equation (9.1) has a solution

$$x_j(t) = (1 + \varepsilon_j(t))\tilde{x}_j(t), \tag{9.43}$$

where $\varepsilon_j(t)$ is continuous for $t \gg 1$ and infinitesimal for $t \to +\infty$.
*   If condition (3) is fulfilled for any j = 1, 2, ..., n then equation (9.1) has an FSS of the form*

$$\{x_j(t) = (1 + o(1))\tilde{x}_j(t)\}.$$

The Theorem is a consequence of the Levinson Theorem. Here we only give the main ideas of the Theorem's proof. Equation (9.1) is equivalent to a system of the form $X' = A(t)X$, where $X = (x, x', ..., x^{(n-1)})^T$ and $A(t)$ is a matrix given in (9.33). Equation (9.1) is reduced to system (9.21) (which satisfies all the conditions of Theorem 9.17) by means of two substitutions. The first one is $X = S(t)Y$ where $S(t)$ is chosen such that the matrix

$$Q^{-1}(t)A(t)Q(t) = \Lambda(t) \equiv \text{diag}(\lambda_1(t), \lambda_2(t), ..., \lambda_n(t))$$

should be diagonal. The matrix $S(t)$ is chosen in the form

$$U(t) = \begin{pmatrix} 1 & 1 & \cdots & 1 \\ \lambda_1(t) & \lambda_2(t) & \cdots & \lambda_n(t) \\ \lambda_1^2(t) & \lambda_2^2(t) & \cdots & \lambda_n^2(t) \\ \cdot & \cdot & & \cdot \\ \lambda_1^{n-1}(t) & \lambda_2^{n-1}(t) & \cdots & \lambda_n^{n-1}(t) \end{pmatrix}. \tag{9.44}$$

We obtain $Y' = (\lambda(t) + U(t))Y$, where the matrix should be diagonal. The matrix $U(t) = (u_{ik}(t))_n$ is equal to $-S^{-1}(t)S'(t))$. We have

$$u_{kk}(t) = -\frac{1}{2}\lambda_k'(t)\frac{\partial^2 H\left(t, \lambda_k(t)\right)/\partial y^2}{\partial H\left(t, \lambda_k(t)\right)/\partial y} \equiv p_k(t) \qquad (9.45)$$

and for $i \neq k$,

$$u_{ik} = \frac{\lambda_k'(t)}{\lambda_k(t) - \lambda_j(t)}\frac{\partial H(t, \lambda_j(t))/\partial y}{\partial H(t, \lambda_k(t)/\partial y}. \qquad (9.46)$$

The second substitution is

$$Y = (E + W(t))Z, \ W(t) = (w_{ik}(t))_n.$$

The system can be written in the form

$$Z' = (\Lambda(t) + U(t) + V(t) + C(t))Z,$$

where

$$V(t) = v_{ik}(t), V(t) = \Lambda(t)W(t) - W(t)\Lambda(t).$$

We choose the matrix $W(t)$ such that $U(t) + V(t)$ will be diagonal. The choice of matrix $W(t)$ is not unique. Since all the diagonal elements of the matrix $V(t)$ equal to zero we may put $w_{ii}(t) = 0$ and for $i \neq k$

$$w_{ik}(t) = \frac{u_{ik}(t)}{\lambda_k(t) - \lambda_i(t)}.$$

Hence

$$Z' = (\Lambda(t) + \mathrm{diag}(p_1(t), p_2(t), ..., p_n(t)) + B(t))Z,$$

where $\int_T^t \|B(t)\|dt < \infty$. Thus, all the conditions of Theorem 9.21 are satisfied which leads to the required.

Remark 9.22. If, in addition to the conditions of Theorem 9.21,

$$a_k'(t) = o\left(a_n^{-(k+1)}(t)\right)$$

as $t \to +\infty$ for all k, then $p_k(t) = o(\lambda_k(t))$ and asymptotics (9.43) can be differentiated n times, i.e.,

$$x_j^{(s)}(t) = (1 + o(1))\tilde{x}_j^{(s)}(t) \qquad (9.47)$$

for $s = 1, 2, ..., n$. In the next theorem, one root have its growth which is less than the growth of all the rest roots for $t \to +\infty$. The theorem is proved in the same way as Theorem 9.21.

Theorem 9.23 (FEDORJUK). *Let all the coefficients of equation* (9.1) *have continuous derivatives and* $a_n(t) \neq 0$ *for* $t \gg 1$. *Besides*

(1) *all the roots of characteristic equation* (9.16) *have the same growth for* $t \to +\infty$ *excluding one of them which has its growth lesser than all the rest roots. For definiteness we assume that*

$$a_n(t) = o(a_{n-1}(t)) \quad for \quad t \to +\infty$$

and there are all finite limits

$$\lim_{t \to +\infty} a_k(t) a_{n-1}^{k/(1-n)}(t) = q_k \quad (k = 1, 2, ..., n-1);$$

the asymptotics of the roots can be written in the form

$$\lambda_k(t) = (\rho_k + o(1)) a_{n-1}^{1/(1-n)}$$

and

$$\lambda_n(t) = -a_n(t) + o\left(\frac{a_n(t)}{a_{n-1}(t)}\right);$$

(2) *the substitution* $y = a_{n-1}^{1/(n-1)}(t) u$ *in* (9.16) *leads to the equation*

$$u^n + (q_1 + o(1)) u^{n-1} + ... + (1 + o(1)) u + o(1) = 0.$$

Let its limiting equation

$$\rho^{n-1} + q_1 \rho^{n-2} + ... + q_{n-2}\rho + 1 = 0 \tag{9.48}$$

have roots

$$\rho_1, \rho_2, ..., \rho_{n-1}.$$

We assume that all the roots are different. Thus, $\rho_i \neq \rho_k$ *for all* $i \neq k$. *Let us mark that all* $\rho_i \neq 0$. *This follows from condition* (1);

(3) *the condition* (3) *of Theorem* 9.21 *is fulfilled;*

(4) $\int_T^{+\infty} \alpha(t) dt < \infty$ (*T is a sufficiently large positive number*) *where*

$$\alpha(t) = \sum_{k=1}^{n} (|a_k'(t)|^2 |a_{n-1}(t)|^{k/(1-n)} + |a_k''(t)| \| a_{n-1}^{k/(1-n)}(t) \|). \tag{9.49}$$

Then equation (9.49) *has a solution of the form* (9.43).

If condition (3) *is fulfilled for any* $j = 1, 2, ..., n$ *then equation* (9.1) *has an FSS of the form*

$$\{x_j(t) = (1 + o(1))\tilde{x}_j(t)\} \quad j = 1, 2, ..., n.$$

Remark 9.24. If in addition to the conditions of Theorem 9.23

$$a_k'(t) = o\left(a_{n-1}^{(k+1)/(n-1)}(t)\right)$$

for all k and $t \to +\infty$ then

$$p_k(t) = o(\lambda_k(t)) \text{ for } k = 1, 2, ..., n - 1$$

and asymptotics (9.43) can be differentiated n times, i.e. relation (9.40) takes place.

If the coefficients of equation (9.1) belong to a field of type M or they are asymptotically close to a field of type M, then the conditions of Theorems 9.21 and 9.23 can be considerably simplified.

Theorem 9.25. *Let all $a_i(t)$ $(i = 1, 2, ..., n)$ belong to a space $[G]$ where G is a field of type M. Besides,*

(1) *let conditions (1) and (2) of Theorem 9.23 be fulfilled;*

(2) *let us denote by $\lambda(t) = a_n^{1/n}(t)$ in the case of fulfillment of condition (1) of Theorem 9.21 and $\lambda(t) = a_{n-1}^{1/(n-1)}(t)$ in the case of Theorem 9.23. Let $\lambda(t)t \to \infty$ for $t \to +\infty$. Then there is a solution of the form (9.43) and relations (9.44) hold.*

Theorem 9.25 is a simple consequence of Theorems 9.21 and 9.23. It is easy to show that all the conditions of the Theorems are fulfilled.

Example 9.26. Consider an equation of the form

$$x^{(n)} + f(t)x = 0, \tag{9.50}$$

where $f(t)$ belongs to $[G]$ (G is a field of type M) and

$$\int_T^{+\infty} t^{-2}|f(t)|^{-1/n}dt < \infty.$$

Its characteristic equation is written in the form

$$H(t, y) \equiv y^n + f(t) = 0.$$

All its roots $\lambda_k(t) = \varepsilon_k[f(t)]^{1/n}$ are asymptotically simple. Here ε_k are distinct roots of nth order of -1 ($k = 1, 2, ..., n$). Consequently all the conditions of Theorem 9.21 are satisfied. We have

$$\lambda_k'(t) = \varepsilon_k(1/n)f'(t)[f(t)]^{(1-n)/n},$$

$$\frac{\partial H(t, \lambda_k(t))}{\partial y} = \lambda_k^{n-1}(t)$$

and

$$\frac{\partial^2 H(t, \lambda_k(t))}{\partial y^2} = n(n-1)\lambda_k^{n-2}(t).$$

Hence

$$p_k(t) = \frac{1-n}{2}\frac{\lambda_k'(t)}{\lambda_k(t)} = \frac{1-n}{2n}\frac{f'(t)}{f(t)}.$$

Consequently equation (9.1) has an FSS of the form

$$\left\{ x_k(t) = (1 + o(1))[f(t)]^{(1-n)/(2n)} \exp\left[\int_T^t \varepsilon_k[f(s)]^{1/n}ds\right] \right\} \qquad (9.51)$$

for $t \to +\infty$ $(k = 1, 2, ..., n)$.

Other examples of formula (9.43) application will be given in the next chapter.

Chapter 10

LINEAR DIFFERENTIAL EQUATIONS WITH POWER ORDER GROWTH COEFFICIENTS ON THE POSITIVE SEMI-AXIS

In chapters 10, 11 we consider linear differential equations on the positive semi-axis with power order growth coefficients of the form

$$F(t, x) \equiv x^{(n)} + a_1(t)x^{(n-1)} + ... + a_n(t)x = 0. \qquad (10.1)$$

where the coefficients $a_i(t)$ belong to a space Q of type M or N (or to $\{Q\}$ which is some more general. See Definitions 6.16, 6.17 and 6.20).

In this chapter we investigate so called *regular case* and a complete set of roots $\Lambda = \{\lambda_1(t), ..., \lambda_n(t)\}$ of (10.1) characteristic equation

$$H(t, y) \equiv y^n + a_1(t)y^{n-1} + ... + a_n(t) = 0 \qquad (10.2)$$

possesses the property of asymptotic separability. The latter, roughly speaking, means that they do not equivalent in pairs for $t \to +\infty$ and for all $i = 1, 2, ..., n$ or may be with exception of no more than one number $j \in \{1, 2, ..., n\}$ there exists a number $\sigma > 0$ such that $|\lambda_i(t)| > t^{-1+\sigma}$ as $t \gg 1$ and $|\lambda_k(t)| < t^{-1+\mu}$ for any fixed positive number μ as $t \gg 1$. More precisely $\Pi\{\lambda_i(t)\} > -1$ and $\Pi\{\lambda_k(t)\} \leq -1$.

Asymptotic solutions to equation (10.1) are obtained by two stages. In the first stage we obtain formal solutions to equation (10.1) (see Definitions 10.12 and 10.13). In the second stage we prove that they form the desired asymptotics.

The first stage is related to equation (10.1) where Q is a field of type N. Equation (10.1) is reduced to a non-linear equation by means of the substitution $x' = yx$. We obtain the equation

$$\Phi(t, y) \equiv \Phi^*(t, y) + H(t, y) = 0$$

(see (8.40), (8.41)) where we have only to investigate n different solutions belonging to the class of the power order functions A_t. For each root $\lambda_j(t) \in$

$\Lambda \; \Pi_j \lambda_j(t) > -1$, there exists a function $g_j(t)$ which is an asymptotic limit of the iterate sequence

$$s_{jm+1}(t) = \lambda_j(t) - \frac{\Phi^*(t, s_{jm}(t))}{H_j^*(t, s_{jm}(t))}, \quad s_{j0}(t) = \lambda_j(t), \quad m = 1, 2, \dots$$

The polynomial $H^*(t, y)$ is of $n-1$ degree. It is obtained from the identity $H(t, y) = H_j^*(t, y)(y - \lambda_j(t))$. That is, $H_j^*(t, y)$ is the result of the division $H(y, \lambda_j)$ by $y - \lambda_j(t)$. In the considered regular cases

$$g_j(t) = \lambda_j(t) - \frac{\lambda_j'(t)}{2} \frac{\partial^2 H(t, \lambda_j(t))}{\partial y^2} \bigg/ \frac{\partial H(t, \lambda_j(t))}{\partial y} + \alpha_j(t),$$

where $\Pi\{a_j(t)\} < -1$. Moreover,

$$\frac{\lambda_j'(t)}{2} \frac{\partial^2 H(t, \lambda_j(t))}{\partial y^2} \bigg/ \frac{\partial H(t, \lambda_j(t))}{\partial y}$$

$$= O\left(\frac{\lambda'(t)}{\lambda_j(t)}\right)$$

$$= O\left(\frac{1}{t}\right) \quad (t \to +\infty).$$

For $\Pi\{\lambda_j(t)\} \leq -1$ the corresponding function $g_j(t) = \lambda_j(t) + \delta_j(t)$, where $\delta_j(t) = o(\lambda_j(t))$ for $t \to +\infty$. More precisely $\Pi\{\delta_j(t)\} < \Pi\{\delta_j(t)\}$. Consequently $\Pi\{\delta_j(t)\} < -1$. This estimate can be made more precisely by means of an iterated sequence given in Chapter 9(4).

The function $\exp[\int g_j(t)dt]$ is a formal solution to equation (10.1) and the collection of such formal solutions (for $j = 1, 2, \dots, n$) forms a formal fundamental system of solutions (FFS).

The second stage consists of the proof that the obtained formal solutions $g_j(t)$ are asymptotic solutions to the equation $\Phi(t, y) = 0$. That is, for each $j = 1, 2, \dots, n$, there is an exact solution $\gamma_j(t)$ to the equation $\Phi(t, y) = 0$ such that $\gamma_j(t) \asymp g_j(t)$ i.e. the difference $\gamma_j(t) - g_j(t)$ has negligible small asymptotics (see Definition 4.7). Equation (10.1) possesses an FSS of the form $\left\{ e^{\int \gamma_j(t)dt} \right\}$ $(j = 1, 2, \dots, n)$. This property holds true if (in addition) Q is a field of type M.

Let us mark that any field of type M is also a field of type N.

The functions $\gamma_j(t)$ and $g_j(t)$ have the same asymptotics. Hence the asymptotic representations of $\gamma_j(t)$ may be obtained from the sequence

$\{s_m(t)\}$ given above. The main term can be obtanind by means of so-called *standard procedure* consisting of the following. Substitute $x' = yx$ in (10.1). This leads to the equation

$$\Phi_n(y) + a_1(t)\Phi_{n-1}(y) + \ldots + a_n(t) = 0 \qquad (10.3)$$

and any $\gamma_j(t)$ is a solution of this equation (the substitution $y = x'/x$ is considered in Chapter 8 in detail). Let $\lambda_j(t) \not\equiv 0$. Then $\lambda_j(t)$ and $g_j(t)$ belong to A_t, and $g_j(t) \sim \lambda_j(t)$ for $t \to +\infty$. Hence (see (5.23)) there are functions $\beta_i(t) \in C_t$ ($\beta_i(t) \to 0$ $t \to +\infty$, $i = 1, 2, \ldots, n-1$) such that

$$g_j(t)^{(m)}(t) = [k_j(k_j - 1)\ldots(k_j - m + 1) + \beta_m(t)]g_j(t)t^{-m},$$

where

$$\begin{aligned} k_j &= P\{\lambda_j(t)\} \\ &= \lim_{t \equiv +\infty} \frac{\ln \lambda_j(t)}{\ln t} \end{aligned}$$

$m = 1, 2, \ldots, n-1$. Substitute the relations obtained in (10.3). We obtain an algebraic equation of $n-1$ order of the form

$$P_j(t, y, \beta_1(t), \ldots, \beta_{n-1}(t)) = 0 \qquad (10.4)$$

containing the functions $\beta_1(t), \ldots, \beta_n(t)$. The functions $\beta_i(t)$ do not know exactly. But it is known that they belong to C_t and, in particular, they are infinitesimal for $t \to +\infty$. Moreover, if all the coefficients $a_i(t)$ have their epensions in the form of generalized power series (see Example 4.13 2°), then the functions $\beta_j(t)$ have negative estimates. That is, $\Pi\{\beta_j(t)\} < 0$.

As a simple example consider the Airy equation $x'' - tx = 0$. $t \in L$ (L is the set of all power logarithmic functions and it is a field of type M). Its characteristic equation $y^2 - t = 0$ has two roots $\lambda_1(t) = \sqrt{t}$ and $\lambda_2(t) = -\sqrt{t}$. Substitute $x' = yx$ we obtain $x'' = (y' + y^2)x$ and (for $x \neq 0$) $y' + y^2 - t = 0$. Clearly, $k_{1,2} = P\{\pm\sqrt{t}\} = 1/2$. For the first root $\sqrt{t}$ we have $y' = (1/2 + \beta_1(t))y/t$, where $\beta_1(t) \in C_t$ which leads to the equation

$$y^2 + \frac{1/2 + \beta_1(t)}{t}y - t = 0.$$

Since $g_1(t) \sim \sqrt{t}$ we obtain

$$g_1(t) = \sqrt{t} - \frac{1}{4t} - \frac{\beta_1(t)}{2t}.$$

where $\Pi\{\beta_1(t)\} < 0$. Finally,

$$x_1(t) = (1 + o(1)))t^{-1/4}e^{(2/3)t^{3/2}} \quad \text{for } t \to +\infty.$$

In the same way the solution corresponding to the second root $-\sqrt{t}$ may be written in the form

$$x_2(t) = (1 + o(1))t^{-1/4}e^{-(2/3)t^{3/2}} \text{ for } t \to +\infty.$$

1. FORMAL SOLUTIONS TO A SINGLE LINEAR DIFFERENTIAL EQUATION OF NTH ORDER

Throughout this paragraph Q *means a field of type* N (which will be not stipulated apart).

Here we look for the formal solutions to equation (10.1). First, we will prove some properties of functions belonging to $\{Q\}$.

Lemma 10.1. *Let* $y(t) \in \{Q\}$. *Then for any* $m = 1, 2, ...$ *if*

(1) $\Pi\{y(t)\} > -1$ *then* $\Phi_m(y(t)) \sim y^m(t)$ *for* $t \to +\infty$;

(2) $\Pi\{y(t)\} \leq -1$ *then* $\Pi\{\Phi_m(y(t))\} \leq \Pi\{y(t)\} - m + 1$.

PROOF. This Lemma is proved by induction on m. Lemma is obvious for $m = 1$ because $\Phi_1(y) = y$. Let it be true for $m-1$. We have (see (8.42))

$$\Phi_m(y(t)) = \Phi_{m-1}(y(t))y(t) + \frac{d\Phi_{m-1}(y(t))}{dt}.$$

In the case $\Pi\{y(t)\} > -1$ we have (by induction) $\Phi_{m-1}(y(t))y(t) \sim y^m(t)$ and (see Proposition 5.15 (5))

$$\begin{aligned}
\frac{d\Phi_{m-1}(y(t)}{dt} &= O\left(\frac{y^{m-1}(t)}{t}\right) \\
&= o(y^m(t))
\end{aligned}$$

consequently $\Phi_m(t) \sim y^m(t)$ ($t \to +\infty$). In the case $\Pi\{y(t)\} \leq -1$ we have (by induction) $\Pi\{\Phi_{m-1}(y(t))y(t)\} \leq \Pi\{\Phi_{m-1}(y(t))\} + \Pi\{y(t)\} \leq \Pi\{\Phi_{m-1}(y(t))\} - 1 \leq \Pi\{y(t)\} - m + 1$, and from the relation

$$d\Phi_{m-1}(y(t))/dt = O(\Phi_{m-1}(y(t))/t) \text{ for } t \to +\infty,$$

clearly,

$$\begin{aligned}
\Pi\{d\Phi_{m-1}(y(t))/dt\} &\leq \Pi\{\Phi_{m-1}(y(t))\} - 1 \\
&\leq \Pi\{y(t)\} - m + 1,
\end{aligned}$$

which leads to the required property.   □

Introduce the following notation:

$$V(y_1(t), \ldots, y_n(t)) = \begin{vmatrix} 1 & 1 & \cdots & 1 \\ \Phi_1(y_1(t)) & \Phi_1(y_2(t)) & \cdots & \Phi_1(y_n(t)) \\ \cdot & \cdot & \cdots & \cdot \\ \Phi_{n-1}(y_1(t)) & \Phi_{n-1}(y_2(t)) & \cdots & \Phi_{n-1}(y_n(t)) \end{vmatrix}.$$

(10.5)

Determinant $V(y_1(t), \ldots, y_n(t))$ has an important significance in the linear differential equations theory, the same as a Wronskian of solutions because the Wronskian of the set $\{\exp\left[\int (y_1(t)dt\right], \ldots, \exp\left[\int y_n(t)dt\right]\}$ is equal to

$$V(y_1(t), \ldots, y_n(t)) \exp\left[\int (y_1(t) + y_2(t) + \ldots + y_n(t))dt\right].$$

Lemma 10.2. *Let $y_1(t), y_2(t) \in \Pi$. Then the following relation holds:*

$$\Phi_m(y_1(t) + y_2(t)) = sum_{p=0}^m \binom{m}{p} \Phi_{m-p}(y_1(t))\Phi_m(y_2(t)).$$

(10.6)

PROOF. Since $x^{(m)} = \Phi_m(y(t))x(t)$ we have

$$\Phi_m(y(t)) = \left\{\exp\left[\int y(t)dt\right]\right\}^{(m)} \exp\left[-\int y(t)dt\right].$$

(10.7)

Hence

$$\Phi_m(y_1(t) + y_2(t))$$

$$= \left\{\exp\left[\int (y_1(t) + y_2(t))dt\right]\right\}^{(m)} \exp\left[-\int (y_1(t) + y_2(t))dt\right]$$

$$= \left\{\exp\left[\int y_1(t)dt\right] \exp\left[\int y_2(t)dt\right]\right\}^{(m)}$$

$$\times \exp\left[-\int y_1(t)dt\right] \exp\left[-\int y_2(t)dt\right]$$

which leads to the required relation. □

Lemma 10.3. *Let $\{y(t), y_1(t), \ldots, y_n(t)\} \subset \Pi$. Then the following identity holds:*

$$V(y_1(t) + y(t), \ldots, y_n(t) + y(t)) = V(y_1(t), \ldots, y_n(t)).$$

(10.8)

Here we may only suppose that $y(t)$ is an $(n-1)$ times differentiable function for $t \gg 1$.

PROOF. By transforming identically determinant (10.5) and using (10.6) we can obtain the left side of identity (10.8). To this end let us multiply the first row of determinant (10.5) by $\Phi_m(y(t))$, the second row by $\binom{m}{1}\Phi_{m-1}(y(t))$, and so on until the mth row which is multiplied by $\binom{m}{m-1}\Phi_1(y(t))$. Identity (10.8) follows by adding all the rows obtained to the $(m+1)$th row ($m = 1, 2, ..., n - 1$). □

Definition 10.4. We say that a set of functions $\{y_1(t), ..., y_n(t)\} \subset \Pi$ possesses the *property of asymptotic separability* for $t \to +\infty$ if the functions are not equivalent in pairs for $t \to +\infty$ and $\Pi\{y_i(t) - y_j(t)\} > -1$ for any $i, j = 1, 2, ..., n, \ i \neq j$.

Proposition 10.5. *Let $\{y_1(t), ..., y_n(t)\} \subset \{Q\}$ be a set of functions possessing the property of asymptotic separability for $t \to +\infty$. Then*

$$V(y_1(t), ..., y_n(t)) \sim \omega(y_1(t), ..., y_n(t)), \qquad (10.9)$$

where $\omega(y_1(t), ..., y_n(t))$ is a Van der Monde determinant for the elements $\{y_1(t), y_2(t), ..., y_n(t)\}$.

PROOF. Assume (for definiteness, otherwise we may change the numeration of the functions $y_j(t)$), that

$$\lim_{t \to +\infty} \left| \frac{y_p(t)}{y_q(t)} \right| \leq 1 \text{ for any } p < q.$$

Hence taking into account the property of asymptotic separability of the set $\{y_j(t)\}$ we conclude that

$$y_q(t) - y_p(t) \sim c_{pq} y_q(t) \text{ for } t \to +\infty,$$

where c_{pq} are non-zero numbers. Hence

$$\omega(y_1(t), ..., y_n(t)) \sim \varphi(t) \equiv c y_2(t) y_3^2(t)...y_n^{n-1}(t) \ (t \to +\infty),$$

where c is a non-zero number.

Consider the case $\Pi\{y_1(t)\} > -1$. Then $\Pi\{y_j(t)\} > -1$ for any $j = 1, 2, ..., n$. Due to Lemma 10.1 (1) we may rewrite $V(y_1(t), ..., y_n(t))$ in the form

$$\begin{vmatrix} 1 & 1 & \cdots & 1 \\ y_1(t) & y_2(t) & \cdots & y_n(t) \\ \cdots & \cdots & \cdots & \cdots \\ y_1^{n-1}(t)(1+\alpha_{n-11}(t)) & y_2^{n-1}(t)(1+\alpha_{n2}(t)) & \cdots & y_n^{n-1}(t)(1+\alpha_{nn}(t)) \end{vmatrix}$$

Here $\alpha_{ij}(t)$ $(i = 2, ..., n; \ j = 1, 2, ..., n)$ are infinitesimal functions for $t \to +\infty$. The derived determinant is equal to a sum of $n!$ members. Each of which is equal to the product of the elements of this determinant located at the different rows and columns. Let us open the brackets and collect all the terms which do not include the functions $\alpha_{ij}(t)$. We obtain the Van der Monde determinant. The terms which include the infinitesimal functions are obviously $o(\varphi(t))$ for $t \to +\infty$ which proves the required property.

Let $\Pi\{y_1(t)\} \le -1$. Choose a function $y(t)$ such that

$$\Pi\{y(t)\} > -1 \text{ and } \Pi\{y(t)\} < \Pi\{y_j(t)\} \text{ for } j = 2, ..., n.$$

Then the set of functions $\{y_1(t)+y(t), ..., y_n(t)+y(t)\}$ possesses the property of asymptotic separability. Thus, we have (according to the previous case)

$$\begin{aligned} V(y_1(t), ..., y_n(t)) &= V(y_1(t) + y(t), ..., y_n(t) + y(t)) \\ &\sim \omega(y_1(t) + y(t), ..., y_n(t) + y(t)) \\ &= \omega(y_1(t), ..., y_n(t)) \text{ for } t \to +\infty. \quad \square \end{aligned}$$

Corollary 10.6. *Let all the conditions of Proposition 10.5 be fulfilled. Then* $V(y_1(t), ..., y_n(t)) \in \{Q\}$ *and*

$$\Pi\{V(y_1(t), ..., y_n(t))\} > \frac{n(1-n)}{2}. \tag{10.10}$$

As a simple consequence $V(y_1(t), ..., y_n(t)) \in A_t$.

PROOF. The inclusion $V(y_1(t), ..., y_n(t)) \in \{Q\}$ is obvious. Let (for definiteness) $\Pi\{y_p(t)\} \le \Pi\{y_q(t)\}$ for any $p < q$. Then $\Pi\{y_j(t)\} > -1$ for $j = 2, ..., n$ and

$$\begin{aligned} \Pi\{V(y_1(t), ..., y_n(t))\} &= \Pi\{\omega(y_1(t), ..., y_n(t))\} \\ &= \Pi\{y_2(t)y_3^2(t)...y_n^{n-1}(t)\} \\ &\le \Pi\{y_2(t)\} + 2\Pi\{y_3(t)\} + ... + (n-1)\Pi\{y_n(t)\} \\ &> \frac{n(1-n)}{2}. \end{aligned}$$

Q is an algebraic field, hence $V(y_1(t), ..., y_n(t)) \in \{Q\}$. That is, it belongs to $A_t \cup O_t$. Since $\Pi\{V(y_1(t), ..., y_n(t))\} > n(1-n)/2 > -\infty$ we conclude that $V(y_1(t), ..., y_n(t)) \in A_t$. $\square$

Consider a set of functions of the form

$$\{y^n(t), a_1(t)y^{n-1}(t), ..., a_n(t)\} \subset \Pi. \qquad (10.11)$$

Let us denote by $\psi(y(t))$ its function of the greatest growth. It means that all the limits

$$\lim_{t \to +\infty} \frac{a_j(t)y^{n-j}(t)}{\psi(y(t))}$$

are finite $(j = 0, 1, ..., n; a_0(t) = 1)$.

Proposition 10.7. *Let $\{y(t), a_1(t), ..., a_n(t)\} \subset \{Q\}$, $\Pi\{y(t)\} > -\infty$. Then for $r = 0, 1, ...$*

$$\frac{\partial^r H(t, y(t))}{\partial y^r} = O\left(\frac{\psi(y(t))}{y^r(t)}\right) \quad for \ t \to +\infty \qquad (10.12)$$

hence

$$\Pi\left\{\frac{\partial^r H(t, y(t))}{\partial y^r}\right\} \le \Pi\{\psi(y(t))\} - r\Pi\{y(t)\}. \qquad (10.13)$$

PROOF. Let $\psi\{y(t)\} = a_s(t)y^{n-s}(t)$. That is,

$$a_j(t)y^{n-j}(t) = O(a_s(t)y^{n-s}(t)) \text{ as } t \to +\infty \text{ for all } j = 1, 2, ..., n.$$

If we differentiate $H(t, y)$ with respect to y, then each term in $H(t, y)$ either vanishes or is divided by y. Then the corresponding function of the greatest grows for $\partial H(t, y(t))/\partial y$ either coincides with $(n-s)a_s(t)y^{n-s-1}(t)$ or is $O(a_s(t)y^{n-s-1}(t))$ for $t \to +\infty$. In both the cases

$$\frac{\partial H(t, y(t))}{\partial y^r} = O\left(\frac{\psi(y(t))}{y(t)}\right).$$

(10.12) is easily proved by induction on r. Relation (10.13) is a simple corollary of (10.12). □

Proposition 10.8. *Let $\{y(t), a_1(t), ..., a_n(t)\} \subset \{Q\}$. Let the function $y(t)$ be not equivalent to any root of equation (10.2) for $t \to +\infty$, and let $\Pi\{y(t)\} > -\infty$. Then there exists a number $c \neq 0$ such that*

$$H(t, y(t)) \sim c\psi(y(t)) \text{ for } t \to +\infty.$$

If $y(t)$ is equivalent to a root of the equation then

$$H(t, y(t)) = o(\psi(y(t))) \text{ for } t \to +\infty.$$

PROOF. The substitution $y = y(t)u$ in (10.2) leads to the relation

$$H(t, y(t)u) \equiv \psi(y(t))[(q_0 + \alpha_0(t))u^n + (q_1 + \alpha_1(t))u^{n-1} + ... + q_n + \alpha_n(t)] = 0,$$

where

$$q_i = \lim_{t \to +\infty} \frac{a_i(t)y^{n-i}(t)}{\psi(y(t))},$$

$\alpha_i(t) \in C_t$ $(i = 0, 1, ..., n)$. Here $a_0(t) = 1$.

Let the function $y(t)$ be equivalent to a root of equation (10.2). Then the polynomial $P(t, u) = H(t, y(t)u)/\psi(y(t))$ has a root $u(t) = 1 + \alpha(t)$, where $\alpha(t) = o(1)$. Hence the limiting polynomial

$$R(v) = q_0 v^n + q_1 v^{n-1} + ... + q_n$$

has a root $v = 1$. Therefore $q_0 + q_1 + ... + q_n = 0$ that is

$$H(t, y(t)) = o(\psi(y(t))) \text{ for } t \to +\infty.$$

If the function $y(t)$ is not equivalent to any of the roots of the polynomial $H(t, y)$, then there are no roots of type $(1 + \alpha(t))$ of the polynomial $P(t, u)$ $(\alpha(t) = o(1))$. Therefore the number 1 is not a root of the polynomial $R(v)$. Hence $c = q_0 + q_1 + ... + q_n \neq 0$. That is

$$H(t, y(t)) \sim c\psi(y(t)) \text{ for } t \to +\infty. \qquad \square$$

Let $\lambda(t)$ be a root of equation (10.2). By $H^*(t, y)$ we denote the polynomial which is obtained from the identity

$$H(t, y) = (y - \lambda(t))H^*(t, y). \qquad (10.14)$$

And put (see (8.41))

$$\Phi^*(t, y) = \Phi(t, y) - H(t, y). \qquad (10.15)$$

The decomposition of expression $\Phi^*(t, y)$ consists of all members of $\Phi(t, y)$ (see (8.52)) belonging to its decomposition containing at least one derivative of y.

Proposition 10.9. *Let* $\{a_1(t), ..., a_n(t)\} \subset \{Q\}$. *Let* $\lambda(t)$ *be an asymptotically simple root of equation (10.2) for* $t \to +\infty$ *and* $\Pi\{\lambda(t)\} > -\infty$. *Let* $y(t)$ *be an arbitrary function such that* $(y(t) \in \{Q\})$ *and*

$$y(t) \sim \lambda(t) \text{ for } t \to +\infty.$$

Then

(1) *there exists a number $p \neq 0$ such that*

$$H^*(t, y(t)) \sim p \frac{\psi(\lambda(t))}{\lambda(t)} \ \textit{for } t \to +\infty, \tag{10.16}$$

where $\psi(y(t))$ is a function of the greatest growth of (10.11) consequently

$$\Pi\{H^*(t, \lambda(t))\} = \Pi\{\psi(\lambda(t))\} - \Pi\{\lambda(t)\}. \tag{10.17}$$

Moreover

$$\Pi\left\{\frac{\partial H^*(t, y(t))}{\partial y}\right\} \leq \Pi\{H^*(t, y(t)\} - \Pi\{y(t)\} \leq \Pi\{\psi(\lambda(t))\} - 2\Pi\{\lambda(t)\};$$
$$\tag{10.18}$$

(2) $$H^*(t, \lambda(t)) = \frac{\partial H(t, \lambda(t))}{\partial y}; \tag{10.19}$$

(3) *there exists a finite limit*

$$\lim_{t \to +\infty} \lambda(t) \frac{\partial^2 H(t, \lambda(t))}{\partial y^2} \bigg/ \frac{\partial H(t, \lambda(t))}{\partial y} = q; \tag{10.20}$$

(4) *if $\Pi\{\lambda(t)\} > -1$ then*

$$\Phi^*(t, \lambda(t)) = \frac{1}{2}\lambda'(t)\frac{\partial^2 H(t, \lambda(t))}{\partial y^2} + \delta(t), \tag{10.21}$$

where

$$\Pi\{\delta(t)\} \leq \Pi\{\psi(\lambda(t))\} - 2\Pi\{\lambda(t)\} - 2,$$

hence

$$\Pi\{\Phi^*(\lambda(t))\} \leq \Pi\{\psi(\lambda(t))\} - \Pi\{\lambda(t)\} - 1. \tag{10.22}$$

Remark . *By Theorem 5.42 and because of the asymptotic simplicity of the root $\lambda(t)$, there exists a field P of type N such that $Q \subset P$ and $\lambda(t) \in \{P\}$.*

Proof *of Proposition 10.9.* Let us prove property (2). We have

$$\frac{\partial H(t, y)}{\partial y} = H^*(t, y) + (y - \lambda(t))\frac{\partial H^*(t, y)}{\partial y}.$$

Substituting $y = \lambda(t)$ in the last relation we obtain (10.19).

Let us prove property (1). Since

$$y(t) \sim \lambda(t) \text{ for } t \to +\infty$$

there are at least two functions belonging to the set (10.11) which we can take as the function $\psi(y(t))$. Therefore without loss we can suppose that $\psi(y(t))$ is one of the functions

$$y^n(t), a_1(t)y^{n-1}(t), ..., a_{n-1}(t)y(t).$$

Let $\psi^*(t)$ be a function of the greatest growth of the set

$$\{ny^{n-1}(t), (n-1)a_1(t)y^{n-2}(t), ..., a_{n-1}(t)\}$$

(which corresponds to the polynomial $\partial H(t,y)/\partial y$). Hence there is a number $a \neq 0$ such that

$$\psi^*(y(t)) \sim a\frac{\psi(y(t))}{y(t)} \text{ for } t \to +\infty.$$

Let

$$\Lambda = \{\lambda(t), \lambda_2(t), ..., \lambda_n(t)\} \tag{10.23}$$

be a complete set of roots of equation (10.2). We have

$$H^*(t,y) = (y - \lambda_2(t))...(y - \lambda_n(t)).$$

Hence

$$\frac{\partial H(t,y(t))}{\partial y} = H^*(t,y(t))\left[1 + \sum_{i=2}^{n} \frac{y(t) - \lambda(t)}{y(t) - \lambda_i(t)}\right].$$

Since $\lambda(t)$ is an asymptotically simple root

$$\frac{y(t) - \lambda(t)}{y(t) - \lambda_i(t)} = o(1) \text{ as } t \to +\infty$$

for any $(i = 2, ..., n)$. Consequently the equation $\partial H(t,y)/\partial y = 0$ has no roots which are equivalent to $\lambda(t)$. Therefore there is a number $b \neq 0$ such that

$$\frac{\partial H(t,y)}{\partial y} \sim b\psi^*(y(t)) \sim ab\frac{\psi(y(t))}{y(t)}.$$

We obtain by putting $p = ab$

$$H^*(t,y) \sim \frac{\partial H(t,y(t))}{\partial y} \sim p\frac{\psi(y(t))}{y(t)}$$

for $t \to +\infty$. Thus, property (1) is proved.

Let us prove property (3). Since the expression under the sign of the limit in (10.17) asymptotically close to a field of type N, q exists (finite or infinite). But

$$\frac{\partial^2 H(t, \lambda(t))}{\partial y^2} = O\left(\frac{\psi(\lambda(t))}{\lambda^2(t)}\right).$$

Hence $q \neq \infty$. Thus, property (3) is proved.

Property (4) follows from decomposition (8.52). Let us substitute $\lambda(t)$ in (8.52) instead of y and estimate the general term under the sign of the sum where $r > 0$. Taking into account the definition of the function $\psi(\lambda(t))$, we conclude that the considered estimate is no more than

$$\Pi\{\psi(\lambda(t))\} - (k_1 + 2k_2 + \ldots + sk_s)(\Pi\{y(t)\} + 1). \qquad (10.24)$$

Taking into account that $\Pi\{y(t)\} > -1$ and $k_1 + 2k_2 + \ldots + sk_s > 0$, maximum of (10.24) is reached when $k_1 = 1, k_2 = \ldots = k_s = 0$ $(r = 2)$. This referred to the member

$$\frac{1}{2}\lambda'(t)\frac{\partial^2 H(t, \lambda(t))}{\partial y^2}.$$

All the rest members have their estimates no more than

$$\Pi\{\psi(\lambda(t))\} - 2(\Pi\{y(t)\} + 1)$$

which leads to the all required properties of the point (4). $\square$

Consider an operator of the form

$$R(y) = -\frac{\Phi^*(t, y)}{H^*(t, y)} \qquad (10.25)$$

which operates from $\{Q\}$ to $\{Q\}$.

Lemma 10.10. *Let $\{a_1(t), \ldots, a_n(t)\} \subset \{Q\}$. Let $\lambda(t)$ be an asymptotically simple root of equation (10.2) and $\Pi\{\lambda(t)\} > -1$. Then $R(y)$ is of power type with a majorant $r \leq -1$ at the point $y(t) \sim \lambda(t)$ for $t \to +\infty$. Moreover*

$$R(y(t)) = O\left(\frac{\lambda'(t)}{\lambda(t)}\right) = O\left(\frac{1}{t}\right) \quad for \ t \to +\infty.$$

PROOF. Without loss of generality we may suppose that $\lambda(t) \in \{Q\}$ (see Remark to Proposition 10.9). Due to Proposition 4.35 $\Phi^*(t, y)$ (in y) is of power type at the point $\lambda(t)$. We have (see (10.21))

$$\Phi^*(t, y(t)) = O\left(\lambda'(t)\frac{\partial^2 H(t, Y(t))}{\partial y^2}\right) \quad for \ t \to +\infty.$$

Hence its majorant $f^* \leq \Pi\{\psi(\lambda(t))\} - \Pi\{\lambda(t)\} - 1$.

$H^*(t, y)$ (in y) is also of power type. Since $\lambda(t)$ is an asymptotically simple root, $H^*(t, y)$ is continuous at this point and (see (10.16)) its majorant is equal to $h^* = \Pi\{H^*(t, y(t))\} = \Pi\{\psi(\lambda(t))\} - \Pi\{\lambda(t)\}$.

By Proposition 5.62 and Example 5.63 $R(y) = -\Phi^*(t, y)/H^*(t, y)$ is of power type with majorant $r = f^* - h^* \leq -1$. Moreover

$$R(y(t)) = O\left(\lambda'(t)\frac{\partial^2 H(t, \lambda(t))}{\partial^2 y} \bigg/ \frac{\partial H(t, \lambda(t))}{\partial y}\right) = O\left(\frac{\lambda'(t)}{\lambda(t)}\right) = O\left(\frac{1}{t}\right)$$

for $t \to +\infty$.   □

Consider an operator of the form

$$\widetilde{R}(y) = -\sum_{j=2}^{n} \frac{a_{n-j}(t)}{a_{n-1}(t)} \Phi(y). \tag{10.26}$$

Lemma 10.11. *Let* $a_1(t), ..., a_n(t) \in \{Q\}$. *Let* $\lambda(t)$ *be an asymptotically simple root of equation* (10.2), *let* $-\infty < \Pi\{\lambda(t)\} \leq -1$ *and let any other root* $\lambda_i(t)$ ($i = 2, ..., n$) *of a complete set of roots of the equation have an estimate* $\Pi\{\lambda_i(t)\} > -1$. *Then* $\widetilde{R}(y)$ *is of power type at the point* $\lambda(t)$ *with majorant* $s = \Pi\{\lambda_i(t)\} - \sigma$ *in the region*

$$U_{\lambda\sigma} = \{y(t) : y(t) \in \Pi, \ \Pi\{y(t) - \lambda(t)\} \leq -\sigma\},$$

where $\sigma = 1 + \min_{i=2,...,n} \Pi\{\lambda_i(t)\} > 0$.

PROOF. We may suppose (see Proof of Lemma 10.10) that $\lambda(t) \in \{Q\}$. We have

$$a_1(t) \quad = \quad \lambda_1(t)\lambda_2(t) + \lambda_2(t)\lambda_3(t) + ... + \lambda_{n-1}(t)\lambda_n(t),$$

$$a_2(t) \quad = \quad -[\lambda_1(t) + \lambda_2(t) + ... + \lambda_n(t)],$$

$$...$$

$$a_{n-1}(t) \quad = \quad (-1)^{n-1}[\lambda_1(t)\lambda_2(t)...\lambda_{n-1}(t) + \lambda_2(t)\lambda_2(t)...\lambda_{n-1}(t)].$$

Since $\lambda_1(t) = o(\lambda_i(t))$ for $t \to +\infty$ ($i = 2, ..., n$) we have

$$a_{n-1}(t) \sim \lambda_2(t)...\lambda_n(t).$$

Hence $a_{n-1}(t) \neq 0$ for $t \gg 1$ and

$$\Pi\{a_{n-1}(t)\} = \Pi\{\lambda_2(t)\} + ... + \Pi\{\lambda_n(t)\} \geq (n-1)\min_{i=2,...,n} \Pi\{\lambda_i(t)\}.$$

It is easy to see that

$$\Pi\{a_{n-j}(t)/a_{n-1}(t)\} \le -(j-1)\min_{i=2,..,n}\Pi\{\lambda_i(t)\}$$

Due to Lemmma 10.1 in $U_{\lambda\sigma}$ taking into account that $\Pi\{y(t)\} < -1$ we have $\Pi\{\Phi_j(y(t))\} = \Pi\{\lambda(t)\} - j + 1$. Consequently

$$\Pi\left\{\frac{a_{n-j}(t)}{a_{n-1}(t)}\Phi_j(y(t))\right\} \le \max_{j=2,...,n}[-(j-1)\min_{i=2,..,n}\Pi\{\lambda_i(t)\}+\Pi\{\lambda(t)\}-j+1]$$

$$= \Pi\{\lambda(t)\} - [1 + \min_{i=2,..,n}\Pi\{\lambda_i(t)\}]$$

$$= \Pi\{\lambda(t)\} - \sigma.$$

The obtained inequality proves this Lemma. $\square$

Definition 10.12. We say that a function $g(t)$ is a *formal variable index* and $G(t) = \exp\left[\int g(t)dt\right]$ is a *formal solution* to equation (10.1) if $\Phi(t, g(t)) \asymp 0$. We say that the formal solution $G(t)$ possesses a *unique asymptotics* for $t \to +\infty$ if $g(t) \in \Pi$ and for any other formal solution

$$\widetilde{G}(t) = \exp\left[\int \tilde{g}(t)dt\right]$$

to the equation such that $\tilde{g}(t) \sim g(t)$ for $t \to +\infty$, the following estimate holds: $\tilde{g}(t) \asymp g(t)$.

Definition 10.13. We say that the set

$$\left\{G_i(t) = \exp\left[\int g_i(t)dt\right]\right\} \quad (i = 1, 2, ..., n)$$

is a *formal fundamental system* (*FFS*) of solutions to equation (10.1) if any function $G_i(t)$ is a formal solution to the equation and

$$\Pi\{V(g_1(t), ..., g_n(t))\} > -\infty.$$

Theorem 10.14. *Let $a_1(t), ..., a_n(t) \in \{Q\}$. Let $\lambda(t)$ be an asymptotically simple root of equation (10.2). Let $\Pi\{\lambda(t)\} > -1$. Then equation (10.1) has a formal solution $G(t) = \exp\left[\int g(t)dt\right]$ where $g(t)$ belongs to $\{Q\}$, $g(t) \sim \lambda(t)$ for $t \to +\infty$ and the function $g(t)$ is a formal solution of the equation $\Phi(t, y) = 0$ in the class of the power order growth functions with unique asymptotics. Besides $g(t)$ is an asymptotic limit of a sequence $\{s_m(t)\}$ where $s_0(t) = \lambda(t)$ and for $m = 1, 2, ...$*

$$s_m(t) = \lambda(t) - \frac{\Phi^*(s_{m-1}(t))}{H^*(t, s_{jm-1}(t))}. \tag{10.27}$$

Hence

$$g(t) = \lambda(t) + \eta(t) + \alpha(t), \tag{10.28}$$

where

$$\eta(t) = -\frac{1}{2}\lambda'(t)\frac{\partial^2 H(t, \lambda(t))}{\partial y^2} \Big/ \frac{\partial H(t, \lambda(t))}{\partial y} \tag{10.29}$$

and

$$\Pi\{\alpha(t)\} \leq -2 - \Pi\{\lambda(t)\} < -1.$$

Let the function $\lambda(t)$ belongs to $\{P\}$, where P is a field of type N and $Q \subset P$ (see remark to Proposition 10.9). Then $g(t) \in \{P\}$.

PROOF. In the considered case, the equation $\Phi(t, y) = 0$ can be rewritten in the equivalent form (see (10.25))

$$y = \lambda(t) + R(y). \tag{10.30}$$

On the basis of Lemma 10.10 and Lemma 4.24 the equation possesses a formal solution $g(t) \sim \lambda(t)$ for $t \to +\infty$ with unique asymptotics. Moreover $g(t)$ is a asymptotic limit of the sequence

$$s_m(t) = \lambda(t) + R(s_{m-1}(t)), \quad s_0(t) = \lambda(t), \quad (m = 1, 2, ...)$$

and $\Pi\{s_m(t) - s_{m-1}(t)\} \leq -\sigma \equiv -1 - \Pi\{\lambda(t)\}$. Clearly, $s_m(t) \in \{Q\}$. We have $g(t) = \lambda(t) + R(\lambda(t)) + \alpha(t)$ where $\Pi\{a(t)\} \leq -\sigma$. Taking into account (10.19) and (10.21) we obtain

$$R(\lambda(t)) = -\frac{\Phi^*(\lambda(t))}{H^*(\lambda(t))} = \eta(t) + \beta(t),$$

$\Pi\{\beta(t)\} \leq -2 - \sigma$. This implies formula (10.28).

Since $g(t)$ is a formal solution we have $g(t) - \lambda(t) - R(g(t) = \Theta(t) \asymp 0$. Hence $\Phi(t, g(t)) \equiv (g(t) - \lambda(t))H^*(g(t)) + \Phi^*(g(t)) = \Theta(t)H^*(g(t))$.

Since $\Pi\{H^*(g(t))\} = \Pi\{\psi(\lambda(t)\} - \Pi\{\lambda(t)\} < +\infty$ clearly $\Theta(t)H^*(g(t)) \asymp 0$. Consequently, $\Phi(t, g(t)) \asymp 0$. That is, $g(t)$ is a variable index of equation (10.1).

Prove the inclusion $g(t) \in \{P\}$. Indeed, $s_m(t) \in \{P\}$ and $\Pi\{g(t) - s_m(t)\} \to -\infty$ for $m \to \infty$. Consequently $g(t) \in \{P\}$. $\square$

Remark 10.15. From (10.27) equation (10.1) has a formal solution which can be written in the form

$$G(t) = (1 + \beta(t)) \exp\left[\int (\lambda(t) + \eta(t))dt\right], \tag{10.31}$$

where

$$\Pi\{\beta(t)\} \leq \Pi\{\alpha(t)\} + 1 < 0.$$

In particular $\beta(t) \to 0$ for $t \to +\infty$. Thus, we obtain the formal solution asymptotics in an explicit form. By Proposition 10.9 (3) we have $\eta(t) = (k + \delta(t))/t$, where

$$k = -\frac{1}{2}qP\{\lambda_j(t)\} \text{ and } \delta(t) \in C_t.$$

Here the number q is given in (10.20). Consequently

$$G(t) = f(t) \exp\left[\int \lambda(t)dt\right], \tag{10.32}$$

where $P\{f(t)\} = k$.

If in addition all the coefficients $a_i(t)$ $(i = 1, 2, ..., n)$ have their asymptotic expansions in the form of generalized power series, then all the infinitesimal functions in the obtaining formulae have negative analytic estimates. That is $\Pi\{\delta(t)\} < 0$. Therefore we can write (instead of (10.32))

$$G(t) = (1 + \theta(t))t^k \exp\left[\int \lambda(t)dt\right], \tag{10.33}$$

where $\Pi\{\theta(t)\} < 0$ (in particular $\theta(t) \to 0$ for $t \to +\infty$).

In the same way on the basis of Definition 10.12 we prove the following:

Theorem 10.16. *Let $\{a_1(t), ..., a_n(t)\} \subset \{Q\}$. Let $\lambda(t)$ be a unique root of equation (10.2) with the estimate $\Pi\{\lambda(t)\} \leq -1$ (each other root of the equation has an estimate more than -1). Then equation (10.1) has a formal solution*

$$G(t) = \exp\left[\int g(t)dt\right]$$

such that if

(1) $\lambda(t) \asymp 0$, *then $g(t)$ is an arbitrary function with the estimate $g(t) \asymp 0$;*

(2) $\Pi\{\lambda(t)\} > -\infty$, *then $g(t) \in \{Q\}$ and*

$$g(t) \sim \lambda(t) \text{ for } t \to +\infty.$$

The function $g(t)$ is a formal solution of the equation $\Phi(t, y) = 0$ in the class of the power order growth functions with unique asymptotics. Besides $g(t)$ is an asymptotic limit of a sequence $\{s_m(t)\}$, where $s_0(t) = -a_n(t)/a_{n-1}(t)$ and for $m = 1, 2, ...$

$$s_m(t) = -\frac{a_n(t)}{a_{n-1}(t)} + \tilde{R}(s_{m-1}(t)), \quad s_0(t) = -a_n(t)/a_{n-1}(t) \quad (m = 1, 2, ...),$$

$$\tag{10.34}$$

where $\tilde{R}(y)$ is defined in (10.26). Hence

$$g(t) = -\frac{a_{n-i}(t)}{a_{n-1}(t)} + \alpha_j(t), \qquad (10.35)$$

where

$$\Pi\{\alpha(t)\} \leq \Pi\{\lambda(t)\} - \sigma, \quad \sigma = 1 + \min_{j=2,\dots,n} \Pi\{\lambda_j(t)\} > 0. \qquad (10.36)$$

Here $\lambda_j(t)$ are all the roots of equation (10.2) excluding the root $\lambda(t)$ ($j = 2, \dots, n$). Besides due to Theorem 5.42 there exists a field P of type N such that $Q \subset P$ and because of the simplicity of the root $\lambda(t)$, we have $\lambda(t) \in \{P\}$. Then $g(t) \in \{P\}$.

Remark 10.17. From (10.36)

$$G_j(t) = (1 + \beta_j(t)) \exp\left[-\int \frac{a_n(t)}{a_{n-1}(t)} dt\right], \qquad (10.37)$$

where

$$\Pi\{\beta_j(t)\} \leq \Pi\{\lambda_j(t)\} - \sigma + 1.$$

Hence $\beta_j(t) \to 0$ for $t \to +\infty$.

Theorem 10.18. *Let $\{a_1(t), \dots, a_n(t)\} \subset \{Q\}$. Let the complete set of roots (10.23) of characteristic equation (10.2) possess the property of asymptotic separability for $t \to +\infty$. Then there exists an FFS of solutions to equation (10.1) of the form*

$$\left\{x_j(t) = \exp\left[\int g_j(t) dt\right]\right\} \quad (j = 1, 2, \dots, n), \qquad (10.38)$$

where $g_j(t) \sim \lambda_j(t)$ for $t \to +\infty$. All the other properties of the functions $g_j(t)$ are obtained in Theorems 10.14 and 10.16, respectively.

PROOF. Because of Theorems 10.14 and 10.16 we have only to prove that the set $\{g_1(t), \dots, \gamma_n(t)\}$ possesses the property of asymptotic separability. The last is a simple consequence of the following properties: $\gamma_j(t) \sim \lambda_j(t)$ for $t \to +\infty$ if $\Pi\{\lambda_j(t)\} > -\infty$, and $\Pi\{\gamma_j(t)\} = -\infty$ if $\Pi\{\lambda_j(t)\} = -\infty$. Hence $g_i(t) \not\sim g_j(t)$ ($t \to +\infty$) and $\Pi\{\gamma_i(t) - \gamma_j(t)\} > -1$ for any $i \neq j$. Consequently the required property is a simple consequence of Proposition 10.9.   □

2. ASYMPTOTIC SOLUTIONS TO A SINGLE LINEAR EQUATION OF NTH ORDER

Beforehand we consider several auxiliary propositions. Consider an expression of the form

$$\eta(t) = \left| \int_{t_0}^{t} \exp\left[\int_{\tau}^{t} \delta(s)ds \right] \alpha(\tau)d\tau \right|, \qquad (10.39)$$

where $\delta(t)$ and $\alpha(t)$ are real continuous functions, $\alpha(t) \geq 0$ for $t \gg 1$ and $h\{\alpha(t)\} = -\infty$.

Lemma 10.19. *Let* $\lim_{t \to +\infty} \delta(t)t = p$, *where* $-\infty < p \leq +\infty$. *Let* $t_0 = +\infty$. *Then* $\eta(t)$ *is determined for* $t \gg 1$ *and* $h\{\eta(t)\} = -\infty$.

PROOF. Let $p < +\infty$. The inequality

$$\int_{t}^{\tau} \left[\delta(s) - \frac{p - \varepsilon}{s} \right] ds \geq 0$$

holds for $t \leq \tau < +\infty, t \gg 1$. Here ε is a positive number. Hence

$$
\begin{aligned}
\eta(t) &= t^{p-\varepsilon} \int_{t}^{+\infty} \exp\left[\int_{\tau}^{t} \left(\delta(s) - \frac{p-\varepsilon}{s} \right) ds \right] \tau^{-p+\varepsilon} \alpha(\tau)d\tau \\
&\leq t^{p-\varepsilon} \int_{t}^{+\infty} \tau^{-p+\varepsilon} \alpha(\tau)d\tau.
\end{aligned}
$$

Since $h\{t^{-p}\alpha(t)\} = -\infty$, the last integral is convergent for $t \gg 1$ and its estimate is equal to $-\infty$. That is, $h\{\eta(t)\} = -\infty$.

Let $p = +\infty$. We have

$$\eta(t) = t^{p} \int_{t}^{+\infty} \exp\left[\int_{\tau}^{t} \left(\delta(s) - \frac{p}{s} \right) ds \right] \tau^{-p} \alpha(\tau)d\tau,$$

where p is a sufficiently large positive number. The inequality

$$\int_{\tau}^{t} \left(\delta(s) - \frac{p}{s} \right) ds \leq 0$$

holds for $t \leq \tau < +\infty, t \gg 1$. Hence

$$\eta(t) \leq t^{p} \int_{t}^{+\infty} \tau^{-p} \alpha(\tau)d\tau.$$

Since $h\{\tau^{-p}\alpha(\tau)\} = -\infty$, the last integral is convergent and its estimate is equal to $-\infty$. That is, $h\{\eta(t)\} = -\infty$. $\qquad \square$

Lemma 10.20. *Let* $\lim_{t \to +\infty} \delta(t)t = -\infty$. *Let* $t_0 = T$ *where* T *is a sufficiently large positive number. Then in* (10.39) $\eta(t)$ *is determined for* $t \gg 1$ *and* $h\{\eta(t)\} = -\infty$.

PROOF. We have

$$0 \leq \eta(t) = t^{-p} \int_T^t \exp\left[\int_\tau^t \left(\delta(s) + \frac{p}{s}\right) ds\right] \tau^p \alpha(\tau) d\tau,$$

where p is a sufficiently large positive number. Let us choose a positive number $T_p > T$ such that

$$\int_\tau^t \left(\delta(s) + \frac{p}{s}\right) ds < 0$$

for $T_p \leq \tau \leq t < +\infty$. Thus, $\eta(t) = \eta_1(t) + \eta_2(t)$ where

$$\eta_1(t) = \int_T^{T_p} \exp\left[\int_\tau^t \delta(s) ds\right] \alpha(\tau) d\tau$$

and

$$\eta_2(t) = \int_{T_p}^t \exp\left[\int_\tau^t \delta(s) ds\right] \alpha(\tau) d\tau.$$

It is obvious that $h\{\eta_1(t)\} = -\infty$. Integral

$$\int_{T_p}^{+\infty} \tau^p \alpha(\tau) d\tau$$

is convergent. Therefore

$$\eta_2(t) = O(t^{-p}) \text{ for } t \to +\infty.$$

Taking into account the arbitrariness of p we conclude that $h\{\eta_2(t)\} = -\infty$ which implies to the required estimate. $\square$

We preserve the hypothesis taken in Chapter 9 and (*in addition*) let G be a field of type M (*which will be not stipulated apart*) (*in the same time* G *is a field of type* N).
  Consider an equation of the form

$$V' = (D(t) + B(t))V, \tag{10.40}$$

where

$$D(t) = \text{diag}(0, \delta_2(t), ..., \delta_n(t)),$$

$B(t) = (b_{ij}(t))_n$ *is a square matrix of nth order.*

Lemma 10.21. *Let* $D(t) \in \{G\}$ *and* $B(t) \asymp 0$ *(this means that* $\delta_i(t) \in \{G\}$ *and* $b_{ij}(t) \asymp 0$ *for* $t \to +\infty$ * $i, j = 1, 2, ..., n$). Then there exists a solution*

$$V(t) = (1 + \alpha_1(t), \alpha_2(t), ..., \alpha_n(t))^T$$

to equation (10.40) where $\alpha_i(t) \asymp 0$ * $(i = 1, 2, ..., n)$, i.e. $V(t) - e_1^* \asymp 0$. Here $e_1^* = (1, 0, ..., 0)^T$.*

PROOF. Since $\delta_i(t) \in \{G\}$, $\Re\delta_i(t)t \in \{G\}$ and thus there exist (finite or infinite) limits

$$\lim_{t \to +\infty} \Re\delta_i(t)t = p_i \ (i = 2, ..., n).$$

Let us prove that there exists a solution $V(t)$ such that $\mathrm{h}\{V(t) - e_1^*\} = -\infty$. Consider an expression of the form

$$y_i(t) = \int_{t_i}^t \exp\left[\int_\tau^t \Re\delta_i(s)ds\right] \|B(\tau)\| d\tau, \tag{10.41}$$

where $t_i = +\infty$ if $p_i > -\infty$ and $t_i = T$ if $p_i = -\infty$. Here T is a sufficiently large positive number. On the bases of Lemma 10.11 it is enough to prove that $\mathrm{h}\{y_i(t)\} = -\infty$. Wile the last follows from Lemmas 10.19 and 10.20.

Let us substitute the obtained solution $V(t)$ to equation (10.40). We have $\mathrm{h}\{V'(t)\} = -\infty$. Differentiating the identity

$$V'(t) = (D(t) + B(t))V(t)$$

$(m-1)$ times we obtain

$$\mathrm{h}\{V^{(m)}(t)\} = -\infty \ (m = 1, 2, ...).$$

This means that $V(t) \asymp 0$.   □

Theorem 10.22. *Let*

$$\left\{G_j(t) = \exp\left[\int g_i(t)dt\right]\right\} \ (j = 1, 2, ..., n)$$

be a formal fundamental system of solutions of the equation

$$F(t, x) \equiv x^{(n)} + a_1(t)x^{(n-1)} + ... + a_n(t)x = 0, \tag{10.1}$$

where all the functions $g_i(t)$ *and* $a_i(t)$ *belong to* $\{G\}$ * $(i = 1, 2, ..., n)$. Then there exists an FSS of the equation of the form*

$$\left\{x_j(t) = \exp\left[\int \gamma_j(t)dt\right]\right\}$$

such that

$$\gamma_j(t) \asymp g_j(t) \quad (j = 1, 2, ..., n).$$

PROOF. We prove this assertion by Lagrange's method of variation of arbitrary constants. Write the solution of equation (10.1) in the form

$$x = u_1 G_1(t) + u_2 G_2(t) + ... + u_n G_n(t), \tag{10.42}$$

where $u_1, u_2, ..., u_n$ are the variable parameters, Put

$$u_1' G_1^{(p)}(t) + u_2' G_2^{(p)}(t) + ... + u_n' G_n^{(p)}(t) = 0 \tag{10.43}$$

for $p = 0, 1, ..., n - 2$. Since

$$G_i^{(p)}(t) = \Phi_p(g_i(t)) G_i(t)$$

relation (10.43) may be rewritten in the form

$$u_1' \Phi_p(g_1(t)) G_1(t) + u_2' \Phi_p(g_2(t)) G_2(t) + ... + u_n' \Phi_p(g_n(t)) G_n(t) = 0. \tag{10.44}$$

Hence

$$x^{(s)} = u_1 \Phi_s(g_1(t)) G_1(t) + u_2' \Phi_s(g_2(t)) G_2(t) + ... + u_n' \Phi_s(g_n(t)) G_n(t)$$

$(s = 0, 1, ..., n - 1)$ and (taking into account that $g_i(t)$ is a formal variable index)

$$
\begin{aligned}
x^{(n)}(t) = {} & u_1 \Phi_n(g_1(t)) G_1(t) + u_2 \Phi_n(g_2(t)) G_2(t) + ... + u_n \Phi_n(g_n(t)) G_n(t) \\
& + u_1' \Phi_{n-1}(g_1(t)) G_1(t) + u_2' \Phi_{n-1}(g_2(t)) G_2(t) + ... \\
& + u_n' \Phi_{n-1}(g_n(t)) G_n(t).
\end{aligned}
$$

On substituting the obtained relations for $x, x', ..., x^{(n)}$ in (10.1) and (taking into account that $g_i(t)$ is a formal variable index) we have $\Phi(t, g_i(t)) \asymp 0$, and

$$u_1' \Phi_{n-1}(g_1(t)) G_1(t) + u_2' \Phi_{n-1}(g_2(t))(t) G_2(t) + ... + u_n' \Phi_{n-1}(g_2(t)) G_n(t)$$

$$= \beta_1(t) G_1(t) u_1 + \beta_2(t) G_2(t) u_2 + ... + \beta_n(t) G_n(t) u_n, \tag{10.45}$$

where $\beta_i(t) \asymp 0$ $(i = 1, 2, ..., n)$. Relations (10.44) and (10.45) may be considered as a system in unknowns $u_1', u_2', ..., u_n'$. Since $\{G_i(t)\}$ is an *FFS* of the equation, the system determinant (denoted by $V(g_1(t), ..., g_n(t))$) has the estimate $\Pi\{V(g_1(t), ..., g_n(t))\} > -\infty$ (see (10.5) and (10.10)). Consequently the system has a unique solution which can be written in the form

$$G_i(t) u_i' = b_{i1}(t) G_1(t) u_1 + b_{i2}(t) G_2(t) u_2 + ... + b_{in}(t) G_n(t) u_n, \tag{10.46}$$

where $b_{ik}(t) \asymp 0$ $(i, k = 1, 2, ..., n)$. Let us put

$$u_s = v_s \frac{G_j(t)}{G_s(t)} \text{ for } s = 1, 2, ..., n.$$

Clearly,

$$v_s' = \delta_{sj}(t)v_s + b_{s1}(t)v_1 + b_{s2}(t)v_2 + ... + b_{sn}(t)v_n,$$

where $\delta_{sj}(t) = \lambda_s(t) - \lambda_j(t)$. For $s = j$ we have

$$v_j' = b_{j1}(t)v_1 + b_{j2}(t)v_2 + ... + b_{jn}(t)v_n).$$

The system satisfies all the conditions of Lemma 10.21 hence it has a solution $v_j(t) = 1 + \delta_j(t)$ and for $s \neq j$ $v_s(t) = \eta_s(t)$, where $\theta_i(t) \asymp 0$ $(i = 1, 2, ..., n)$. Consequently

$$u_j(t) = 1 + \theta_1(t) \text{ and } u_s(t) = \theta_s(t)\frac{G_1(t)}{G_2(t)}.$$

On substituting these functions in (10.1) we obtain the solution $x_j(t) = (1 + \eta_j(t))G_j(t)$, where

$$\eta_j(t) = \theta_1(t) + \theta_2(t) + ... + \theta_n(t) \asymp 0.$$

Hence

$$x_j(t) = \exp\left[\int \gamma_j(t)dt\right],$$

where $\gamma_j(t) = g_j(t) + \eta_j(t)$. Here

$$\eta_j(t) = \frac{\eta_j'(t)}{1 + \eta_j(t)} \in O_t.$$

The solutions form a fundamental system because

$$W(x_1(t), ..., x_n(t)) = V(\gamma_1(t), ..., \gamma_n(t))G_1(t)...G_n(t) \neq 0$$

for $t \gg 1$. The last inequality follows from the relation

$$\Pi\{V(\gamma_1(t), ..., \gamma_n(t))\} = \Pi\{V(g_1(t), ..., g_n(t))\}$$
$$= \Pi\{V(\lambda_1(t), ..., \lambda_n(t))\}$$
$$> \frac{(1-n)n}{2}. \qquad \square$$

The following theorem is a simple consequence of Theorem 10.22.

Theorem 10.23. *Let* $\{a_1(t), ..., a_n(t)\} \in \{G\}$. *Let a complete set of roots* $\Lambda = \{\lambda_1(t), ..., \lambda_n(t)\}$ *of characteristic equation* (10.2) *possess the property of asymptotic separability. Let*

$$\left\{ G_j(t) = \exp\left[\int g_i(t)dt \right] \right\} \quad (j = 1, 2, ..., n)$$

be a formal fundamental system of solutions where each function $g_j(t)$ *possesses all the properties obtained in Theorems* 10.14, 10.16 *and* 10.18 *respectively. Then equation* (10.1) *has a fundamental system of solutions*

$$\left\{ x_j(t) = \exp\left[\int \gamma_j(t)dt \right] \right\}$$

such that

$$\Pi\{\gamma_j(t) - g_j(t)\} = -\infty \quad (j = 1, 2, ..., n).$$

Consider three simple examples illustrating Theorems 10.22 and 10.23.

Examples 10.24. 1° Given the equation

$$x''' - tx'' - 4t^2 x' + 4t^3 x = 0. \tag{10.47}$$

Its characteristic equation $H(t, y) \equiv y^3 - ty^2 - 4t^2 y + 4t^3 = 0$ has three simple roots $\lambda_1(t) = t$, $\lambda_2(t) = -2t$ and $\lambda_3(t) = 2t$. The coefficients of the equation and the characteristic roots belong to L which is a field of type M. Besides the roots form a complete system possessing the property of asymptotic separability. Indeed they do not equivalent in pairs and $\Pi\{\lambda_i(t)\} = \Pi\{\lambda_i(t) - \lambda_j(t)\} = \Pi\{t\} = 1 > -1$ for any $i, j = 1, 2, 3$, $i \neq j$. First, on the basis of Theorems 10.14 and 10.18 (see also Remark 10.15) equation (10.47) has a formal fundamental system of solutions (FFS) in the form $\{G_j(t) = (1 + \beta_j(t))t^{k_j} e^{\int \lambda_j(t)dt}\}$, where $\Pi\{\beta_j(t)\} \leq -1$. More exactly $\beta_j(t) = O(1/t)$ for $t \to +\infty$. $\partial H(t, y)/\partial y = 3y^2 - 2ty - 4t^2$ and $\partial^2 H(t, y)/\partial y^2 = 6y - 2t$. For $\lambda_1(t) = t$ we have

$$q_1 = \lim_{t \to +\infty} \lambda_1(t) \frac{\partial^2 H(\lambda_1(t), t)}{\partial y^2} \Big/ \frac{\partial H(\lambda_1(t), t)}{\partial y} = -4/3$$

and $k_1 = -\frac{1}{2}q_1 \Pi\{\lambda_1(t)\} = 2/3$. That is,

$$G_1(t) = \left(1 + O\left(\frac{1}{t}\right)\right) t^{2/3} e^{t^2/2}.$$

We have for $\lambda_2(t) = -2t$

$$q_2 = \lim_{t \to +\infty} \lambda_2(t) \frac{\partial^2 H(\lambda_2(t), t)}{\partial y^2} \Big/ \frac{\partial H(\lambda_2(t), t)}{\partial y} = 7/3$$

and $k_2 = -7/6$. In the same way we have $q_3 = 5$ and $k_3 = -5/2$ for $\lambda_3(t) = 2t$. This gives the following formal solutions

$$G_2(t) = \left(1 + O\left(\frac{1}{t}\right)\right) t^{-7/6} e^{-t^2} \text{ and } G_3(t) = \left(1 + O\left(\frac{1}{t}\right)\right) t^{-5/2} e^{t^2}.$$

On the basis of Theorem 10.23 the system has an FSS of the form, $\{x_j(t) = (1 + \theta_j(t))G_j(t)\}$, where $\theta_j(t) \asymp 0$. Consequently the solution possesses an FSS of the same form as the obtained FFS. That is, for $t \to +\infty$

$$\left\{ x_1(t) = \left(1 + O\left(\frac{1}{t}\right)\right) t^{2/3} e^{t^2/2}, \quad x_2(t) = \left(1 + O\left(\frac{1}{t}\right)\right) t^{-7/6} e^{-t^2}, \right.$$

$$\left. x_3 = \left(1 + O\left(\frac{1}{t}\right)\right) t^{-5/2} e^{t^2} \right\}.$$

2° Consider the equation

$$x''' + \sqrt{t}x'' + \frac{6x'}{5\sqrt{t}} + x = 0. \tag{10.48}$$

Its characteristic equation

$$H(t, y) \equiv y^3 + \sqrt{t}y^2 + \frac{6y}{5\sqrt{t}} + 1 = 0$$

as it is easy to see has a set of roots

$$\lambda_1(t) \sim -t^{1/2} \text{ and } \lambda_{2,3}(t) \sim \pm it^{-1/4} \text{ for } t \to +\infty$$

which possesses the property of asymptotic separability. The roots have to be obtained more accurate. For the first root let us rewrite the characteristic equation in the form

$$y = -t^{1/2} - \left(\frac{6}{5}t^{-1/2}y + 1\right)y^{-2}.$$

Substitute the approximate value of the root to the right side of the last relation we obtain

$$\lambda_1(t) \sim -t^{1/2} + \frac{1}{5}t^{-1}.$$

Let us substitute the new approximate value hence

$$\lambda_1(t) = -t^{1/2} + \frac{1}{5}t^{-1} + O\left(t^{-5/2}\right) \text{ for } t \to +\infty.$$

We have

$$\frac{\partial H(t, \lambda_1(t))}{\partial y} = 3\lambda_1^2(t) + 2t^{1/2}\lambda_1^2(t) + \frac{6}{5}t^{-1/2} \sim t,$$

$$\frac{\partial^2 H(t, \lambda_1(t))}{\partial y^2} = 6\lambda_1(t) + 2t^{1/2} \sim -4t^{1/2}$$

$(t \to +\infty)$. Consequently

$$q_1 = \lim_{t \to +\infty} \lambda_1(t)\frac{\partial^2 H(t, \lambda_1(t))}{\partial y^2} \left/ \frac{\partial H(t, \lambda_1(t))}{\partial y} = 4. \right.$$

We have $k_1 = -\frac{1}{2}q_1 P\{\lambda_1(t)\} = -1$ and

$$x_1(t) \sim t^{k_1} \exp\left[\int \lambda_1(t)dt\right].$$

Thus,

$$x_1(t) \sim t^{-1} \exp\left[\int \left(-t^{1/2} + \frac{1}{5}t^{-1}\right) dt\right].$$

Finally

$$x_1(t) \sim t^{-4/5}e^{-(2/3)t^{3/2}} \quad \text{for } t \to +\infty.$$

In the same way we have

$$\lambda_{2,3}(t) = \pm it^{-1/4} - \frac{1}{10}t^{-1} + O\left(t^{-7/4}\right)$$

for $t \to +\infty$

$$\frac{\partial H(t, \lambda_{2,3}(t))}{\partial y} \sim \pm it^{1/4}, \quad \frac{\partial^2 H(t, \lambda_{2,3}(t))}{\partial y^2} \sim 2t^{1/2}$$

and $P\{\lambda_{2,3}(t)\} = \frac{1}{4}$. Hence $q_{2,3} = 1$, $k_{2,3} = 1/8$ and

$$x_{2,3}(t) \sim t^{k_{2,3}}e^{\int \lambda_{2,3}(t)dt} \sim t^{1/40}e^{\pm i\frac{4}{3}t^{3/4}} \quad \text{for } t \to +\infty.$$

3° The roots of the characteristic equation may not posses the property of asymptotic separability. Consider the equation

$$x''' + tx'' - t^2x' - t^3x = 0. \tag{10.49}$$

Its characteristic equation $y^3 + ty^2 - t^2y - t^3 = 0$ has one simple root $\lambda_1(t) = t$ and one double root $\lambda_2(t) = -t$. Substitution $x' = yx$ leads to the equation

$$y^3 + ty^2 - t^2y - t^3 + 3yy' + ty' + y'' = 0.$$

We rewrite this equation in the form

$$(y + t)^2(y - t) + 3yy' + ty' + y'' = 0. \tag{10.50}$$

For the first root $\lambda_1(t) = t$ we represent the equation in the form

$$y = t - \frac{3yy' + ty' + y''}{(y + t)^2}$$

and correspondingly form the iteration sequence $\{s_m(t)\}$, where $x_0(t) = t$ and

$$s_m(t) = t - \frac{3s_{m-1}(t)s_{m-1}(t)' + ts_{m-1}(t)' + s_{m-1}(t)''}{(s_{m-1}(t) + t)^2}$$

for $m = 1, 2, \ldots$ We have $s_1(t) = t - 1/t$, and so on. Clearly, we obtain a formal series of the form

$$t - \frac{1}{t} + \frac{a_2}{t^3} + \ldots + \frac{a_m}{t^{2m-1}} + \ldots$$

and

$$s_m(t) = t - \frac{1}{t} + \frac{a_2}{t^3} + \ldots + \frac{a_m}{t^{2m-1}}$$

(a_m are constants). The series has an asymptotic limit

$$g_1(t) = t - \frac{1}{t} + O(t^{-3}) \text{ for } t \to +\infty$$

which is a formal index to the considered equation. For the root $\lambda_2(t) = -t$ we have two suitable asymptotic series. The first formal index is obtained from the following representation of equation (10.49):

$$y = -t + \sqrt{-\frac{3yy' + ty' + y''}{y - t}}.$$

Thus, we form the iteration sequence $\{s_m(t)\}$, where $x_0(t) = -t$ and for $m = 1, 2, \ldots$

$$s_m(t) = -t + \sqrt{-\frac{3s_{m-1}(t)s_{m-1}(t)' + ts_{m-1}(t)' + s_{m-1}(t)''}{s_{m-1}(t) - t}}.$$

Thus, we obtain a series of the form

$$-t + 1 - \frac{1}{2t} + \frac{b_3}{t^3} + \ldots + \frac{b_m}{t^m} + \ldots$$

(b_m are constants). The index $g_2(t)$ is equal to an asymptotic sum of the series and

$$g_2(t) = -t + 1 - \frac{1}{2t} + O(t^{-2}) \text{ for } t \to +\infty$$

which is a formal index to the considered equation. In the same way the representation

$$y = -t - \sqrt{-\frac{3yy' + ty' + y''}{y - t}}$$

leads to the following formal index

$$g_3(t) = -t - 1 - \frac{1}{2t} + O(t^{-2}) \text{ for } t \to +\infty.$$

We have to prove that the obtained formal solutions

$$G_1(t) = \exp\left[\int g_1(t)dt\right], \quad G_2(t) = \exp\left[\int g_1(t)dt\right],$$

$$G_3(t) = \exp\left[\int g_3(t)dt\right]$$

form an FFS. Clearly, $\Phi_2(g_1(t)) \equiv g_1^2(t) + g'(t) = t^2 - 1 + O(t^{-2})$, $\Phi_2(g_2(t)) = (t-1)^2 + O(t^{-2})$ and $\Phi_2(g_3(t)) = (t+1)^2 + O(t^{-2})$. Hence

$$V(g_1(t), g_2(t), g_3(t))$$

$$= \begin{vmatrix} 1 & 1 & 1 \\ t - 1/t + O(t^{-3}) & -t + 1 + O(t^{-2}) & -t - 1 + O(t^{-2}) \\ t^2 - 1 + O(t^{-2}) & (t-1)^2 + O(t^{-2}) & (t+1)^2 + O(t^{-2}) \end{vmatrix}$$

$$\sim \begin{vmatrix} 1 & 1 & 1 \\ t - 1/t & -t + 1 & -t - 1 \\ t^2 - 1 & (t-1)^2 & (t+1)^2 \end{vmatrix} \sim -8t^2$$

for $t \to +\infty$. Consequently $\Pi\{V(g_1(t), g_2(t), g_3(t))\} = 2 > -\infty$. On the basis of Theorem 10.22 we conclude that there exists an FSS of the considered equation in the form $\{x_1(t), x_2(t), x_3(t)\}$ where the functions $x_i(t)$ have the same asymptotics as the functions $G_i(t)$ respectively $(i = 1, 2, 3)$. This leads to the following representation of the functions $x_i(t)$:

$$x_1(t) = (1 + O(t^{-2}))t^{-1}e^{t^2/2}, \quad x_2(t) = (1 + O(t^{-1}))t^{-1/2}e^{t^2/2+t},$$

$$x_3(t) = (1 + O(t^{-1}))t^{-1/2}e^{t^2/2-t}.$$

3.   QUALITATIVE CHARACTERISTICS OF LINEAR DIFFERENTIAL EQUATIONS

Here we consider the main important characteristics of linear differential equations. Namely their asymptotic stability, instability, and oscillation.

Definition 10.25. Equation (10.1) is said to be *asymptotically stable* (for $t \to +\infty$) if each its solution $x(t) \to 0$ for $t \to +\infty$. The equation is *instable* if there exists at least one unbounded solution of the equation for $t \to +\infty$.

Since equation (10.1) may have complex valued solutions we consider the notion of a solution of an oscillating type (instead of an oscillating solution) which is close to the notion of an oscillating function but more natural for our investigation.

Definition 10.26. A function $f(t)$ is said to be of an *oscillating type* (on the positive semi-axis) if the function $\Re f(t)$ has zeros for $t \gg 1$ (this means that there is a sequence of points $\{t_m\}$ $m = 1, 2, ...$ such that $t_m \to +\infty$ for $m \to \infty$ and $\Re f(t_m) = 0$).

Equation (10.1) is called of an *oscillating type* for $t \to +\infty$ if there exists at least one solution $x(t)$ of an oscillating type (on the positive semi-axis) to the equation.

Theorem 10.27. *Let all $a_i(t) \in \{G\}$ where G is a field o type M and let the complete set of roots $\Lambda = \{\lambda_1(t), ..., \lambda_n(t)\}$ possesses the property of asymptotic separability for $t \to +\infty$. Then equation (10.1) is*

(1) asymptotically stable for $t \to +\infty$, if

$$\lim_{t \to +\infty} \Re \lambda_j(t) t < \frac{1}{2} \Re q_j P\{\lambda_j(t)\} \text{ for any } j = 1, 2, ..., n, \qquad (10.51)$$

where

$$q_j = \lim_{t \to +\infty} \lambda_j(t) \frac{\partial^2 H(t, \lambda_j(t))}{\partial y^2} \bigg/ \frac{\partial H(t, \lambda_j(t))}{\partial y} ; \qquad (10.52)$$

(2) instable if

$$\lim_{t \to +\infty} \Re \lambda_j(t) t > \frac{1}{2} \Re q_j P\{\lambda_j(t)\} \text{ for at least one } j \in \{1, 2, ..., n\};$$
$$\qquad (10.53)$$

(3) of an oscillating type if

$$\lim_{t \to +\infty} \Im \lambda_j(t) t \neq \frac{1}{2} \Im q_j P\{\lambda_j(t)\} \text{ for at least one } j \in \{1, 2, ..., n\}. \quad (10.54)$$

PROOF. If $\Pi\{\lambda_j(t)\} \leq -1$ then the required properties follow from Theorems 10.14, 10.18 and 10.23. If $\Pi\{\lambda_j(t)\} \leq -1$, let us mark that on the basis of the Vieté theorem we have for the corresponding solution $x_j(t)$

$$a_n(t)/a_{n-1}(t) = \lambda_j(t) + \alpha_j(t),$$

where $\Pi\{\alpha_j(t)\} < -1$. Hence

$$\gamma_j(t) \equiv x'_j(t)/x_j(t) = \lambda_j(t) + \beta_j(t)$$

where $\Pi\{\beta_j(t)\} < -1$. Moreover the number q_j is equal to zero which leads to the required.   □

Example 10.28. $1°$ Consider the Airy equation $x'' + tx = 0$. Its characteristic equation $y^2 + t = 0$ has roots $\lambda_{1,2}(t) = \pm i\sqrt{t}$. Here $P\{\lambda_{1,2}(t)\} = P\{\sqrt{t}\} = \frac{1}{2}$ and $q_{1,2} = 1$. From (10.51) the equation is asymptotically stable for $t \to +\infty$. For (10.53) taking into account that $\lim_{t\to+\infty} \Im\lambda_{1,2}i(t)t = \infty$, the equation is of oscillating type.

$2°$ Consider equation (10.48). Its characteristic equation has roots $\lambda_1(t) \sim -t^{1/2}$ and

$$\lambda_{2,3}(t) = \pm it^{-1/4} - \frac{1}{10}t^{-1} + O(t^{-7/4}) \text{ for } t \to +\infty.$$

We have $\lim_{t\to+\infty} \Re\lambda_1(t)t = -\infty$, $\lim_{t\to+\infty} \Re\lambda_{2,3}(t)t = -\frac{1}{10}$, $P\{\lambda_{3.4}(t) = P\{t^{-1/4}\} = -1/4$ and (see Example 10.24.$2°$) $q_{2,3} = 1$. From (10.53) it follows that the solutions related to the roots $\lambda_{2,3}(t)$ are instable. Hence the equation is instable for $t \to +\infty$.

4.   SECOND ORDER DIFFERENTIAL EQUATIONS

In this subsection we consider in detail a single linear differential equation of the second order of the form

$$F(t, x) \equiv x'' + a_1(t)x' + a_2(t)x = 0. \tag{10.55}$$

Throughout this paragraph we suppose that G *is a field of type* M *which will be not stipulated apart.*

Let the characteristic equation

$$H(t, y) \equiv y^2 + a_1(t)y + a_2(t) = 0 \tag{10.56}$$

have a complete set of roots $\Lambda = \{\lambda_1(t), \lambda_2(t)\}$.

Let us substitute $x' = yx$ in (10.55). We obtain (for $x \neq 0$) the equation

$$\Phi(t, y) \equiv y' + y^2 + a_1(t)y + a_2(t) = 0. \tag{10.57}$$

The following Theorem is a particular case of Theorem 10.23.

Theorem 10.29. *Let Λ belong to $\{G\}$ and possess the property of asymptotic separability. Then equation (10.55) has a fundamental system of solutions (FSS)*

$$\left\{ x_1(t) = \exp\left[\int \gamma_1(t)dt\right], \quad x_2(t) = \exp\left[\int \gamma_2(t)dt\right] \right\} \qquad (10.58)$$

such that for $i = 1,2$ $\gamma_i(t) \in \{G\}$ and if $\Pi\{\lambda_i(t)\} > -\infty$ then

$$\gamma_i(t) \sim \lambda_i(t) \text{ for } t \to +\infty;$$

if $\lambda_i(t) \asymp 0$ then $\gamma_i(t) \asymp 0$. Moreover

(1) *if $\Pi\{\lambda_i(t)\} > -1$ then*

$$\gamma_i(t) = \lambda_i(t) - \frac{\lambda_i'(t)}{\lambda_i(t) - \lambda_j(t)} + \alpha_i(t), \qquad (10.59)$$

where

$$\Pi\{\alpha_i(t)\} \le -2 - \Pi\{\lambda_1(t) - \lambda_2(t)\} < -1$$

$j \ne i$, $i,j \in \{1,2\}$. Besides $\gamma_i(t)$ is an asymptotic limit of the iteration sequence $\{s_{mi}(t)\}$ where $s_{0i}(t) = \lambda_i(t)$ and

$$s_{mi}(t) = \lambda_i(t) - \frac{s_{m-1i}'(t)}{s_{m-1i}(t) - \lambda_j(t)} \text{ for } m = 1,2,... \qquad (10.60)$$

(2) *if $\Pi\{\lambda_i(t)\} \le -1$ then*

$$\gamma_i(t) = -\frac{a_1(t)}{a_2(t)} + \alpha_i(t), \qquad (10.61)$$

where

$$\Pi\{\alpha_i(t)\} \le \Pi\{\lambda_1(t)\} - \Pi\{\lambda_2(t)\} - 1 < -1$$

$(j \ne i, j \in \{1,2\})$. Moreover $\gamma_i(t)$ is an asymptotic limit of he iterate sequence $\{s_{mi}(t)\}$, where $s_{0i}(t) = -a_1(t)/a_2(t)$ and

$$s_{mi}(t) = -\frac{a_2(t)}{a_1(t)} - \frac{s_{m-1i}'(t) + s_{m-1i}^2(t)}{a_1(t)} \text{ for } m = 1,2,... \qquad (10.62)$$

Example 10.30. Consider an equation of the form

$$x'' + at^\alpha x' + bt^\beta x = 0, \qquad (10.63)$$

where α and β are real numbers, a, b are complex numbers. Its characteristic equation is in the form

$$H(t, y) \equiv y^2 + at^\alpha y + bt^\beta = 0.$$

Its roots are

$$\lambda_{1,2}(t) = -\frac{at^\alpha}{2} \pm \sqrt{\frac{a^2 t^{2\alpha}}{4} - bt^\beta}.$$

Consider the case when $b \neq 0$, $\beta > -2$ and $2\alpha < \beta$. We have

$$\lambda_{1,2}(t) \sim \pm\sqrt{-bt^\beta} \quad \text{for } t \to +\infty.$$

The roots are not equivalent in pairs and

$$\Pi\{\lambda_1(t) - \lambda_2(t)\} = \Pi\left\{2\sqrt{-bt^\beta}\right\} = \frac{\beta}{2} > -1.$$

On the basis of Theorem 10.29 the logarithmic derivatives to solutions of equation (10.63) may be written in the form

$$\gamma_{1,2}(t) = \lambda_{1,2}(t) - \frac{\beta}{4t} + O(t^{-1-\varepsilon}),$$

where ε is a positive number. Consequently

$$x_{1,2}(t) \sim t^{-\beta/4} \exp\left[\pm \int \lambda_{1,2}(t)dt\right]. \tag{10.64}$$

For instance consider the equation $x'' + 2x' - tx = 0$. Here $\alpha = 0$ and $\beta = 1$. Its characteristic equation $y^2 + 2y - t = 0$ has the roots

$$\lambda_{1,2}(t) = -1 \pm \sqrt{1+t} = \pm\sqrt{t} - 1 \pm \frac{1}{2\sqrt{t}} + O\left(\frac{1}{t\sqrt{t}}\right)$$

for $t \to +\infty$. Consequently the considered equation has the solutions

$$x_1(t) \sim t^{-1/4} \exp\left[\frac{2}{3}t\sqrt{t} - t + \sqrt{t}\right],$$

$$x_2(t) \sim t^{-1/4} \exp\left[-\frac{2}{3}t\sqrt{t} - t - \sqrt{t}\right] \quad \text{for } t \to +\infty.$$

Let $a \neq 0$, $\alpha > -1$ and $2\alpha > \beta$. We have

$$\lambda_{1,2}(t) = -\frac{a}{2}t^\alpha \pm \frac{a}{2}t^\alpha\sqrt{1 - \frac{4b}{a^2}t^{\beta-2\alpha}}.$$

Hence

$$\lambda_1(t) \sim -\frac{b}{a}t^{\beta-\alpha}, \lambda_2(t) \sim -at^{\alpha} \text{ for } t \to +\infty,$$

and

$$\Pi\{\lambda_1(t) - \lambda_2(t)\} = \Pi\{t^{\alpha}\} = \alpha > -1.$$

It is easy to show (on the basis of Theorem 10.29) that

$$\gamma_1(t) = \lambda_1(t) + O(t^{-1-\varepsilon_1}) \text{ and } \gamma_2(t) = \lambda_2(t) - \frac{\alpha}{t} + O(t^{-1-\varepsilon_2}),$$

where $\varepsilon_1, \varepsilon_2$ are positive numbers. Consequently

$$x_1(t) \sim \exp\left[\int \lambda_1(t)dt\right] \text{ and } x_2(t) \sim t^{-\alpha}\exp\left[\int \lambda_2(t)dt\right]$$

for $t \to +\infty$.

Example 10.31. Consider the equation $x'' - tx' + tx = 0$. Here $\alpha = 1, \beta = 1$. Its characteristic equation has the roots

$$\begin{aligned}\lambda_1(t) &= \frac{t}{2} - \sqrt{\frac{t^2}{4} - t} \\ &= 1 + \frac{1}{t} + O(t^{-2})\end{aligned}$$

and

$$\begin{aligned}\lambda_2(t) &= \frac{t}{2} + \sqrt{\frac{t^2}{4} - t} \\ &= t - 1 - \frac{1}{t} + O(t^{-2})\end{aligned}$$

for $t \to +\infty$. Consequently, $x_1(t) \sim te^t$ and $x_2(t) \sim t^{-2}e^{t^2/2-t}$.

Now we examine the so called singular cases of equation (10.55) when the complete set of roots of the characteristic equation does not possess the property of asymptotic separability. Let us make the substitution

$$x = ue^{-\frac{1}{2}\int a_1(t)dt}$$

in equation (10.55) where u is a new unknown. We obtain the equation

$$u'' + q(t)u = 0. \tag{10.65}$$

Here

$$q(t) = a_2(t) - \frac{a_1^2(t)}{4} - \frac{a_1'(t)}{2}. \tag{10.66}$$

Let us suppose that $\sqrt{q(t)}$ belongs to $\{G\}$. Mark that the function $q(t)$ also belongs to the space $\{G\}$ because $q(t) = \left(\sqrt{q(t)}\right)^2$. Consider all the possible cases.

(1) Let $\Pi\{q(t)\} > -2$. Then the characteristic equation $y^2 + q(t) = 0$ to equation (10.65) has a complete set of roots $\mu_{1,2}(t) = \pm\sqrt{-q(t)} \in \{G\}$ possessing the property of asymptotic separability. On the basis of Theorem 10.29 it easy to show that equation (10.65) has an FSS $\{u_1(t), u_2(t)\}$, where

$$u_{1,2}(t) = (1 + \alpha_{1,2}(t))(-q(t))^{-1/4} \exp\left[\pm \int \sqrt{-q(t)}dt\right]. \tag{10.67}$$

Here $\Pi\{\alpha_{1,2}(t) < -1 - \frac{1}{2}\Pi\{q(t)\} < 0$. And finally equation (10.55) has an FSS consisting of the functions $x_{1,2} = u_{1,2}(t)e^{-\frac{1}{2}\int a_1(t)dt}$.

(2) let $\Pi\{q(t)\} < -2$. Then in (10.65) carry the member $q(t)u$ to the right side and twice integrate between t and $+\infty$. Supposing that $\lim_{t\to+\infty} u(t) = 1$ and $\lim_{t\to+\infty} u'(t) = 0$ we come to the integral equation

$$u(t) = 1 - \int_t^{+\infty} d\tau_1 \int_{\tau_1}^{+\infty} q(\tau)u(\tau)d\tau. \tag{10.68}$$

It is obvious that any solution to equation (10.68) is also a solution to equation (10.65). Taking into account that $\Pi\{q(t)\} < -2$ we conclude that the integral in the right side of (10.68) is absolutely convergent for $t \gg 1$ for any bounded continuous function $u(t)$, and it is $o(1)$ for $t \to +\infty$. On the basis of the principle of contractive mappings it is easy to show that equation (10.68) has a unique continuous solution $u(t) \sim 1$ for $t \to +\infty$. Let us substitute the obtained solution in (10.68). Let $\alpha_1(t) = u(t) - 1$. Hence

$$|\alpha_1(t)| \le \sup_{t \ge T} |u(t)| \int_t^{+\infty} d\tau_1 \int_{\tau_1}^{+\infty} |q(\tau)|d\tau$$

which leads to the estimate $\Pi\{\alpha_1(t)\} \le \Pi\{q(t)\} + 2 < 0$.

To obtain the second solution let us make the substitution $u = (1 + \alpha_1(t))v$ in (10.68) where v is a new unknown. We obtain $v'' + b(t)v' = 0$, where $b(t) = 2\alpha_1'(t)/(1 + \alpha_1(t))$. Hence

$$\Pi\{b(t)\} = \Pi\{\alpha_1(t)\} - 1 < -1.$$

It is easy to show that the last equation has a solution

$$v(t) = \int_{t_0}^t \exp\left[\int_{\tau_1}^{+\infty} b(\tau)d\tau\right] d\tau_1 = t(1 + \alpha_2(t)),$$

where
$$\alpha_2(t) \sim \frac{1}{t} \int_{t_0}^{t} d\tau_1 \int_{\tau_1}^{+\infty} b(\tau)d\tau.$$

Here $t_0 = +\infty$ if $\Pi\{b(t)\} < -2$, and t_0 is a number, $t_0 \gg 1$ if $\Pi\{b(t)\} \geq -2$. Consequently $\Pi\{\alpha_2(t)\} \leq \Pi\{q(t)\} + 2 < 0$;

(3) let $\Pi\{q(t)\} = -2$. This case we consider with an additional condition. We suppose that $q(t) \in \{L\}$ where L is the set of all power-logarithmic functions. We prove the following

Proposition 10.32. *Equation* (10.65) *(where* $\Pi\{q(t)\} = -2$ *and* $q(t) \in \{L\}$*) has an FSS of the form*

$$\left\{ u_i(t) = \exp\left[\int \frac{\gamma_i^*(t)}{t} dt \right] \right\},$$

where $\gamma_i^*(t) \in A_t$ *and* $\Pi\{\gamma_i^*(t)\} = 0$, $(i = 1, 2)$.

PROOF. The function $q(t)$ may be written in the form
$$q(t) = l(t)t^{-2} + \delta(t),$$
where
$$l(t) \in L \ \delta(t) \in \{L\}, \ \Pi\{l(t)\} = 0 \ \text{and} \ \Pi\{\delta(t)\} < -2.$$

Let the length of the logarithmic chain (see Definitions 6.26 and 6.36) of the function $l(t)$ is equal to m. We prove this Proposition by induction with respect to $m = -1, 0, ...$

Let $m = -1$. Then $l(t) \equiv c \neq 0$, where c is a number. Equation (10.65) is in the form
$$u'' + (ct^{-2} + \delta(t))u = 0. \tag{10.69}$$

We may presuppose that the solutions to equation (10.65) are close to the corresponding solutions of the equation $v'' + cvt^{-2} = 0$. The last equation is of Euler's type. It has solutions of the form t^σ (and may be $t^\sigma \ln t$), where σ is a number. And since $c \neq 0$ then $\sigma \neq 0$. Let us substitute $u = yt^\sigma$ in (10.65). We obtain the equation

$$y'' + 2\sigma \frac{y'}{t} + \delta(t)y = 0. \tag{10.70}$$

The substitution $w = y't^{-2\sigma}$ leads to the equation $w' + t^{2\sigma}\delta(t)y = 0$. The last enable us to consider the following integral equation

$$y(t) = 1 + \int_{t}^{+\infty} \tau_1^{-2\sigma} d\tau_1 \int_{t_0}^{\tau_1} \tau^{2\sigma}\delta(\tau)y(\tau)d\tau. \tag{10.71}$$

Here $t_0 = +\infty$ if $\Pi\{\delta(t)\} + 2\sigma < -1$ and t_0 is a sufficiently large positive number if $\Pi\{\delta(t)\} + 2\sigma \geq -1$. Any solution to equation (10.71) is also a solution to equation (10.70). The integral in the right side of (10.71) is absolutely convergent for $t \gg 1$ for any bounded continuous function $y(t)$ and it is $o(1)$ for $t \to +\infty$. On the basis of the principle of contractive mappings it is easy to show that equation (10.71) has a unique continuous solution $y(t) \sim 1$ for $t \to +\infty$. Let us substitute the obtained solution in (10.71). Let $\alpha(t) = y(t) - 1$. Hence

$$|\alpha(t)| \leq \sup_{t \geq T} |y(t)| \int_t^{+\infty} \tau_1^{-2\Re\sigma} d\tau_1 \int_{t_0}^{\tau_1} \tau^{-2\Re\sigma} |\delta(\tau)| d\tau$$

which leads to the estimate

$$h\{\alpha(t)\} \leq \Pi\{\delta(t)\} + 2 < 0.$$

On substituting the obtained solution in (10.70) and differentiating n times $(n = 1, 2, ...)$ by induction with respect to n we obtain

$$\alpha^{(n)}(t) = y^{(n)}(t) = o(t^{-n})$$

for $t \to +\infty$. This means that $\alpha(t) \in C_t$. Moreover it is easy to show, that $\Pi\{\alpha(t)\} \leq 2 + \Pi\{\delta(t)\}$. Hence equation (10.65) has a solution

$$u(t) = t^\sigma (1 + \alpha(t))$$

and the function

$$\gamma^*(t) \equiv t\frac{u'(t)}{u(t)} = \sigma + \beta(t),$$

where

$$\beta(t) = \frac{\alpha'(t)t}{1 + \alpha(t)} \in C_t.$$

Since $\sigma \neq 0$ $P\{\gamma^*(t)\} = 0$. If equation $v'' + cvt^{-2} = 0$ has a solution of the form $t^\sigma \ln t$ we can prove in the same way that equation (10.65) has the corresponding solution

$$u(t) = t^\sigma (1 + \alpha(t)) \ln t, \quad \text{where } \Pi\{\alpha(t)\} < 0.$$

Hence

$$\gamma^*(t) \equiv t\frac{u'(t)}{u(t)} = \sigma + \frac{1}{\ln t} + \beta(t),$$

where $\beta(t) \in C_t$ and hence $P\{\gamma^*(t)\} = 0$. On the basis of the proved it is easy to conclude that equation (10.65) has an FSS of the desired form. Thus, the case $m = -1$ is proved. Let this Proposition be true for $m - 1$

$(m \geq 0)$ and $\mathrm{dim}l(t) = m$. Hence $l(t) = q_1(\ln t)/t^2$, where $q_1(t) \in \{L\}$, $q_1(t) \not\equiv 0$ and

$$q_1(t) = l_1(\ln t) + \delta_1(\ln t).$$

Here $l_1(t) \in L$, $\mathrm{dim}l_0(t) = \max[-1, m-1]$. Let us make the substitution $\tau = \ln t$ in (10.65). We obtain the equation

$$u''_\tau - u'_\tau + q_1(\tau) + \delta_2(\tau) = 0, \tag{10.72}$$

where $\Pi\{\delta_2(\tau)\} < -2$ (by τ).

On the basis of properties (1), (2) and of the supposition of induction we conclude that equation (10.72) has an FSS of the form

$$\left\{v_i(\tau) = \exp\left[\int \gamma_{i*}(\tau)d\tau\right]\right\},$$

where $\gamma_{i*}(\tau) \in A_\tau$ $(i = 1,2)$. Hence FSS of equation (10.72) may be represented in the form

$$\left\{u_i(t) \equiv v_i(\ln t) = \exp\left[\int \frac{\gamma_{i*}(\ln t)}{t}dt\right]\right\}.$$

Hence $\gamma_i^*(t) = \gamma_{i*}(\ln t)$. Since

$$\gamma_{i*}(\tau) \in A_\tau, \quad P\{\gamma_{i*}(\ln t)\} = 0$$

which proves this Proposition. $\qquad\square$

Example 10.33. Given the equation

$$x'' + x\frac{\ln^2 t}{t^2} = 0.$$

Its characteristic equation $y^2 + t^{-2}\ln^2 t = 0$ has two roots $\lambda_{1,2}(t) = \pm t^{-1}i(\ln t)$ $(i = \sqrt{-1})$. Hence $\Pi\{\lambda_{1,2}(t)\} = -1$. Thus, here we have a singular case. Let us make the substitution $\tau = \ln t$. We obtain the equation

$$x''_\tau - x'_\tau + \tau^2 = 0.$$

Its characteristic equation

$$H(\tau, y) \equiv y^2 - y + \tau^2 = 0$$

has roots

$$\lambda^*_{1,2}(\tau) = \frac{1}{2} \pm \sqrt{\frac{1}{4} - \tau^2} \sim \pm i\tau.$$

More precisely

$$\lambda_{1,2}^*(\tau) = \frac{1}{2} \pm i\tau \mp \frac{1}{8}\tau^{-2} + O(\tau^{-2})$$

for $\tau \to +\infty$. We have

$$\lambda_1^*(\tau) \not\sim \lambda_2^*(\tau)$$

and

$$\Pi\{\lambda_1^*(\tau) - \lambda_2^*(\tau)\} = \Pi\{\tau\} = 1.$$

The last differential equation satisfies (by τ) all the conditions of Theorem 10.23 and consequently has the *FSS*

$$\left\{ x_{1,2}^*(t) = \exp\left[\int \gamma_{1,2*}(\tau)d\tau\right] \right\},$$

where

$$\gamma_{1,2*}(\tau) = \lambda_{1,2}^*(\tau) \mp \frac{\gamma_{1,2}^{*'}(\tau)}{\lambda_1^*(\tau) - \lambda_2^*(\tau)}\frac{1}{2} \pm i\tau \mp \frac{1i}{8\tau} - \frac{1}{2\tau} + O(\tau^{-2})$$

for $\tau \to +\infty$ which leads to the *FSS* of the original equation

$$\left\{ x_1(t) \sim \left(\frac{t}{\ln t}\right)^{1/2} \cos\left[\ln t - \frac{1}{8\ln t} + O\left(\ln^{-2} t\right)\right], \right.$$

$$\left. x_2(t) \sim \left(\frac{t}{\ln t}\right)^{1/2} \sin\left[\ln t - \frac{1}{8\ln t} + O(\ln^{-2} t))\right]\right\}$$

for $t \to +\infty$.

4.1   STANDARD PROCEDURE

The main terms of the solutions to equation (10.55) may be obtained by means of so called *standard procedure*.

Let us make the substitution $x' = yx$ in (10.55). Hence (for $x \neq 0$) we obtain the equation

$$y' + y^2 + a_1(t)y + a_2(t) = 0. \qquad (10.73)$$

Let the equation have a solution $y(t) \in A_t$ then $y'(t) = (k + \alpha(t))y(t)/t$, where k is a number

$$k = \lim_{t\to+\infty} \frac{\ln y(t)}{\ln t}$$

and $\alpha(t) \in C_t$ consequently $\alpha(t)$ is an infinitesimal function for $t \to +\infty$. Substitute $y'(t)$ in (10.73) we obtain an algebraic equation of the form

$$y^2 + \left(a_1(t) + \frac{k + \alpha(t)}{t}\right)y + a_2(t) = 0. \qquad (10.74)$$

In reality (10.74) is a family of equations depending on a numerical parameter k and on an arbitrary function $\alpha(t) \in C_t$. Equation (10.74) is said to be the *precise characteristic equation* of equation (10.55). It has two families of solution

$$y_{1,2}(t) = -\frac{1}{2}\left(a_1(t) + \frac{k+\alpha(t)}{t}\right) \pm \sqrt{\frac{1}{4}\left[a_1(t) + (k+\alpha(t))/t\right]^2 - a_2(t)}.$$

Let it be possible to represent the family $y_j(t)$ $(j = 1, 2)$ in the following form

$$y_j(t) = \Phi_j(k, t) + f_j(k, t, \alpha(t)). \tag{10.75}$$

The function $\Phi_j(k_0, t)$ is called a *kernel* of the family $y_j(t)$ or *standard approximation* to solution of equation (10.73) if

(1) $k_0 = \lim_{t \to +\infty}[\ln \Phi_j(k_0, t)]/ \ln t$;

(2) $f_j(k_0, t, \alpha(t)) = o(\Phi_j(k_0, t))$ as $t \to +\infty$ for any function $\alpha(t) \in C_t$;

(3) either $f_j(k_0, t, \alpha(t)) \in O_t$ for any $\alpha(t) \in C_t$, or there are two functions $\alpha_1(t), \alpha_2(t) \in C_t$ such that

$$f_j(k_0, t, \alpha_1(t)) \not\sim f_j(k_0, t, \alpha_2(t)) \text{ for } t \to +\infty. \tag{10.76}$$

The standard procedure consists of *obtaining of all standard approximations to solutions of equation* (10.73). In all the proved cases in this subsection any standard approximation to solution of equation (10.73) is an asymptotic approximation to a solution of this equation for $t \to +\infty$. Notice that if the coefficients $a_1(t), a_2(t)$ have their asymptotic decompositions in he form of generalized asymptotic power series, then any function $\alpha(t)$ in the precise characteristic equation may be considered as a function with the estimate $\Pi\{\alpha(t)\} < 0$.

Examples 10.34. 1° Consider the equation

$$x'' + 2x' - tx = 0$$

which we investigate above. Substitute $x' = yx$. We obtain the equation $y' + y^2 + 2y - t = 0$ and

$$y^2 + \left(2 + \frac{k+\alpha(t)}{t}\right)y - t = 0$$

is its precise characteristic equation, where $\Pi\{\alpha(t)\} < 0$. Hence

$$y_{1,2}(t) = -\frac{1}{2}\left[2 + \frac{k+\alpha(t)}{t}\right] \pm \sqrt{\frac{1}{4}\left[2 + \frac{k+\alpha(t)}{t}\right]^2 + t} \sim \pm\sqrt{t}.$$

Consequently $k = \Pi\left\{\sqrt{t}\right\} = \frac{1}{2}$ for any standard approximation. We have

$$y_{1,2}(t) = -1 - \frac{1}{4t} \pm \sqrt{t}\left(1 + \frac{1}{2t}\right) + \beta_{1,2}(t),$$

where $\Pi\{\beta_{1,2}(t)\} < -1$. Finally

$$y_{1,2}(t) = \pm\sqrt{t} - 1 \pm \frac{1}{2\sqrt{t}} - \frac{1}{4t} + \beta_{1,2}(t).$$

$2°$ Consider the equation

$$x'' + \frac{\ln^2 t}{t^2}x = 0. \tag{10.77}$$

Substitution $x' = yx$ leads to the equation $y' + y^2 + (\ln^2 t)/t^2 = 0$ and

$$y^2 + \frac{k + \alpha(t)}{t}y + \frac{\ln^2 t}{t^2} = 0$$

is its precise characteristic equation. Hence

$$
\begin{aligned}
y_{1,2}(t) &= -\frac{k + \alpha(t)}{2t} \pm \sqrt{\frac{(k + \alpha^2(t))^2}{4t^2} - \frac{\ln^2 t}{t^2}} \\
&= \frac{1}{t}\left(-\frac{k}{2} \pm \sqrt{k^2 - \ln^2 t}\right) + o\left(\frac{1}{t}\right) \\
&\sim \pm i\frac{\ln t}{t}
\end{aligned}
$$

as $t \to +\infty$ for any $\alpha(t) \in C_t$ $(i = \sqrt{-1})$. Since

$$k = P\left\{i\frac{\ln t}{t}\right\} = -1$$

we have two following standard approximations:

$$\varphi_{1,2}(t) = \frac{1/2 \pm i\ln t}{t} + o\left(\frac{1}{t}\right). \tag{10.78}$$

The functions $\pm\frac{1}{2}(\ln t)/t$ are the main terms of the functions $y_{1,2}(t) = x'_{1,2}/x_{1,2}(t)$, where $x_{1,2}(t)$ are the corresponding solutions to the considered equation. There are many ways to obtain the approximations more accurate. One of them was shown above in the example 10.24. Another way is to form an appropriate iteration sequences of approximations. In our case we may write

$$y' + \left(y - i\frac{\ln t}{t}\right)\left(y + i\frac{\ln t}{t}\right) = 0.$$

The first solution satisfy the relation $y_1(t) \sim (i \ln t)/t$. Hence

$$y_1(t) = i\frac{\ln t}{t} - \frac{y_1'(t)}{y_1(t) + i(\ln t)/t}$$

and from (10.78)

$$y_1(t) = i\frac{\ln t}{t} - \frac{1}{2t} + o\left(\frac{1}{t}\right).$$

Consequently

$$y_1(t) = i\frac{\ln t}{t} - \frac{[(i \ln t)/t - 1/(2t) + o(1/t)]'}{2i(\ln t)/t + 1/(2t) + o(1/t)}. \tag{10.79}$$

In fact the function given by the relation $o(1/t)$ belong to A_t and their order $p = -1$. That is $o(1/t)' = o(1/t^2)$ for $t \to +\infty$. Consequently we obtain

$$y_1(t) = i\frac{\ln t}{t} + \frac{1}{2t} - \frac{1 + i/4}{2t \ln t} + \beta_1(t),$$

where $\beta_1(t) = o(1/(t \ln t)$ for $t \to +\infty$. Reasoning in the same way we obtain the iterate sequence

$$s_m(t) = i\frac{\ln t}{t} - \frac{s_{m-1}'(t)}{s_{m-1}(t) + i(\ln t)/t}, \quad m = 1, 2, ...,$$

where from (10.78)

$$s_0(t) = \frac{i \ln t}{t} + \frac{1}{2t}.$$

This results in a formal series of the form

$$\frac{1}{t}\left[i \ln t + \frac{1}{2} - \frac{i}{2 \ln t} + \frac{c_2}{\ln^2 t} + ... + \frac{c_m}{\ln^m t} + ...\right]$$

where each its particular sum is an asymptotic approximation to the function $y_1(t)$. Here $c_2, c_3, ...$ are constants. In particular we obtain the following relation

$$y_1(t) = i\frac{\ln t}{t} + \frac{1}{2t} - \frac{1 + 4i}{2t \ln t} + O\left(\frac{1}{t \ln^2 t}\right) \quad \text{for } t \to +\infty.$$

Thus, the function $x_1(t)$ satisfy the following relation

$$x_1(t) = \left(1 + O\left(\frac{1}{\ln t}\right)\right)\sqrt{\frac{t}{\ln t}} \exp\left[i\frac{\ln^2 t}{2} - i\frac{\ln \ln t}{8}\right].$$

In the same way, we have

$$x_2(t) = \left(1 + O\left(\frac{1}{\ln t}\right)\right) \sqrt{\frac{t}{\ln t}} \exp\left[i\frac{\ln^2 t}{2} + i\frac{\ln \ln t}{8}\right].$$

Thus, the general solution of the considered equation may be written in the form

$$x = \sqrt{\frac{t}{\ln t}} \left[(1 + \delta_1(t))A \cos\left(\frac{\ln^2 t}{2} - \frac{\ln \ln t}{8}\right)\right.$$
$$\left. + (1 + \delta_2(t))B \sin\left(\frac{\ln^2 t}{2} - \frac{\ln \ln t}{8}\right)\right], \qquad (10.80)$$

where A, B are arbitrary constants and $\delta_{1,2}(t) = O(1/lnt))$ for $t \to +\infty$. If we look for real solutions then the constants A and B have to be real numbers and $\beta_{1,2}(t)$ real functions. Relation (10.80) may be differentiate any number times where $\delta_{1,2}^{(m)}(t) = o(t^{-m}/\ln t)$ for $m = 1, 2, ...$ ($t \to +\infty$).

5.   SYSTEMS OF TWO DIFFERENTIAL EQUATIONS WITH TWO UNKNOWNS

Here we investigate a system of linear differential equations of the form

$$\begin{cases} x_1' = a_{11}(t)x_1 + a_{12}(t)x_2 \\ x_2' = a_{21}(t)x_1 + a_{22}(t)x_2. \end{cases} \qquad (10.81)$$

Here $\{a_1(t), a_2(t)\} \subset \{G\}$ ($i, j = 1, 2$), where G is a field of type M. Consider all the possible cases. Let at least one of the coefficients $a_{12}(t)$, $a_{21}(t)$ belong to the class A_t. Then system (10.81) may be reduced to one linear equation of second order by means of a linear non-singular substitution. For definiteness let $\Pi\{a_{12}(t)\} > -\infty$. We use the simplest linear form $x = x_1$ and differentiate it according to system (10.81). We have

$$x' = a_{11}(t)x_1 + a_{12}(t)x_2.$$

This leads to the desired linear substitution $\tilde{X} = Q(t)X$ where $\tilde{X} = (x, x')^T$ and

$$Q(t) = \begin{pmatrix} 1 & 0 \\ a_{11}(t) & a_{12}(t) \end{pmatrix}. \qquad (10.82)$$

For the coordinate x_1 we obtain the equation

$$x_1'' - \left(\mathrm{Sp}A(t) + \frac{a_{12}'(t)}{a_{12}(t)}\right)x_1' + \left(\det A(t) + \frac{a_{11}(t)a_{12}'(t)}{a_{12}(t)} - a_{11}'(t)\right)x_1 = 0,$$
$$(10.83)$$

where $A(t)$ is the matrix of system (10.81)

$$A(t) = \begin{pmatrix} a_{11}(t) & a_{12}(t) \\ a_{21}(t) & a_{22}(t) \end{pmatrix}. \qquad (10.84)$$

The coordinate x_2 may be obtained from the first line of system (10.81).

$$x_2(t) = \frac{\gamma(t) - a_{11}(t)}{a_{12}(t)} x_1,$$

where (for $x_1(t) \neq 0$) $\gamma(t) = x_1'(t)/x_1(t)$.

In the case when $a_{12}(t)$ and $a_{21}(t)$ belong to the class O_t the substitution

$$x_1 = u_1 e^{\int a_{11}(t)dt}, \quad x_2 = u_2 e^{\int a_{11}(t)dt}$$

leads to the system

$$\begin{cases} u_1' = & a_{12}(t)u_2, \\ u_2' = a_{21}(t)u_1 + [a_{22}(t) - a_{11}(t)]u_2. \end{cases}$$

It is easy to show that the last system has a solution

$$u_1(t) = 1 + \zeta_1(t), \quad u_2(t) = \zeta_2(t),$$

where $\Pi\{\zeta_{1,2}(t)\} = -\infty$. Consequently system (10.81) has a solution

$$\begin{aligned} X_1(t) &= (x_{11}(t), x_{21}(t))^T \\ &= e^{\int a_{11}(t)dt}(1 + \zeta_1(t), \zeta_2(t))^T. \end{aligned}$$

In the same way we obtain another solution

$$X_2(t) = e^{\int a_{22}(t)dt}(\tilde{\zeta}_1(t), 1 + \tilde{\zeta}_2(t))^T,$$

where $\Pi\{\tilde{\zeta}_{1,2}(t)\} = -\infty$. Clearly, the set $\{X_1(t), X_2(t)\}$ forms an *FSS* of system (10.81).

Example 10.35. Given the system

$$\begin{cases} x_1' = & tx_1 + x_2, \\ x_2' = & x_1 - tx_2. \end{cases} \qquad (10.85)$$

Here $a_{12}(t) = 1$ and $\Pi\{1\} = 0 > -\infty$. We obtain $x_1'' - (t^2 + 2)x_1 = 0$ from (10.82) . Its characteristic equation $H(t, y) \equiv y^2 - t^2 - 2 = 0$ has roots

$$\lambda_{1,2}(t) = \pm\sqrt{t^2 + 2} = \pm\left(t + \frac{1}{t} + O(t^{-3})\right) \quad \text{for } t \to +\infty,$$

$$\Pi\{\lambda_{1,2}(t)\} = \Pi\{\pm t\} = 1.$$

Hence the set of roots possesses the property of asymptotic separability. This equation has an *FSS*

$$\{x_{11}(t) = \exp[\int \gamma_1(t)dt], \ x_{12}(t) = \exp[\int \gamma_2(t)dt]\},$$

where

$$\gamma_{1,2}(t) \sim \lambda_{1.2}(t) \ \text{for} \ t \to +\infty.$$

System (10.85) has correspondingly two solutions

$$X_1(t) = (x_{11}(t), x_{21}(t))^T \ \text{and} \ X_2(t) = (x_{12}(t), x_{22}(t))^T,$$

where (see the first row of system (10.85))

$$x_{2i}(t) = x'_{1i}(t) - tx_{1i}(t) \ (i = 1, 2).$$

We have

$$\gamma_{1,2}(t) = \lambda_{1,2}(t) - \frac{1}{2}\frac{\lambda'_{1,2}(t)}{\lambda_{1,2}(t)} + \alpha_{1,2}(t)$$

from Theorems 10.22 and 10.23, where $\Pi\{\alpha_{1,2}(t)\} \leq -3$. Consequently

$$\gamma_1(t) = t + 1/(2t) + \beta_1(t) \ \text{and} \ \gamma_2(t) - 3/(2t) + \beta_2(t),$$

where $\Pi\{\beta_{1,2}(t)\} \leq -3$. Therefore

$$x_{11}(t) \sim t^{1/2}e^{t^2/2} \ \text{and} \ x_{21}(t) \sim \frac{1}{2}t^{-1/2}e^{t^2/2}$$

for $t \to +\infty$. In the same way

$$x_{12}(t) \sim t^{-3/2}e^{-t^2/2} \ \text{and} \ x_{22}(t) \sim -2t^{-1/2}e^{-t^2/2} \ (t \to +\infty).$$

Chapter 11

LINEAR DIFFERENTIAL EQUATIONS IN SINGULAR CASES ON THE POSITIVE SEMI-AXIS

In this chapter we considered the equation

$$F(t, x) \equiv x^{(n)} + a_1(t)x^{(n-1)} + \ldots + a_n(t)x = 0 \qquad (11.1)$$

with coefficients belonging to a field Q of type N or to $\{Q\}$ in the regular case when a complete set of roots

$$\Lambda = \{\lambda_1(t), \lambda_2(t), \ldots, \lambda_n(t)\} \qquad (11.2)$$

of its characteristic equation

$$H(t, y) \equiv y^n + a_1(t)y^{n-1} + \ldots + a_n(t) = 0 \qquad (11.3)$$

possesses the property of asymptotic separability. Now we discuss the general case (when Λ may contains non-asymptotically separable roots).

Here in the preamble we give the results of the chapter in a simplified form.

In section 11.1 we somewhat spread the method of investigation used in Chapter 10 to the equation where Λ is a set of so called asymptotically independent functions. The set may contain equivalent roots. The demand of the equivalence is changed by conditions imposed on their derivatives (see Definition 11.1). The asymptotic representations of the solutions to equation (11.1) also can be represented in an implicit form. But the formulae are some more complicated.

In sections 11.2 and 11.3 we discuss the conditions of asymptotic proximity of logarithmic derivatives of solutions and their corresponding characteristic roots and the possibility to reduce equation (11.1) to equations of orders less than n.

In 11.3 the case when the set Λ consists of two asymptotically separable subsets of roots is considered . Two sets of functions $F = \{f_1(t), ..., f_p(t)\}$ and $G = \{g_1(t), ..., g_q(t)\}$ are asymptotically separable if any pair $f_i(t), g_j(t)$ is asymptotically separable. That is $f_i(t) \not\sim g_j(t)$ for $t \to +\infty$ and $\Pi\{f_i(t) - g_j(t)\} > -1$ $(i = 1, 2, ..., p; j = 1, 2, ..., q)$. And (in the same way) $F_1 = \{f_{11}(t), ..., f_{1m}(t)\}$ and $F_2 = \{f_{21}(t), ..., f_{2m}(t)\}$ are asymptotically similar if for each (fixed) $j = 1, 2, ..., m$, either $f_{1j}(t) \sim f_{2j}(t)$ for $t \to +\infty$ or $\Pi\{f_{1j}(t) - f_{2j}(t)|\} \leq -1$. If all the coefficients of equation (11.1) belong to a field Q of type N (or to $\{Q\}$) and the set Λ may break up into two asymptotically separable subsets Λ_1 and Λ_2 [containing p and q roots respectively $(p+q = n)$], then asymptotic integration of the equation can be reduced to asymptotic integration of two linear differential equations with sets of characteristic roots similar to Λ_1 and Λ_2 respectively. The procedure is very complicated and therefore it has purely theoretical meaning. For example in Chapter 13 (by help of this procedure) we may investigate all the main asymptotic properties of linear differential equations with power-logarithmic coefficients.

In 11.2 equation (11.1) is considered in the case when its characteristic equation has asymptotically similar subsets of roots. It is proved that there is a substitution of the form $x = y \exp[\int g(t)dt]$ which leads to an equation of the form $\widetilde{F}(t,y) \equiv y^{(n)} + b_1(t)y^{(n-1)} + ... + b_n(t)y = 0$ such that its characteristic equation either has a complet set of roots $\widetilde{\Lambda}$ possessing the property of asymptotic separability or $\widetilde{\Lambda}$ may contain subsets of asymptotically similar roots, but any such subset (say $\widetilde{\Lambda}$) consists of functions such that $\Pi\{f_i(t) - f_j(t)\} \leq -1$ for each two functions $f_i(t), f_j(t) \in \widetilde{\Lambda}$.

1.   ASYMPTOTICALLY INDEPENDENT FUNCTIONS

Let $Y = \{y_1(t), y_2(t), ..., y_n(t))\} \subset \Pi$. Consider a function $y(t) \in Y$. We also suppose (for definiteness) that $y(t) = y_1(t)$ (otherwise we may change the numeration of the functions).

Definition 11.1. We say that $y_1(t)$ is an *asymptotically independent function* in Y (for $t \to +\infty$) if

$$\Pi\{y_1(t) - y_i(t)\} > -1 \text{ and } \Pi\{y_1'(t)\} < 2\Pi\{y_1(t) - y_i(t)\} \qquad (11.4)$$

for any $i = 2, ..., n$.

We say that the set Y possesses the property of *asymptotic independence* (or simply Y is an *independent set* of functions) if any $y_j(t) \in Y$ is an independent function in Y (for $t \to +\infty$; $j = 1, 2, ..., n$).

For instance the set

$$\left\{ t, \sqrt{t} + 1, \sqrt{t} - 1, \frac{1}{\sqrt{t}}, \frac{1}{t} \right\}$$

possesses the property of asymptotic independence.

Consider a determinant of the form

$$V(y_1(t), \ldots, y_n(t)) = \begin{vmatrix} 1 & 1 & \cdots & 1 \\ \Phi_1(y_1(t)) & \Phi_1(y_2(t)) & \cdots & \Phi_1(y_n(t)) \\ \cdots & \cdots & \cdots & \cdots \\ \Phi_{n-1}(y_1(t)) & \Phi_{n-1}(y_2(t)) & \cdots & \Phi_{n-1}(y_n(t)) \end{vmatrix},$$

(11.5)

where $\Phi_m(t)$ are defined in Chapter 8(3) (see (8.50)).

Proposition 11.2. *Let Q be a field of type N. Let the set $Y = \{y_1(t), y_2(t), ..., y_n(t)\}$ belong to $\{Q\}$ and possess the property of asymptotic independence. Then*

$$V(y_1(t), ..., y_n(t)) \sim \omega(y_1(t), ..., y_n(t)) \ \text{for} \ t \to +\infty. \tag{11.6}$$

As a simple consequence

$$\Pi\{V(y_1(t), ..., y_n(t))\} > \frac{n(1-n)}{2}. \tag{11.7}$$

Here $\omega(y_1(t), ..., y_n(t))$ is the Van der Monde determinant for the set of functions $y_1(t), y_2(t), ..., y_n(t)$.

PROOF. For simplicity throughout the proof we will write y_i instead of $y_i(t)$ and y instead y_1; we put $d_{ij} = y_i - y_j$ and $d_j = y_1 - y_j$ $(i, j = 1, 2, ..., n)$.

(1) Since $y^{(m)} = (y')^{(m-1)}$ $(m \geq 1)$ we obtain the following estimate $\Pi\{y^{(m)}\} \leq \Pi\{y'\} - m + 1$. Hence

$$\Pi\{(y')^{k_1}(y'')^{k_2}...(y^{(s)})^{k_s}\} \leq \sigma\Pi\{y'\} - r + 2\sigma,$$

where $\sigma = k_1 + k_2 + ... + k_s$ and $r = 2k_1 + 3k_2 + ... + (s+1)k_s$;

(2)
$$\frac{\partial^r \omega(y_1, ..., y_n)}{\partial y^r} = r! \omega(y_1, ..., y_n) \sum_{1 < j_1 < ... < j_r \leq n} \frac{1}{d_{j_1} d_{j_2} ... d_{j_r}}.$$

Let $p = \min_{j=2,...,n} \Pi\{d_j\}$ hence

$$\Pi\left\{\partial^r \frac{\omega(y_1, ..., y_n)}{\partial y^r}\right\} \leq \Pi\{\omega(y_1, ..., y_n)\} - pr;$$

(3) from the relations obtained in (1) and (2), taking into account that $\Pi\{y'\} < 2p$ and $p > -1$, for $k > 0$, we may write the following inequality

$$\Pi\left\{(y')^{k_1}(y'')^{k_2}...(y^{(s)})^{k_s} \partial^r \frac{\omega(y_1, ..., y_n)}{\partial y^r}\right\}$$
$$\leq \quad \Pi\{\omega(y_1, ..., y_n)\} + \sigma(\Pi\{y'\} - 2p) - (r - 2\sigma)(p + 1)$$
$$< \quad \Pi\{\omega(y_1, ..., y_n)\}.$$

On the basis of formula (8.45) taking into account that the decomposition of the differential expression $\Phi_m(y)$ is explicitly independent of m, $V(y_1, ..., y_n)$ may be represented as a linear combination of members of the form

$$[(y_1')^{k_{11}}...(y_1^{(n-2)})^{k_{n-21}}]...[(y_n')^{k_{1n}}...(y_n^{(n-2)})^{k_{n-2n}}] \frac{\partial^{r_1 + ... r_n} \omega(y_1, ..., y_n)}{\partial y_1^{r_1}...\partial y_n^{r_n}},$$

(11.8)

where $k_{11}, ..., k_{n-2n}, r_1, ..., r_n$ are non-negative integers and

$$r_i = 2k_{1i} + 3k_{2i} + ... + (n-1)k_{n-2i}. \tag{11.9}$$

Member (11.8), where $r_1 = r_2 = ... = r_n = 0$ is equal to $\omega(y_1, ..., y_n)$ and its coefficient in the decomposition is equal to 1.

If at least on number $k_{ij} \neq 0$, on the basis of the inequality obtained in (3) we may assert that the estimate of member (11.8) is less than $\Pi\{\omega(y_1, ..., y_n)\}$. This leads to the desired properties. $\qquad \square$

2.   FORMAL ASYMPTOTIC SOLUTIONS

Throughout this paragraph we suppose that *Q is a field of type N.*

Consider a ball of the following form

$$U_{h\lambda} = \{y(t) : y(t) \in \{Q\}, \Pi\{y(t) - \lambda(t)\} \leq h\} \tag{11.10}$$

in $\{Q\}$ where h is a real number ($\lambda(t) \in \{Q\}$). The expressions $H^*(t, y)$ and $\Phi^*(t, y)$ are given in (10.14) and (10.15) respectively;

$$\Phi(t, y) \equiv H(t, y) + \Phi^*(y) = \Phi_n(y) + a_1(t)\Phi_{n-1}(y) + \ldots + a_n(t).$$

$\Phi(t, y)$ are obtained with the help of substitution $y = x'/x$ in (11.1) where $x \neq 0$. So that if equation (11.1) has a solution $x(t) \neq 0$, then $y(t) = x'(t)/x(t)$ is a solution of the equation $\Phi(t, y) = 0$ and vice versa, if the equation $\Phi(t, y) = 0$ has a solution $y(t)$ then (11.1) has a solution $x(t) = e^{\int y(t)dt}$.

Let $\lambda(t) \in \Lambda$, where Λ is a complete set of roots of characteristic equation (11.3).

Remark. *If all $a_i(t) \in Q$ ($i = 1, 2, ..., n$) then there is a field P of type N such that $Q \subset P$ and $\Lambda \in Q$. This implies that if $a_i(t) \in \{Q\}$ and $\lambda_i(t)$ does not equivalent to any other root belonging to Λ, then $\lambda_i(t) \in \{P\}$. So that in the subsequent consideration (without loss of generality) we may suppose that the coefficients $a_i(t)$ and the set Λ belong to the same set $\{Q\}$. Moreover if $\Lambda subset\{Q\}$ then (clearly) all coefficients $a_i(t)$ belong to $\{Q\}$.*

Throughout this paragraph we suppose that $\lambda(t) = \lambda_1(t)$ (otherwise we can change the numeration of the roots).

Set

$$p = \min_{i=2,...,n} \Pi\{\lambda_i(t) - \lambda(t)\} \text{ and } h = \Pi\{\lambda'(t)\} - p.$$

Lemma 11.3. *Let (see (11.2)) $\Lambda \subset \{Q\}$, and let $\lambda(t)$ be an asymptotically independent function of Λ for $t \to +\infty$. Then the equation*

$$y = \lambda(t) - \frac{\Phi^*(t, y)}{H(t, y)} \tag{11.11}$$

has a formal solution $g(t)$ in the class of the power order growth functions such that $g(t) \in U_{h\lambda}$ (see (11.10). So that

$$\Pi\{g(t) - \lambda(t)\} \leq h \text{ and } \lambda(t) \sim \lambda(t) \text{ for } t \to +\infty.$$

Moreover the function $g(t)$ is an asymptotic limit of any sequence of the form $\{s_m(t)\}$ where $s_0(t)$ is an arbitrary function belonging to $U_{h\lambda}$ and for $m = 1, 2, ...$

$$s_m(t) = \lambda(t) - \frac{\Phi^*(t, s_{m-1}(t))}{H^*(t, s_{m-1}(t)} \tag{11.12}$$

and $\Pi\{s_m(t) - s_{m-1}(t)\} \leq -\sigma$, where $\sigma = p - h > 0$. A function $\tilde{g}(t) \in U_{h\lambda}$ is a formal solution of the considered equation if and only if $g(t) \asymp \{\tilde{g}(t)\}$.

PROOF. This Lemma follows from Lemma 4.24 and Proposition 4.29. Since $H(t, y) = (y - \lambda_1(t))...(y - \lambda_n(t))$ we have

$$\frac{\partial^r H(t, y)}{\partial y^r} = \sum (y - \lambda_{i_1}(t))...(y - \lambda_{i_{n-r}}(t)),$$

where each term of the sum is a product of $n-r$ different expressions of the form $y - \lambda_s(t)$. The sum is taken over all such terms. Hence

$$\frac{\partial^r H(t, y)}{\partial y^r} \bigg/ H^*(t, y)$$

is the sum of all possible members of the form

$$\frac{1}{(y - \lambda_{j_1}(t))...(y - \lambda_{j_{r-1}}(t))} \quad \text{and} \quad \frac{y - \lambda(t)}{(y - \lambda_{j_1}(t))...(y - \lambda_{j_r}(t))},$$

where all the indices $j_1, ..., j_r$ are different in pairs and are not equal to 1. On the basis of formula (8.45) the function $\Phi^*(t, y)/H^*(t, y)$ is a decomposition of the following members

$$c(k_1)...c(k_q) \frac{(y')^{k_1}...(y^{(q)})^{k_q}}{(y - \lambda_{j_1}(t))...(y - \lambda_{j_{r-1}}(t))} \tag{11.13}$$

and

$$c(k_1)...c(k_q) \frac{(y')^{k_1}...(y^{(q)})^{k_q}(y - \lambda(t))}{(y - \lambda_{j_1}(t))...(y - \lambda_{j_r}(t))}, \tag{11.14}$$

where all numbers j_s are different in pairs and are not equal to 1. Let us introduce the designations:

$$\Sigma \equiv k_1 + k_2 + ... + k_q, \quad I = k_1 + 2k_2 + ... + qk_q.$$

Then

$$r = \Sigma_q + I_q = 2k_1 + 3k_2 + ... + (q + 1)k_q,$$

that is $r = \Sigma + I$. Because of $\lambda(t)$ is asymptotically independent this implies the inequalities $p > -1$ and $h < p$.

We will estimate the general terms (11.13) and (11.14) when $y = y(t) \in U_{h\lambda}$. We have the following relations:

$$\Pi\{y(t) - \lambda_i(t)\} = \Pi\{\lambda(t) - \lambda_i(t)\} \geq p$$

for any $i = 2, ..., n$. Indeed since $h < p$, $y(t) - \lambda(t) = o(\lambda(t) - \lambda_i(t))$ hence $y(t) - \lambda_i(t) \sim \lambda(t) - \lambda_i(t)$ for $t \to +\infty$. Consequently

$$\Pi\{y(t) - \lambda_i(t)\} = \Pi\{\lambda(t) - \lambda_i(t)\} \geq p.$$

Besides
$$\Pi\{y'(t)\} = \Pi\{\lambda'(t)\}.$$

Indeed $y(t) = \lambda(t) + \alpha(t)$ where $\Pi\{\alpha(t)\} \leq h$. Hence $\Pi\{\alpha'(t)\} \leq h - 1 = \Pi\{\lambda'(t)\} - (p+1)$. Consequently if $\Pi\{\lambda'(t)\} > -\infty$ then $\Pi\{\alpha'(t)\} < \Pi\{\lambda'(t)\}$ and hence $\Pi\{y'(t)\} = \Pi\{\lambda'(t)\}$. If $\Pi\{\lambda'(t)\} = -\infty$ then clearly $\Pi\{\alpha'(t)\} = -\infty$ and $\Pi\{y'(t)\} = -\infty$. Let

$$R_1(y) = c(k_1)...c(k_q)(y')^{k_1}...(y^{(q)})^{k_q},$$

$$R_2(y) = \frac{1}{(y - \lambda_{j_1}(t))...(y - \lambda_{j_{r-1}}(t))}$$

and

$$\widetilde{R}_2(y) = \frac{y - \lambda(t)}{(y - \lambda_{j_1}(t))...(y - \lambda_{j_r}(t))}.$$

It is easy to check that $\Pi\{R_2(y(t))\} \leq -(r-1)p$ and $\Pi\{\widetilde{R}_2(y(t))\} \leq -rp + h \leq -(r-1)p$. Taking into account that

$$\Pi\{y^{(s)}(t)\} \leq \Pi\{y'(t)\} - s + 1$$

for any integer $s = 1, 2, ..$ we have

$$
\begin{aligned}
\Pi\{R_1(t)\} &\leq \Pi\{y'(t)\}(k_1 + ... + k_q) - k_2 - 2k_3 - ... - (q-1)k_q \\
&= \Pi\{y'(t)\}\Sigma + \Sigma - I.
\end{aligned}
$$

Consequently taking into account that $I \geq \Sigma \geq 1$, $h < p$ and $\Pi\{y'(t)\} = h + p$, we obtain

$$
\begin{aligned}
\Pi\{R_1(y(t))R_2(y(t))\} &= \Pi\{R_1(y(t))\} + \Pi\{R_2(y(t))\} \\
&\leq \Pi\{y'(t)\}\Sigma + \Sigma - I - (I + \Sigma - 1)p \\
&\leq (h - p)\Sigma + p \\
&\leq h.
\end{aligned}
$$

In the same way we obtain the relation $\Pi\{R_1(y(t))\widetilde{R}_2(y(t))\} \leq h$. Thus

$$A(y(t)) \equiv \lambda(t) - \frac{\Phi^*(t, y(t))}{H^*(t, y(t))} \in U_{h\lambda}.$$

Consider the equation

$$y = A(y) \equiv \lambda(t) - \frac{\Phi^*(t, y)}{H^*(t, y)}$$

in the ball $U_{h\lambda}$. We proved that $A(U_{h\lambda}) \subset U_{h\lambda}$. Let $y(t) \in U_{h\lambda}$. The derivative estimates of the operators $R_1(y)R_2(y)$ and $R_1(y)\tilde{R}_2(y)$ are easily considered. Namely

$$\Pi'_h\{R_1(\lambda(t))R_2(\lambda(t))\} \leq h - p \text{ and } \Pi'_h\{R_1(\lambda(t))\tilde{R}_2(\lambda(t))\} \leq h - p.$$

This leads to the estimate $\Pi'_h\{A(\lambda(t))\} \leq -\sigma$, where $\sigma = p - h > 0$. $\square$

Theorem 11.4. *Let all conditions of Lemma 11.3 be fulfilled. Then equation (11.1) has a formal solution*

$$G(t) = \exp\left[\int g(t)dt\right], \qquad (11.15)$$

where the function $g(t)$ is obtained in Lemma 11.3.

PROOF. Indeed since $g(t)$ is a formal solution of equation (11.11) we have

$$g(t) - \lambda(t) + \frac{\Phi^*(t, g(t))}{H(t, g(t))} \asymp 0$$

and because of $g(t) \in \Pi$ then $H^*(t, g(t)) \in \Pi$. Consequently

$$\Phi(t, g(t)) = H^*(g(t))\left[\lambda(t) - \frac{\Phi^*(t, g(t))}{H(t, g(t))}\right] \asymp 0. \qquad \square$$

Theorem 11.5. *Let all the coefficients $a_i(t)$ and the complete set of roots Λ (given in (11.2)) of characteristic equation (11.3) belong to $\{Q\}$ and possess the property of asymptotic independence ($i = 1, 2, ..., n$). Then equation (11.1) has a formal fundamental system of solutions (FFS) of the form*

$$\left\{G_j(t) = \exp\left[\int g_j(t)dt\right]\right\} \quad (j = 1, 2, ..., n), \qquad (11.16)$$

where the functions $g_j(t)$ possess all the properties obtained in Theorem 11.4 (where for application of the Theorem for each fixed $j \in \{1, .., n\}$ we have renumber all the roots such that $\lambda_j(t)$ is renumbered in $\lambda(t) = \lambda_1(t)$).

PROOF. Proof of this theorem is based on inequality (11.7) obtained in Proposition 11.2. From the properties of $g_j(t)$ it easily follows that the set $\{g_1(t), g_2(t), ..., g_n(t)\}$ possesses the property of asymptotic independence. Consequently $\Pi\{V(g_1(t), ..., g_n(t))\} > n(1-n)/2$. $\square$

The following theorem is proved in the same way as Theorem 11.5.

Theorem 11.6. *Let the complete set of roots Λ (given in (11.2)) of characteristic equation (11.3) belong to a space $\{G\}$ where G is a field of type M, and Λ possesses the property of asymptotic independence. Then equation (11.1) has a fundamental system of solutions (FSS) of the form*

$$\left\{ G_j(t) = (1 + \theta_j(t)) \exp\left[\int g_j(t)dt \right] \right\} \quad (i = 1, 2, ..., n), \qquad (11.17)$$

where the functions $g_j(t)$ possess all the properties obtained in Theorem 11.4 and $\{\theta_j(t)\} \asymp 0$ $(j = 1, 2, ..., n)$.

Example 11.7. Consider the equation

$$x''' - 3(t + \sqrt{t})x'' + (3t^2 + 6\sqrt{t} + 2t)x' - t(t + \sqrt{t})(t + 2\sqrt{t})x = 0. \quad (11.18)$$

Its characteristic equation

$$y^3 - 3(t + \sqrt{t})y^2 + (3t^2 + 6\sqrt{t} + 2t)y - t(t + \sqrt{t})(t + 2\sqrt{t}) = 0$$

has the complete set of roots

$$\Lambda = \{\lambda_1(t) = t, \quad \lambda_2(t) = t + \sqrt{t}, \quad \lambda_3(t) = t + 2\sqrt{t}\}.$$

We have

$$\Pi\{\lambda_i(t) - \lambda_j(t)\} = \Pi\{\sqrt{t}\}$$
$$= 1/2$$
$$> -1$$

and $\Pi\{\lambda_i'(t)\} = 0 < 2\Pi\{\lambda_i(t) - \lambda_j(t)\} = 1$ for any $i \neq j$ $(i, j = 1, 2, 3)$. Here Q is a field of all rational fraction in $\sqrt{t}$ which is a field of type M. $p = 1/2$, $h = -1/2$ for any solution. Thus all the conditions of Theorem 11.6 are satisfied. This means that equation (11.18) has an FSS of the form

$$\left\{ x_i(t) = \exp\left[\int \gamma_i(t)dt \right] \right\},$$

where $\gamma_i(t) = \lambda_i(t) + \alpha_i(t)$. Here $\Pi\{\alpha_i(t)\} \leq h = -1/2$.

If $i = 1$ we have $g_1(t) = t + \alpha_1(t)$ where $\Pi\{\alpha_1(t)\} \leq -1/2$.

Calculate the function $\gamma_1(t)$ more accurate. We have

$$H_1^*(t, y) = (y - t - \sqrt{t})(y - t - 2\sqrt{t}) \text{ and } \Phi^*(t, y) = y'' + 3y'(y - t - \sqrt{t}).$$

Hence

$$\gamma_1(t) = t - \frac{3\gamma_1'(t)}{\gamma_1(t) - t - 2\sqrt{t}} - \frac{\gamma_1''(t)}{(\gamma_1 - t - \sqrt{t})(\gamma_1 - t - 2\sqrt{t})}. \qquad (11.19)$$

Substitute $\gamma_1(t) = t + \alpha_1(t)$ to the right side of (11.17) we obtain the relation

$$\gamma_1(t) = t + 3/(2\sqrt{t}) + \beta(t),$$

where $\Pi\{\beta(t)\} = -3/2$. Since the function $\gamma_i(t)$ has its asymptotic representation in the form of asymptotic series in $1/\sqrt{t}$ we conclude that $\beta(t) = O(1/(t\sqrt{t}))$. The obtained relation leads to the required asymptotic representation

$$x_1(t) = \left(1 + O\left(\frac{1}{\sqrt{t}}\right)\right) \exp\left[\frac{t^2}{2} + 3\sqrt{t}\right] \quad \text{for } t \to +\infty.$$

In the same way we obtain

$$x_2(t) = \left(1 + O\left(\frac{1}{\sqrt{t}}\right)\right) \exp\left[\frac{t^2}{2} + \frac{2}{3}t\sqrt{t}\right]$$

and

$$x_3(t) = \frac{1 + O(1/\sqrt{t})}{t\sqrt{t}} \exp\left[\frac{t^2}{2} + \frac{4}{3}t\sqrt{t} - 3\sqrt{t}\right]$$

for $t \to +\infty$.

3. CONNECTION BETWEEN CHARACTERISTIC ROOTS AND SOLUTIONS OF A LINEAR DIFFERENTIAL EQUATION

If we have a linear homogeneous differential system with constant coefficients the connection between its solutions and the corresponding roots of the characteristic equation is very simple. But the problem is far from being trivial for equations with variable coefficients. Moreover comparatively simple systems of linear differential equations with variable coefficients may have characteristic roots distinguished from the corresponding logarithmic derivatives of the solutions as the following example shows: consider the system

$$\begin{cases} x_1' &= tx_1 + x_2, \\ x_2' &= (3 - t^2)x_1 - tx_2, \end{cases}$$

the characteristic roots of the system matrix are equal to $\pm\sqrt{3}$. But (as it is easy to check) the system has the general solution

$$x_1 = C_1 e^{2t} + C_2 e^{-2t}, \quad x_2 = C(1 - t)e^{2t} - C_2(2 + t)e^{-2t}$$

(C_1, C_2 are arbitrary constants) which shows that there are not functions among the components of the solution with logarithmic derivatives equivalent to $\pm\sqrt{3}$ for $t \to +\infty$. In the same time an intimate connection (similar, in some sense, to the connection as for the differential equations with constant coefficients) is preserved for a single linear differential equation with coefficients of the power order of growth.

Consider an equation of the form

$$P(t, y) \equiv P_p(t, y)P_q(t, y) + \alpha_0(t) + \alpha_1(t)y + \ldots + \alpha_{n-1}(t)y^{n-1} = 0. \quad (11.20)$$

Here $P_p(t, y)$ and $P_q(t, y)$ are polynomials of p and q degree (respectively) and the coefficients of their highest degree terms are equal to 1, $p + q = n$.

Lemma 11.8. *Let*

$$\{\lambda_1(t), \ldots, \lambda_p(t)\} \quad and \quad \{\delta_1(t), \ldots, \delta_q(t)\}$$

be complete sets of roots for the equations $P_p(t, y) = 0$ and $P_q(t, y) = 0$ respectively and let they and the functions $\alpha_0(t), \ldots, \alpha_{n-1}(t)$ belong to an algebraically closed field U of type CF. Let there be a function $\lambda(t) \in U$ such that

$$\lim_{t \to +\infty} \frac{\lambda(t)}{\lambda_i(t)} \neq \infty \text{ for any } i = 1, 2, \ldots, p.$$

Let (for definiteness)

$$\lim_{t \to +\infty} \frac{\lambda_r(t)}{\lambda_{r-1}(t)} = c_r \neq \infty \ (r = 2, \ldots, p - 1)$$

and

$$\lim_{t \to +\infty} \frac{\delta_j(t)}{\lambda(t)} = 0 \ (j = 1, 2, \ldots, q).$$

Let

$$
\begin{aligned}
\alpha_0(t) &= o(\lambda_1(t) \ldots \lambda_p(t) \lambda^q(t)), \\
\alpha_1(t) &= o(\lambda_1(t) \ldots \lambda_p(t) \lambda^{q-1}(t)), \\
& \cdots\cdots\cdots\cdots\cdots \\
\alpha_q(t) &= o(\lambda_1(t) \ldots \lambda_p(t)), \\
\alpha_{q+1}(t) &= o(\lambda_1(t) \ldots \lambda_{p-1}(t)), \\
& \cdots\cdots\cdots\cdots\cdots \\
\alpha_{n-1}(t) &= o(\lambda_1(t))
\end{aligned}
$$

for $t \to +\infty$. Then the polynomial $P(t, y)$ has a complete set of roots of the form

$$\{\hat{\lambda}_1(t), \ldots, \hat{\lambda}_p(t), \hat{\delta}_1(t), \ldots, \hat{\delta}_q(t)\}$$

(*belonging to* U) *such that*

$$\lim_{t \to +\infty} \frac{\hat{\lambda}_i(t)}{\lambda_i(t)} = 1 \ (i = 1, 2, ..., p)$$

and

$$\lim_{t \to +\infty} \frac{\hat{\delta}_j(t)}{\lambda(t)} = 0 \ (j = 1, 2, ..., q).$$

PROOF. Since U is an algebraically closed field, all the roots of poly-nomial $P(t, y)$ belong to U. We are going to prove that equation (11.20) has exactly p roots $\hat{\lambda}_m(t)$ equivalent (respectively) to the roots $\lambda_m(t)$ ($m = 1, 2, ..., p$). Let there be a natural s such that $c_s \neq 0$ and $c_{s+1} = 0$. Clearly, all the following limits are finite and do not vanish:

$$\lim_{t \to +\infty} \frac{\lambda_k(t)}{\lambda_1(t)} = \mu_k \neq 0 \text{ for } k = 1, 2, ..., s.$$

All the other limits are equal to zero that is

$$\lim_{t \to +\infty} \frac{\lambda_m(t)}{\lambda_1(t)} = 0 \text{ for } m = s + 1, ..., p.$$

Moreover

$$\lim_{t \to +\infty} \frac{\delta_j(t)}{\lambda_1(t)} = 0 \text{ for } j = 1, 2, ..., q.$$

Substitute $y = \lambda_1(t)\tilde{y}$ in (11.20). This leads to the equation

$$(\tilde{y} - \mu_1 + \beta_1(t))...(\tilde{y} - \mu_s + \beta_s(t))(\tilde{y} + \beta_{s+1}(t))$$

$$...(\tilde{y} + \beta_n(t)) + \gamma_0(t) + \gamma_1(t)\tilde{y} + ... + \gamma_{n-1}(t)\tilde{y}^{n-1} = 0, \qquad (11.21)$$

where all the numbers $\mu_1, ..., \mu_s$ do not equal to zero and the functions

$$\beta_0(t), ..., \beta_n(t), \gamma_0(t), ..., \gamma_{n-1}(t)$$

are infinitesimal for $t \to +\infty$. Clearly, the last equation has exactly s roots equivalent to μ_m for $t \to +\infty$, respectively, and its other roots are infinitesimal functions. Thus, equation (11.20) has exactly s roots $\hat{\lambda}_m(t) \sim \lambda_m(t)$ for $t \to +\infty$ ($m = 1, 2, ..., s$). Divide the equation $P(t, y) = 0$ by the polynomial

$$(y - \hat{\lambda}_1(t))...(y - \hat{\lambda}_s(t)).$$

As the result we obtain an algebraic equation with its polynomial of $n - s$ degree and the equation is equivalent to the equation

$$(y - \lambda_{s+1}(t))...(y - \lambda_p(t))(y - \delta_1(t))...(y - \delta_q(t))$$
$$+ \frac{\alpha_0(t) + \alpha_1(t)y + ... + \alpha_{n-1}(t)y^{n-1}}{(y - \lambda_1(t))...(y - \lambda_s(t))} = 0 \qquad (11.22)$$

for any function $y(t) = o(\lambda_1(t))$ $(t \to +\infty)$. On substituting $y = \lambda_{s+1}(t)\tilde{y}$ in the last equation it is easy to show that the equation has a root $\hat{\lambda}_{s+1}(t) \sim \lambda_{s+1}(t)$ for $t \to +\infty$. Continuing in the same way (by finite steps) we may assert that equation (11.20) has the required roots $\hat{\lambda}_1(t), ..., \hat{\lambda}_p(t)$ and all other roots are $o(\lambda_p(t))$ for $t \to +\infty$. Moreover for any such root the equation is equivalent to the equation

$$(y - \delta_1(t))...(y - \delta_q(t)) + \frac{\alpha_0(t) + \alpha_1(t)y + ... + \alpha_{n-1}(t)y^{n-1}}{(y - \lambda_1(t))...(y - \lambda_p(t))} = 0. \quad (11.23)$$

Let us substitute $y = \lambda(t)\tilde{y}$ in the last equation. This leads to the equation

$$(\tilde{y} - \eta_1(t))...(\tilde{y} - \eta_q(t)) + \varphi_0(t, y) + \varphi_1(t, \tilde{y})\tilde{y} + ... + \varphi_{n-1}(t, \tilde{y})\tilde{y}^{n-1} = 0, \qquad (11.24)$$

where $\eta_1(t), ..., \eta_q(t)$ are infinitesimal functions and $\varphi_1(t, \tilde{y}), ..., \varphi_{n-1}(t, \tilde{y})$ are infinitesimal functions for any

$$\tilde{y} = o(\lambda_p(t)) \text{ for } t \to +\infty$$

and continuous in $\tilde{y}$. Clearly, the last equation has q infinitesimal roots. This means that the complete set of roots of equation (11.20) possesses q roots which are $o(\lambda(t))$ for $t \to +\infty$.   □

Remark 11.9. On the basis of Lemma 11.8 and taking into account that the roots $\hat{\delta}_j(t) = o(\lambda(t))$ for $t \to +\infty$, from (11.23) it follows that

$$\hat{\delta}_j(t) = \delta_j(t) + o\left(\left[\frac{|\alpha_0(t)| + |\alpha_1(t)||\lambda(t)| + ... + |\alpha_{n-1}(t)||\lambda^{n-1}(t)|}{\lambda_1(t)...\lambda_q(t)}\right]^{1/q}\right) \qquad (11.25)$$

for $t \to +\infty$ $(j = 1, 2, ..., q)$.

Theorem 11.10. *Let all the roots of complete set of roots (11.2) of characteristic equation (11.3) belong to $A_t \cup O_t$. Let equation (11.1) has a formal solution of the form*

$$x(t) = \exp\left[\int g(t)dt\right], \qquad (11.26)$$

where $g(t) \in A_t \cup O_t$ and there exists at least one finite limit

$$\lim_{t \to +\infty}[g(t)/\lambda_j(t)] \text{ or } \lim_{t \to +\infty}[\lambda_j(t)/g(t)] \text{ for any (fixed) } j = 1, 2, ..., n.$$

Then (i) *If* $\Pi\{g(t)\} > -1$ *then there exists a root* $\lambda(t) \in \Lambda$ *such that* $\lambda(t) \sim g(t)$ *for* $t \to +\infty.$

(ii) *If* $\Pi\{g(t)\} \leq -1$ *then there exists a root* $\lambda(t) \in \Lambda$ *with the estimate* $\Pi\{\lambda(t)\} \leq -1.$

PROOF. Let (for definiteness)

$$\Pi\{\lambda_{j+1}(t)\} \leq \Pi\{\lambda_j(t)\} \text{ for } j = 1, 2, ..., n - 1.$$

Let $\Pi\{g(t)\} > -1$. Suppose the contradiction. That is $g(t) \not\sim \lambda_i(t)$ for any $i = 1, 2, ..., n$. Then $g(t) - \lambda(t) \in A_t \cup O_t$ and

$$\Pi\{P(t, g(t))\} = \Pi\{g(t) - \lambda_1(t)\} + ... + \Pi\{g(t) - \lambda_n(t)\}.$$

Consider all the possible cases:

(1) Let $\Pi\{\lambda_1(t)\} \leq \Pi\{g(t)\}$. Hence $\Pi\{g(t) - \lambda_i(t)\} = \Pi\{g(t)\}$. We are going to estimate the expression $\Phi(t, g(t))$. Consider decomposition (8.52) of the expression $\Phi(t, y)$. For $y = g(t)$ it consists of terms of the form

$$c(g'(t))^{k_1}(g''(t))^{k_2}...(g^{(q)}(t))^{k_q} \frac{\partial^r H(t, g(t))}{\partial y^r},$$

where c is a positive number (depending on the considered term), $r = 2k_1 + 3k_2 + ... + (q + 1)k_q$. If $r = 0$ the term is equal to $H(t, g(t))$. Clearly, $\Pi\{H(t, g(t))\} = n\Pi\{g(t)\}$. Let $r > 0$ (then $I = k_1 + 2k_2 + ... + qk_q > 0$). Evidently

$$\Pi\left\{\frac{\partial^r H(t, g(t))}{\partial y^r\}}\right\} \leq (n - r)\Pi\{g(t)\}.$$

Denote the considered term by $R(t, g(t))$. We have $\Pi\{g(t)\} + 1 > 0$ and

$$\begin{aligned}
\Pi\{R(t, g(t))\} &\leq (k_1 + k_2 + ... + k_q + n - r)\Pi\{g(t)\} - k_1 - 2k_2 - ... - qk_q \\
&= n\Pi\{g(t)\} - I(1 + \Pi\{g(t)\}) \\
&< n\Pi\{g(t)\}.
\end{aligned}$$

Thus

$$\Pi\{\Phi(t, g(t))\} = n\Pi\{g(t)\} > -\infty.$$

On the other hand $g(t)$ is a formal solution to equation $\Phi(t, y) = 0$ which implies $\Phi(t, g(t)) \asymp 0$. The obtained contradiction proves the considered case.

Let there be a number j such that $1 < j < n$, $\Pi\{g(t)\} \leq \Pi\{\lambda_p(t)\}$ if $p \leq j$ and

$$\Pi\{\lambda_q(t)\} \leq \Pi\{g(t)\} \text{ for } q > j.$$

Hence

$$\Pi\{H(t, g(t))\} = \Pi\{\lambda_1(t)\} + \dots + \Pi\{\lambda_j(t)\} + (n - j)\Pi\{g(t)\}.$$

Besides we have

$$\Pi\{\partial H^r(t, g(t))/\partial y^r\} \leq \Pi\{\lambda_1(t)\} + \dots + \Pi\{\lambda_j(t)\} + (n - j - r)\Pi\{g(t)\}$$

for $r > 0$ and $r \leq j$. Hence

$$
\begin{aligned}
\Pi\{R(t.g(t))\} \quad &\leq \quad (k_1 + k_2 + \dots + k_q)\Pi\{\lambda_1(t)\} + \dots + \Pi\{\lambda_j(t)\} \\
&\quad + (n - j - r)\Pi\{g(t)\} - (k_1 + 2k_2 + \dots + qk_q) \\
&= \quad \Pi\{H(t, g(t))\} - I(\Pi\{g(t)\} + 1) \\
&< \quad \Pi\{H(t, g(t))\}
\end{aligned}
$$

which leads to a contradiction. If $r > j$ the same consideration also leads to a contradiction. Thus the case is proved.

Let $\Pi\{g(t)\} \leq \Pi\{\lambda_n(t)\}$. Proceeding from formula (8.52), it is easy to show that

$$\Pi\{\Phi(t, g(t))\} = \Pi\{H(t, g(t))\} = \Pi\{\lambda_1(t)\} + \dots + \Pi\{\lambda_n(t)\} > -\infty$$

and case (1) is proved.

(2) Let $\Pi\{g(t)\} \leq -1$. Suppose the contradiction. That is, all characteristic roots have the estimates $\Pi\{\lambda_i(t)\} > -1$. This leads to the estimate

$$\Pi\{H(t, g(t))\} = \Pi\{\lambda_1(t)\} + \dots + \Pi\{\lambda_n(t)\}.$$

But as is easy to show the rest terms of decomposition (8.52) have the estimate

$$\Pi\{g(t)\} - 1 + \Pi\{\lambda_1(t)\} + \dots + \Pi\{\lambda_{n-2}(t)\} < \Pi\{H(t, g(t))\}$$

which leads to a contradiction. $\qquad\qquad\square$

The inverse theorem may be proved with some additional alterations.

Consider a substitution of the form

$$x = \tilde{x} \exp\left[\int g(t)dt\right] \tag{11.27}$$

in (11.1). As the result we obtain the equation

$$\widetilde{F}(t, \tilde{x}) \equiv \tilde{x}^{(n)} + b_1(t)\tilde{x}^{(n-1)} + \dots + b_n(t)\tilde{x} = 0. \tag{11.28}$$

Denote its characteristic equation by

$$\widetilde{H}(t, \tilde{y}) \equiv \tilde{y}^n + b_1(t)\tilde{y}^{n-1} + \dots + b_n(t) = 0. \tag{11.29}$$

Let $\tilde{\Lambda} = \{\tilde{\lambda}_i(t)\}$ $(i = 1, 2, ..., n)$ be its complete set of roots.

Theorem 11.11. *Let all $\lambda_i(t) \in \Lambda$ (see (11.2)) belong to Q which is an algebraic closed field of type N. Let $\lambda(t) \in \Lambda$ be an $q-$asymptotically multiple root of equation (11.2) for $t \to +\infty$ and $\Pi\{\lambda(t)\} > -1$. Then there exists a substitution of the form (11.27) in (11.1) where $g(t) \in Q$ and $g(t) \sim \lambda(t)$ for $t \to +\infty$ such that the set $\tilde{\Lambda}$ contains exactly q roots each of which is either an asymptotically simple and $\tilde{\lambda}(t) = o(\lambda(t))$ for $t \to +\infty$, or $\Pi\{\tilde{\lambda}(t)\} \leq -1$.*

PROOF. First we substitute $x = x_1 \exp\left[\int \lambda(t)dt\right]$ in (11.1). As the result we obtain a linear differential equation of the form

$$F_1(t, x_1) \equiv x_1^{(n)} + c_{11}(t)x_1^{(n-1)} + ... + c_{1n}(t)x_1 = 0. \qquad (11.30)$$

Its characteristic equation is written in the form

$$H_1(t, y_1) \equiv y_1^n + c_{11}(t)y_1^{n-1} + ... + c_{1n}(t) = 0. \qquad (11.31)$$

Let

$$\Lambda_1 = \{\lambda_{11}(t), ..., \lambda_{1p}(t), \delta_{11}(t), ..., \delta_{1q}(t)\} \qquad (11.32)$$

be its complete set of roots. The substitution is equivalent to the substitution $y = y_1 + \lambda(t)$ in the equation $\Phi(t, y) \equiv H(t, y) + \Phi^*(y) = 0$. We have

$$\Phi_1(t, y_1) \equiv H(t, y_1 + \lambda(t)) + \Phi^*(y_1 + \lambda(t)) = 0. \qquad (11.33)$$

Its decomposition may be obtained from decomposition (8.52) by means of corresponding arithmetical operations and differentiation. Polynomial $H(t, y_1 + \lambda)$ may be written as a product of polynomials $P_{1p}(t, y_1)$ and $P_{1q}(t, y_1)$ of p and q degree with their complete sets of roots

$$\Lambda_{1p} = \{\lambda_{1i}^*(t)\} \ (i = 1, 2, ..., p) \ \text{and} \ \Lambda_{1q} = \{\delta_{1j}^*(t)\} \ (j = 1, 2, ..., q),$$

respectively. Here $\lambda_{1i}^*(t) = \lambda_i(t) - \lambda(t)$ and (for definiteness) we suppose that

$$\lim_{t \to +\infty} \frac{\lambda_{1i+1}^*(t)}{\lambda_{1i}^*(t)} \neq \infty.$$

Clearly,

$$\lim_{t \to +\infty} \frac{\lambda(t)}{\lambda_{1p}^*(t)} \neq \infty \ \text{and} \ \lim_{t \to +\infty} \frac{\delta_{1j}^*(t)}{\lambda(t)} = 0.$$

Polynomial $H_1(t, y_1)$ (see (11.33)) consists of the sum of all members of polynomials $H(t, y_1 + \lambda(t))$ and $P_{1n-2}(t, y_1)$. The last expression is a polynomial of $n-2$ degree composed from the terms obtaining from $\Phi^*(y_1 + \lambda)$

decomposition where each such term must not contain any derivative of y_1. Consider the terms obtained from the expression

$$U(t, y_1) \equiv \frac{1}{2}\lambda'(t)\frac{\partial^2 H(t, y_1 + \lambda(t))}{\partial y^2}$$

belonging to $P_{1n-2}(t, y_1)$ decomposition. We have

$$U(t, y_1) = \lambda'(t)[\beta_{10}(t) + \beta_{11}(t)y_1 + \dots + \beta_{1n-2}(t)y^{n-2}].$$

Here

$$\beta_{1m}(t) = \frac{1}{2m!}\frac{\partial^{m+2}H(t, \lambda(t))}{\partial y^{m+2}} \quad (m = 0, 1, \dots, n-2).$$

Clearly,

$$\beta_{10}(t) = O(\lambda_{11}^*(t)\dots\lambda_{1p}^*(t)\lambda^{q-2}(t)),$$
$$\beta_{11}(t) = O(\lambda_{11}^*(t)\dots\lambda_{1p}^*(t)\lambda^{q-3}(t)),$$

$$\cdot \quad \cdot \quad \cdot \quad \cdot \quad \cdot \quad \cdot \quad \cdot \quad \cdot$$

$$\beta_{1q-2}(t) = O(\lambda_{11}^*(t)\dots\lambda_{1p}^*(t)),$$
$$\beta_{1q-1}(t) = O(\lambda_1^*(t)\dots\lambda_{1p-1}^*(t)),$$

$$\cdot \quad \cdot \quad \cdot \quad \cdot \quad \cdot \quad \cdot \quad \cdot \quad \cdot$$

$$\beta_{1n-2}(t) = O(1).$$

Taking into account that $\lambda^{(s)}(t) = t^{-s+1}\lambda'(t)O(1)$ as $t \to +\infty$ (for any $s > 1$), it is easy to show that any member of the considered decomposition has the desired form and estimate. Thus (preserving, for convenience, the notation for all the terms as in the expression $U(t, y_1)$) we conclude that

$$\begin{aligned} H_1(t, y_1) &= (y - \lambda_1^*(t))\dots(y_1 - \lambda_p^*(t))(y_1 - \delta_1^*(t))\dots(y_1 - \delta_q^*(t)) \\ &\quad + \lambda'(t)\left[\beta_0(t) + \beta_1(t)y_1 + \dots + \beta_{n-2}(t)y_1^{n-2}\right]. \end{aligned}$$

On the basis of Lemma 11.8 we conclude that complete set of roots (11.32) possesses the following properties:

$$\lim_{t \to +\infty}[\lambda(t)/\lambda_{1i}(t)] \neq \infty \quad (i = 1, 2, \dots, p),$$

and
$$\lim_{t \to +\infty}[\delta_{1j}(t)/\lambda(t)] = 0 \quad (j = 1, 2, \dots, q)$$
$$\lim_{t \to +\infty}[\lambda_{1r+1}(t)/\lambda_{1r}(t)] \neq \infty \text{ for } r = 1, 2, \dots, p-1.$$

We have more precise estimates for the functions $\delta_{1j}(t)$. Due to (11.25)

$$\delta_{1j}(t) = \lambda(t)O(1/[\lambda(t)t]^{1/q}) \text{ for } t \to +\infty \quad (j = 1, 2, \dots, q).$$

Thus taking into account that $\Pi\{\lambda(t)\} > -1$ clearly

$$\Pi\{\delta_{1j}(t)\} \leq \Pi\{\lambda(t)\} - \sigma_\lambda$$

where

$$\sigma_\lambda = \frac{1}{q}[\Pi\{\lambda(t)\} + 1] > 0.$$

Being trivial for $q = 1$ we prove this Theorem by induction with respect to q. Thus we suppose that Theorem is true for any r-asymptotically multiple root where $r < p$. We construct the proof by contradiction. Then we have to prove that the roots $\delta_{1j}(t)$ are q-asymptotically multiple and $\Pi\{\delta_{1j}(t)\} > -1$. Hence

$$\Sigma_1(t) = \delta_{11}(t) + ... + \delta_{1q}(t) \sim n\delta_{1j}(t)$$

for $t \to +\infty$ and for any $j = 1, 2, ..., q$. In particular $\Pi\{\Sigma_1(t)\} \leq \Pi\{\lambda(t)\} - \sigma_\lambda$. Denote

$$\Delta_1(t) = \frac{1}{q}[\delta_{11}(t) + ... + \delta_{1q}(t)]$$

and make the second substitution

$$x_1 = x_2 \exp\left[\int \Delta_1(t)dt\right].$$

As the result we obtain a linear differential equation of the form

$$F_2(t, x_2) \equiv x_2^{(n)} + c_{21}(t)x_2^{(n-1)} + ... + c_{2n}(t)x_2 = 0. \tag{11.34}$$

Its characteristic equation is written in the form

$$H_2(t, y_2) \equiv y_2^n + c_{21}(t)y_2^{n-1} + ... + c_{2n}(t) = 0. \tag{11.35}$$

Let

$$\Lambda_2 = \{\lambda_{21}(t), ..., \lambda_{2p}(t), \delta_{21}(t), ..., \delta_{2q}(t)\} \tag{11.36}$$

be its complete set of roots. In the same way as it was shown in the previous substitution, the estimates of the roots are obtained from the following equation

$$(y - \lambda_{21}^*(t))...(y - \lambda_{2p}^*(t)) + \Delta_1'(t)[\beta_{20}(t) + \beta_{21}(t)y + ... + \beta_{2n-2}(t)y^{n-2}] = 0,$$

where

$$\lambda_{2i}^*(t) = \lambda_{1i}(t) - \Delta_1(t) \quad (i = 1, 2, ..., p)$$

and $\delta_{2j}^*(t) = \delta_{1j}(t) - \Delta_1(t) \quad (j = 1, 2, ..., q),$

$$\beta_{20}(t) = O(\lambda_{21}^*(t)...\lambda_{2p}^*(t)\Delta_1^{q-2}(t)),$$

$$\beta_{21}(t) = O(\lambda_{21}^*(t)...\lambda_{2p}^*(t)\Delta_1^{q-3}(t)),$$

$$\cdots\cdots\cdots\cdots\cdots$$

$$\beta_{2q-2}(t) = O(\lambda_{21}^*(t)...\lambda_{2p}^*(t)),$$

$$\beta_{2q-1}(t) = O(\lambda_1^*(t)...\lambda_{2p-1}^*(t)),$$

$$\cdots\cdots\cdots\cdots\cdots$$

$$\beta_{2n-2}(t) = O(1).$$

Clearly $\delta_{21}^*(t) + ... + \delta_{2q}^*(t) = 0$. Moreover

$$\Pi\{\delta_{2j}(t) - \delta_{2j}^*(t)\} \le \Pi\{\Delta_1(t)\} - \sigma_{\Delta_1},$$

where

$$\sigma_{\Delta_1} = \frac{1}{q}[\Pi\{\Delta_1(t)\} + 1] > 0.$$

Besides we must believe that all the functions $\delta_{2j}(t)$ are q−asymptotically multiple and $\Pi\{\delta_{2j}(t)\} > -1$ $(j = 1, 2, ..., q)$.

Unfortunately the obtained estimates are insufficient to prove the theorem. Making the estimates more precisely we have to estimate the sum

$$\Sigma_2(t) = \delta_{21}(t) + ... + \delta_{2q}(t).$$

Since all the roots $\delta_{2j}(t)$ are equivalent in pairs, $\delta_{2j}(t) \sim \frac{1}{q}\Sigma_2(t)$. Let us rewrite the last equation in the form

$$Q(t, y_2) \equiv (y_2 - \delta_{21}^*)...(y_2 - \delta_{2q}^*)$$
$$+ \frac{\Delta_1'(t)\left[\beta_{20}(t) + \beta_{21}(t)y + ... + \beta_{2n-2}(t)y^{n-2}\right]}{(y - \lambda_{21}^*(t))...(y - \lambda_{2p}^*(t))} = 0$$

$$(11.37)$$

which we rewrite in the following form:

$$Q(t, y_2) \equiv (y_2 - \delta_{21}^*)...(y_2 - \delta_{2q}^*) + \eta_0(t) + \eta_1(t)y_2 + ...$$
$$+ \eta_{q-2}(t)y_2^{q-2} + \varphi_{q-1}(t, y_2)y^{q-2} = 0. \qquad (11.38)$$

Here $\varphi_{q-1}(t, y_2)y^{q-2}$ is the remainder of Taylor's formula for the function

$$V(t, y_2) = \frac{\Delta_1'(t)\left[\beta_{20}(t) + \beta_{21}(t)y_2 + ... + \beta_{2n-2}(t)y_2^{n-2}\right]}{(y - \lambda_{21}^*(t))...(y - \lambda_{2p}^*(t))},$$

where $y_2 = y_2(t) = O(\Delta_1(t))$ as $t \gg 1$. Thus $\eta_0(t) = V(t, 0)$. Hence

$$\eta_0(t) = (-1)^p \frac{\delta_1'(t)\beta_{20}(t)}{\lambda_{21}^*(t)...\lambda_{2p}^*(t)} = O(\Delta_1'(t)\Delta_1^{q-2}(t))$$

for $t \to +\infty$. We can obtain expression (11.38) expanding

$$\frac{1}{(1 - y_2/\lambda_{21}^*(t))...(1 - y_2/\lambda_{2p}^*(t))}$$

into a power series in y_2. Consequently

$$\eta_j(t) = O\left(\frac{|\Delta_1'(t)|}{\lambda_{21}^*(t)...\lambda_{2p}^*(t)}[|\beta_{20}(t)/[\lambda_{2p}^*(t)]^j|\right.$$

$$\left. + \frac{|\Delta_1'(t)|}{\lambda_{21}^*}|\beta_{21}(t)/[\lambda_{2p}^*(t)]^{j-1}| + ... + |\beta_{2j}(t)|]\right)$$

$$= O(\Delta_1'(t)\Delta_1^{q-2-j})$$

$(j = 1, 2, ..., q - 2)$. In the same way we have

$$\varphi(t, y_2(t)) = O(\Delta_1'(t)/\lambda_{1p}^*(t))$$

for any $y_2 = y_2(t) = O(\Delta_1(t))$. Consider the equation

$$(y_2 - \delta_{21}^*(t))...(y_2 - \delta_{2q}^*(t)) + \eta_0(t) + \eta_1(t)y_2 + ... + \eta_{q-2}(t)y^{q-2} = 0$$

and let $\{\hat{\delta}_1(t), ..., \hat{\delta}_q(t)\}$ be its complete set of roots. Clearly

$$\hat{\delta}_1(t) + ... + \hat{\delta}_q(t) = \delta_{21}^*(t) + ... + \delta_{2q}^*(t) = 0.$$

Due to Lemma 11.8 we may conclude that equation (11.31) can be written in the form

$$\left(y_2 - \hat{\delta}_1(t) + O\left([\Delta_1'(t)/[\Delta_1(t)\lambda_p(t)]]^{1/q}\right)\right)$$

$$... \left(y_2 - \hat{\delta}_q(t) + O\left([\Delta_1'(t)/[\Delta_1(t)\lambda_p(t)]]^{1/q}\right)\right) = 0.$$

Consequently

$$\Sigma_2(t) \equiv \delta_{21}(t) + ... + \delta_{2q}(t)$$

$$= \hat{\delta}_1(t) + ... + \hat{\delta}_q(t) + O\left([y_2^{p-1}\Delta_1'(t)/[\Delta_1(t)\lambda_p(t)]]^{1/q}\right)$$

$$= O\left([\Delta_1(t)[\Delta_1'(t)/[\Delta_1(t)\lambda_p(t)]]^{1/q}\right)$$

for $t \to +\infty$. Because of the asymptotic multiplicity of the roots $\delta_{2j}(t)$ (taking into account that $\Delta_1'(t)/\Delta_1(t) = O(1/t)$) we obtain the following estimates:

$$\delta_{2j}(t) = O\left(\frac{\Delta_1(t)}{t\lambda_p(t)}\right) \quad \text{for } t \to +\infty.$$

Consequently

$$\Pi\{\delta_{21}(t)\} \le \Pi\{\Delta_1(t)\} - \sigma \text{ and } \Pi\{\delta_{21}(t)\} > -1 \qquad (11.39)$$

where $\sigma = \frac{1}{q}[\Pi\{\lambda_p(t)\} + 1] > 0$. By making in succession substitutions like the one in the above case, we obtain the sequence $\{\delta_{m1}(t)\}$ $(m = 2, ...)$, where

$$\Pi\{\delta_{m1}(t)\} \le \Pi\{\Delta_1(t)\} - (m-1)\sigma$$

and (by the supposition) $\Pi\{\delta_{m1}(t)\} > -1$. Since $m\sigma \to +\infty$, we have $\Pi\{\delta_{m1}(t)\} \to -\infty$ for $m \to \infty$. The obtained contradiction proves this Theorem. $\qquad\qquad\square$

Remark 11.12. Assume the hypothesis and notation of Theorem 11.11. As it is proved in the Theorem the obtained substitution (11.27) leads to equation (11.28) where its characteristic equation has exactly q roots $\tilde{\lambda}_{j1}(t), ..., \tilde{\lambda}_{jp}(t)$ with the estimates

$$\tilde{\lambda}_{jk} = o(\lambda(t)) \text{ for } t \to +\infty.$$

Then there exist numbers r and s such that $r + s = p$ and (for definiteness) the roots $\tilde{\lambda}_{jl}(t)$ are asymptotically simple and

$$\Pi\{\tilde{\lambda}_{jl}(t)\} > -1 \quad (j = 1, 2, ..., r).$$

The rest s roots $\{\tilde{\lambda}_{jr+m}(t)\}$ $(m = 1, 2, ..., s)$ have the estimates

$$\Pi\{\tilde{\lambda}_{jr+m}(t)\} \le -1.$$

Thus, on the basis of Theorem 11.5, equation (11.1) possesses a formal fundamental set of solutions where exactly r of the solutions may be written in the form

$$\zeta_{jl}(t) = \exp\left[\int g_{jl}(t)dt\right], \quad g_{jl}(t) \sim \lambda(t) \text{ fort } t \to +\infty$$

and

$$\Pi\{g_{jm}(t) - g_{ju}(t)\} > -1 \ (m \ne u, l, m, \ u = 1, 2, ..., r).$$

The functions $g_{jl}(t)$ may be obtained by means of so called *standard procedure*. The standard procedure was applied to the second order equations (see subsection 10.4.1). Indeed, it has a more wide application.

Let the considered formal solution be in the form

$$\zeta(t) = \exp\left[\int y(t)dt\right]$$

(corresponding to the root $\lambda(t)$ of characteristic equation (11.3) with the estimate $\Pi\{\lambda(t)\} > -1$). Then $y(t) \in A_t$, $y(t) \sim y(t)$ for $t \to +\infty$, and hence $k \equiv P\{y(t)\} = P\{\lambda(t)\}$. Substitute $x' = yx$ in (11.1). We obtain the equation $\Phi(t, y) \equiv H(t, y) + \Phi^*(t, y) = 0$. Let us substitute $y = y(t)$ in the last equation. Since $y(t)$ is a formal solution to the equation we have $\Phi(t, y(t)) \in O_t$. By Proposition 5.15 (see Chapter 5(2)) there exist functions $\alpha_j(t) \in C_t$ such that

$$y^j(t) = t^{-j}[k(k-1)...(k-j+1) + \alpha_j(t)]y(t) \quad (j = 1, 2, ...).$$

This leads to the equation

$$\Phi(t, y) \equiv P(t, k, \alpha_1(t), ..., \alpha_n(t)) + \theta(t) = 0.$$

Here $\theta(t) \asymp 0$, $P(t, k, \alpha_1(t), ..., \alpha_n(t))$ is a polynomial on nth degree which contains the infinitesimal functions $\alpha_1(t), ..., \alpha_n(t)$. To obtain the formal solutions we may omit the function $\theta(t)$. Consequently we obtain an equation of the form

$$P(t, k, \alpha_1(t), ..., \alpha_n(t)) = 0. \tag{11.40}$$

Equation (11.40) contains n beforehand unknown infinitesimal functions $\alpha_i(t)$ (moreover $\alpha_i(t) \in C_t$), but it is an algebraic equation which is more convenient for investigation. On the basis of (11.40) we may obtain principal terms for an asymptotic representation of the function $y(t)$. The equation is said to be the *accurate characteristic equation* for equation (11.1). This method may be considered as an extension of the Euler method of solution of linear differential equations with constant coefficients to the linear equations with power order growth coefficients.

Example 11.13. Given the equation

$$x''' - 3tx'' + 3t^2 x' - t^3 x = 0. \tag{11.41}$$

Its characteristic equation has an 3-multiple root $\lambda(t) = t$. Clearly, $P\{\lambda(t)\} = 1$. Substitution $x' = yx$ leads to the equation

$$(y - t)^3 + y'' + 3yy' - 3ty' = 0. \tag{11.42}$$

Hence the accurate characteristic equation is written in the form

$$(y - t)^3 - \{3t(k + \alpha_1(t))/t + [k(k-1) + \alpha_2(t)]/t^2\}y + [k + \alpha_1(t)]y^2/t = 0.$$

Since $k = 1$ we have

$$(y - t)^3 - [3 + o(1)]y + (1 + o(1))y^2/t = 0.$$

Rewrite this equation in the form

$$y = t + \varepsilon_i \sqrt[3]{[3 + o(1)]y - (1 + o(1))y^2/t}.$$

Here ε_i are the different cube roots of unit. Since $y(t) \sim t$ we have

$$y_i(t) = t + \varepsilon_i \sqrt[3]{2t}(1 + o(1)) \text{ for } t \to +\infty. \tag{11.43}$$

We may obtain the asymptotic approximations more precisely by means of the substitution $y = y_1 + t + \varepsilon_i \sqrt[3]{2t}$ (in (11.42)) for each fixed $i = 1, 2, 3$.

Let us mark that the coefficients of equation (11.41) belong to a field consisting of fractional rational functions in t which is a field of type M. Besides due to (11.43) the substitution $x = \tilde{x} \exp\left[t^2/2\right]$ leads to a linear differential equation where its characteristic polynomial possesses a complete set of roots $\tilde{\Lambda} = \{\tilde{\lambda}_j(t)\}$ where

$$\tilde{\lambda}_j(t) \sim \varepsilon_j \sqrt[3]{2t} \text{ for } t \to +\infty \ (i = 1, 2, 3).$$

Hence $\tilde{\Lambda}$ is a set of function possessing the property of asymptotic separability. Hence any formal solution (11.43) to equation (11.42) is an asymptotic solution to the equation.

4.   ANALYTIC TRANSFORMATIONS

Here we consider a theoretical possibility to reduce the problem of asymptotic solution of linear differential equation of the form

$$F(t, x) \equiv x^{(n)} + a_1(t)x^{(n-1)} + \dots + a_n(t)x = 0, \tag{11.1}$$

to the same problem for linear equations of the same form but with their orders less then the order of the original equation. This may be obtained in the case when the characteristic equation

$$H(t, y) = y^n + a_1(t)y^{n-1} + \dots + a_n(t) = 0 \tag{11.3}$$

possesses its complete set of roots $\Lambda = \{\lambda_i(t)\}$ (see (11.2)) which contains asymptotically separable subsets.

Definition 11.14. Consider two sets

$$Y_1 = \{y_{11}(t), y_{12}(t), \dots, y_{1p}(t)\} \text{ and } Y_2 = \{y_{21}(t), y_{22}(t), \dots, y_{2q}(t)\}.$$

We say that Y_1 and Y_2 are asymptotically separable if

$$y_{1i}(t) \not\sim y_{2j}(t) \text{ for } t \to +\infty \text{ and } h\{y_{1i}(t) - y_{2j}(t)\} > -1$$

for any $i = 1, 2, \dots, p$ and $j = 1, 2, \dots, q$.

The required reduction is made by means of identical transformations of equation (11.1).

4.1   *SEVERAL PROPERTIES OF MATRIX EQUATIONS*

We proceed with a simple proposition. *For simplicity we will consider all the properties in a field Q of type N.*

Lemma 11.15. *Given a linear matrix equation of the form*

$$D_1(t)X - XD_2(t) = B(t), \tag{11.44}$$

where $D_1(t)$ and $D_2(t)$ are square matrices of order p and q respectively. The unknown X and $B(t)$ are $p \times q$ matrices. Let the matrices $D_1(t)$, $D_2(t)$, and $B(t)$ belong to Q. Let the limits

$$\lim_{t\to+\infty} D_1(t) = D_1(\infty) \quad and \quad \lim_{t\to+\infty} D_2(t) = D_2(\infty)$$

be finite. Denote the sets of characteristic roots of the matrices $D_1(\infty)$ and $D_2(\infty)$ by $J_1 = \{\lambda_1, \lambda_2, ..., \lambda_p\}$ and $J_2 = \{\mu_1, \mu_2, ..., \mu_q, \}$ respectively (each root of J_1 and J_2 is written so many times as its multiplicity). Let

$$\lambda_i \neq \mu_j \quad for \ any \ i = 1, 2, ..., p \ and \ j = 1, 2, ..., q,$$

e.i. the sets J_1 and J_2 are disjoint. Then equation (11.44) possesses a unique solution $X_0(t)$. Moreover

$$X_0(t) \in Q \quad and \quad \Pi\{\|X_0(t)\|\} \leq \Pi\{\|B(t)\|\}.$$

The proof of the theorem is based on the following algebraic theorem:

Lemma 11.16. *Let D_1 and D_2 be constant square matrices of the orders p and q, respectively, let their characteristic sets of roots be not intersected, and let B be a $p \times q$ matrix. Then the matrix equation $D_1 X - X D_2 = B$ has a unique solution X_0. Each element of the solution is a linear combination of elements of the matrix B, where the coefficients of the linear combination are various products of matrices D_1 and D_2 elements.*

PROOF. The equation $D_1 X - X D_2 = B$ is equivalent to a linear system $AY = F$ where A is a square (constant) matrix of pq order with elements composed of various elements of the matrices D_1 and D_2. Moreover since the sets of characteristic roots of the matrices do not intersect, $\det A \neq 0$; F is a column matrix consisting from the corresponding elements of the matrix B, and Y is a column matrix consisting of the corresponding elements of the matrix X. Thus there exists the inverse matrix A^{-1} and $Y = A^{-1}F$ which leads to the required properties.   □

Lemma 11.15 is a simple consequence of Lemma 11.16. Indeed equation (11.44) is equivalent to a system of the form $A(t)Y = F(t)$, where $A(t) \in$

Q and $\det A(t) \to \det A(\infty) \neq 0$; $F(t)$ consists of the elements of the matrix $B(t)$. The last system has a unique solution $Y_0(t) = D^{-1}(t)F(t)$, where evidently $\det A^{-1}(t) \to \det A^{-1}(\infty) \neq \infty$ which leads to the required properties.

Consider a matrix expression $\Delta(t) + B(t)$, where

$$\Delta(t) = \operatorname{diag}(D_1(t), D_2(t)).$$

Here $D_1(t)$ and $D_2(t)$ are square matrices of orders p and q (respectively)

$$B(t) = \begin{pmatrix} 0 & B_1^*(t) \\ B_2^*(t) & 0 \end{pmatrix}.$$

Here $B_1^*(t)$ and $B_2^*(t)$ are $p \times q$ and $q \times p$ matrices, respectively.

Lemma 11.17. *Let $D_1(t)$ and $D_2(t)$ possess all the properties given in Lemma 11.15. Let the matrices $\Delta(t)$ and $B(t)$ belong to Q and $B(t) \in C_t$. Then there exists a matrix*

$$R(t) = \begin{pmatrix} 0 & R_1^*(t) \\ R_2^*(t) & 0 \end{pmatrix}$$

of the same structure as the matrix $B(t)$ such that $\Pi\{R(t)\} \leq \Pi\{B(t)\}$ and

$$(E + R(t))^{-1}(\Delta(t) + B(t))(E + R(t))$$

$$= \operatorname{diag}(D_1(t) + B_1^*(t)R_2^*(t), D_2(t) + B_2^*(t)R_1^*(t)), \qquad (11.45)$$

where E is a unit matrix of order $p + q$. Moreover the matrix $R(t)$ belongs to a field $\widetilde{Q}$ which is an algebraic extension of Q.

PROOF. Relation (11.45) is equivalent to the following two relations:

$$\begin{aligned} D_1(t)R_1^*(t) - R_1^*(t)D_2(t) &= -B_1^*(t) + R_1^*(t)B_2^*(t)R_1^*(t), \\ D_2(t)R_2^*(t) - R_2^*(t)D_1(t) &= -B_2^*(t) + R_2^*(t)B_1^*(t)R_2^*(t). \end{aligned}$$

Prove the matrix $R_1^*(t)$ existence (the proof of matrix $R_2^*(t)$ existence is made in the same way). Matrix $R_1^*(t)$ is a solution of the matrix equation

$$D_1(t)X - XD_2(t) = -B_1^*(t) + XB_2^*(t)X. \qquad (11.46)$$

Consider a sequence of matrices $\{X_m(t)\}$ where $X_0(t) = \theta$ and for $m = 1, 2, \ldots$ the matrix $X_m(t)$ is a solution to the equation

$$D_1(t)X_m - X_m D_2(t) = -B_1^*(t) + X_{m-1}(t)B_2^*(t)X_{m-1}(t). \qquad (11.47)$$

On the basis of Lemma 11.15 there exists a unique solution $X_m(t)$. It belongs to Q and (since $X_m(t)$ is a linear combination of the elements disposed on the right hand side with bounded coefficients) we conclude that there is a number $C > 0$ (independent of m) such that

$$\|X_m(t)\| \leq C\|B_1^*(t) - X_{m-1}(t)B_2^*(t)X_{m-1}(t)\| \qquad . \qquad (11.48)$$

for $t \geq 1$. Choose a number $T > 0$ so large that inequality (11.48) is valid for $t > T$

$$\|B_1^*(t)\| < \frac{1}{4}C^{-2}, \text{ and } \|B_2^*(t)\| < 1.$$

Prove that

$$\|X_m(t)\| \leq 2C\|B_1^*(t)\| \text{ for any } m.$$

Being trivial for $m = 1$, we suppose that it is valid for $m - 1$. Hence

$$\|X_m(t)\| \leq C\|B_1^*(t)\| + 4C^2\|B_1^*(t)\|^2 \leq 2C\|B_1^*(t)\|.$$

Thus proof of the considered inequality is complete by induction with respect to m. Estimate the norm of the difference

$$\delta_m(t) = X_{m+1}(t) - X_m(t).$$

We have

$$D_1(t)\delta_m(t) - \delta_m(t)D_2(t) = \delta_{m-1}(t)B_2^*(t)X_m(t) + X_{m-1}(t)B_2^*(t)\delta_{m-1}(t).$$

Choose the number $T_1 > T$ so large that for $t > T_1$, $\|B_2^*(t)\| < \frac{1}{8}C^{-1}$. Hence

$$\|\delta_m(t)\| \leq \frac{1}{2}\|\delta_{m-1}(t)\|.$$

Thus the matrix series

$$(X_1(t) - X_0(t)) + (X_2(t) - X_1(t)) + ... + (X_{m+1}(t) - X_m(t)) + ...$$

is uniformly convergent to a infinitesimal matrix $R_1^*(t)$ which is obviously a solution to equation (11.46).

Let us prove that equation (11.46) has not any other sufficiently small solution for $t \to +\infty$. Suppose the contrary. That is, for a sufficiently small number $\varepsilon > 0$, there is another solution $R_\varepsilon(t)$ such that $\|R_\varepsilon(t)\| < \varepsilon$. On putting $\delta(t) = R_1^*(t) - R_\varepsilon(t)$ we obtain the relation

$$\Delta_1(t)\delta(t) - \delta(t)\Delta_2(t) = \delta(t)B_2^*(t)R_1^*(t) - R_\varepsilon(t)B_2^*(t)\delta(t).$$

Consequently in the same way as it is proved above for $t \gg 1$ leads to the inequality

$$\|\delta(t)\| \leq C[\varepsilon\|B_2^*(t)\|\|\delta(t)\| + \|R_1^*(t)\|\|B_2^*(t)\|\|\delta(t)\|]$$
$$\leq 2C\varepsilon\|\delta(t)\| \leq \frac{1}{2}\|\delta(t)\|$$

which leads to a contradiction. Thus, we proved that the considered matrix equation has an isolated infinitesimal solution $R_1^*(t)$. To prove the inclusion $R_1^*(t) \in \tilde{Q}$, it is sufficient to prove that any element of the matrix $R_1^*(t)$ is an algebraic element under the field Q. Indeed matrix equation (11.46) is equivalent to a system of pq algebraic equations with pq unknown elements which in the same time are elements of the matrix X in equation (11.46). Thus the assertion follows from Theorem 8.35. $\qquad\square$

Consider a matrix equation of the form

$$X' = g(t)P(t)X \qquad (11.49)$$

Proposition 11.18. *Let all elements of the matrix $P(t)$ and the function $g(t)$ belong to an algebraic closed field Q of type N. Let $\Pi\{g(t)\} > -1$ and let there be a finite limit $\lim_{t\to+\infty} P(t) = P(\infty)$. However let the complete set of characteristic roots of the matrix $P(\infty)$ be the union $J_1 \cup J_2$ (determined in Lemma 11.15), and let J_1 and J_2 be disjoint. Let D_1^* and D_2^* be constant matrices possessing the complete sets of roots J_1 and J_2, respectively, and there be a (constant) non-singular matrix H such that*

$$H^{-1}P(\infty)H = \operatorname{diag}(D_1^*, D_2^*).$$

Then there exist matrices $U(t) \in \{Q\}$ and $\Upsilon(t) \in O_t$ such that

$$\lim_{t\to+\infty} U(t) = H$$

and the linear transformation $X = U(t)V$ of system (11.49) leads to the system

$$V' = [g(t)\Delta(t) + \Upsilon(t)]V, \qquad (11.50)$$

where $\Delta(t) = \operatorname{diag}(D_1(t), D_2(t)) \in \{Q\}$. Here $D_1(t)$ and $D_2(t)$ are square matrices of orders p and q, respectively, and

$$\lim_{t\to+\infty} \operatorname{diag}(D_1(t), D_2(t)) = \operatorname{diag}(D_1(\infty), D_2(\infty)) \equiv \operatorname{diag}(D_1^*, D_2^*).$$

PROOF. The transformation to be found is a result of the following substitutions. The first substitution $X = HY_0$ leads to the system $Y_0' = g(t)\Delta_0^*(t)Y_0$, where $\Delta_0^*(t) = H^{-1}\Delta(t)H$. The last matrix can be written in the form $D_0^*(t) = D_0(t) + B_0(t)$, where

$$\Delta_0(t) = \Delta_0^*(t)\operatorname{diag}(E_p, E_q) + B_0(t).$$

The matrix $\Delta_0(t)$ is quasi-diagonal and it is written in the form

$$\Delta_0(t) = \operatorname{diag}(D_{01}(t), D_{02}(t)),$$

where $D_{01}(t)$, $D_{02}(t)$ are square matrices of orders p and q, respectively and

$$\lim_{t \to +\infty} \text{diag}(D_{01}(t), D_{02}(t)) = \text{diag}(D_1^*, D_2^*).$$

$B_0(t)$ is infinitesimal for $t \to +\infty$ and it has the same form as the matrix $B(t)$, that is

$$B_0(t) = \begin{pmatrix} 0 & B_{01}(t) \\ B_{02}(t) & 0 \end{pmatrix},$$

where E_p and E_q are unit matrices of orders p and q, respectively, B_{01} and $B_{02}(t)$ are $p \times q$ and $q \times p$ matrices, respectively. Thus, the first substitution leads to the system

$$Y_0' = g(t)[D_0(t) + B_0(t)]Y_0.$$

The second substitution is

$$Y_1 = (E + R_0(t))Y_0, \quad R_0(t) \in Q.$$

$R_0(t)$ is chosen (according to 11.16) such that $\Pi\{R_0(t)\} \leq \Pi\{B_0(t)\}$ and

$$(E + R_0(t))^{-1}(\Delta_0(t) + B_0(t))(E + R_0(t))$$

have to be a quasi-diagonal matrix according to Lemma 11.15. The second substitution leads to the following system

$$\begin{aligned} Y_1' &= g(t)[(E + R_0(t))^{-1}(\Delta_0(t) + B_0(t))(E + R_0(t)) \\ &\quad - \frac{1}{g(t)}(E + R_0(t))^{-1}R_0'(t)]. \end{aligned}$$

Since $\Pi\{\|(E + R_0(t)\|\} \leq -1$, $\Pi\{1/g(t)\} = -\Pi\{g(t)\} < 1$ and $\Pi\{\|R_0'(t)\|\} \leq -1$ the last equation may be written in the form

$$Y_1' = (\Delta_1(t) + B_1(t))Y_1, \tag{11.51}$$

where $\Delta_1(t)$ is a quasi-diagonal matrix of the same structure as the matrix $\Delta_0(t)$, it belong to Q and

$$\lim_{t \to +\infty} \Delta_1(t) = \text{diag}(D_1^*, D_2^*).$$

$B_1(t)$ is a matrix of the form

$$B_1(t) = \begin{pmatrix} 0 & B_{11}(t) \\ B_{12}(t) & 0 \end{pmatrix}$$

and

$$\Pi\{\|B_1(t)\|\} \leq \Pi\{\|(1/g(t))R_0'(t)\|\} \leq \Pi\{\|R_0(t)\|\} - \sigma.$$

Here $\sigma = 1 - \Pi\{g(t)\} > 0$. Making a succession of substitutions analogous to the second case, we obtain a sequence of matrices

$$\{U_m(t) = H(E + R_0(t))...(E + R_m(t))\} \quad (m = 1, 2, ...),$$

where $\Pi\{R_m(t)\} \le \Pi\{R_0(t)\} - m\sigma$. Owing to Theorem 9.21 we can conclude that there exists a matrix $U(t) \in \{Q\}$ such that

$$\Pi\{U(t) - U_m(t)\} \to -\infty \text{ for } m \to \infty$$

which is the matrix to be found. $\square$

Remark 11.19. Evidently Proposition 11.18 is valid also in the case where $P(t) \in \{Q\}$ (instead of the inclusion $P(t) \in Q$).

Now we consider one more transformation of similarity. Given a (constant Frobenius) matrix F in the form

$$F = \begin{pmatrix} 0 & 1 & 0 & 0 & ... & 0 & 0 \\ 0 & 0 & 1 & 0 & ... & 0 & 0 \\ \cdot & \cdot & \cdot & \cdot & & \cdot & \cdot \\ 0 & 0 & 0 & 0 & ... & 0 & 1 \\ -a_n & -a_{n-1} & -a_{n-2} & -a_{n-3} & ... & -a_2 & -a_1 \end{pmatrix}. \quad (11.52)$$

Lemma 11.20. *Let the complete set of characteristic roots J of matrix (11.52) be $J_1 \cup J_2$ (which are determined in Lemma 11.15) and let J_1 and J_2 be disjoint. Then there exists a non-singular constant matrix H such that $H^{-1}FH = \text{diag}(F_1, F_2)$. Here F_1 and F_2 are constant matrices of form (11.52) of orders p and q respectively. Moreover the first row of the matrix H has the following form: the first and $(p+1)-$th elements are equal to 1 and the remaining elements are all zeros.* [S_*] (*on* Γ).

PROOF. Let us consider the system

$$X' = \text{diag}(F_1, F_2)X, \quad (11.53)$$

where $X = (x_1, ..., x_n)^T$ and $X' = (x_1', ..., x_n')^T$. A non-singular transformation $Y = HX$ with constant matrix H reduces system (11.53) to a system $Y' = FY$ where $F = H\text{diag}(F_1, F_2)$ or (which is the same) $\text{diag}(F_1, F_2) = H^{-1}FH$. We obtain the required transformation by means of the following procedure: Put $y_1 \equiv y = x_1 + x_{p+1}$ and successively differentiate this relation $(n-1)$ times according to system (11.53). We have $y_2 \equiv y' = x_1' + x_{p+1}' = x_2 + x_{p+2}$ and so forth. The identities $y_1' = y_2, y_2' = y_3, ..., y_{n-1}' = y_n$ hold for all the solutions to the obtained system which means that F is a Frobenius matrix. Its complete set of characteristic roots coincides with one of the matrix $\operatorname{diag}(F_1, F_2)$. Moreover the matrix H is non-singular. Indeed the set of all functions $y_1(t)$ consists of all functions $x_1(t) + x_p(t)$ which a linear n–dimensional space. Hence the substitution $Y = HX$ is not singular. Hence $\det H \neq 0$. Besides since $y_1 = x_1 + x_p$ we obtain the first row of the matrix R is in the required form. Thus we obtained the transformation $Y = HX$ with the matrix H to be found.   □

4.2   *THEOREM OF QUASI-DIAGONALIZATION*

The results obtained above can be applied to equation (11.1) which should be reduced to a system putting $x = y_1, x' = y_2, ..., x^{(n-1)} = y_n$. Hence for $Y = (y_1, ..., y_n)$ we have the system

$$X' = A(t)X \tag{11.54}$$

with the matrix

$$A(t) =$$

$$\begin{pmatrix} 0 & 1 & 0 & 0 & \cdots & 0 & 0 \\ 0 & 0 & 1 & 0 & \cdots & 0 & 0 \\ \cdots & \cdots & \cdots & \cdots & & \cdots & \cdots \\ 0 & 0 & 0 & 0 & \cdots & 0 & 1 \\ -a_n(t) & -a_{n-1}(t) & -a_{n-2}(t) & -a_{n-3}(t) & \cdots & -a_2(t) & -a_1(t) \end{pmatrix}$$

The characteristic equation of the matrix $A(t)$ coincides with the characteristic equation

$$H(t, y) \equiv y^n + a_1(t)y^{n-1} + ... + a_n(t) = 0. \tag{11.3}$$

Let the set of functions $\{[a_j(t)]^{1/j}\}$ have a function of the greatest growth $g(t)$ with the estimate $\Pi\{g(t)\} > -\infty$. Substitute $y = g(t)u$ in (11.3). We obtain the so called reduced equation

$$P(t, x) \equiv u^n + p_1(t)u^{n-1} + ... + p_n(t) = 0, \tag{11.55}$$

where $p_j(t) = a_j(t)[g(t)]^{-j}$. Let there exist the limits

$$\lim_{t \to +\infty} p_j(t) = p_j(\infty) \equiv \tilde{p}_j \neq \infty. \tag{11.56}$$

Pass to the limit in (11.55). As the result we obtain the *limiting equation*

$$\tilde{P}(v) = v^n + \tilde{p}_1 v^{n-1} + ... + \tilde{p}_n = 0. \tag{11.57}$$

Lemma 11.21. *Let $a_i(t) \in \{Q\}$ where Q is an algebraically closed field of type N $(i = 1, 2, ..., n)$. Let $\Pi\{g(t)\} > -1$. Then there exists the limiting equation for equation (11.3) (where at least one root does not vanish). Let the complete set of roots of equation (11.57) be $J = J_1 \cup J_2$ determined in Lemma 11.15 (the sets J_1 and J_2 are disjoint). Then there exist the following differential expressions*

$$
\begin{aligned}
F_p(t,x) &= x^{(p)} + b_1(t)x^{(p-1)} + ... + b_p(t)x, \\
F_q(t,x) &= x^{(q)} + c_1(t)x^{(q-1)} + ... + c_q(t)x, \\
F_p^*(t,x) &= x^{(p)} + b_1^*(t)x^{(q-1)} + ... + b_p^*(t)x, \\
F_q^*(t,x) &= x^{(q)} + c_1^*(t)x^{(q-1)} + ... + c_q^*(t)x, \\
\Delta_{p-1}(t,x) &= \alpha_1(t)x^{(p-1)} + ... + \alpha_p(t)x, \\
\Delta_{q-1}^*(t,x) &= \alpha_1^*(t)x^{(q-1)} + ... + \alpha_q^*(t)x
\end{aligned}
$$

such that

$$
F(t,x) = F_q(t, F_p(t,x)) + \Delta_{p-1}(t,x)
$$
$$
F(t,x) = F_p^*(t, F_q^*(t,x)) + \Delta_{q-1}^*(t,x)
$$

and

$$
a_1(t) = b_1(t) + c_1^*(t) - \frac{pq}{t}P\{g(t)\} + \frac{\alpha(t)}{t}. \tag{11.58}
$$

Here the functions $b_i(t)$, $b_i^*(t)$, $c_j(t)$, and $\alpha(t)$ belong to $\{Q\}$, $\alpha(t) \in C_t$, and $\alpha_i(t)$, $\alpha_j^*(t)$ belong to O_t. Moreover the union of FFS of equations $F_p(t,x) = 0$ and $F_q^*(t,x) = 0$ forms an FFS of the equation $F(t,x) = 0$.

PROOF. Let us prove the existence of the expressions $F_p(t,x)$, $F_q(t,x)$ and $\Delta_{p-1}(t,x)$. The proof of the existence of the expressions $F_p^*(t,x)$, $F_q^*(t,x)$ and $\Delta_{p-1}^*(t,x)$ is similar.

In (11.54) let us make the substitutions:

$$
y_1 = x_1, \quad y_j = \frac{dy_{j-1}}{d\tau}, \quad \text{where } \tau = \int g(t)dt, \quad j = 2, ..., n
$$

and the differentiation is made according to system (11.54), that is

$$
x_2 = \frac{dx_1}{dt} = \frac{dy_1}{d\tau}\frac{d\tau}{dt} = g(t)y_2,
$$
$$
x_3 = \frac{dx_2}{dt} = g'(t)y_2 + g(t)\frac{dy_2}{dt} = g'(t)y_2 + g^2(t)y_3
$$

and so forth. This results in the transformation $X = G(t)Y$, where $G(t)$ is a triangular matrix with the elements $1, g(t), ..., g^{n-1}(t)$ on the main diagonal.

The considered substitution is equivalent to the substitution $\tau = \int g(t)dt$ of the independent variable t in equation (11.1) which leads to the equation

$$x_\tau^{(n)} + q_1(t)x_\tau^{(n-1)} + ... + q_n(t)x = 0, \qquad (11.59)$$

where $q_j(t) \in \{Q\},\ j = 1, 2, ..., n$.

Since $g(t) \in A_t$ and by Proposition 6.7 we have

$$g^{(s)}(t) = t^{-s}[k(k-1)...(k-s+1) + \beta_s(t)]g(t),$$

where $k = P\{g(t)\}$ and $\beta_s(t) \in C_t$ ($s = 1, 2, ...$). Then it is easy to show that

$$x^{(s)} = g^s(t)\left[x_\tau^{(s)} + \frac{\rho_{s1} + \beta_{s1}}{\tau}x_\tau^{(s-1)} + ... + \frac{(\rho_{ss-1} + \beta_{ss-1}(\tau)}{\tau^{s-1}}x_\tau'\right],$$
$$(11.60)$$

where $\rho_{si} = \text{const}$, $\beta_{si} \in C_t$ ($i = 1, 2, ..., s-1$) so that (see (11.56)) $\lim_{t\to+\infty} p_j(t) = \lim_{t\to+\infty} a_j(t)g^{-j}(t) = \tilde{p}_j$. Thus we obtain system (11.49) where $P(t)$ is a Frobenius matrix of with its last row

$$-p_n(t), -p_{n-1}(t), ..., -p_1(t).$$

The resulting system satisfies Proposition 11.18. Hence system (11.50) may be obtained by means of the transformation $Y = U(t)V$. Moreover $\lim_{t\to+\infty} U(t) = H$, where H is determined in Proposition 11.18.

Let $\Upsilon(t)$ (see (11.50)) be a null matrix. This means that the system is divided into two systems of orders p and q:

$$V_1' = g(t)\Delta_1(t)V_1 \text{ and } V_2' = g(t)\Delta_2(t)V_2, \qquad (11.61)$$

where $V_1 = (v_1, v_2, ..., v_p)^T$ and $V_2 = (v_{p+1}, v_{p+2}, ..., v_n)^T$. It should be noted that the variable x in (11.1) is equal to

$$x = v_1 + v_{p+1} + \delta_1(t)v_1 + ... + \delta_n(t)v_n$$

where the functions $\delta_j(t)$ are elements of the first row of the matrix $U(t) - H$, and $\delta_j(t) \in C_t$ ($j = 1, 2, ..., n$). Moreover, equation (11.1) has all the solutions determined by the functions

$$x(t) = v_1(t) + v_{p+1}(t) + \delta_1(t)v_1(t) + ... + \delta_n(t)v_n(t)$$

where $v_i(t)$ are components of solutions of systems (11.61). Since $V_2(0, ..., 0)^T$ is a solution to the system $V_2' = g(t)\Delta_2(t)V_2$ then the functions of the form

$$x(t) = v_1(t) + \delta_1(t)v_1(t) + ... + \delta_p(t)v_p(t)$$

are also solutions to equation (11.1). Let us put

$$w = v_1 + \delta_1(t)v_1 + \ldots + \delta_p(t)$$

and differentiate in succession up to the $(p-1)$th derivative according to the system $V_1' = g(t)\Delta_1(t)V_1$. As the result we obtain a non-singular transformation $W = Q(t)V_1$, where $W = (w_1, \ldots, w_p)^T$. Differentiating the obtained $w^{(p-1)}$ according to the system considered and replacing the variables $v_1, \ldots, v_p$ in the expression using the relation $V_1 = Q^{-1}(t)W$, where $W = (w, w', \ldots, w^{(p)})^T$ we obtain the relation

$$F_p(t, w) \equiv w^{(p)} + b_1(t)w^{(p-1)} + \ldots + b_p(t)w = 0, \tag{11.62}$$

where $b_i(t) \in \{Q\}$ $(i = 1, 2, \ldots, p)$. Obviously any solution of the last equation is a solution to equation (11.1). Let us substitute expression $L_p(t, x)$ in (11.1). To do this we rewrite expression (11.62) in the form

$$x^{(p)} = F_p(t, x) - b_1(t)x^{(p-1)} - \ldots - b_p(t)x. \tag{11.63}$$

On differentiating (11.63) q times substituting the derivatives in (11.1) and replacing each derivative $x^{(p)}$ which appears in the differentiation by the right side of expression (11.63), we rewrite (11.1) in the following form:

$$F(t, x) = F_q(t, F_p(t, x)) + \Delta_{p-1}(t, x),$$

where $F_p(t, x)$ and $F_q(t, x)$ have properties still to be found. The order of the differential expression $\Delta_{p-1}(t, x) = 0$ is equal to $p-1$ and this equation has all the solutions to equation $F_p(t, x) = 0$ which is possible only if $\Delta_{p-1}(t, x) \equiv 0$. Thus $F(t, x) = F_q(t, F_p(t, x))$ which proves the case under consideration.

Let $\Upsilon(t)$ be a non-null matrix. Then we consider the truncated system $\tilde{V}' = g(t)\Delta(t)\tilde{V}$, where $\tilde{V} = (\tilde{v}_1, \ldots, \tilde{v}_n)^T$. We wish to recover a differential equation of nth order. To make this we put

$$w = \tilde{v}_1 + \tilde{v}_{p+1} + \delta_1(t)\tilde{w}_1 + \ldots + \delta_n\tilde{w}_n,$$

and differentiate in succession up to the $(n-1)$th derivative inclusive according to the truncated system. We obtain a differential equation in w which can be written (because of $\Upsilon(t) \in O_t$) in the following form

$$\tilde{F}(t, w) \equiv w^{(n)} + (a_1 + \nu_1(t))w^{(n-1)} + \ldots + (a_n + \nu_n(t))w = 0,$$

where $\nu_i(t) \asymp 0$ $(i = 1, 2, \ldots, n)$. On the basis of the previous case, we can write $\tilde{F}(t, x) = F_q(t, L_p(t, x))$. Hence

$$F(t, x) = F_q(t, F_p(t, x)) - (\nu_1(t)x^{(n-1)} + \ldots + \nu_n(t)x).$$

On substituting expression (11.62) (as in the previous case) in the sum $\nu_1(t)x^{(n-1)} + \ldots + \nu_n(t)x$ (and making the necessary change of the designations) we obtain the required form of the expression

$$F(t, x) = F_q(t, F_q(t.x)) + \Delta_{p-1}(t, x),$$

where the obtained differential expressions possess all the required properties.

The proof of relation (11.58) is based on the property that the sum Sp of diagonal elements of similar matrices is invariant. Let us examine the construction of the matrix $\Delta(t)$ in the previous proof (see (11.50)). Below in the proof we suppose that the functions $\eta_i(t)$ satisfy the relations $\eta_i(t)t \in C_t$ $(i = 1, 2, \ldots, 6)$. We have $SpP(t) = -p_1(t)$. By induction

$$p_1(t) = -\frac{a_1(t)}{g(t)} + \frac{n(n-1)}{2}\frac{g'(t)}{g(t)}.$$

Since $g(t) \in A_t$ we have

$$\frac{g'(t)}{g(t)} = t^{-1}P\{g(t)\} + \eta_1(t).$$

Consequently

$$g(t)p_1(t) = a_1(t) + \frac{n(n-1)}{2t}P\{g(t)\} + \eta_2(t).$$

Clearly

$$Sp(g(t)\Delta(t) + \Upsilon(t)) = Sp(g(t)P(t) - SpU^{-1}(t)U'(t)),$$

but since $tU^{-1}(t)U'(t) \in C_t$ we obtain

$$Sp(g(t)\Delta(t)) = -g(t)p_1(t) + \eta_3(t),$$

that is

$$a_1(t) = -Sp(g(t)\Delta(t)) - \frac{n(n-1)}{2t}P\{g(t)\} + \eta_4(t).$$

Similarly

$$b_1(t) = -Sp(g(t)\Delta_1(t)) - \frac{p(p-1)}{2t}P\{g(t)\} + \eta_5(t)$$

and

$$c_1^*(t) = -Sp(g(t)\Delta_2(t)) - \frac{q(q-1)}{2t}P\{g(t)\} + \eta_6(t).$$

Hence on taking into account that

$$\Delta(t) = \text{diag}(\Delta_1(t), \Delta_2(t))$$

and that $p + q = n$, we obtain (11.58).   □

Chapter 12

LINEAR DIFFERENTIAL EQUATIONS IN A SECTOR OF THE COMPLEX PLANE

In this chapter we consider a single linear differential equation of nth order of the form

$$F(x, z) \equiv x^{(n)} + a_1(z)x^{(n-1)} + \ldots + a_n(z)x = 0 \qquad (12.1)$$

in a central sector S of the complex plane for sufficiently large argument z where $a_1(z), \ldots, a_n(z)$ are holomorphic functions.

It is convenient to consider the equation in $[S]$ (recall, consideration in $[S]$ means that the equation is examined only in any fixed closed central sector S^* whose every point $z \neq 0$ is an interior point of S). This follows from a cardinal property that (as a rule) there are so called *regular sectors* for the equation. In such a sector (let, for simplicity, it is also denote by S) the formulae of analytic asymptotic representations for any (fixed) solution are preserved in any sector S^*. But the asymptotics may be considerably changed if z tends to a boundary ray of the sector S. So that the asymptotic formulae (and, clearly, the character of asymptotic behaviour) for the same solution may be different in adjacent regular sectors of the sector S. This property is called *the Stokes phenomenon*.

In this Chapter we only consider the *regular* case when the characteristic equation

$$H(y, z) \equiv y^n + a_1(z)y^{n-1} + \ldots + a_n(z) = 0. \qquad (12.2)$$

has a complete set of roots

$$\Lambda = \{\lambda_1(z), \lambda_2(z), \ldots, \lambda_n(z)\}, \qquad (12.3)$$

which possesses the property of asymptotic separability (see Definition 12.1) which is similar to the definition for real argument discussed in Chapter 11 and the results (in a regular sector) is also similar to the case of reel

307

argument, respectively. Thus if the coefficients belong to a field of type N_S, $\lambda_j(z) \in \Lambda$ and $\Pi_S\{\lambda_j(z)\} > -1$ then in a fixed sector $\widetilde{S}$ whose every point $z \neq 0$ is an interior point of S possesses a formal solution (which general speaking may depend on $\widetilde{S}$) $G_j(z) = e^{\int g_j(z)dz}$ such that $g_j(z) \sim \lambda_j(z)$ for $z \to \infty$ in $\widetilde{S}$, the function $g_j(z)$ is an asymptotic limit of the sequence

$$s_{jm+1}(z) = \lambda_j(z) - \frac{\Phi^*(s_{jm}(z), z)}{H_j(s_{jm}(z), z)} \quad s_{j0}(z) = \lambda_j(z), \quad m = 0, 1, \ldots$$

Here $\Phi^*(y, z) = \Phi(y, z) - H(y, z)$, $\Phi(y, z) = F(x, z)/x$, where $y = x'/x$ for $x \neq 0$ and $H_j(y, z)$ is determined from the identity $H(y, z) = H_j(y, z)(y - \lambda_j(z))$.

In particular

$$g_j(z) = \lambda_j(z) - \frac{1}{2}\lambda_j'(z)\frac{\partial^2 H(\lambda_j(z), z)}{\partial y^2} \bigg/ \frac{\partial H(\lambda_j(z), z)}{\partial y} + \alpha_j(s),$$

where $\Pi_{\widetilde{S}}\{\alpha_j(s)\} < -1$.

Let (in addition) Q be a normal field of type N_S and let S' be a regular sector of the sector S. Then there exists an exact solution $x_j(z) = e^{\int \gamma_j(z)dz}$ of (12.1) such that $\gamma_j(z) \asymp g_j(z)$ in any (fixed) sector S'^*. And so on.

The formulae to calculate the angles of the regular sectors are given and the Stokes phenomenon is discussed. In particular the conditions for a solution to preserve its analytic asymptotic formulae in an adjacent regular sector are given.

1.   FORMAL SOLUTIONS

The formal theory of linear differential equations on the positive semi-axis described in chapter 11, takes place for equation (12.1) with some evident alterations. We give the main Definitions and Propositions in the necessary form.

Throughout this paragraph Q *means a field of type* N_S.

Definition 12.1. We say that a set of functions $\{y_1(z), \ldots, y_n(z)\} \subset \Pi_S$ possesses the property of asymptotic separability for $z \to \infty, z \in [S]$ if the functions are not equivalent in pairs for $z \to \infty, z \in [S]$ and

$$\Pi_S\{y_i(z) - y_j(z)\} > -1 \quad \text{for any } i, j = 1, 2, \ldots, n, \quad i \neq j. \qquad (12.4)$$

Proposition 12.2. *Let* $\{y_1(z), \ldots, y_n(z)\} \subset \{Q\}_S$ *and be a set of functions possessing the property of asymptotic separability for* $z \to \infty, z \in [S]$.

Then (see (11.6))

$$V(y_1(z), ..., y_n(z)) \sim \omega(y_1(z), ..., y_n(z)) \ for \ z \to \infty, z \in [S],$$

where $\omega(y_1(z), ..., y_n(z))$ is Van der Monde's determinant for the set $\{y_1(z), ..., y_n(z)\}$. Hence

$$V(y_1(z), ..., y_n(z)) \in A_S$$

and

$$\Pi_S\{V(y_1(z), ..., y_n(z))\} > \frac{n(1-n)}{2}. \tag{12.5}$$

Let $\{a_1(t), ..., a_n(t), y(t)\} \subset \{Q\}_S$. Consider a set of functions

$$\{y^n(z), a_1(z)y^{n-1}(z), ..., a_n(z)\}. \tag{12.6}$$

Denote by $\psi(y(z))$ a function of the greatest growth of the set. It means that all the limits

$$\lim_{z \to \infty, z \in [S]} \frac{a_j(z)y^{n-j}(z)}{\psi(y(z))} \tag{12.7}$$

are finite $(j = 0, 1, ..., n; a_0(z) = 1)$.

Proposition 12.3. *Let $\{a_1(z), ..., a_n(z), y(z)\} \subset \{Q\}_S$. Let $\Pi_S\{y(z)\} > -\infty$. Then for $r = 0, 1, ...$ (see (12.2))*

$$\frac{\partial^r H(y(z), z)}{\partial y^r} = O\left(\frac{\psi(y(z))}{y^r(z)}\right) \tag{12.8}$$

for $z \to \infty, z \in [S]$, hence

$$\Pi_S\left\{\frac{\partial^r H(y(z), z)}{\partial y^r}\right\} \le \Pi_S\{\psi(y(z))\} - r\Pi_S\{y(z)\}. \tag{12.9}$$

Proposition 12.4. *Let $\{a_1(z), ..., a_n(z), y(z)\} \subset \{Q\}$. If the function $y(z)$ is not equivalent to any root of equation (12.2) and $\Pi_S\{y(z)\} > -\infty$ then there is a number $c \ne 0$ such that*

$$H(y(z), z) \sim c\psi(y(z)) \ for \ z \to \infty, z \in [S].$$

If $y(z)$ is equivalent to a root of the polynomial, then

$$H(y(z), z) = o(\psi(y(z))) \ for \ z \to \infty, z \in [S].$$

Proposition 12.5. *Let $\{a_1(z), ..., a_n(z)\} \subset \{Q\}_S$. Let $\lambda(z)$ be an asymptotically simple root of equation (12.2) for $z \to \infty, z \in [S]$ and $\Pi_S\{\lambda(z)\} > -\infty$. The polynomial $H^*(y, z)$ is defined from the identity*

$$H(y, z) = (y - \lambda(z))H^*(y, z). \tag{12.10}$$

Let $y(z)$ be an arbitrary function belonging to $\{Q\}_S$ such that

$$y(z) \sim \lambda(z) \text{ for } z \to \infty, z \in [S].$$

Then
   (1) *there is a number $p \neq 0$ such that*

$$H^*(y(z), z) \sim p\frac{\psi(\lambda(z))}{\lambda(z)} \text{ for } z \to \infty, z \in [S], \tag{12.11}$$

where $\psi(y(z))$ is a function of the greatest growth of (12.6), consequently,

$$\Pi_S\{H^*(\lambda(z), z)\} = \Pi_S\{\psi(\lambda(z))\} - \Pi_S\{\lambda(z)\}; \tag{12.12}$$

(2) $$H^*(\lambda(z), z) = \frac{\partial H(\lambda(z), z)}{\partial y}; \tag{12.13}$$

   (3) *there exists a finite limit*

$$\lim_{z \to \infty, z \in [S]} \lambda(z)\frac{\partial^2 H(\lambda(z), z)}{\partial y^2} \bigg/ \frac{\partial H(\lambda(z), z)}{\partial y} = q; \tag{12.14}$$

   (4) *introduce the designations*

$$\Phi(y, z) = \Phi_n(y) + a_1(z)\Phi_{(n-1)}(y) + ... + a_n(z)\Phi_0(y), \tag{12.15}$$

where $\Phi_0(y) = 1$ and for $m > 0$ the expression is determined from the substitution $x' = yx$ in the expression $x^{(m)}$ so that (for $x \neq 0$) $\Phi_m(y) = x^{(m)}/x$,

$$\Phi^*(y, z) = \Phi(y, z) - H(y, z). \tag{12.16}$$

If $\Pi_S\{\lambda(z)\} > -1$ then

$$\Phi^*(\lambda(z), z) = \frac{1}{2}\lambda'(z)\frac{\partial^2 H(\lambda(z), z)}{\partial y^2} + \delta(z), \tag{12.17}$$

where

$$\Pi_S\{\delta(z)\} \leq \Pi_S\{\psi(\lambda(z))\} - 2\Pi_S\{\lambda(z)\} - 2,$$

hence

$$\Pi_S\{\Phi^*(\lambda(z), z)\} \leq \Pi_S\{\psi(\lambda(z))\} - \Pi_S\{\lambda(z)\} - 1. \tag{12.18}$$

Definition 12.6. The function

$$G(z) = \exp\left[\int g(z)dz\right]$$

is said to be a *formal solution* to equation (12.1) in the class of power order growth functions in the sector S (or simply, formal solution), and the function $g(z)$ is said to be a *formal variable index* (in the class of the power order functions) to the equation in the sector S if $\Pi_S\{\Phi(z, g(z)\} = -\infty$.

The set

$$\left\{G_i(z) = \exp\left[\int g_i(z)dz\right]\right\}$$

is called a *formal fundamental system* (FFS) of solutions in the sector S (in the class of the power order growth functions) to equation (12.1) if any function $G_i(z)$ is a formal solution to the equation and

$$\Pi_S\{V(g_1(z), ..., g_n(z))\} > -\infty,$$

$i = 1, ..., n;$ (see (11.7)).

Consider two following central sectors of the complex plane.

$$S = \{z : \varphi_1 < \arg z < \varphi_2, z \neq \infty\} \text{ and } \widetilde{S} = \{z : \tilde{\varphi}_1 \leq \arg z < \tilde{\varphi}_2, z \neq \infty\} \tag{12.19}$$

where $\varphi_1, \varphi_2, \tilde{\varphi}_1, \tilde{\varphi}_2$ are constants and $\tilde{\varphi}_1 < \varphi_1 < \varphi_2 < \tilde{\varphi}_2$. Thus $\widetilde{S}$ is a closed central sector such that every its point $\neq 0$ is an interior point of the sector S.

Theorem 12.7. *Let* $\{a_1(z), ..., a_n(z)\} \subset \{Q\}_S$. *Let* $\lambda(z)$ *be asymptotically simple root of equation* (12.2), $\lambda(z) \in \{Q\}_S$ *and* $\Pi_{\widetilde{S}}\{\lambda(z)\} > -1$. *Then equation* (12.1) *has a formal solution* $G(z) = \exp\left[\int g(z)dz\right]$, *where* $g(z)$ *belongs to* $\{Q\}_S$, $g(z) \sim \lambda(z)$ *for* $z \to +\infty, z \in \widetilde{S}$ *and the function* $g(z)$ *is a formal solution of the equation* $\Phi(y, z) = 0$ *in the class of the power order growth functions with unique asymptotics in the sector* $\widetilde{S}$. *In addition* $g(z)$ *is an asymptotic limit of a sequence* $\{s_m(z)\}$ *where* $s_0(z) = \lambda(z)$ *and for* $m = 1, 2, ...$

$$s_m(z) = \lambda(z) - \frac{\Phi^*(s_{m-1}(z))}{H^*(t, s_{m-1}(z))}. \tag{12.20}$$

Hence

$$g(z) = \lambda(z) + \eta(z) + \alpha(z), \tag{12.21}$$

where

$$\eta(z) = -\frac{1}{2}\lambda'(z)\frac{\partial^2 H(\lambda(z), z)}{\partial y^2} \Big/ \frac{\partial H(\lambda(z), z)}{\partial y} \tag{12.22}$$

and

$$\Pi\{\alpha(z)\} \le -2 - \Pi\{\lambda(z)\} < -1. \tag{12.23}$$

Moreover $g(z) \in \{Q\}_{\widetilde{S}}$.

Remark 12.8. *Because of* (12.21) *equation* (12.1) *has a formal solution written in the form*

$$G(z) = (1 + \beta(z)) \exp\left[\int (\lambda(z) + q(z))dz\right], \tag{12.24}$$

where

$$\Pi_S\{\beta(z)\} \le \Pi_S\{\alpha(z)\} + 1 < 0$$

and

$$\beta(z) \to 0 \text{ for } z \to \infty, z \in \widetilde{S}.$$

Thus we obtain the asymptotics for the formal solution in an explicit form. Moreover $q(t) = (k + \delta(z))/z$, where $k = -\frac{1}{2}qP_S\{\lambda(z)\}$ and $\delta(t) \in C_{tilde S}$.

$$\lim_{z\to\infty, z\in[S]} \lambda(z)\frac{\partial^2 H(\lambda(z), z)}{\partial y^2} \Big/ \frac{\partial H(\lambda(z), z)}{\partial y} = q; \tag{12.14}$$

Consequently

$$G(z) = f(z) \exp\left[\int \lambda(z)dz\right], \tag{12.25}$$

where $P_S\{f(z)\} = k$. If in addition all the coefficients $a_1(z), ..., a_n(z)$ have their asymptotic expansions in the form of generalized power series, then all infinitesimal functions in the obtained formulas have negative analytic estimates. Consequently we may write

$$G(z) = (1 + \theta(z))z^k \exp\left[\int \lambda(z)dz\right] \tag{12.26}$$

(instead of (12.25)*) where $\Pi_{\widetilde{S}}\{\theta(z)\} < 0$.*

Theorem 12.9. *Let $\{a_1(z), ..., a_n(z)\} \subset \{Q\}_S$. Let $\lambda(z)$ be a unique root of polynomial* (12.2) *with the estimate*

$$\Pi_S\{\lambda(z)\} \le -1 \text{ and } \lambda(z) \in \{Q\}_S.$$

Then equation (12.1) has a formal solution

$$G(z) = \exp\left[\int g(z)dz\right] \text{ where } g(z) \in \{Q\}_{\widetilde{S}}$$

and if

(i) $\Pi_S\{\lambda(z)\} = -\infty$ *then* $g(z)$ *is an arbitrary function with the estimate* $\Pi_{\widetilde{S}}\{g(z)\} = -\infty$;

(ii) $\Pi_S\{\lambda(z)\} > -\infty$ *then*

$$g(z) \sim \lambda(z) \text{ for } z \to \infty, z \in S.$$

The function $g(z)$ *is a formal solution of the equation* $\Phi(y, z) = 0$ *in the class of the power order growth function with unique asymptotics in the sector* $\widetilde{S}$. *Besides* $g(z)$ *is an asymptotic limit of a sequence* $\{s_m(z)\}$ *where* $s_0(z) = \lambda(z)$ *and for* $m = 1, 2, ...$

$$s_m(z) = \lambda(z) + R(s_{m-1}(z)), \tag{12.27}$$

where

$$R(s_{m-1}(z)) = -\sum_{i=2}^{n} \frac{a_{n-i}(z)}{a_{n-1}(z)} \Phi_i(s_{m-1}(z)). \tag{12.28}$$

Hence

$$g(z) = -\frac{a_n(z)}{a_{n-1}(z)} + \alpha(z). \tag{12.29}$$

Here

$$\Pi_{\widetilde{S}}\{\alpha(z)\} \le \Pi_S\{\lambda(z)\} - \sigma, \sigma = 1 + \min_{i=1,...,j-1,j+1,...,n} \Pi_S\{\lambda(z)\} > 0. \tag{12.30}$$

Theorem 12.10. *Let* $\{a_1(z), ..., a_n(z)\} \subset \{Q\}_S$. *Let the complete set of roots* $\Lambda(z)$ *(see (12.3)) belong to* $\{Q\}_S$ *and possess the property of asymptotic separability for* $z \to \infty, z \in [\widetilde{S}]$. *Then equation (12.1) has an FFS of solutions*

$$\left\{ G_i(z) = \exp\left[\int g_i(z)dz\right] \right\} \quad (i = 1, 2, ..., n),$$

where the functions $g_i(t)$ *possess all the properties obtained in Theorems 12.7 and 12.9 respectively.*

PROOF *of Theorems 12.7, 12.9, and 12.10 is based on Lemma 4.24 and almost no differ from the corresponding proof of Theorems 10.14, 10.16, and 10.18, respectively. Therefore we omit them.*

2.   SOME AUXILIARY PROPOSITIONS

In this paragraph G is a *normal field of type* N_S.

Definition 12.11. Let $p(z) \in \{G\}_S$ and $k = \Pi_S\{p(z)\} > -1$. We say that a direction φ (ray $\arg z = \varphi$, angle) is an *extremal direction (ray, angle)* of the function $p(z)$ in S if φ is an interior angle of the sector S and there is an integer $m \in \{0, \pm 1, ...\}$ such that

$$\varphi_0 + (k+1)\varphi = \pi m, \tag{12.31}$$

where

$$\varphi_0 = \lim_{z \to \infty, z \in [S]} \Im[\ln p(z) - k \ln z]. \tag{12.32}$$

If m is an even number we say that φ is a maximal direction (ray, angle);

If m is odd we say that φ is a minimal direction (ray, angle);

If φ satisfy the relation

$$\varphi_0 + (k+1)\varphi = \frac{\pi}{2} + \pi m \tag{12.33}$$

we say that φ is a *singular direction (ray, angle)* of the function. If φ is not singular, we say that it is a *regular direction (ray, angle)*.

Mark that the order $P_S\{p(z)\}$ is a real number which is equal to $\Pi_S\{p(z)\} = k$.

Definition 12.12. An unbounded and connected set Γ of points $z \in [S], |z| \gg 1$ is called a *Stokes element* of the function $p(z)$ in S if the relation

$$\Im(p(z)z) = 0 \tag{12.34}$$

holds for any point $z \in \Gamma$.

Proposition 12.13. *Let*

$$p(z) \in \{G\}_S \ \ and \ \ k = \Pi_S\{p(z)\} > -1.$$

Then

(1) *for each extremal direction φ and only for it, there exists a unique Stokes element Γ_φ such that*

$$\arg z \to \varphi \ \ for \ \ z \to \infty, z \in \Gamma_\varphi;$$

on the Stokes element

$$dz \sim e^{i\varphi}d|z| \; for \; z \to \infty, z \in \Gamma_\varphi \;\; (i = \sqrt{-1});$$

(2) *the function*

$$\zeta(z) = \int \Re(p(z)dz)$$

has extremums on the circumcircle $z = re^{it}$ for $r \gg 1$ only at the points of intersection with the Stokes elements. And if φ is a maximal direction we have a maximum. If φ is a minimal direction we have a minimum.

PROOF. The function $p(z)$ belongs to the class A_S and has a real order $k > -1$. Let us put $z = re^{it}$ and designate by

$$h(z) = \Im[\ln p(z) - k \ln z] - \varphi_0$$

(see (12.32)). Hence $h(z) \to 0$ and

$$\frac{\partial h(re^{it})}{\partial t} = iz\Im\left(\frac{p'(z)}{p(z)} - \frac{k}{z}\right) \to 0 \; for \; z \to \infty, \; z \in [S].$$

Let Γ_φ be a Stokes element corresponding to an extremal direction φ. Its equation (see (12.34)) may be rewritten in the form

$$P(r, t) \equiv \Im \ln(p(z)z) - \pi m = 0,$$

where m is an integer given in Definition 12.11. The last equation can be represented in the form

$$P(r, t) \equiv \varphi_0 + (k + 1)t - \pi m + h(re^{it}) = 0.$$

We have

$$\partial P(r, t)/\partial t \sim k + 1 \neq 0 \; for \; r \gg 1.$$

By the well known implicit function theorem there exists one and only one function $t = \varphi + \delta(r)$ which is continuous for $r \gg 1$ and $\delta(r) \to 0$ for $r \to \infty$. On the Stokes element the relation $\Im[(p(z)z)'dz] = 0$ holds. Since $p(z) \in A_S$ clearly

$$p'(z)/p(z) = z^{-1}(k + \beta(z)), \; where \; \beta(z) \in C_S.$$

Taking into account that $p(z)z$ is a real value on the Stokes element we have

$$\Im[(k + 1 + o(1))z^{-1}dz] = 0.$$

Hence
$$(k + 1 + o(1))|z|dt + o(1)d|z| = 0.$$
Consequently $dt = O(|z|^{-1})d|z|$ which leads to the relation $dz \sim e^{it}d|z|$. Thus the first property is proved.

Let us prove the second property. Mark that $p(z)z \to \infty$ for $z \to \infty, z \in [S]$. Therefore on the Stokes element $p(z)z \to +\infty$ if φ is a maximal direction and $p(z)z \to -\infty$ if φ is a minimal direction. Let us put $\gamma(t) = \zeta(re^{it})$. We have
$$\gamma(t) = - \int \Im[p(re^{it})re^{it}]dt.$$
Consequently the extrema are achieved only if
$$\Im(p(re^{it})re^{it}) = 0,$$

that is, the extrema are achieved at the points of intersection of the circumcircle $z = re^{it}$ with the Stokes elements. Let us estimate the second order derivative of the function $\gamma(t)$ at the point of intersection with the curve Γ_φ. Because here $p(re^{it})re^{it}$ is a real value we have
$$\gamma''(t) = -\Im[ip'(re^{it})r^2e^{2it} + p(re^{it})re^{it}]$$
$$\sim -(k+1)\Re[p(re^{it})re^{it}]$$
$$= (-(k+1)p(re^{it})re^{it}.$$

Let φ be a maximal direction (the case when φ is a minimal direction is proved in the same way). Since $k + 1 > 0$ and
$$p(re^{it})re^{it} > 0 \text{ for } r \gg 1, \quad \gamma''(t) < 0,$$

that is, maximum takes place. □

Definition 12.14. A closed central sector S' is said to be a *normal sector* of the function $p(z) \in \{G\}_S$ in S if

(1) both the boundary rays of S' are regular rays of the function $p(z)$ in S;

(2) there is no more than one extremal direction of the function $p(z)$ in S', and if it is minimal, there are not singular directions of the function $p(z)$ in S'.

Consider an expression of the form
$$\int_{z_0}^{z} e^{\int_\tau^z p(s)ds}\theta(\tau)d\tau. \tag{12.35}$$

Proposition 12.15. *Let*

$$p(z) \in \{G\}_S \text{ and } \Pi_S\{p(z)\} = k > -1.$$

Let S' be a normal sector of the function $p(z)$ in S. Let the function $\theta(z)$ be holomorphic inside of the sector S' and continuous on its boundary rays for $|z| \gg 1$. Let $\Pi_S\{\theta(z)\} = -\infty$. Then there are a point $z_o \in S' \cup \infty$ and a single valued branch of (generally speaking, multi valued) integral (12.35), which we denote by $v(z_o, z)$, such that $\Pi_S\{v(z_o, z)\} = -\infty$.

PROOF. Let us put

$$W(z_1, z) = \exp\left[\int_{z_1}^{z} p(s)ds\right].$$

Then

$$v(z_o, z) = W(z_1, z)v(z_o, z_1) + v(z_1, z). \tag{12.36}$$

We denote by φ_1' and φ_2' the boundary angles of the sector S' ($\varphi_1' < \varphi_2'$).

Let in S' there be not extremal directions of the function $p(z)$. Then in S' the function

$$\zeta(z) = \int \Re(p(z)dz)$$

changes monophonically on the circumcircle $z = re^{it}$ with sufficiently large radius r. Let for definiteness $\zeta(z)$ decrease if the point z is moved from the boundary

$$\Gamma_1 = \{z : \arg z = \varphi_1', z \neq \infty\}.$$

Since φ_1' is a regular direction and $\Pi_S\{p(z)\} > -1$,

$$\lim_{z \to \infty, z \in \Gamma_1} \Re(p(z)z) = q,$$

where $q = +\infty$ or $-\infty$. Let

(a) $q = +\infty$. Put $z_o = \infty$. The function $v(z_o, z)$ is determined uniquely if the integration path between the points z_0 and z is given. We choose the

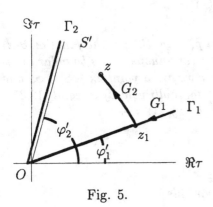

Fig. 5.

path consisting of two chains: the first one (see Fig. 5), G_1, is a segment of the ray Γ_1 which connects the points z_o and $z_1 = |z|e^{i\varphi_1}$. The second chain G_2 is a segment of a central circumcircle which connects points z_1 and z in S'. Let us make the necessary estimates of the members belonging to expression (12.35). We have

$$|W(z_1, z)| = \left|\exp\left[\int_{G_2} p(s)ds\right]\right| \le \exp\left[\int_{G_2} \Re(p(s))ds\right] \le 1;$$

$$|v(z_1, z)| = \left|\int_{G_2} \exp\left[\int_\tau^z p(s)ds\right]\theta(\tau)d\tau\right|$$

$$\le \int_{G_2} |\theta(\tau)||d\tau|$$

$$\le (\varphi_2' - \varphi_1')|z| \max_{\tau \in G_2} |\theta(\tau)|.$$

Hence $h_{\tau \in T_2} v(z_1, z) = -\infty$. In (12.34) let us substitute $\tau = re^{i\varphi_1'}$ and $s = s_1 e^{i\varphi_1'}$. Consequently

$$|v(z_o, z_1)| \le \int_{|z|}^{+\infty} \exp\left[-\int_{|z|}^r \Re(p(s_1 e^{i\varphi_1'})e^{i\varphi'_1})ds_1\right] |\theta(re^{i\varphi'})|dr.$$

On the basis of Lemma 10.19 we conclude that $h_{S'}\{v(z_o, z)\} = -\infty$. Taking into account that $v(z_o, z)$ is a holomorphic function in S' for $|z| \gg 1$, we obtain $\Pi_{S'}\{v(z_o, z)\} = -\infty$;

(b) $q = -\infty$. Let us put $z_o = t_o e^{i\varphi_1}$ where $t_o \in J_+, t_o \gg 1$. The estimate of the expressions $W(z_1, z)$ and $v(z_1, z)$ is the same just as in case (a). In the expression $v(z_o, z_1)$ we also put $t = re^{i\varphi_1'}$ and $s = s_1 e^{i\varphi_1'}$. Hence

$$v(z_o, z) \le \int_{t_o}^{|z|} \exp\left[\int_r^{|z|} \Re(p(s_1 e^{i\varphi_1'})e^{i\varphi_1'})ds_1\right] |\theta(re^{i\varphi_1'})|dr,$$

which leads to the estimate $\Pi_{S'}\{v(z_o, z)\} = -\infty$.

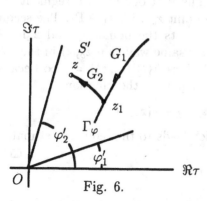

Fig. 6.

In S' let there be a maximal direction φ of the function $p(z)$. Then in S' there is a Stokes element Γ_φ where the function $p(z)z$ is positive and $p(z)z \to +\infty$ for $z \to \infty$. Put $z_0 = \infty$ and choose the path of integration consisting of two chains (see Fig. 6). The first chain G_1 is a segment of the Stokes element Γ_φ which connects the points $z_0 = \infty$ and z_1, where $|z_1| = |z|$. The second chain, G_2, is a segment of a central circumcircle which connects the points z_1 and z in S'. Since the function $\zeta(z) = \int \Re(p(z)dz)$ achieves its minimum at the point z_1, the inequality $\Re(p(\tau)d\tau) \leq 0$ is satisfied on the arc G_2. Hence $h_{G_2}\{v(z_1,z)\} = -\infty$. We have

$$|v(z_1,z)| \leq \int_{|z|}^{+\infty} \exp\left[\int_\tau^{|z|} \Re\left[p(s)\frac{ds}{d|s|}\right]d|s|\right]|\theta(\tau)d\tau|$$

on G_1. Since $dz \sim e^{it}d|z|$ for $z \to \infty$ on the Stokes element we have

$$\lim_{z\to\infty, z\in\Gamma_\varphi} \frac{\Re[g(z)dz]}{d|z|}|z| = +\infty$$

then on the basis of Lemma 10.19 $h_{G_1}\{v(z_0,z)\} = -\infty$ which leads to the required estimate $\Pi_{S'}\{v(z_0,z)\} = -\infty$.

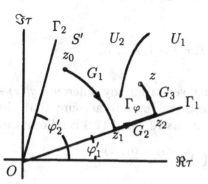

Fig. 7.

In S' let there be a minimal direction φ. Then in S' there are not singular directions. Consequently $\Re(g(z)z) \to -\infty$ for $z \to \infty, z \in S'$. In the sector S' there exists a Stokes element which divides it by two regions U_1 and U_2 (see Fig. 7) where in each region, the function

$$\zeta(z) = \int \Re(p(z)dz)$$

decreases on a central circumcircle of sufficiently large radius if the point moves from a boundary to the Stokes element. Put $z_0 \in S'$ $[z_0, z \in U_1]$

(the case $z \in U_2$ is proved in the same way). We choose the path of integration consisting of three chains. The first one G_1 is a segment of a central circumcircle which connects the point z_o with $z_1 \in \Gamma_1$. The second chain G_2 is a segment of Γ_1 which connects the points z_2 and z in S'. The expression $v(z_1, z)$ is estimated in the same way as $v(z_o, z)$ in the case when S' has not extremal directions. Besides $h\{W(z_o, z)\} = -\infty$ because $\Re(p(z)z) \to -\infty$ for $z \to \infty, z \in S'$. Consequently the function

$$v(z_o, z) = v(z_o, z_1)W(z_o, z) + v(z_1, z)$$

has the estimate $h\{v(z_o, z)\} = -\infty$ which leads to the required estimate.
$\square$

3.   MAIN THEOREMS

Theorem 12.16. *Let $\{a_1(z), ..., a_n(z)\} \subset \{G\}_S$ where G is a normal field of type N_S. Let the characteristic polynomial $H(y, z)$ (see (12.2)) have a complete set of roots Λ (see (12.3)) which possesses the property of asymptotic separability for $z \to \infty, z \in [S]$. Let $\widetilde{S}$ (see (12.9)) be a normal sector for any difference*

$$\lambda_i(z) - \lambda_j(z) \text{ where } i \neq j, i, j = 1, 2, ..., n.$$

Let

$$\left\{ G_i(z) = \exp\left[\int g_i(z)dz\right] \right\} \quad (i = 1, 2, ..., n)$$

be an FFS of solutions obtained in Theorem 12.10. Then there exists a fundamental system of solutions

$$\left\{ x_i(z) = \exp\left[\int \gamma_i(z)dz\right] \right\}$$

to equation (12.1) such that

$$\Pi_{\widetilde{S}}\{\gamma_i(z) - g_i(z)\} = -\infty \ for \ i = 1, 2, ..., n.$$

PROOF. Let us prove that $\widetilde{S}$ is a normal sector for any difference $g_i(z) - g_j(z)$ where $i \neq j$; $i, j = 1, 2, ..., n$. Indeed (taking into account that also G_S is a field of type N_S in S) from Theorems 12.7 and 12.9 it follows that

$$g_i(z) - g_j(z) = \lambda_i(z) - \lambda_j(z) + \frac{c_{ij} + \alpha_{ij}(z)}{z},$$

where $c_{ij} = \text{const}$ and $\alpha_{ij}(z) \in C_{\widetilde{S}}$. Since

$$\Pi_S\{\lambda_i(z) - \lambda_j(z)\} > -1,$$
$$\lim_{z \to \infty, z \in [S]} (\lambda_i(z) - \lambda_j(z))z = \infty.$$

Hence

$$k = P_{\widetilde{S}}\{g_i(z) - g_j(z)\} = P_S\{\lambda_i(z) - \lambda_j(z)\}$$

and

$$\varphi_0 = \lim_{z \to \infty, z \in \widetilde{S}} \Im[\ln(g_i(z) - g_j(z)) - k \ln z]$$

$$= \lim_{z \to \infty, z \in \widetilde{S}} \Im\{[\ln(\lambda_i(z) - \lambda_j(z)) - k \ln z$$

$$+ \ln[1 + \frac{c_{ij} + o(1)}{z}(\lambda_i(z) - \lambda_j(z))]\}$$

$$= \lim_{z \to \infty, z \in \widetilde{S}} \Im[\ln(\lambda_i(z) - \lambda_j(z)) - k \ln z]$$

which leads to the required property. Subsequent proof of this proposition based on Proposition 12.15 and practically it is not distinguished from the proof of Theorem 10.22. Therefore we omit it. □

Remark 12.17. Let φ be an interior ray of the sector S. By

$$S_{\varphi\varepsilon} = \{z : |\arg z - \varphi| < \varepsilon, 0 < \varepsilon \ll 1, z \neq \infty\}$$

we denote a sufficiently small sector with its bisector $\{z : \arg z = \varphi, z \neq \infty\}$. Since such a sector is normal for any difference $\lambda_i(z) - \lambda_j(z)$. Theorem 12.16 is valid in any interior sufficiently simal sector $S_\varepsilon \subset S$.

Definition 12.18. Let $\{a_1(z), ..., a_n(z)\} \subset \{G\}_S$ where G is a normal field of type N_S. Let the $H(y, z)$ (see (12.2)) have complete set of roots Λ (see (12.3)) such that

$$\Pi\{\lambda_i(z) - \lambda_j(z)\} > -1 \text{ for } i \neq j, \quad i, j = 1, 2, ..., n.$$

Then

(1) an extremal or singular direction (ray, angle) for at least one of the differences $\lambda_i(z) - \lambda_j(z)$ in S is said to be an *extremal* or *singular direction* (*ray, angle*) of equation (12.1) in S, respectively.

An interior direction (ray, angle) of the sector S which is not singular is called a *regular direction* (*ray, angle*) of the equation in the sector S;

(2) an open central sector $S^o \subset S$ is called a *regular sector* of equation (12.1) in S if all its directions are regular and its both boundary rays are singular.

Definition 12.19. Let all the conditions of Definition 12.18 be fulfilled. Let S_1 be a closed central subsector of S. Let $\lambda_i(z), \lambda_j(z)$ belong to complete set of roots (12.4) of equation (12.2). We say that the *root* $\lambda_i(z)$ *is* *on the left hand side* ($\lambda_j(z)$ *is on the right hand side*) with respect of the root $\lambda_j(z)(\lambda_i(z))$ in their *natural arrangement* in the sector S if

$$\exp\left[\int (\lambda_j(z) - \lambda_i(z))dz\right] \to 0 \text{ for } z \to \infty, z \in [S].$$

We also say that the root $\lambda_i(z)$ has a *dominant asymptotic behavior* with respect to $\lambda_j(z)$. We use a similar terminology for the corresponding formal solutions and variable indices to equation (12.1).

Example 12.20. Consider the Airy equation $x'' - zx = 0$. Its characteristic equation $y^2 - z = 0$ has roots $\lambda_{1,2}(z) = \pm\sqrt{z}$. The difference

$$\lambda_1(z) - \lambda_2(z) = 2\sqrt{z}$$

has singular angles $\varphi = \frac{\pi}{3}(2m+1)$. Thus in the complex plane we have the following singular angles $\varphi = -\pi, -\frac{\pi}{3}, \frac{\pi}{3}, \pi$ and the corresponding three regular sectors

$$S_1 = \{z : |\arg z| < \frac{\pi}{3}, z \neq \infty\}, \quad S_2 = \{z : \frac{\pi}{3} < \arg z < \pi, z \neq \infty\}$$

and $S_3 = \{z : -\pi < \arg z < -\frac{\pi}{3}, z \neq \infty\}$. Extremal angles are $\varphi = 0, \pm\frac{2}{3}\pi$. The natural arrangement of the roots is $\lambda_1(z), \lambda_2(z)$ in S_1, and $\lambda_2(z), \lambda_1(z)$ in S_2, S_3. The regular sectors and extremal rays of the equation are shown on the diagram 1. The singular lines are given by full lines and the regular lines are given by dotted lines. The natural arrangement of the roots is shown on the diagram in any regular sector where the roots are replaced by their indices.

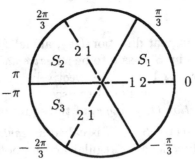

Diagram 1.

Theorem 12.21. *Let* $\{a_1(z), ..., a_n(z)\} \subset \{G\}_S$ *where* G *is a normal field of type* N_S. *Let the complete set of roots* Λ *of characteristic equation*

(12.2) *belong to* $\{G\}_S$ *and possess the property of asymptotic separability for* $z \to \infty, z \in [\tilde{S}]$. *Let*

$$\left\{ G_i(z) = \exp\left[\int g_i(z)dz \right] \right\} \quad (i = 1, 2, ..., n)$$

be an FFS of solutions obtained in Theorem 12.10. Let S_1, S_2 *be two regular adjacent sectors of equation* (12.1) *which belong to* $\tilde{S}$ (*see* (12.9)) *and let* Γ *be their common boundary ray. Then*

(1) *there exists a fundamental system of solutions*

$$\left\{ x_i(z) = \exp\left[\int \gamma_i(z)dz \right] \right\}$$

of equation (12.1) *such that*

$$\Pi_{S_1}\{\gamma_i(z) - g_i(z)\} = -\infty \quad (i = 1, 2, ..., n);$$

(2) *let* $j \in \{1, 2, ..., n\}$ *be a fixed number and let the following conditions be fulfilled: each root* $\lambda_s(z) \in \Lambda$ *which is on the right-hand side with respect to the root* $\lambda_j(z)$ *in their natural arrangement in the sector* S_1 *preserve its arrangement in the sector* S_2 *with respect to* $\lambda_j(z)$. *Then*

$$\Pi_{S_1 \cup \Gamma \cup S_2}\{\gamma_j(z) - g_j(z)\} = -\infty.$$

To prove this Theorem we have preliminarily to prove the following Lemma:

Lemma 12.22. *Let all the conditions of Theorem 12.21 be fulfilled. Let* S' *and* S'' *be central open intersecting sectors belonging to* S. *Let*

$$x(z) = \exp\left[\int \gamma(z)dz \right]$$

be a solution to equation (12.1) *which satisfies the relation* $\Pi_{S'}\{\gamma(z) - g_j(z)\} = -\infty$. *Let* S'' *be a normal sector in* S *for any difference* $\lambda_r(z) - \lambda_s(z)$ $(r \neq s;\ r, s = 1, 2, ..., n)$ *such that each root* $\lambda_s(z)$ *which is on the right-hand side with respect of the root* $\lambda_j(z)$ *in their natural arrangement in the sector* S' *remains on the right-hand side in the sector* S''. *Then* $\Pi_{S' \cup S''}\{\gamma(z) - g_j(z)\} = -\infty$.

PROOF. On the basis of Theorem 12.16 there exists an *FSS*

$$\left\{ x_i^*(z) = \exp\left[\int \gamma_i^*(z)dz \right] \right\}, \quad (i = 1, 2, ..., n)$$

where
$$\Pi_{S''}\{\gamma_i^*(z) - g_i(z)\} = -\infty.$$

Consequently

$$x(z) = x_j^*(z) + \sum_{i=1}^n c_i x_i^*(z). \tag{12.37}$$

Let us choose a sufficiently small central sector $S^o \subset S' \cap S''$ which does not contain singular directions of equation (12.1). It is obviously a regular sector and all the roots of equation (12.2) preserve their natural arrangement in it. For definiteness let it be in the form

$$\lambda_1(t), \lambda_2(t), ..., \lambda_n(t).$$

Divide expression (12.37) by $x_1^*(z)$ and pass to the limit for $z \to \infty, z \in S^o$. Hence $c_1 = 0$. In the same way, dividing (12.37) simultaneously by $x_2^*(z), ..., x_j^*(z)$ we obtain $c_2 = c_3 = ... = c_j = 0$. Therefore $x(z) = x_j(z) + \theta(z)$, where

$$\theta(z) = \sum_{i=j+1}^n \frac{x_i^*(z)}{x_j^*(z)}.$$

It is obvious that $\Pi_{S''}\{\theta(z)\} = -\infty$. Hence $\Pi_{S''}\{\gamma(z)g_j(z)\} = -\infty$. Since $\Pi_{S'}\{\gamma(z) - g_j(z)\} = -\infty$ we obtain $\Pi_{S' \cup S''}\{\gamma(z) - g_j(z)\} = -\infty$. □

Continuation of the Theorem proof. Let us fix a sufficiently small sector $S_\varepsilon \subset S_1$. On the basis of Theorem 12.16 there exists an *FSS*

$$\left\{ x_i(z) = \exp\left[\int \gamma_i(z)dz\right] \right\},$$

where
$$\Pi_{S_\varepsilon}\{\gamma_i(z) - g_i(z)\} = -\infty.$$

Let S_* be a central closed subsector of S_1 which contains the sector S_ε. Since in S_* there exists only finite number of extremal directions of equation (12.1) and there are not singular directions, it may be covered with a finite system of sufficiently small central opened sectors belonging to S_1 such that each of them is a normal for any difference

$$\lambda_r(z) - \lambda_s(z), \quad r \neq s, r, s = 1, 2, ..., n,$$

and at least one of this covering, say $S^{(1)}$, contains the sector S_o. Since the natural arrangement of the roots is preserved in $S^{(1)}$, on the basis of Lemma 12.22 we have

$$\Pi_{S^{(1)}}\{\gamma_i(z) - g_i(z)\} = -\infty.$$

Let $S^{(2)}$ be a sector of the considered covering which is intersected with $S^{(1)}$. In the same way we may conclude that

$$\Pi_{S^{(2)}}\{\gamma_i(z) - g_i(z)\} = -\infty.$$

On going from one sector of this covering to another adjacent sector we obtain $\Pi_{S_*}\{\gamma_i(z) - g_i(z)\} = -\infty$. Since S_* is an arbitrary closed sector we obtain $\Pi_{S_1}\{\gamma_i(z) - g_i(z)\} = -\infty$. Thus property (1) of the Theorem is proved.

Prove property (2). Let us choose a sufficiently small central open sector S_ε which contains ray Γ. On the basis of Lemma 12.22

$$\Pi_{S_\varepsilon}\{\gamma_j(z) - g_j(z)\} = -\infty.$$

Reasoning in the same way as in the Proof of property (1) we obtain

$$\Pi_{S_2}\{\gamma_j(z) - g_j(z)\} = -\infty.$$

Consequently

$$\Pi_{S_1 \cup \Gamma \cup S_2}\{\gamma_j(z) - g_j(z)\} = -\infty. \qquad \square$$

The so called Stokes phenomenon takes place for the class of equations considered in Theorem 12.21. The Stokes phenomenon consists of the following. Let all the conditions of Theorem 12.21 be fulfilled. Let S_1 and S_2 are two regular sectors and $x(z)$ be a solution of equation (12.1). Then there are numbers $c_i^{(1)}$ and functions $\theta_i^{(1)}(z)$ with the estimates $\Pi_{S_1}\{\theta_i^{(1)}(z)\} = -\infty$ such that

$$x(z) = \sum_{i=1}^{n} c_i^{(1)}(1 + \theta_i^{(1)}(z)) \exp\left[\int g_i(z)dz\right].$$

Here $\{G_i(z) = \exp\left[\int g_i(z)dz\right]\}$ is an FFS obtained in Theorem 12.10 Thus in the sector S_1 we have an asymptotic representation of the solution $x(z)$ in the following form

$$x(z) \sim \sum_{i=1}^{n} c_i^{(1)} \exp\left[\int g_i(z)dz\right].$$

For the sector S_2 we also have an asymptotic representation in the form

$$x(z) \sim \sum_{i=1}^{n} c_i^{(2)} \exp\left[\int g_i(z)dz\right],$$

where $c_i^{(2)}$ are constants.

The Stokes phenomenon lies in the fact that the numbers $c_i^{(1)}$ and $c_i^{(2)}$ may not coincide. This means that

$$x(z) \sim \sum_{i=1}^{n} c_i(t) \exp\left[\int g_i(z)dz\right],$$

where $c_i(t)$ are piecewise constant functions depending on $t = \arg z$.

Proposition 12.23. *Let equation (12.1) satisfy all the condition of Theorem 12.16. Let S_1 and S_2 be two adjacent regular sectors of equation (12.1) and*
$$\Gamma = \{z : \arg z = \varphi, z \neq \infty\}$$
be their common boundary ray. Let $W_j(z)$ be a dominant asymptotics in S_1 and $W_k(z)$ be a dominant asymptotics in S_2. Let $x(z)$ be a solution of equation (12.1) such that

(a) $x(z) \sim c_j W_j(z)$ for $z \to \infty, z \in [S_1]$ and $x(z) \sim c_k W_k(z)$ for $z \to \infty, z \in [S_2]$, where c_j and c_k, are non-zero constants;

(b) let $W_m(z) = o(W_j(z))$ and $W_m(z) = o(W_k(z))$ as $z \to \infty, z \in \Gamma$ for any $m \neq j, k$. Then

$$x(z) \sim c_j W_j(te^{i\varphi}) + c_k W_k(te^{i\varphi}) \qquad (12.38)$$

for $t \to +\infty$ $(i = \sqrt{-1})$.

The proof of this proposition is obvious.

Remark 12.24. We suppose that at least one of the numbers c_j, c_k is not equal to zero; if for instance $c_j = 0$ it means that the solution $x(z)$ has a non-dominant asymptotic in S_1.

Remark 12.25. If all the condition of Proposition 12.23 are fulfilled we say that there is a *normal situation* about the solution $x(z)$ on Γ. Then the following assertion takes place: *the asymptotic representation of the function $x(z)$ on Γ consists on a sum of asymptotic solutions in the adjacent sectors of $x(z)$ extended on the ray Γ.*

Examples 12.26. 1°. Given the Airy equation $x'' - zx = 0$. Its regular sectors and extremal directions were obtained in the example 12.20. Here we preserve the accepted terminology. By Theorem 12.21 (1) the equation has two linearly independent solutions

$$x_{1,2}^*(z) = (1 + \alpha_{1,2}^*(z))W_{1,2}(z),$$

where

$$\Pi_{S_1}\{\alpha^*_{1,2}(z)\} \le -\frac{3}{2}$$

and

$$W_{1,2}(z) = z^{-1/4} \exp\left[\pm\frac{2}{3}z^{3/2}\right]. \tag{12.39}$$

The solution $x^*_2(z)$ has a minor asymptotics with respect to the solution $x^*_1(z)$ in S_1. Due to property (2) of Theorem 12.21 it follows that the solution $x^*_2(z)$ preserves its asymptotics in the sectors S_2 and S_3. Thus the solution $x^*_2(z))$ has the asymptotics $W_2(z)$ in the entire complex plane except the ray $\{z : \arg z = -\pi, z \ne \infty\}$. Renumber $x^*_2(z)$ by $x_2(z)$. In the same way we may consider another solution $x_1(z)$ which has a minor asymptotics $W_1(z)$ in the sector S_2 and preserve its asymptotic in the sectors S_1 and S_3. Thus $x_1(z)$ has the asymptotic $W_1(z)$ in the entire complex plane except the ray $\{z : \arg z = -\frac{\pi}{3}, z \ne \infty\}$. Hence there exists an FSS $\{x_1(z), x_2(z)\}$ which preserves its asymptotics in the entire complex plane except two rays determined by the angles $-\pi$ and $-\frac{\pi}{3}$.

Let $x(z)$ be a solution of the Airy equations. Then there are two constants c_1 and c_2 such that

$$x(z) = c_1 x_1(z) + c_2 x_2(z).$$

Hence (if $c_1 \ne 0$ and $c_2 \ne 0$) $x(z) \sim c_1 x_1(z)$ for $z \to \infty, z \in [S_1]$ and $x(z) \sim c_2 x_2(z)$ for $z \to \infty, z \in [S_2]$. On the singular ray $z = te^{i\pi/3}$ we have

$$
\begin{aligned}
x(te^{i\pi/3}) &= c_1 x_1(te^{i\pi/3}) + c_2 x_2(te^{i\pi/3}) \\
&\sim c_1 W_1(te^{i\pi/3}) + c_2 W_2(te^{i\pi/3})
\end{aligned}
$$

for $t \to +\infty$. Hence we may conclude that there is a normal situation for every solution $x(z)$ on any singular ray $z = te^{i\pi/3}$ of the equation. In the same way it is easy to show that a normal situation takes place for every solution of the equation on any singular ray. On the basis of the rule obtaining in Remark 12.25 let us calculate the missing asymptotics of the solutions $x_{1,2}(z)$ on the rays $z = te^{-i\pi/3}$ and $z = te^{-i\pi}$. We have

$$x_1(te^{-i\pi/3}) \sim W_1(te^{-i\pi/3}) + W_1(te^{5i\pi/3})$$

$$= 2e^{-i\pi/6}\,(1/\sqrt[4]{t})\cos((2/3)t\sqrt{t} - \pi/4);$$

$$x_2(-t) \sim W_2(te^{i\pi}) + W_2(te^{-i\pi}) = (1/\sqrt[4]{t})\cos((2/3)t\sqrt{t} - \pi/4)$$

for $t \to +\infty$.

It is possible to calculate the asymptotic approximations of solutions to the Airy equation proceeding from the Cauchy problem. Consider for example the FSS $\{x_{10}(z), x_{01}(z)\}$ with the initial conditions

$$x_{10}(0) = x'_{01}(0) = 1 \text{ and } x'_{10}(0) = x_{01}(0) = 0.$$

They have a dominant asymptotics in any regular sector. Thus

$$x_{10}(z) \sim pW_1(z) \text{ and } x_{01}(z) \sim qW_1(z) \text{ for } z \to \infty, z \in [S_1].$$

It is possible to calculate approximately the coefficient p (analogously for q) proceeding from the limiting relation

$$\lim_{t \to +\infty} \frac{x_{10}(t)}{W_1(t)} = 1,$$

if we take advantage of the approximation

$$p \sim \frac{x_{10}(t)}{W_{10}(t)} \text{ for } t \gg 1$$

$x_{10}(t)$ may be calculate by means of any known computational method. In this simple case (expanding the solution $x_{10}(t)$ into a power series and applying the known asymptotic method of power series summation) we can give the exact values of p and q :

$$p = \frac{\sqrt{\pi}}{3^{1/3}\Gamma(1/3)} \text{ and } q = \frac{\sqrt{\pi}}{9^{1/3}\Gamma(2/3)},$$

where $\Gamma(t)$ is the Euler Gamma function.

Let us calculate the asymptotics of solutions $x_{10}(z)$ and $x_{01}(z)$ on the adjacent sectors S_2. In S_2 we can write

$$x_{10}(z) \sim p^*W_2(z) \text{ and } x_{01}(z) \sim q^*W_2(z)$$

for $z \to \infty, z \in [S_2]$. The Airy equation is invariant under the substitution $z = \tau e^{2i\pi/3}$. Hence we have

$$x_{10}(te^{2i\pi/3}) = x_{10}(t) \text{ and } x_{01}(te^{2i\pi/3}) = e^{2i\pi/3}x_{01}(t)$$

on the extremal ray $\{z : \arg z = \frac{2}{3}\pi, z \neq \infty\}$. Consequently $pW_1(t) = p^*W_2(te^{2i\pi/3})$. Hence $p^* = pe^{i\pi/3}$. In the same way $q^* = qe^{(5/6)i\pi}$.
On the singular ray $z = te^{i\pi/3}$ for $t \to +\infty$

$$
\begin{aligned}
x_{10}(te^{i\frac{\pi}{3}}) &\sim pW_1(te^{i\pi/3}) + p^*W_2(te^{i\pi/3}) \\
&= 2pt^{-1/4}\cos\left(\frac{2}{3}t^{3/2} - \frac{\pi}{12}\right);
\end{aligned}
$$

$$
\begin{aligned}
x_{01}(te^{i\pi/3}) &\sim qW_1(te^{i\pi/3}) + q^*W_2(te^{i\pi/3}) \\
&= 2qt^{-1/4}\cos\left(\frac{2}{3}t^{3/2} - \frac{5\pi}{12}\right).
\end{aligned}
$$

In the same way it is possible to find the asymptotics of the solutions in the sector S_3 and on the negative ray $z = -t$.

In conclusion consider ones more solution

$$x(z) = \frac{x_{10}(z)}{p} - \frac{x_{01}(z)}{q}.$$

Clearly $x(z)$ has a minor asymptotics in the sector S_1. Hence there is a number $c \neq 0$ such that $x(z) \sim cW_2(z)$ for $z \to \infty, z \in [S_1]$.

It should be noted that it is impossible to calculate c proceeding from the relation

$$c = \lim_{t \to +\infty} \frac{x(t)}{W_2(t)}$$

and replacing c by the approximate value $x(t)/W_2(t)$ for $t \gg 1$. Indeed if we calculate (by means of an approximate method) the value $x(t)$ proceeding from the initial conditions $x(0) = 1/p$ and $x'(0) = -1/q$, then the inevitable errors in the calculations and in writing of the initial conditions may lead to a solution which have a dominant asymptotics. Consequently the errors of the number c calculation do not lend itself to control and may unlimited grow as t increases. To correct calculate the coefficient c, we should consider the solution $x(t)$ in an adjacent regular sector and then establish the connection of asymptotic behavior of the solution in the sector S_1 and in the adjacent sector. On the basis of Theorem 12.21 (2) the solution $x(z)$ considered in the sector S_1 preserves its asymptotic on the ray $z = te^{2i\pi/3} \subset S_2$. Hence

$$x(te^{2i\pi/3}) \sim cW_2(te^{2i\pi/3}).$$

In the other hand

$$x(te^{2i\pi/3}) = \frac{1}{p}x_{10}(te^{2i\pi/3}) - \frac{1}{q}x_{01}(te^{2i\pi/3}) \sim \left(\frac{p^*}{p} - \frac{q^*}{q}\right) W_2(te^{2i\pi/3})$$

for $t \to +\infty$. Consequently $c = p^*/p - q^*/q = \sqrt{3}$.

Now we can rewrite the initial conditions of the solution $x_2(z)$ (belonging to the previous considered FSS). We have $x_2(z) = x(z)/c$. Hence $x_2(0) = \sqrt{3}/p$ and $x_2'(0) = -\sqrt{3}/q$.

2°. Given the equation

$$x''' - z^2 x = 0. \tag{12.40}$$

Its characteristic equation $H(y, z) \equiv y^3 - z^2 = 0$ has roots

$$\lambda_1(z) = z^{2/3} \quad \text{and} \quad \lambda_{2,3}(z) = e^{\pm 2i\pi/3} z^{2/3}$$

which form a complete set of roots possessing the property of asymptotic separability (in any sector S). The regular sectors and extremal rays of the equation are shown in the diagram 2. The singular lines are given by full lines and the regular lines are given by dotted lines. The natural arrangement of the roots is shown in the diagram in any regular sector where the roots are replaced by their indices. On the basis of Theorem 12.21 it easy to show that in an arbitrary (fixed) regular sector S_j there exists an FSS

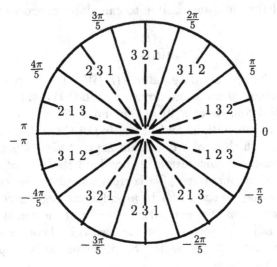

Diagram 2.

$$\left\{ x_{js}(z) = (1 + \alpha_{js}(z))z^{-2/3} \exp\left[\frac{3}{5}\varepsilon_s z^{5/3}\right] \right\}, \quad s = 1, 2, 3,$$

where $\Pi_{S_j}\{\alpha_{js}(z)\} \leq -\frac{5}{3}$, $\varepsilon_1 = 1$ and $\varepsilon_{2,3} = e^{\pm 2i\pi/3}$. Let us introduce the designations of asymptotics

$$W_s(z) = z^{-2/3} \exp\left[\frac{3}{5}\varepsilon_s z^{5/3}\right]. \tag{12.41}$$

For instance let us consider a solution

$$x(z) \sim W_1(z) \text{ for } z \to \infty, z \in [S_1],$$

where

$$S_1 = \left\{ z : -\frac{3\pi}{5} < \arg z < -\frac{2\pi}{5}, z \neq \infty \right\}.$$

Considering the diagram we see that $x(z)$ preserves its asymptotics in the sector $\{z : -\pi < \arg z < \phi/5, z \neq \infty\}$.

Consider the fundamental system of solutions which consists of functions $x_{100}(z), x_{010}(z)$ and $x_{001}(z)$ which satisfy the following initial conditions:

$$x_{100}(0) = 1, \quad x'_{100}(0) = 0, \quad x''_{100}(0) = 0;$$

$$x_{101}(0) = 0, \quad x'_{010}(0) = 1, \quad x''_{010}(0) = 0;$$

$$x_{001}(0) = 0, \quad x'_{010}(0) = 0, \quad x''_{010}(0) = 1.$$

On the positive semi-axis J_+, $W_1(z)$ is a dominant asymptotics. Hence

$$x_{100}(t) \sim pW_1(t), x_{010}(t) \sim qW_1(t) \text{ and } x_{001}(t) \sim rW_1(t)$$

for $t \to +\infty$. For information

$$p = \frac{\Gamma(3/5)\Gamma(4/5)}{2\pi\sqrt{3}}, \quad q = \frac{\sqrt{3}}{2\sin\frac{\pi}{10}}, \quad r = \frac{\Gamma(1/5)\Gamma(2/5)}{2\pi\sqrt{3}\sqrt[5]{25}}.$$

In conclusion let us calculate the asymptotics of the solution

$$x(t) = \frac{x_{100}(t)}{p} - \frac{x_{010}(t)}{q}$$

for $t \to +\infty$. On the positive semi-axis the considered asymptotics is not dominant. Hence there are constants A and B such that (for $t \to +\infty$)

$$x(t) \sim AW_2(t) + BW_3(t). \tag{12.42}$$

To obtain the constants let us consider the diagram. We see that by turn beginning from positive semi-axis to the rays with the angles $\varphi = \pm\frac{2\pi}{5}$, the asymptotics of the solution still is determined by expression (12.42). On the ray $z = te^{2i\pi/5}$, $W_3(z)$ is the dominant asymptotics. Hence on this ray $x_{100}(z) \sim p^*W_3(z)$ and $x_{010}(z) \sim q^*W_3(z)$. The equation is invariant under the substitution $z = \tau e^{2i\pi/5}$. Consequently it is possible to obtain the exact dependence between p and p^*, q and q^*. We have

$$x_{100}(te^{2i\pi/5}) = x_{100}(t) \text{ and } x_{010}(te^{2t\pi/5}) = e^{2i\pi/5}x_{010}(t).$$

This gives

$$p^*W_3(te^{2i\pi/5}) = pW_1(t)$$

and

$$q^*W_3(te^{2i\pi/5}) = qe^{2i\pi/5}W_1(t).$$

Hence $p^* = pe^{4i\pi/15}$ and $q^* = qe^{2i\pi/3}$. Thus

$$B = \frac{p^*}{p} - \frac{q^*}{q} = e^{4i\pi/15} - e^{2i\pi/3}.$$

In the same way, considering the solution $x(z)$ on the ray with the angle $\varphi = -2\pi/5$ we obtain

$$A = e^{-4i\pi/15} - e^{-2i\pi/3}.$$

Hence for $t \to +\infty$

$$x(t) \sim (e^{-4i\pi/15} - e^{-2i\pi/3})t^{-2/3} \exp\left[\frac{3}{5}e^{2i\pi/3}t^{5/3}\right]$$

$$+(e^{4i\pi/15} - e^{2i\pi/3})t^{-2/3} \exp\left[\frac{3}{5}e^{2i\pi/3}t^{5/3}\right]$$

$$= 4(\sin\frac{2\pi}{15})t^{-2/3}e^{-(3/10)t^{5/3}} \cos\left(3\sqrt{3}\,t^{5/3} - \frac{2\pi}{15}\right).$$

4.   QUALITATIVE CHARACTERISTICS OF LINEAR DIFFERENTIAL EQUATIONS IN A SECTOR OF THE COMPLEX PLANE

In the same way as in section 11.3 we can obtain the criteria of asymptotic stability and instability of equation (12.1) in the complex plane.

Equation (12.1) is said to be *asymptotically stable* in S (on a ray Γ) for $z \to \infty$ if each its solution $x(z) \to 0$ for $z \to \infty, z \in S$ $(z \in \Gamma)$.

The equation is *unstable* in S (on Γ) if there exists at least one unlimited solution of the equation for $z \to \infty, z \in S (z \in \Gamma)$.

The following proposition is a simple consequence of Theorem 12.21

Theorem 12.27. *Let $\{a_1(z), ..., a_n(z)\} \subset \{G\}_S$, where G is a normal field of type N_S, and characteristic equation (12.2) have a complete set of roots possessing the property of asymptotic separability for $z \to \infty, z \in [S]$. Let $\lambda(z)$ be a root of equation (12.2) and $\Pi_S\{\lambda(z)\} > -\infty$. Let Γ be a central interior ray of the sector $\widetilde{S}$ (see (12.9)). Let*

$$x(z) = \exp\left[\int \gamma(z)dz\right]$$

be a solution to equation (12.1) in the sector $\widetilde{S}$ and

$$\gamma(z) \sim \lambda(z) \text{ for } z \to \infty, z \in [\widetilde{S}] \ (z \in \Gamma).$$

Let

$$q = \lim_{z\to\infty, z\in[S]} \lambda(z)\frac{\partial^2 H(\lambda(z), z)}{\partial y^2} \Big/ \frac{\partial H(\lambda(z), z)}{\partial y}. \tag{12.43}$$

Let in $[S]$ there exist a finite or (of a definite sign) infinite limit

$$p = \lim_{z \to \infty} \Re(\lambda(z)z). \tag{12.44}$$

If (1)

$$p < \frac{1}{2} \Re q P_S\{\lambda(z)\} \tag{12.45}$$

then $x(z) \to 0$ for $z \to \infty, z \in [\tilde{S}]$ (Γ). If relation (12.45) holds for any root of equation (12.2) then equation (12.1) is asymptotically stable in $[\tilde{S}]$ (on Γ);

(2)

$$p > \frac{1}{2} \Re q P_S\{\lambda(z)\} \tag{12.46}$$

then $x(z) \to \infty$ for $z \to \infty, z \in [\tilde{S}]$ (Γ). If relation (12.46) holds for at least one root of equation (12.2) then equation (12.1) is unstable in $[\tilde{S}]$ (on Γ).

Examples 12.28. 1°. Consider the Airy equation $x'' - zx = 0$. Its characteristic equation $y^2 - z = 0$ has roots $\lambda_1(z) = \sqrt{z}$ and $\lambda_2(z) = -\sqrt{z}$. Consequently the equation is nstable in any sector of the complex plane. The equation may be stable on separate rays where

$$p = \Re(\lambda_{1,2}(z)z) = 0.$$

That is, it may be stable only on the singular rays of the equation. We have

$$P_S\{\lambda_{1,2}(z)\} = P_S\{\sqrt{z}\} = 1/2.$$

The number q is equal to 1 (see (12.17)) for both roots.
Hence

$$p = 0 < \frac{1}{2} \Re q P_S\{\lambda_{1,2}(z)\} = 1/4.$$

Thus the equation is stable on the rays $\varphi = \pm\frac{\pi}{3}$ and $\pm\pi$.

2°. Let us estimate stability of the equation

$$x''' + z^{1/2}x'' + \frac{6}{5}z^{-1/2}x' + x = 0. \tag{12.47}$$

Its characteristic equation

$$y^3 + z^{1/2}y^2 + \frac{6}{5}z^{-1/2}y + 1 = 0$$

has roots $\lambda_1(z) \sim -z^{1/2}$, $\lambda_{2,3}(z) \sim \pm iz^{-1/4}$ for $z \to \infty$. As it is easy to see in any sector of the complex plane there is at least one direction where either $\Re(iz\sqrt{z}) \to +\infty$ or $-\Re(iz\sqrt{z}) \to +\infty$ for $z \to \infty$. This means that the equation is unstable in any sector of the complex plane. First consider the equation on its separate rays where simultaneously $\Re(\lambda_1(z)z) \leq 0$ and $\Re(\lambda_{2,3}(z)) = 0$. Put $z = re^{it}$. We obtain $\cos \frac{3t}{2} \geq 0$ from the first inequality and $\sin \frac{3t}{4} = 0$ from the second one. Hence $t = \frac{4}{3}\pi m$ ($m = 0, \pm 1, ...,$) and $\cos \frac{3t}{2} = 1$. Consequently the partial solutions corresponding to the root $\lambda_1(z)$ tend to zero for $z \to \infty$ on the considered rays. On the obtained rays let us estimate the asymptotic behavior of partial solutions corresponding to the roots $\lambda_{2,3}(z)$. It is easy to obtain the asymptotic representations of the roots. They are given by the following relations:

$$\lambda_{2,3}(z) = \pm iz - \frac{1}{10z} + O(z^{-7/4}) \text{ for } z \to \infty.$$

Hence $\Re(\lambda_{2,3}(z)z) \to -1/10$. We have

$$Ps\{\lambda_{2,3}(z)\} = Ps\{\pm iz^{-1/4}\} = -\frac{1}{4}$$

and $q = 1$. Consequently from (12.46) the equation is unstable on the rays considered.

Chapter 13

LINEAR DIFFERENTIAL EQUATIONS WITH POWER-LOGARITHMIC COEFFICIENTS

The most complete results can be obtained for linear differential equations of the form

$$F(t, x) \equiv x^{(n)} + a_1(t)x^{(n-1)} + \ldots + a_n(t)x = 0 \qquad (13.1)$$

with power-logarithmic coefficients. Its characteristic equation we denote by

$$H(t, y) \equiv y^n + a_1(t)y^{n-1} + \ldots + a_n(t) = 0 \qquad (13.2)$$

and let $\Lambda = \{\lambda_1(t), \lambda_2(t), \ldots, \lambda_n(t)\}$ be a complete set of roots of the last equation.

The main result is given in Theorem 13.6.

In section 4 we consider a linear differential system of the form $X' = A(z)X$ where $A(z)$ is a square n-order matrix belonging to a field of type N_S. We discuss the methods to reduce the system to a single linear differential equation with coefficients belonging to the same field.

1. SOME AUXILIARY PROPOSITIONS

Consider an expression of the form

$$I(t) = \int_{t_0}^{t} \exp\left[\int_{\tau}^{t} p(s)ds\right] \alpha(\tau)d\tau. \qquad (13.3)$$

Lemma 13.1. *Let $p(t)$ belong to L. Denote the limit $\lim_{t \to +\infty} p(t)t = q$ [since L is a field of type M, the limit (finite or of definite signh infinite) exists]. Let $\alpha(t)$ be continuous functions for $t \gg 1$ and $h\{\alpha(t)\} = r < -1$. Then it is possible to choose $t_0 \in J_+ \cup +\infty$ such that $h\{I(t)\} \leq r + 1$. In particular if $r = -\infty$ then $h\{I(t)\} = -\infty$.*

335

PROOF.   (1) Let $q = +\infty$. In this case $t_0 = +\infty$ and $\tau \geq t$. Clearly $p(s) > 0$ for $s \gg 1$. Consequently $\int_\tau^t p(s)ds \leq 0$ and

$$|I(t)| \leq \int_t^{+\infty} |\alpha(\tau)|d\tau.$$

Since $r < -1$ the last integral is convergent for $t \gg 1$ and $h\{I(t)\} \leq r+1$.

(2) Let q be a number and $q > r+1$. In this case $t_0 = +\infty$ and $\tau \geq t$. Clearly $p(s)s - q + \varepsilon < 0$ for for any sufficiently small number $\varepsilon > 0$ as $s \gg 1$. Hence

$$|I(t)| \leq t^{q-\varepsilon} \int_t^{+\infty} \exp\left[\int_t^\tau \frac{q - \varepsilon - p(s)s}{s}ds\right] \tau^{-q+\varepsilon}|\alpha(\tau)|d\tau$$

$$\leq t^{-q+\varepsilon} \int_t^{+\infty} \tau^{-q+\varepsilon}|\alpha(\tau)|d\tau.$$

We have $h\{t^{-q+\varepsilon}|\alpha(t)|\} = r - q + \varepsilon < -1$ for $\epsilon \ll 1$ and hence

$$h\left\{\int_t^{+\infty} \tau^{-q+\varepsilon}|\alpha(\tau)|d\tau\right\} \leq r - q + 1 + \varepsilon.$$

Consequently $h\{I(t)\} \leq r+1$.

(3) Let q be a number and $q \leq r+1$. In this case $t_0 = T$ is a sufficiently large positive number and $T \leq \tau \leq t$. There is an infinitesimal function $\beta(t)$ (for $t \to +\infty$) such that $p(t) = (q + \beta(t))/t$. Expression (13.3) may be estimate as follows. We have

$$I(t) = t^{q+\gamma(t)} \int_T^t \tau^{-q-\gamma(\tau)}\alpha(\tau)d\tau.$$

Here

$$\gamma(t) = \frac{1}{\ln t} \int_T^t \beta(s)ds \to 0 \ \ for \ t \to +\infty.$$

Clearly

$$h\{t^{-q-\gamma(t)}\alpha(t)\} = r - q \geq -1.$$

Hence

$$h\left\{\int_T^t \tau^{-q-\gamma(\tau)}\alpha(\tau)d\tau\right\} \leq r - q + 1.$$

Consequently $h\{I(t)\} \leq r+1$.

(4) Let $q = -\infty$. In this case $t_0 = T$ is a sufficiently large positive number and $T \leq \tau \leq t$. Clearly for any number $K > 0$ there is a number

$T_K \geq T$ such that $p(s) + K/s < 0$ for $s > T_K$. We have $|I(t)| \leq I_1(t) + I_2(t)$, where

$$I_1(t) = \int_T^{T_K} \exp\left[\int_\tau^t p(s)ds\right] |\alpha(\tau)|d\tau$$

and

$$I_2(t) = \int_{T_K}^t \exp\left[\int_\tau^t p(s)ds\right] |\alpha(\tau)|d\tau.$$

Since $p(t)t \to -\infty$ clearly $h\{I_1(t)\} = -\infty$. Besides

$$I_2(t) = t^{-K} \int_{T_K}^t \exp\left[\int_\tau^t (p(s) + K/s)ds\right] \tau^K |\alpha(\tau)|d\tau \leq t^{-K} \int_{T_K}^t |\tau^n \alpha(\tau)|d\tau.$$

Clearly, for $K > h\{\alpha(t)\}$, the last integral has a simple estimate no more than $K + h\{\alpha(t)\} + 1$ so that $h\{I_2(t)\} \leq r + 1$. Consequently $h\{I(t)\} \leq r + 1$.

$\square$

Lemma 13.2. *Let $p(t) \in L$. Let $\Pi\{\alpha(t)\} = r < -1$. Then it is possible to choose the value $t_0 \in J_+ \cup +\infty$ such that $\Pi\{I(t)\} \leq r + 1$. In particular if $r = -\infty$ then $I(t) \asymp 0$.*

PROOF. Since $p(t)$ belongs to L, there is a (finite or infinite) limit $\lim_{t \to +\infty} \Re p(t)t$. Due to Lemma 13.1 we have the estimate $h\{I(t)\} \leq r + 1$. On differentiating (13.3) we obtain the following identity

$$I'(t) = p(t)I(t) + \alpha(t). \tag{13.4}$$

(1) Let $\Pi\{p(t)\} \leq -1$. Clearly

$$h\{I'(t)\} \leq \max[\Pi\{p(t)\} + \Pi\{I(t)\}, \Pi\{\alpha(t)\}] = r.$$

On differentiating in succession relation (13.4) n times we conclude (by induction with respect to $n = 1, 2, ...$) that $h\{I^{(n)}(t)\} \leq r + 1 - n$. This implies the estimate $\Pi\{I(t)\} \leq r + 1$.

(2) Let $\Pi\{p(t)\} > -1$. Form a sequence of functions $\{I_k(t)\}$, where $I_0(t) = 0$ and

$$I_k(t) = I'_{k-1}(t)/p(t) + \alpha(t)/p(t) \text{ for } k = 1, 2, ...$$

We have $I_k(t) \in \Pi$ and

$$\Pi\{I_k(t) - I_{k-1}(t)\} \leq \Pi\{I_{k-1}(t) - I_{k-2}(t)\} - \sigma,$$

where $\sigma = 1 - \Pi\{p(t)\} > 0$. Hence

$$\Pi\{I_k(t) - I_{k-1}(t)\} \to -\infty \text{ for } k \to +\infty$$

and the series

$$I_1(t) + (I_2(t) - I_1(t)) + ... + (I_k(t) - I_{k-1}(t)) + ...$$

has an asymptotic sum $I^*(t) \in \Pi$. Clearly $h\{I(t) - I^*(t)\} = -\infty$ and

$$I^{*\prime}(t) - q(t)I^*(t) - \alpha(t) \asymp 0.$$

Substitute $I(t) = I^*(t) + y(t)$ in identity (13.4). We have $y'(t) = p(t)y(t) + \theta(t)$, where $\theta(t) \asymp 0$. Consequently $h\{y'(t)\} = -\infty$. On differentiating $y'(t)$ $(n-1)$ times (by induction) we obtain the estimations

$$h\{y^{(n)}(t)\} = -\infty \quad (n = 1, 2, ...).$$

The last means that $y(t) \asymp 0$. Thus $I(t) \asymp I^*(t)$. Hence $\Pi\{I'(t)\} \leq \Pi\{I(t)\} - 1$ and (from (13.4))

$$\Pi\{I(t)\} \leq \max[\Pi\{I(t)\} - \sigma, \Pi\{\alpha(t)\} - \Pi\{p(t)\}],$$

where $\sigma = 1 - \Pi\{p(t)\} > 0$. Hence

$$\Pi\{I(t)\} \leq \Pi\{\alpha(t)\} - \Pi\{p(t)\} < \Pi\{\alpha(t)\} + 1. \qquad \square$$

As a simple consequence of the last Lemma we have the following proposition.

Lemma 13.3. *Let $p(t) = l(t) + \beta(t)$, where $l(t) \in L$, $\beta(t) \in C_t$ and let the integral $\int_T^{+\infty} \beta(t)dt$ be convergent for a sufficiently large positive number T. Let $\Pi\{\alpha(t)\} = r < -1$. Then it is possible to choose the value $t_0 \in J_+ \cup +\infty$ such that $\Pi\{I(t)\} \leq r+1$. In particular if $\Pi\{\alpha(t)\} = -\infty$ then $I(t) \asymp 0$.*

PROOF. Expression (13.3) can be rewritten in the form

$$I(t) = (1 + \delta(t)) \int_{t_0}^{t} \exp\left[\int_{\tau}^{t} l(s)ds\right] \alpha^*(\tau)d\tau, \qquad (13.5)$$

where

$$\delta(t) = \exp\left[\int_{t}^{+\infty} \beta(s)ds\right] - 1 \in C_t$$

and
$$\alpha^*(t) = \alpha(t)/(1 + \delta(t)).$$

Clearly the function $1+\delta(t)$ does not change the estimates of the function $\alpha(t)$ and of the integral (13.3). That is, $\Pi\{I(t)\} = \Pi\{I^*(t)\}$, where

$$I^*(t) = \int_{t_0}^t \exp\left[\int_\tau^t l(s)ds\right] \alpha^*(\tau)d\tau.$$

On the basis of Lemma 13.1 we have $\Pi\{I^*(t)\} \leq r + 1.$ $\qquad\square$

Consider a linear differential equation with constant coefficients of the form
$$x^{(n)} + a_1 x^{(n-1)} + \dots + a_n x = 0, \tag{13.6}$$

where $a_1, \dots, a_n$ are numbers.

Lemma 13.4. *Equation (13.6) has an FSS of the form*

$$\{x_i(t) = e^{y_i t} t^{q_i}\} \ (i = 1, 2, \dots, n),$$

where y_i are all the roots of the characteristic equation

$$y^n + a_1 y^{n-1} + \dots + a_n = 0,$$

q_i are non-negative integers. Besides

$$\Pi\{V(\gamma_1(t), \dots, \gamma_n(t))\} \geq \frac{n(1-n)}{2}, \tag{13.7}$$

where $\gamma_i(t) = y_i + q_i/t$.

Proof. The existence of the considered *FSS* is well known. It is easy to show that $\Phi_s(y_i + q_i/t) = t^{-s}P_i(t)$, where $\Phi_i(y)$ is determined in Chapter 8(3). Here $P_i(t)$ is a polynomial in t with constant coefficients. Hence
$$V(y_1 + q_1/t, \dots, y_n + q_n/t) = t^{n(1-n)/2}P(t),$$

where $P(t)$ is a polynomial with constant coefficients. Since $\{x_i(t)\}$ is an *FSS*, $P(t) \not\equiv 0$. Consequently $\Pi\{P(t)\} \geq 0$, which implies the required inequality. $\qquad\square$

Lemma 13.5. *Let $\{a_1(t), \dots, a_n(t)\} \subset L$ and let $\Pi\{a_j(t)\} > -j$ for at least one $j \in \{1, 2, \dots, n\}$. Then the polynomial*

$$H(t, y) = y^n + a_1(t)y^{n-1} + \dots + a_n(t)$$

has at least one root $\lambda(t)$ with the estimate $\Pi\{\lambda(t)\} > -1$.

PROOF. Evidently the function $g(t)$ of the greatest growth of the set $\{[a_j(t)]^{1/j}\}$ $(j = 1, 2, ..., n)$ has the estimate $\Pi\{g(t)\} > -1$. Besides there exists at least one root $\lambda(t) \sim \mu g(t)$ for $t \to +\infty$, where $\mu \neq 0$ is a (complex) number. This leads to the required estimate. □

2.   MAIN THEOREMS

Theorem 13.6. *Let $\{a_1(t), ..., a_n(t)\} \subset \{L\}$ and let $\Lambda = \{\lambda_i(t)\}$ be a complete set of roots of characteristic equation*

$$H(t, y) \equiv y^n + a_1(t)y^{n-1} + ... + a_n(t) = 0 \qquad (13.2).$$

Then there exists a fundamental set of solutions to equation

$$F(t, x) \equiv x^{(n)} + a_1(t)x^{(n-1)} + ... + a_n(t)x = 0 \qquad (13.1)$$

of the following form

$$\left\{ x_j(t) = \exp\left[\int \gamma_j(t)dt\right] \right\} \quad (j = 1, 2, ..., n) \qquad (13.8)$$

such that for any $\gamma_j(t) \in \{\gamma_1(t), ..., \gamma_n(t)\}$ there exists a function $l_j(t) \in L$ such that $\gamma_j(t) = l_j(t) + \alpha_j(t)$, where $\alpha_j(t) \in C_t$ and there is a number T such that $\int_T^{+\infty} \alpha_j(t)dt$ is convergent. Moreover either $\Pi\{l_j(t)\} > -1$ then $\gamma_j(t) \sim \lambda_j(t)$ for $t \to +\infty$. If $\Pi\{l(t)\} \leq -1$ then characteristic equation (13.2) has a root $\lambda_j(t)$ with the estimate $\Pi\{\lambda_j(t)\} \leq -1$. For any $j \neq k$ $(j, k = 1, 2, ..., n)$ $\Pi\{g_j(t) - g_k(t)\} \geq -1$. Besides the integral

$$\int_t^{+\infty} |g_i(\tau) - g_k(\tau)|d\tau$$

is divergent for $t \gg 1$. There exists the following estimate

$$\Pi\{V(g_1(t), ..., g_n(t))\} \geq \frac{(1-n)n}{2}. \qquad (13.9)$$

PROOF. Without loss of generality we may suppose that all coefficients $a_j(t)$ in (13.1) belong to L (instead of $\{L\}$). Indeed let the theorem be true for this case. Choose a sufficiently large positive number N and (according to Proposition 6.37) the coefficients may be written in the form $a_j(t) = a_{jN}(t) + \alpha_{jN}(t)$, where $a_{jN}(t) \in L$ and $\Pi\{\alpha_{jN}(t)\} < -N$ $(j = 1, 2, ..., n)$. Equation (13.1) may be presented in the form

$$F_N(t, x) \equiv x^{(n)} + a_{1N}(t)x^{(n-1)} + ... + a_{nN}(t)x$$

$$= -\sum_{j=1}^{n} \alpha_{jN}(t) x^{(n-j)}. \tag{13.10}$$

To obtain the solutions to the last equation we use the Lagrange method of variation of arbitrary constants putting

$$x = u_1 G_{1N}(t) + u_2 G_{2N}(t) + ... + u_n G_{nN}(t),$$

where u_j are the variable parameters and $\{G_{jN}(t) = \exp[\int \gamma_{jN}(t)dt]\}$ is an *FSS* of the truncated equation $F_N(t,x) = 0$ possessing all the properties to be proved. Due to the properties of the *FSS* of the equation $F_N(t,x) = 0$ and on the basis of Lemma 13.3, it is easy to show that equation (13.1) has an *FSS* of the form

$$x_j(t) = \exp\left[\int \left(\gamma_{jn}(t) + \theta_{jN}(t)\right) dt\right],$$

where $\Pi\{\theta_{jN}(t)\} \ll -1$ for $N \gg 1$, and this implies the desired properties of *FSS* of equation (13.1).

We prove the theorem by induction with respect to $n = 1, 2, ...$

If $n = 1$ (13.1) is in the form $x' + a(t)x = 0$, where $a(t) \in L$. Hence its *FSS* consists of functions of the form $x(t) = \exp[\int \gamma(t)dt]$, $\gamma(t) = -a(t)$ and $V(\gamma(t)) = 1$. Clearly $\Pi\{V(\gamma(t)\} = 0$. Thus the case $n = 1$ is proved.

For an arbitrary $n > 1$ only two cases are possible:

(1)
$$\Pi\{a_j(t)\} \le -j \quad \text{for any} \quad j = 1, 2, ..., n;$$

(2) $\Pi\{a_j(t)\} > -1$ for at least one number $j \in \{1, 2, ..., n\}$.

Let
$$\max_{i=1,2,...,n} \{\dim a_i(t)\} = m.$$

In addition we prove the Theorem by induction with respect to $m = -1, 0, 1, ...$ If $m = -1$, (13.1) is an equation with constant coefficients which leads (see Lemma 13.4)) to the required properties. Therefore we suppose that the theorem is proved for any order n where the maximal length of the logarithmic chain of the coefficients is less than m.

Consider case (1). Each root $\lambda_i(t)$ of equation (13.2) has the estimate $\Pi\{\lambda_i(t)\} \le -1$. As it follows from Proposition 6.37 (see (6.18)) each function $a_j(t)$ can be represented in the form

$$a_j(t) = t^{-j} q_j(\ln t) + \delta_j(t),$$

where $q_j(t) \in L$ and $\Pi\{\delta_j(t)\} < -j$. We have $\dim q_j(t) \le m$, and (13.1) is in the form

$$L(t,x) \equiv x^{(n)} + \frac{q_1(\ln t)}{t} x^{(n-1)} + ... + \frac{q_n(\ln t)}{t^n} x$$

$$= [-\delta_1(t)]x^{(n-1)} - ...[-\delta_n(t)]x. \qquad (13.11)$$

Substitute $\tau = \ln t$, then the truncated equation $L(t, x) = 0$ is written in the form

$$x_\tau^{(n)} + c_1(\tau)x_\tau^{(n-1)} + ... + c_n(\tau)x = 0, \qquad (13.12)$$

where $c_i(\tau) \in L$ and (by τ) $\dim c_j(\tau) \leq m - 1$. By induction equation (13.12) has an FSS of the form

$$\left\{ \widetilde{G}_i(\tau) = \exp\left[\int \widetilde{\gamma}_i(\tau)d\tau\right]\right\} \quad (i = 1, 2, ..., n), \qquad (13.13)$$

where the functions $\widetilde{\gamma}_i(\tau)$ possess all the properties having to be proved. In particular

$$\widetilde{\gamma}_i(\tau) = \widetilde{l}_i(\tau) + \widetilde{\alpha}(\tau),$$

where $\widetilde{l}_i(\tau) \in L$, $\widetilde{\alpha}(\tau) \in C_\tau$ and $\int_\tau^\infty \alpha(s)ds$ is convergent for $\tau \gg 1$ ($i = 1, 2, ..., n$). Hence the corresponding FSS to equation $L(t, x) = 0$ may be rewritten in the form

$$\left\{ G_i(t) = \exp\left[\int \gamma_i^*(t)dt, \right]\right\} \quad (i = 1, 2, ..., n),$$

where

$$\gamma_i^*(t) = \frac{\widetilde{l}_i(\ln t)}{t} + \frac{\widetilde{\alpha}_i(\ln t)}{t}.$$

Clearly, the obtained FSS of the truncated equation possesses all the properties to be proved. In particular

$$\Pi\{V(\gamma_1^*(t), ..., \gamma_n^*(t))\} = n(1 - n)/2.$$

Taking into account that $\delta_j(e^\tau) \asymp 0$ (by τ, $j = 1, 2, ..., n$) we conclude that equation (13.1) possesses all the properties have to be proved.

Consider case (2). First we suppose that the asymptotic multiplicity of any root of equation (13.1) is less than n. Then the set of roots of the limiting equation (11.57) may be divided into two disjoint sets J_1 and J_2 considered in Lemma 11.21, thus equation (13.1) satisfies all the conditions of this Lemma. Consequently (see Lemma 11.21), all the functions of the FSS of the equations $F_p(t, x) = 0$ and $F_q^*(t, x) = 0$ (which possesses (by induction) all the properties to be proved) are formal solutions to equation (13.1). Their union $\{G_i(t) = exp[\int g_i(t)dt]\}$ form an FFS of solutions to equation (13.1). On the basis of Liouvile formula and (11.58) its Wronskian determinant equals to

$$W(G_1(t), ..., G_n(t)) \equiv \exp\left[-\int a_1(t)dt\right]$$

$$= \exp\left[-\int (b_1(t) + c_1(t) - \frac{pq}{t}\Pi\{g(t)\} + \frac{\alpha(t)}{t})dt\right]$$

$$= W(G_1(t), ..., G_p(t))W(G_{p+1}(t), ..., G_n(t))t^{-pq\Pi\{g(t)\}+\beta(t)},$$

where $\alpha(t)t \in C_t$ and $\beta(t) = (\ln t)^{-1}\int \alpha(t)dt$. Hence $P\{t^{\beta(t)}\} = 0$. Consequently

$$\begin{aligned}
\Pi\{V(g_1(t), ..., g_n(t))\} &= \Pi\{V(g_1(t), ..., g_p(t))\} + \Pi\{V(g_{p+1}(t), ..., g_n(t))\} \\
&\geq \frac{p(1-p)}{2} + \frac{q(1-q)}{2} + pq\Pi\{g(t)\} \\
&> \frac{p(1-p)}{2} + \frac{q(1-q)}{2} - pq \\
&= \frac{n(1-n)}{2}.
\end{aligned}$$

Now we may apply Lagrange's method of variation of arbitrary constants to the equation (13.1) considering the solutions in the form

$$x = u_1 G_1(t) + u_2 G_2(t) + ... + u_n G_n(t)$$

(u_i are the variable parameters). The well known procedure leads to *FSS* of equation (13.1) in the form (13.8) where $\gamma_j(t) = g_j(t) + \zeta_j(t)$. Here $\zeta_i(t) \asymp 0$ $(j = 1, 2, ..., n)$.

To complete the proof we have to consider the last case when the characteristic equation has n−asymptotically multiple root $\lambda(t)$. $(\Pi\{\lambda(t)\} > -1)$. Then each root of characteristic equation (13.2) is equivalent to $-a_n(t)/n$ for $t \to +\infty$. On putting

$$x = u\exp\left[-\int \frac{a_1(t)}{n}dt\right]$$

we obtain

$$u^{(n)} + b_2(n)(t)x^{(n-2)} + ... + b_n(t)x = 0, \tag{13.14}$$

where $b_i(t) \in L$ (i=1,2,...,n). The logarithmic derivatives $\gamma_*(t)$ of the solutions to (13.14) connect with the corresponding logarithmic derivatives $\gamma(t)$ to equation (13.1) by the relationship $\gamma(t) = \gamma_*(t) - a_1(t)/n$, and all the roots of the characteristic equation to (13.14) are $o(\lambda(t))$ for $t \to +\infty$. Let

$$\left\{u_i(t) = \exp\left[\int \gamma_{*i}(t)dt\right]\right\}$$

be an *FSS* of equation (13.14) which possesses all the required properties $(i = 1, 2, ..., n)$. Then equation (13.1) has the *FSS*

$$\left\{x_i(t) = \exp\left[\int (\gamma_{*i}(t) - a_1(t)/n)dt\right]\right\} \quad (i = 1, 2, ..., p).$$

This implies all the required to be proved. In particular

$$\Pi\{V(\gamma_1(t), ..., \gamma_n(t))\} = \Pi\{V(\gamma_{*1}(t), ..., \gamma_{*n}(t))\} \geq n(1-n)/2. \quad \square$$

Example 13.7. Given the equation

$$x^{(4)} - t^{-4}\ln^4 t \cdot x = 0. \tag{13.15}$$

We have

$$-t^{-4}\ln^4 t \in L \text{ and } \Pi\{-t^{-4}\ln^4 t\} = -4.$$

Hence the *FSS* of this equation consists of four functions of the form

$$x_i(t) = \exp\left[\int \gamma_i(t)dt\right] \quad (i = 1, ..., 4),$$

where $\Pi\{\gamma_i(z)\} \leq -1$. To find the asymptotic approximations to the functions $\gamma_i(t)$ let us substitute $x' = yx$. Hence

$$y^4 + 6y^2y' + 4yy'' + 3(y')^2 + y''' - t^{-4}\ln^4 t = 0. \tag{13.16}$$

We have

$$y' = t^{-1}(k + \alpha_1(t))y, \quad y'' = t^{-2}(k(k-1) + \alpha_2(t))y$$

and

$$y''' = t^{-3}(k(k-1)(k-2) + \alpha_3(t))y,$$

where $\alpha_i(t)\beta_t$ $(i = 1, 2, 3)$ and $k = P\{y(t)\}$ for any solution $y(t) \in A_t$. On substituting the derived asymptotic expression in (13.16) we obtain

$$y^4 + 6t^{-1}y^3(k + \alpha_1(t)) + t^{-2}y^2[4k(k-1) + 3k^2 + \tilde{\alpha}(t)]$$

$$+t^{-3}y[k(k-1)(k-2) + \alpha_3(t)] - t^{-4}\ln^4 = 0 \tag{13.17}$$

$(\tilde{\alpha}(t) \in C_t)$. It is easy to see that (for any k) the first approximations of solutions may be obtained from the equation $y^4 - t^{-4}\ln^4 t = 0$. Hence $\gamma_j(t) \sim \epsilon_j t^{-1}\ln t$ $(t \to +\infty)$ where ϵ_j are the distinct fourth roots of 1 $(j = 1, ..., 4)$. We can obtain the second approximation from the last equation. The next approximations to the functions $\gamma_j(t)$ may be obtained from equation (13.16) taking into account that $k = P\{t^{-1}\ln t\} = -1$. But we may find the approximations to the solution $\gamma_j(t)$ from the relations

$$\gamma_j(t)_j = \epsilon_j \sqrt[4]{t^{-4}\ln^4 t - 6\gamma_j(t)_j^2\gamma_j'(t) - 4\gamma_j(t)\gamma_j''(t) - 3(\gamma_j'(t))^2 - \gamma_j'''(t)}$$

$$\tag{13.18}$$

for each (fixed) j where

$$\gamma_j(t) = \epsilon_j t^{-1} \ln t (1 + \beta_{j1}(t)).$$

Here $\beta_{j1}(t) \in C_t$. Consequently

$$\gamma_j(t) \sim \epsilon_j t^{-1} \ln t \cdot \sqrt[4]{1 - 6\epsilon_j^3 t^{-1}/\ln t}.$$

Hence $\gamma_j(t) \sim \epsilon_j t^{-1} \ln t - 3/(2t)$. More precisely

$$\gamma_j(t) = \epsilon_j t^{-1} \ln t - (3 + \beta_{j2}(t))/(2t),$$

where $\beta_{j2}(t) \in C_t$. On substituting the obtained relation to the right side in (13.18) we obtain

$$\gamma_j(t) = \epsilon_j t^{-1} \ln t + \frac{3}{2t} + \left(\frac{5}{8}\epsilon_j^3 - \frac{3}{2}\right)\frac{1}{t \ln t} + O\left(\frac{1}{t \ln^2 t}\right)$$

for $t \to +\infty$ which leads to the following FSS :

$$\{x_j(t) = (1 + \delta_j(t))[\ln t]^{5/8\epsilon_j^2 - 3/2} t^{3/2} \exp\left[\frac{1}{2}\ln^2 t\right],$$

where

$$\delta_j(t) = O(1/\ln t) \text{ for } t \to \infty, t \in S_t \ (j = 1, ..., 4).$$

Theorem 13.8. *Let all the suppositions of Theorem 13.6 be fulfilled, and let, in addition, characteristic equation (13.2) possess a root $\lambda(t)$ which is either asymptotically simple and has the estimate $\Pi\{\lambda(t)\} > -1$, or it is a unique root with the estimate $\Pi\{\lambda(t) \le -1$. Then equation (13.1) has a unique solution $x(t) = \exp[\int \gamma(t)dt]$, where $\gamma(t) \in \{L\}$ and $\gamma(t) \sim \lambda(t)$ for $t \to +\infty$.*

PROOF. As it was proved earlier there exists a formal variable index $g(t) \in \{L\}$ and $g(t) \sim \gamma(t)$ for $t \to +\infty$. Substitute $x = u \exp[\int g(t)dt]$ in (13.1). We obtain the equation

$$u^{(n)} + b_1(t)u^{(n-1)} + ... + b_{n-1}(t)u' + \theta(t)u = 0,$$

where $b_i(t) \in \{L\}$, $\Pi\{b_{n-1}(t)\} > -\infty$ and $\theta(t) \asymp 0$. It is easy to show that the last equation has (a unique) solution $u(t) = \exp[\int \delta(t)dt]$ where $\delta(t) \asymp 0$. $\square$

3.   DEFICIENCY INDICES OF CERTAIN DIFFERENTIAL OPERATORS

In this section (to illustrate the results obtained) we consider asymptotic properties of certain fourth-order differential operators associated with self-adjoin differential expressions of the form

$$l(t, x) \equiv x^{(4)} - (a(t)x')' + b(t)x$$

considered on the positive semi-axis J_+. We shall look for the dimension m of the subspace for the set of solutions of the equation

$$x^{(4)} - (a(t)x')' + b(t)x = \mu x \tag{13.19}$$

which belong to the space $L_2(t_0, +\infty)$. Here $a(t)$ and $b(t)$ are real functions $\mu =$const, $\Im\mu \neq 0$, $t_0 \gg 1$. The value m is also called the deficiency index of the minimal operator L_0 associated with the differential expression $l(t, x)$. Thus we consider the equation

$$F(t, x) \equiv x^{(4)} - (a(t)x')' + (b(t) - \mu)x = 0 \tag{13.20}$$

for $t \gg 1$. If a complete set of roots of the characteristic equation

$$H(t, y, \mu) \equiv y^4 - a(t)y^2 - a'(t)y + b(t) - \mu = 0 \tag{13.21}$$

possesses the property of asymptotic separability we consider the coefficients $a(t)$ and $b(t)$ to belong to $\{Q\}$ where Q is a field of type M. In all other cases we suppose that the coefficients belong to $\{L\}$. It will be recalled that a solution $x(t)$ to the equation (13.20) belongs to the space $L_2(t_0, +\infty)$ if $\int_{t_0}^{+\infty} |x(t)|^2 dt < \infty$.

We introduce the following notation: $P_a = \Pi\{a(t)\}$ and $P_b = \Pi\{b(t)\}$. The signs of the functions $a(t)$ and $b(t)$ on the semi-axis J_+ are given by sign a and sign b respectively (for $t \gg 1$).

This investigation shows that the number m depends with the exception of some special cases on the values P_a, P_b and on the signs sign a and sign b. Therefore we consider a coordinate plane with its abscissa P_a and ordinate P_b (see Fig. 8).

All special cases are possible only on the thick lines of the figure. Thus, the plane is divided into several open regions separated by the thick lines. In each such region the number m depends only on sign a and sign b. Here we restrict ourselves to consideration of equation (13.20) in the above mentioned regions.

Let us consider the case when sign b is positive $P_b > 2P_a$ and $P_b > 4/3$. Here $a^2(t)/b(t) \to 0$ for $t \to +\infty$ and characteristic equation (13.21) has four roots $\lambda_i(t) \sim \varepsilon_j(\beta(t) - \mu)^{1/4}$, where ε_j are distinct roots of

-1, $j = 1, 2, 3, 4$. That is, all the roots do not equivalent in pairs and $\Pi\{\lambda_j(t)\} = (1/4)P_a > -1$. Consequently the set $\{\lambda_j(t)\}$ forms a complete set of roots of equation (13.21) and possesses the property of asymptotic separability. Consequently there exists an FSS $\{x_j(t) = \exp[\int \gamma_j(t)dt]\}$ of equation (13.20) where $\gamma_j(t) \sim \lambda_j(t)$ for $t \to +\infty$. As it is easy to show, $m = 2$.

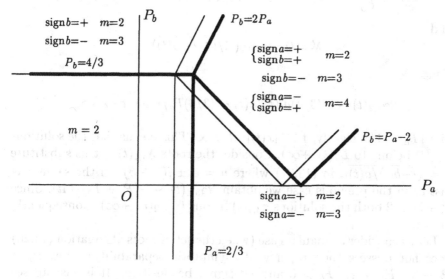

Fig. 8.

Consider the case when sign a is negative, sign b is positive and the region is specified by inequalities $P_b < 2P_a$, $P_a + P_b > 2$ and $P_b > P_a - 2$. To solve characteristic equation (13.21) let us substitute $y = \sqrt{-a(t)}u$. Hence

$$u^4 + u^2 + \frac{a'(t)}{a(t)\sqrt{-a(t)}}u + \frac{b(t) - \mu}{a^2(t)} = 0. \tag{13.22}$$

The last equation has four solutions $u_{1,2}(t) \sim \pm i$ and $u_{3,4}(t) \sim \pm i\sqrt{b(t)/a^2(t)}$ ($i = \sqrt{-1}$) for $t \to +\infty$. It is obvious that the set of roots possesses the property of asymptotic separability. However the main terms of the asymptotic relations have purely imaginary values. Therefore we must find the asymptotic representations of the functions $\lambda_j(t)$ more precisely. Let us substitute $u = \alpha \pm i(1 + \beta)$ in (13.22). The resulting equations have roots $\alpha(t) \to 0$ and $\beta(t) \to 0$ for $t \to +\infty$. We have only to estimate the real part of the function $u(t)$. We have

$$\Re u_{1,2}(t) = \alpha_{1,2}(t) \sim a'(t)/(2a(t)\sqrt{-a(t)}).$$

Hence $\Re\lambda_{1,2}(t) \sim a'(t)/(2a(t))$ for $t \to +\infty$. Also

$$\frac{\partial^2 H(t, \lambda_{1,2}(t), \mu)}{\partial y^2} = 12\lambda_{1,2}^2(t) - 2a(t) \sim 10a(t),$$

$$\frac{\partial H(t, \lambda_{1,2}(t), \mu)}{\partial y} \sim 4\lambda_{1,2}^3(t) - 2a(t)\lambda_{1,2}(t) \sim \pm 2ia(t)\sqrt{-a(t)}$$

and

$$\lambda_{1,2}'(t) \sim \pm ia'(t)/(2\sqrt{-a(t)}).$$

Hence

$$\Re\gamma_{1,2}(t) \sim -(3/4)a'(t)/a(t) \sim (3/4)P_a/t \text{ for } t \to +\infty.$$

Since $P_a > 2/3$, clearly, $\int^{+\infty} |x(t)|^2 dt < \infty$. This means that the solutions $x_{1,2}(t)$ belong to $L_2(t_0, +\infty)$. Consider the roots $\lambda_{3,4}(t)$, let us substitute $y = \sqrt{-b(t)/a(t)}u$ in (13.21) where $u = \alpha \pm i(1 + \beta)$. In the same way just as in the previous case we obtain $\Re\gamma_{3,4}(t) \sim -(P_a + P_b)/(4t)$. Since $P_a + P_b > 2$ both the solutions $x_{3,4}(t)$ belong to $L_2(t_0, +\infty)$. Consequently $m = 4$.

Let us consider a singular case (when the set of roots of equation (13.20) does not possess the property of asymptotic separability). Let $P_b < P_a - 2$, $P_a > 2$, $P_b > 0$ and let sign a be positive. It is easy to see that equation (13.21) has two roots $\lambda_{1,2}(t) \sim \pm\sqrt{a(t)}$, a root $\lambda_3(t) \sim -a'(t)/a(t)$ and a root $\lambda_4(t) \sim b(t)/a'(t)$ for $t \to +\infty$. Thus the roots $\lambda_{3,4}(t)$ have estimates $\Pi\{\lambda_{3,4}(t)\} \leq -1$. In this case we should suppose that the functions $a(t)$ and $b(t)$ belong to $\{L\}$. We conclude (in the same way just as in the previous case) that $x_1(t)$ does not belong to $L_2(t_0, +\infty)$ and $x_2(t)$ belongs to $L_2(t_0, +\infty)$ (where $x_1(t)$ and $x_2(t)$ are solutions corresponding to the roots $\lambda_1(t)$ and $\lambda_2(t)$, respectively). The functions $\gamma_{3,4}(t)$ which corresponds to the roots $\lambda_{3,4}(t)$, satisfy the equation

$$\Phi_4(y) - a(t)\Phi_2(y) - a'(t)y + b(t) - \mu = 0,$$

where the unknowns belong to the class A_t. Hence $y' \sim P_y y/t$,

$$y'' \sim P_y(P_y - 1)y/t^2 \text{ and } y''' \sim P_y(P_y - 1)(P_y - 2)y/t^3 \text{ for } t \to +\infty.$$

Here $P_y = P\{y\}$. Therefore as it is easily seen we can obtain the required asymptotic approximations of the functions $\gamma_{3,4}(t)$ from the equation

$$y^2 + \left(\frac{a'(t)}{a(t)} + \frac{P_y}{t}\right)y - (b(t) - \mu) = 0.$$

Hence $\gamma_3(t) \sim -(P_a + P_b)/t$. Because in this case

$$P_y = \lim_{t \to +\infty} \frac{\ln[(P_a + P_b)/t]}{\ln t} = -1$$

we have $\gamma_3(t) \sim (1 - P_a)/t$ $(t \to +\infty)$. Since $P_a > 2$ the solution $x_3(t)$ belongs to $L_2(t_o, +\infty)$. Furthermore $\gamma_4(t) \sim b(t)t/[(P_a + P_b)a(t)]$. Consequently $P\{\gamma_4(t)\} < -1$ and the solution $x_4(t)$ does not belong to $L_2(t_o, +\infty)$. Thus $m = 2$.

It is possible to investigate all the other cases in the same way.

Similar results were obtained by A. DEVINATZ where $a(t) = at^\alpha$ and $b(t) = bt^\beta$. Here a, b, α and β are real numbers.

4. SYSTEMS OF LINEAR DIFFERENTIAL EQUATIONS

Here we consider a system of the form

$$X' = A(z)X \tag{13.23}$$

where $A(z) = (a_{ij}(z))_n$ is a square matrix belonging to a space G of type N_S. We consider the possibility to reduce system (13.23) to a single linear differential equation of nth order with coefficients belonging to the field V, and then the system may be solved by the already known methods. To this end we apply the method described in Chapter 8.

The case when $A(z) \equiv A = (a_{ij})_n$ is a constant (complex) matrix is the most simple and it was considered in Chapter 8. Here we investigate the general case.

We look for a substitution $U = P(z)X$ obtained by means of a linear form u taken in the form

$$u = c_1 z^{\sigma_1} x_1 + c_2 z^{\sigma_2} x_2 + ... + c_n z^{\sigma_n} x_n, \tag{13.24}$$

where c_i are constants equal to 1 or 0 for each (fixed) i, σ_i are positive numbers $(i = 1, 2, ..., n)$. To obtain the required substitution we differentiate form (13.24) $(n-1)$ times according to system (13.23); $U = (u_1, u_2, ..., u_n)^T$ where $u_1 = u, u_2 = u', ..., u_n = u^{(n-1)}$. Consequently the following relations must be fulfilled $u_1'(z) = u_2(z), ..., u_{n-1}'(z) = u_n(z)$ for any function $u(z)$ belonging to the resulting space of system (13.23) and the form (13.24). So that if the matrix $P(z)$ is non-singular we obtain the resulting system $U' = F(z)U$ where $F(z)$ has to be a Frobenius matrix. The necessary and sufficient condition for $P(z)$ to be non-singular (for $|z| \gg 1, z \in [S]$) is the following: the dimension of the resulting space (which is, clearly, linear) has to be equal to n. Consider the last requirement in detail. System (13.23) has a set of solutions where any its coordinate x_i is given in the form

$$x_i = q_{i1}C_1\phi_{i1}(z) + q_{i2}C_2\phi_{i2}(z) + ... + q_{in}C_n\phi_{in}(z). \tag{13.25}$$

Here $\phi_{ij}(z)$ are the corresponding elements of the fundamental matrix of solutions $\Phi(z) = (\phi_{ij}(z))_n$, C_i are arbitrary constants, and q_{ij} is equal to 0 or 1 for any (fixed) $i, j = 1, 2, ..., n$. Without loss of generality we may suppose that each $\phi_{ij}(z) \not\equiv 0$ (otherwise we can determine $\phi_{ij}(z)$ by arbitrariness and taking $c_{ij} = 0$). If all $q_{ij} = 1$, then we can take $u = x_i$. Unfortunately some numbers in (3) may be equal to zero. This situation can be repeated even for $u = x_1 + x_2 + ... + x_n$. Indeed the resulting space will be of n dimension if any arbitrary constant C_i is essential. For example consider the system $X' = \left(\begin{smallmatrix} \lambda & 0 \\ 0 & \lambda \end{smallmatrix}\right) X$. If the resulting space is taken in the form $u = x_1 + x_2$, then $u(t)$ consist of the functions of the form $(C_1 + C_2)e^{\lambda z}$. Therefore $C_1 + C_2$ may be rewritten as a single arbitrary constant C. Similar situation may arise for the form (13.25). Let us take the following procedure. Examine the form $u_1 = x_1$ and let the resulting space contains the essential arbitrary constants $C_1, ..., C_m$ and do not contain the constants $C_{m+1}, ..., C_n$. Clearly there exists at least one form $u_j = x_j$ where its resulting space contains the constant C_{m+1}. Then it is possible to choose a number $\sigma_j > 0$ such that the resulting space $u^* = x_1 + z^{\sigma_j} x_j$ will essentially contains all the constants $C_1, ..., C_m$ and the constant C_{m+1}. This is made by permissible arbitrariness of the number σ_j. If the resulting space of the form u^* is not sufficient we may consider a new space which is formed of three members and so on. Clearly by finitely many steps we obtain the desired form.

Remark 13.9. As it follows in the given proof the numbers σ_i may be chosen as integers. Consequently the matrix $P(z)$ may belong to the same field V because any field of type N_S contains the variable z.

Remark 13.10. Since the coefficients are power functions the matrix $P(z)$ belong to a field of type N_S. And since $\det P(z) \not\equiv 0$, then $\det P(z) \in A_S$. Consequently $\Pi\{\det P(z)\} > -\infty$.

Example 13.11. Given the system

$$\begin{cases} x_1' = & zx_1 + & x_2 + & x_3 \\ x_2' = & x_1 + & zx_2 & \\ x_3' = & -2x_1 - & x_2 + & zx_3 \end{cases}.$$

Here $X' = A(z)X$ where $X = (x_1, x_2, x_3)^T$, $X' = (x_1', x_2', x_3')^T$ and

$$A(z) = \begin{pmatrix} z & 1 & 1 \\ 1 & z & 0 \\ -2 & -1 & z \end{pmatrix}.$$

Put $u = x_1$ and differentiate according the sistem. We have $u' = zx_1 + x_2 + x_3$ and $u'' = z^2 x_1 + (2z - 1)x_2 + 2zx_3$. Thus we obtained the substitution $U = P(z)X$ where $U = (u, u', u'')^T$ and

$$P(z) = \begin{pmatrix} 1 & 0 & 0 \\ z & 1 & 1 \\ z^2 & 2z - 1 & 2z \end{pmatrix}.$$

We have $\det P(z) = 1$. Besides

$$P^{-1}(z) = \begin{pmatrix} 1 & 0 & 0 \\ -z^2 & 2z & -1 \\ z^2 - z & 1 - 2z & 1 \end{pmatrix}, \quad P'(z) = \begin{pmatrix} 0 & 0 & 0 \\ 1 & 0 & 0 \\ 2z & 2 & 2 \end{pmatrix}.$$

Hence $U' = B(z)U$, where

$$B(z) = P'(z)P^{-1} + P(z)A(z)P^{-}(z)$$

$$= \begin{pmatrix} 0 & 1 & 0 \\ 0 & 0 & 1 \\ z^3 - 2z - 1 & 2 - 3z^2 & 3z \end{pmatrix}.$$

Here $B(z)$ is a Frobenius matrix. Then all the solutions $u(z)$ coincide with all the solutions of the equation

$$u''' - 3zu'' + (3z^2 - 2)u' - (z^3 - 2z - 1)u = 0.$$

Let us substitute $u = ve^{z^2/2}$ in the last equation. We have $v''' - 2v' + v = 0$. Its general solution is in the form

$$u = C_1 e^z + C_2 e^{q_1 z} + C_3 e^{q_2 z},$$

where $q_{1,2} = -\frac{1}{2} \pm \frac{\sqrt{5}}{2}$ (C_1, C_2 and C_3 are arbitrary constants). So that

$$u = e^{z^2/2}(C_1 e^z + C_2 e^{q_1 z} + C_3 e^{q_2 z}),$$

$$u' = e^{z^2/2}[C_1(z + 1)e^z + C_2(z + q_1)e^{q_1 z} + C_3(z + q_2)e^{q_2 z}]$$

and

$$u'' = e^{z^2/2}[C_1(z^2 + 2z + 2)e^z + C_2(z^2 + 2q_1 z + 2q_1)e^{q_1 z}$$
$$+ C_3(z^2 + 2q_2 z + 2q_2)e^{q_2 z}]$$

and $X = P^{-1}(z)U;\ U = (u, u', u'')^T$.

Since system (13.23) is reduced to a single equation with coefficients belonging to a field of type N_S, we may apply to the system the standard procedure. It consists of the following: we suppose, for each $i = 1, 2, ..., n$, $x_i = u_i(z) \exp[\int \gamma(z)dz]$ where $u_i(z) \in V$ and

$$u_i' = z^{-1}(k_i + \alpha_i(z))u_i, \tag{13.26}$$

where $\alpha_i(z) \in C_S$. Substitute relations (13.26) in (13.23). We obtain

$$\left(a_{11}(z) - \frac{k_1 + \alpha_1(z)}{z} - \gamma(z)\right) u_1 + a_{12}(z)u_2 + ... + a_{1n}(z)u_n = 0,$$

$$a_{21}(z)u_1 + \left(a_{22}(z) - \frac{k_2 + \alpha_2(z)}{z} - \gamma(z)\right) u_1 + ... + a_{2n}(z)u_n = 0,$$

$$. \quad . \quad . \quad . \quad . \quad . \quad . \quad . \quad . \quad . \quad . \quad . \quad . \quad . \quad . \quad . \quad .$$

$$a_{n1}(z)u_1 + a_{n2}u_2 + ... + \left(a_{nn}(z) - \frac{k_n + \alpha_i(z)}{z} - \gamma(z)\right) u_n = 0 \tag{13.27}$$

which can be considered as a system in unknowns $u_1, u_2, ..., u_n$. We suppose that the system has a non-zero solution. It is possible if and only if the determinant of the system is equal to zero. That is,

$$\begin{vmatrix} a_{11}(z) - \frac{k_1+\alpha_1(z)}{z} - \gamma(z) & a_{12}(z) & \cdots & a_{1n}(z) \\ a_{21}(z) & a_{22}(z) - \frac{k_2+\alpha_2(z)}{z} - \gamma(z) & \cdots & a_{2n}(z) \\ . \quad . \quad . \quad . & . \quad . \quad . & . \quad . & . \quad . \\ a_{n1}(z) & a_{n2}(z) & \cdots & a_{nn}(z) - \frac{k_n+\alpha_n(z)}{z} - \gamma(z) \end{vmatrix} = 0. \tag{13.28}$$

This expression may be considered as a family of equation in γ with numerical parameters $k_1, k_2, ..., k_n$ and unknown functions $\alpha_1(z), \alpha_2(z), ..., \alpha_n(z)$ belonging to the class C_S. Let one of the solutions of the family be $\gamma(z) = f(z, k_1, ...k_n, \alpha_1(z), ..., \alpha_n(z))$. Substitute the solution to the system (13.27). Since determinant of the system vanishes the system possesses a non-trivial family of solutions $u_1, ...u_n$, where

$$u_i = u_i(z, k_1, ..., k_n, \alpha_1(z), ..., \alpha_n(z)) \quad (i = 1, 2, ..., n).$$

The numbers k_i are determined from the relations

$$\lim_{z\to\infty,z\in[S]} (\ln \phi_i(z))/\ln z,$$

where $\phi_i(z)$ is the kernel of the family u_i. The formal procedure consists of obtaining all the functions $\gamma(t)$ and the corresponding functions $u_1(z), u_2(z), u_n(z)$. For definiteness we may put $u_1(z) = 1$ then $x_1(t) = \exp[\int \gamma(t)dt]$ and $(k_1 + \alpha_1(t))/t = 0$.

Chapter 14

LINEAR DIFFERENCE EQUATIONS. GENERAL THEORY

In two subsequent chapters we mainly consider a linear difference equation of the form

$$\Phi(t, x(t)) \equiv x(t+n) + a_1(t)x(t+n-1) + ... + a_n(t)x(t) = 0 \quad (14.1)$$

on the positive semi-axis $J_+ = \{t : t > 0, t \neq +\infty\}$, n is a natural number, $a_i(t)$ are functions of the real argument t $(i = 1, 2, ..., n)$, $a_n(t) \not\equiv 0$ for $t \gg 1$. We investigate the asymptotic behavior of the solutions to equation (14.1) for $t \to +\infty$. Asymptotic investigation of such equations is a classical problem which goes back to H. POINCARÉ and O. PERRON.

In many problems the asymptotic behavior of solutions of the equation depends on asymptotic behavior of roots of the characteristic equation

$$H(t, y) \equiv y^n + a_1(t)y^{n-1} + .. + a_n(t) = 0. \quad (14.2)$$

In this chapter we give a short outline of the general theory of systems and higher-order linear difference equations.

In Section 1 we investigate the asymptotic properties of linear difference equation of the first order written in the form $x(t+1) = q(t)x(t) + \alpha(t)$, where $x(t)$ is a unknown functions, $\alpha(t)$ and $q(t)$ are continuous functions on the positive semi-axis J_+.

Previous the equation $x(t+1) - x(t) = \alpha(t)$ is examined. If $h\{\alpha(t)\} = p < -1$ then it is proved that the equation has a unique infinitesimal solution $x^*(t)$ satisfying the inequality

$$|x^*(t)| \leq |\alpha(t)| + \int_t^{+\infty} \sup_{s \in [\tau, +\infty[} |\alpha(s)| d\tau$$

and $h\{x^*(t)\} \leq p + 1$. If $\alpha(t) \in \Pi$ then the equation has solutions which are given by the Euler–Maclaurin Formula (see (14.11)).

For the equation $x(t + 1) = q(t)y(t)$ if $q(t)$ is a power order growth function ($q(t) \in A_t$) we obtain formulae (14.18) and (14.19) (which are a simple consequence of the Euler–Maclaurin formula). Thus we may obtain the asymptotic representation of a solution (say $\tilde{x}(t)$) of equation (14.1) if it can be reduced (for the considered solution) to an equation of the first order of the form $x(t + 1) = q(t)x(t)$ where $q(t) \in A_t$.

For a non-homogeneous equation $x(t + 1) = q(t)x(t) + \alpha(t)$ there were investigate the cases: (1) $|q(t)| \geq 1$ and $h\{\alpha(t)\} < -1$; (2) $\lim_{t \to +\infty} t|q(t)| = p < +\infty$ and $h\{\alpha(t)\} = -\infty$; and (3) $h\{q(t)\} = -\infty$ or $\lim_{t \to +\infty} t \ln |q(t)| = -\infty$ and $h\{\alpha(t)\} = -\infty$.

In the case (1) it is proved that the equation has a unique infinitesimal solution $x^*(t)$ and it has the estimate $h\{x^*(t)\} \leq h\{\alpha(t)\} + 1$. In the cases (2), (3) it is proved that the equation has a solution $x^*(t) = O(t^{-\infty})$.

1. FIRST ORDER EQUATIONS

1.1 SOME AUXILIARY PROPOSITIONS

In this paragraph we consider an equation of the form

$$x^-(t) \equiv x(t + 1) - x(t) = \alpha(t), \qquad (14.3)$$

where $\alpha(t)$ is a function defined for $t \gg 1$ on the positive semi-axes. First, let us mark two simple propositions.

Proposition 14.1. *Let* $\Pi\{x(t)\} = p$. *Then the asymptotic representation of the difference* $x^-(t)$ *can be written in the form*

$$x^-(t) \asymp \frac{x'(t)}{1!} + \frac{x''(t)}{2!} + \ldots + \frac{x^{(m)}(t)}{m!} + \ldots \qquad (14.4)$$

This means that

$$\Pi\{x^-(t) - s_m(t)\} \to -\infty \ \ for \ m \to \infty,$$

where

$$s_m(t) = \frac{x'(t)}{1!} + \frac{x''(t)}{2!} + \ldots + \frac{x^{(m)}(t)}{m!}.$$

Hence $\Pi\{x^-(t)\} \leq p - 1$.

PROOF. The required property immediately follows from Taylor's formula applied to $x(t + 1)$ at the point t with the increment 1. $\qquad \square$

Proposition 14.2. *Let* $p = h\{\alpha(t)\} < -1$. *Then equation* (14.3) *has a unique infinitesimal solution* $x^*(t)$ *which is represented as a sum of the* *absolutely and uniformly convergent series*

$$-\alpha(t) - \alpha(t+1) - \dots - \alpha(t+m) - \dots \qquad (14.5)$$

on $[T, +\infty[$, where T is a sufficiently large positive number. The solution satisfies the following inequality

$$|x^*(t)| \le |\alpha(t)| + \int_t^{+\infty} \sup_{s \in [\tau, +\infty[} |\alpha(s)| d\tau \qquad (14.6)$$

and (as a consequence) $h\{x^*(t)\} \le p+1$. In particular if $\alpha(t) = O(t^{-\infty})$ (i.e. $h\{\alpha(t)\} = -\infty$) then $x^*(t) = O(t^{-\infty})$.

PROOF. For any positive number $\varepsilon < -1 - p$ there exists a positive number A (which generally speaking depends on ε) such that $|\alpha(t+m)| \le A(t+m)^{p+\varepsilon}$ uniformly on $[T, +\infty[$ ($m = 0, 1, \dots$). Since $p + \varepsilon < -1$, the series with the general term $(t+m)^{p+\varepsilon}$ is absolutely and uniformly convergent, and therefore series (14.5) is absolutely and uniformly convergent to a function $x^*(t)$ continuous on $[T, +\infty[$. Substitute series (14.5) in (14.3) and clearly $x^*(t)$ is a solution to the equation.

Let us prove the uniqueness of an infinitesimal solution. Let it be $x^*(t)$. The general solution of (14.3) can be represented in the form

$$x(t) = C(t) + x^*(t), \qquad (14.7)$$

where $C(t)$ is an arbitrary periodic function with period 1. Since a periodic function (which does not vanish identically) is not infinitesimal due to (14.7) equation (14.3) can have no more than one infinitesimal solution.

Remark. The last expression is called a *general solution to equation* (14.3).

Formula (14.6) is a consequence of (14.5). Indeed

$$|\alpha(t+m)| \le \sup_{\zeta \le s \le m} |\alpha(t+s)|,$$

$m = 1, 2, \dots$ Clearly

$$
\begin{aligned}
|\alpha(t+m)| &= \int_{m-1}^m |\alpha(t+m)| d\tau \\
&\le \int_{m-1}^m \sup_{\zeta \le s \le m} |\alpha(t+s)| d\zeta \\
&\le \int_{m-1}^m \sup_{\zeta \le s < +\infty} |\alpha(t+s)| d\zeta.
\end{aligned}
$$

Hence

$$|x^*(t)| \leq |\alpha(t)| + \sum_{m=1}^{\infty} |\alpha(t+m)|$$

$$\leq |\alpha(t)| + \sum_{m=0}^{\infty} \int_m^{m+1} \sup_{\zeta \leq s < \infty} |\alpha(t+s)| d\zeta$$

$$= |\alpha(t)| + \int_0^{+\infty} \sup_{\zeta < s < +\infty} |\alpha(s+t)| d\zeta.$$

On substitution $\tau = t + \zeta$ we obtain relation (14.6). We have

$$|x^*(t)| \leq |\alpha(t)| + A \int_t^{\infty} \tau^{p+\varepsilon} d\tau = O(t^{p+1+\varepsilon}) \text{ for } t \to +\infty.$$

Hence $h\{x^*(t)\} \leq p + 1 + \varepsilon$. Taking into account the arbitrariness of ε we conclude that $h\{x^*(t)\} \leq p + 1$.   □

Proposition 14.3. *Let all the conditions of Proposition 14.2 be fulfilled and let (in addition)* $\Pi\{\alpha(t)\} = p < -1$. *Then the solution* $x^*(t)$ *has the estimate* $\Pi\{x(t)\} \leq p + 1$.

PROOF. On differentiating series (14.5) we obtain $x^{*\prime}(t)$ as a sum of the series

$$-\alpha'(t) - \alpha'(t+1) - \dots - \alpha'(t+m) - \dots$$

Hence

$$|x^{*\prime}(t)| \leq |\alpha'(t)| + \int_0^{+\infty} \sup_{\tau < s < +\infty} |\alpha'(t+s)| d\tau.$$

Consequently

$$h\{x^{*\prime}(t)\} \leq h\{\alpha'(t)\} + 1 \leq p.$$

On differentiating (14.5) m times (by induction with respect to $m = 0, 1, \dots$) we conclude that $h\{x^{*(m)}(t)\} \leq p - m + 1$. This implies $\Pi\{x^*(t)\} \leq p + 1$.   □

1.2   BERNOULLI NUMBERS

The so called BERNOULLI *numbers* are defined by the following relation:

$$B_k = -\left(\frac{B_0}{k+1} + B_1 + \dots + \frac{B_s k!}{s!(k-s+1)!} + \dots + k\frac{B_{k-1}}{2} \right) \qquad (14.8)$$

for $k = 1, 2, ...$ where $B_0 = 1$. Thus $B_1 = -1/2$, $B_2 = 1/6$, $B_3 = 0, ...$ Bernoulli numbers are used to write Taylor's and asymptotic series for some functions. We consider the expansion into a power series of the function

$$\varphi(t) = \frac{t}{e^t - 1}, \quad |t| < 2\pi.$$

Let

$$\varphi(t) = \sum_{k=0}^{+\infty} b_k \frac{t^k}{k!} .$$

Taking into account that $\varphi(t)(e^t - 1) = t$ we obtain

$$\sum_{k=0}^{+\infty} b_k \frac{t^k}{k!} \sum_{k=1}^{+\infty} \frac{t^k}{k!} = t. \tag{14.9}$$

Multiply the series and collect all like members. The result can be represented in the following form $t + \sum_{k=2}^{+\infty} c_k t^k$. We have

$$c_{k+1} = \frac{1}{(k+1)!} + \frac{b_1}{k!} + ... + \frac{b_s}{s!(k+1-s)!} + ... + \frac{b_k}{k!} .$$

But $c_k = 0$ from (14.9). Hence multiplying the both sides by $k!$ we obtain

$$b_k = -\left(\frac{1}{k+1} + b_1 + ... + \frac{b_s k!}{s!(k-s+1)!} + ... + \frac{k b_{k-1}}{2} \right).$$

It means (see (14.8)) that $b_k = B_k$ $(k = 2, ...)$ and we may write

$$\varphi(t) \equiv \frac{t}{e^t - 1} = \sum_{k=0}^{+\infty} B_k \frac{t^k}{k!} . \tag{14.10}$$

Let us mark that all odd numbers B_{2p+1}, excluding B_1, are equal to zero. Indeed

$$\varphi(-t) = \frac{t}{1 - e^{-t}} = t + \frac{t}{e^t - 1},$$

i.e. $\varphi(-t) = t + \varphi(t)$. Hence

$$\sum_{k=2}^{+\infty} (-1)^k B_k \frac{t^k}{k!} = \sum_{k=2}^{+\infty} B_k \frac{t^k}{k!}$$

which leads to the relation $B_k = (-1)^k B_k$.

The relation turns into identity for even k and $B_{2p+1} = -B_{2p+1}$ holds for any $p = 1, 2, ...$ Hence $B_{2p+1} = 0$.

1.3 *THE EULER–MACLAURIN FORMULA*

Theorem 14.4. *Let* $\Pi\{\alpha(t)\} = p$. *Then equation* (14.3) *has a family of solutions continuous on* $[T, +\infty[$, *where* T *is a sufficiently large positive number. Their asymptotic representation can be written in the form*

$$x(t) \asymp \int \alpha(t)dt - \frac{1}{2}\alpha(t) + \sum_{k=1}^{\infty} \frac{B_{2k}}{(2k)!}\alpha^{(2k-1)}(t)\,, \qquad (14.11)$$

where B_{2k} *are Bernoulli numbers* $(k = 1, 2, ...)$. *This means that for any fixed antiderivative* $\int \alpha(t)dt$ *there exists a solution* $x(t)$ *of equation* (14.3) *such that* $\Pi\{x(t) - s_m(t)\} \to -\infty$ *for* $m \to \infty$, *where*

$$s_m(t) = \int \alpha(t)dt - \frac{1}{2}\alpha(t) + \sum_{k=1}^{m} \frac{B_{2k}}{(2k)!}\alpha^{(2k-1)}(t)\,. \qquad (14.12)$$

Hence $\Pi\{x(t)\} \le \max[0, \Pi\{\alpha(t)\} + 1]$, *if* $p < -1$ *then equation* (14.3) *has a unique infinitesimal solution* $x^*(t)$ *for* $t \to +\infty$, $\int \alpha(t)dt$ *is considered as* $\int_{+\infty}^{t} \alpha(\tau)d\tau$ *and* $\Pi\{x^*(t)\} \le p + 1$.

PROOF. Let us find a formal series in the form

$$\beta(t) = \int \alpha(t)dt + \frac{c_1}{1!}\alpha(t) + ... + \frac{c_k}{k!}\alpha^{(k-1)}(t) + ... \qquad (14.13)$$

which formally satisfies equation (14.3). Here $c_1, c_2, ...$ are numbers. Clearly

$$\beta^-(t) = \int_t^{t+1} \alpha(\tau)d\tau + \frac{c_1}{1!}\alpha^-(t) + ... + \frac{c_k}{k!}[\alpha^{(k-1)}(t)]^- + ...$$

and it is (formally) equal to $\alpha(t)$. On expanding the terms into power series we obtain:

$$\int_t^{t+1} \alpha(\tau)d\tau = \int_0^1 \alpha(t+s)ds = \int_0^1 \sum_{p=0}^{\infty} \alpha^{(p)}(t)\frac{s^p}{p!}ds = \sum_{p=0}^{\infty} \frac{\alpha^p(t)}{(p+1)!}\,;$$

$$[\alpha^{(k)}(t)]^- = \sum_{p=1}^{\infty} \frac{\alpha^{(k+p)}(t)}{p!} \quad (k = 0, 1, ...).$$

Consequently

$$\sum_{p=1}^{\infty} \frac{\alpha^{(p)}(t)}{(p+1)!} + \frac{c_1}{1!p!}\alpha^{(p)}(t) + ... + \frac{c_k}{k!p!}\alpha^{(k+p-1)}(t) + ... = 0.$$

Let us collect the similar terms containing $\alpha^{(k)}(t)$. Its coefficient should be equal to zero. Hence

$$\frac{1}{(k+1)!} + \frac{c_1}{1!k!} + ... + \frac{c_s}{s!(k-s+1)!} + ... + \frac{c_{k-1}}{(k-1)!2!} + \frac{c_k}{k!} = 0.$$

Consequently

$$c_k = -\left(\frac{1}{k+1} + \frac{c_1}{1!} + ... + \frac{c_s k!}{(s1(k-s+1)!} + ... + \frac{c_{k-1}}{2}\right).$$

From (14.8) $c_k = B_k$ $k = 1, 2, ...$ Taking into account that $B_{2p+1} = 0$ for $p = 1, 2, ...$ we conclude that the series in (14.13) formally satisfy equation (14.4). To prove relation (14.11) make the substitution $x(t) = y(t) + S_k(t)$ in (14.3) where

$$S_k(t) = \int \alpha(t)dt + \frac{B_1}{1!}\alpha(t) + ... + \frac{B_k}{k!}\alpha^{(k-1)}(t). \tag{14.14}$$

The substitution leads to the equation

$$y^-(t) = \alpha(t) - S_k^-(t) \equiv k(t).$$

Clearly

$$\left[\int \alpha(t)dt\right]^- \equiv \int_t^{t+1} \alpha(s)ds = \int_0^1 \alpha(t+s)ds,$$

hence

$$\alpha(t) - \left[\int \alpha(t)dt\right]^- = -\int_0^1 [\alpha(t+s) - \alpha(t)]ds$$

and

$$\alpha_k(t) = -\int_0^1 [\alpha(t+s) - \alpha(t)]ds - \frac{B_1}{1!}[\alpha(t)]^- - ... - \frac{B_k}{k!}[\alpha^{(k-1)}(t)]^-.$$

We have

$$\int_0^1 [\alpha(t+s) - \alpha(t)]ds = \sum_{p=1}^k \frac{\alpha^{(p)}(t)}{(p+1)!} + R_k(t), \tag{14.15}$$

where

$$R_k(t) = \int_0^1 \alpha^{(k+1)}(t+\theta(s))\frac{s^{k+1}}{(k+1)!}ds \quad (0 < \theta(s) < 1).$$

Estimate $R_k(t)$. Clearly

$$|R_k(t)| \leq \frac{1}{(k+2)!} \sup_{0 < \theta < 1} |\alpha^{(k+1)}(t + \theta)|.$$

Consequently $h\{R_k(t)\} \leq \Pi\{\alpha(t)\} - k - 1$. On differentiating relation (14.15) n times we obtain

$$R_k^{(n)}(t) = \int_0^1 \alpha^{(k+n+1)}(t + \theta_n(s)) \frac{s^{k+1}}{(k+1)!} ds,$$

$0 < \theta_n(s) < 1$, $n = 1, 2, ...$, which leads to the estimate

$$h\{R_k^{(n)}(t)\} \leq \Pi\{\alpha(t)\} - k - n - 1.$$

This means that

$$\Pi\{R_k(t)\} \leq \Pi\{\alpha(t)\} - k - 1.$$

In the same way for $m = 0, 1, ..., k$

$$[\alpha^{(m)}(t)]^- = \sum_{p=1}^{k-p} \frac{\alpha^{(p+m)}}{p!} + R_{k,m+1}(t),$$

where

$$\Pi\{R_{k,m+1}(t)\} \leq \Pi\{\alpha(t)\} - k - 1.$$

Thus the substitution leads to the equation $y^-(t) = \alpha_k(t)$, where

$$\Pi\{\alpha_k(t)\} < \Pi\{\alpha(t)\} - k.$$

Let us choose $k \gg \Pi\{\alpha(t)\}$. Then $\Pi\{\alpha_k(t)\} < -1$. Consequently on the basis of Proposition 14.2 there exists a unique infinitesimal solution $y_k(t)$ of the last equation where $\Pi\{y_k(t)\} \leq \Pi\{\alpha(t)\} - k$. Taking into account the uniqueness of the solution, tending k to infinity we prove relation (14.11). All the other required properties follows from Proposition 14.2. $\quad\square$

Consider an equation of the form

$$x(t + 1) = q(t)x(t) + \alpha(t), \tag{14.16}$$

where $q(t)$ and $\alpha(t)$ are complex valued functions defined for $t \gg 1$. Consider a homogeneous equation of the form

$$x(t + 1) = q(t)x(t). \tag{14.17}$$

Theorem 14.5. *Let $q(t) \in A_t$. Then equation (14.16) has a solution*

$$x(t) = \exp[y(t)], \qquad (14.18)$$

where the asymptotic representation of $y(t)$ can be written in the form

$$y(t) \asymp \int \ln q(t) dt - \frac{1}{2} \ln q(t) + \sum_{k=1}^{\infty} \frac{B_{2k}}{(2k)!} \left(\frac{q'(t)}{q(t)} \right)^{(2k-2)}. \qquad (14.19)$$

Theorem 14.5 is a simple consequence of Proposition 14.3. Indeed let us put $y(t) = \ln x(t)$ and $\alpha(t) = \ln q(t)$ in (14.18), this leads to (14.3). Owing to the condition $q(t) \in A_t$, we have $\Pi\{\ln q(t)\} \leq 0$. In addition

$$(\ln q(t))^{(2k-1)} = \left(\frac{q'(t)}{q(t)} \right)^{(2k-2)}$$

which leads to the required relation. ☐

1.4   *NON-HOMOGENEOUS EQUATIONS*

Lemma 14.6. Let $|q(t)| \geq 1$ for $t \gg 1$ and let $h\{\alpha(t)\} < -1$. Then equation (14.16) has a unique infinitesimal solution $x^*(t)$. It is continuous and

$$|x^*(t)| \leq \int_0^{+\infty} \sup_{\tau \leq s < +\infty} |\alpha(t + s - 1)| d\tau \qquad (14.20)$$

for $t \gg 1$, hence

$$h\{x^*(t)\} \leq h\{\alpha(t)\} + 1.$$

In particular if $h\{\alpha(t)\} = -\infty$ then $h\{x^*(t)\} = -\infty$.

PROOF. We have $1/|q(t)| \leq 1$ for $\gg 1$. Since $h\{\alpha(t)\} < -1$ there exists a number $\sigma > 1$ such that $|\alpha(t + m)| < m^{-\sigma}$ for $t \gg 1$ uniformly in $m = 1, 2, ...$, hence the series

$$-\frac{\alpha(t)}{q(t)} - \frac{\alpha(t + 1)}{q(t)q(t + 1)} - \cdots - \frac{\alpha(t + m)}{q(t)q(t + 1)...q(t + m)} - \cdots \qquad (14.21)$$

absolutely and uniformly converges to a continuous function $x^*(t)$ which is a solution to equation (14.16) (for $t \gg 1$). We have

$$|x^*(t)| \leq |\alpha(t)| + |\alpha(t + 1)| + ... + |\alpha(t + m)| + ... \qquad (14.22)$$

which leads to (14.20).

Let us prove the uniqueness of the infinitesimal solution. The general solution of equation (14.16) can be written in the form

$$x(t) = C(t)u(t) + x^*(t), \tag{14.23}$$

where $C(t)$ is an arbitrary periodic function with period 1, $u(t)$ is a fixed non-trivial solution of the homogeneous equation $u(t+1) = q(t)u(t)$. It is easy to see that any sequence $|C(t+m)u(t+m)|$ $(t = \text{const}, \ m = 1, 2, ...)$ does not decrease and hence $C(t)u(t)$ is not an infinitesimal function for $t \to +\infty$ (if $C(t)u(t) \not\equiv 0$). Consequently $x^*(t)$ is an unique infinitesimal solution to equation (14.16). $\qquad\square$

Lemma 14.7. *Let there exist a finite limit*

$$\lim_{t\to+\infty} t \ln|q(t)| = p$$

and let there exist a function $q_m(t) \in A_t$ for any natural m such that $h\{q(t) - q_m(t)\} < -m$. Let $h\{\alpha(t)\} = -\infty$. Then there exists a continuous solution $x^*(t)$ to equation (14.16) (for $t \gg 1$) with the estimate $h\{x^*(t)\} = -\infty$.

PROOF. Let $q(t) \in A_t$. On the basis of (14.19) equation $y(t+1) = q(t)y(t)$ has a solution $y(t)$ continuous for $t \gg 1$ such that

$$\ln|y(t)| = \int \ln|q(t)|dt + O\left(\frac{1}{t}\right) \quad \text{for } t \to +\infty.$$

Hence $|y(t)| = t^{p+o(1)}$ for $t \to +\infty$. Consequently $h\{1/y(t)\} = -p$. Let us substitute $x(t) = y(t)u(t)$ in (14.16). We have $u(t+1) = u(t) + \beta(t)$, where $\beta(t) = \alpha(t)/[q(t)y(t)]$ and hence $h\{\beta(t)\} = -\infty$. Thus the last equation has a solution $u^*(t)$ continuous for $t \gg 1$ and $h\{u^*(t)\} = -\infty$. Consequently the solution $x^*(t) = y(t)u^*(t)$ to equation (14.16) has the estimate $h\{x^*(t)\} = -\infty$.

In the general case it is sufficient to prove that the equation $y(t+1) = q(t)y(t)$ has a solution $y(t)$ continuous for $t \gg 1$ and $y(t) = t^{p+o(1)}$ for $t \to +\infty$. To this end, put $u(t) = \ln y(t)$. We have

$$u(t+1) = u(t) + \ln q(t). \tag{14.24}$$

It is obvious that $q(t) = q_m(t)(1 + \alpha_m(t))$ for sufficiently large natural number m. Here $q_m(t) \in A_t$ and $h\{\alpha_m(t)\} \ll -1$. The equation $u(t+1) = u(t) + \ln q_m(t)$ has a continuous solution (for $t \gg 1$)

$$u_m^*(t) = \int \ln q_m(t)dt + O\left(\frac{1}{t}\right) \quad \text{for } t \to +\infty.$$

Let us substitute $u(t) = u_m^*(t) + v(t)$ in (14.24). We have

$$v(t+1) = v(t) + \ln(1 + \alpha_m(t)),$$

where

$$h\{\ln(1 + \alpha_m(t))\} = h\{\alpha_m(t))\} \ll -1.$$

On the basis of Theorem 14.4 the last equation has a continuous solution $v^*(t)$ for $t \gg 1$ and

$$h\{v^*(t)\} \leq h\{\alpha_m(t)\} + 1 \ll -1.$$

Thus we proved that equation (14.24) has a continuous solution

$$u^*(t) = \int \ln q_m(t)dt + O\left(\frac{1}{t}\right) \quad \text{for } t \to +\infty \; (t \gg 1)$$

which leads to the required relation.   □

Lemma 14.8. *Let* $h\{q(t)\} = -\infty$ *or* $\lim_{t \to +\infty} t \ln |q(t)| = -\infty$. *Then any continuous (for $t \gg 1$) solution $x(t)$ to equation (14.16) has the estimate* $h\{x(t)\} = -\infty$.

PROOF. Let us choose a solution $x(t)$ to equation (14.16) $(t \gg 1)$. Consider an equation of the form

$$y(t+1) = \left(1 + \frac{1}{t}\right)^{-\sigma} y(t) + |\alpha(t)|, \qquad (14.25)$$

where σ is a positive number. Let us fix a number σ and choose a positive number T_σ such that the functions $q(t)$ and $\alpha(t)$ are continuous and $(1 + 1/t)^{-\sigma} > |q(t)|$ for $t > T_\sigma$. Let $y(t)$ be a solution to equation (14.25) such that it coincides with $|x(t)|$ on the interval $]T_\sigma, T_\sigma + 1[$. The inequality $y(t) \geq |x(t)|$ for $t > T_\sigma + 1$ is obvious. Equation (14.25) has a general solution

$$y(t) = C(t)t^{-\sigma} + y^*(t),$$

where $C(t)$ is an arbitrary periodic function with period 1, $y^*(t)$ (see Lemma 14.6) is a function continuous for $t \gg 1$ and $h\{y^*(t)\} = -\infty$. Hence $h\{x(t)\} \leq -\sigma$. Taking into account the arbitrariness of σ we conclude that $h\{x(t)\} = -\infty$.   □

2.   SYSTEMS OF NTH ORDER EQUATIONS

Here we consider a *normal system* of n ordinary linear difference equations which is a system of the form

$$\begin{cases} x_1(t+1) & = & a_{11}(t)x_1(t) + a_{12}(t)x_2(t) + \ldots + a_{1n}(t)x_n(t) + f_1(t), \\ x_2(t+1) & = & a_{21}(t)x_1(t) + a_{22}(t)x_2(t) + \ldots + a_{2n}(t)x_n(t) + f_2(t), \\ \quad \cdots \cdots \cdots \cdots \cdots \cdots \cdots \cdots \cdots \cdots \\ x_n(t+1) & = & a_{n1}(t)x_1(t) + a_{n2}(t)x_2(t) + \ldots + a_{nn}(t)x_n(t) + f_n(t). \end{cases}$$

$$(14.26)$$

Let τ be a fixed number belonging to the interval $[0, 1[$ and $D_\tau = \{t : t = \tau + m,\ m = 1, 2, \ldots\}$. First, we suppose that the functions $a_i(t)$ and $f_i(t)$ specified on the set D_τ. In some problems it is convenient to consider the system on the positive semi-axis $J_+ = \{t : t \geq 0, t \neq \infty\}$, on the real axis and even in the entire complex plane, where $a_i(t)$ and $f_i(t)$ are functions defined in the considered domains respectively. But the case of the set D_τ is the most important. Below we limit ourself to consideration of the cases of the set D_τ and semi-axis J_+.

System (14.26) is equivalent to a single (matrix) equation of the form

$$X(t+1) = A(t)X(t) + F(t). \qquad (14.27)$$

Here $A(t) = (a_{ij}(t))_n$ is a square matrix of nth order consisting of the elements $a_{ij}(t)$;

$$F(t) = (f_1(t), f_2(t), \ldots, f_n(t))^T$$

is a column matrix (column vector);

$$X(t) = (x_1(t), x_2(t), \ldots, x_n(t))^T.$$

We also say that (14.27) is a *system of n* (scalar) *linear difference equations.*

The general theory of systems (14.27) is similar (in many respects) to the general theory of systems of linear differential equation (see Chapter 8).

Definition 14.9. A point $t_0 \neq \infty$ is said to be a *regular point* of equation (14.27) if the matrix-functions $A(t)$ and $F(t)$ are defined at this point and each of their elements does not become infinity. Any other point is said to be a *singular point* of the equation.

Throughout this subsection (if it is not stipulated apart) *we suppose $A(t)$ and $F(t)$, to be finite, and* $\det A(t) \neq 0$ *for any t in the considered domain.*

Clearly the solution $X(t)$ to system (14.27) exists and is unique on D_τ if the value $X_\tau = X(\tau)$ is given; X_τ is called an *initial condition* for the system. The solution exists and unique on J_+ if $X(t)$ is specified at any point $t \in [0,1[$. This means that it is given an *initial matrix function* $\varphi_0(t)$ on the interval $[0.1[$ and $X(t) = \varphi_0(t)$ on this interval.

Remark 14.10. *Let $A(t)$, $F(t)$ be real matrices and the initial condition X_τ consists only of real components. Then the solution $X(\tau + m)$ is a real matrix for any $m = 0, 1, ...$*

If $F(t) = \theta$ (θ is a null matrix) equation (14.27) turns into the equation

$$X(t+1) = A(t)X(t) \qquad (14.28)$$

which is called a *homogeneous linear difference (matrix) equation* (or *system* of n scalar equations).

A zero column matrix is obviously a solution of the equation (which is called *trivial*).

Due to the supposition and because of $\det A(t) \neq 0$, $X(\tau + m) = \theta$ if and only if $X(\tau) = \theta$. Indeed if $X(\tau) = \theta$ then obviously $X(\tau + m) = \theta$. And if $X(\tau + m) = \theta$ taking into account that there exists the inverse matrix $A^{-1}(\tau + m - 1)$ we have $X(\tau + m - 1) = A^{-1}(\tau + m - 1)X(\tau + m) = \theta$. Reasoning in the same way we obtain the relation $X(\tau) = \theta$ by finitely many steps.

Theorem 14.11. *The set of all solutions to equation (14.28) (on D_τ) (for a fixed $\tau \in [0,1[$) forms n-dimensional linear space over the field $\mathbb{C}_n$ of all complex numbers.*

PROOF. Let $X_1(t)$ and $X_2(t)$ be solutions to (14.28) and let c_1 and c_2 be (complex) numbers. Then obviously the function $c_1 X_1(t) + c_2 X_2(t)$ is a solution to the equation. Thus the considered space is linear. Consider the system of column matrices (vectors)

$$\{E_1 = (1,0,...,0)^T, E_2 = (0,1,0,...,0)^T, ..., E_n = (0,...,0,1)^T\}$$

which is linearly independent in the space $\mathbb{C}_n$. There exist unique solutions $X_i(t)$ where $X_i(\tau) = E_i$ ($i = 1, 2, ..., n$). They form a linearly independent system. Indeed let us consider a linear combination

$$X(t) = c_1 X_1(t) + c_2 X_2(t) + ... + c_n X_n(t),$$

where $c_1, c_2, ..., c_n$ are (complex) numbers. Then $X(t)$ is a solution to equation (14.28) with the initial condition $X(\tau) = (c_1, c_2, ..., c_n)^T$. If $X(t)$ is a zero matrix then $X(\tau)$ is a zero matrix. Hence $c_1 = c_2 = \ldots = c_n = 0$. Thus the solutions $X_1(t), X_2(t), \ldots, X_n(t)$ form a linearly independent system. If $X(t)$ is a solution to the equation and $X(\tau) = (c_1, c_2, \ldots, c_n)^T$, where c_i are (complex) numbers ($i = 1, 2, \ldots, n$), then we have $X(t) = c_1 X_1(t) + c_2 X_2(t) + \ldots + c_n X_n(t)$. Consequently the set of the solutions $\{X_1(t), X_2(t), \ldots, X_n(t)\}$ forms a basis of the space. □

Definition 14.12. Any linearly independent system of solutions

$$\{X_1(t), X_2(t), \ldots, X_n(t)\} \tag{14.29}$$

to equation (14.28) for $t \in D_\tau$ is said to be a *basis* or a *fundamental system* (*fundamental set*) of solutions (briefly *FSS*) of system (14.28) on D_τ, and the expression

$$x = C_1 X_1(t) + C_2 X_2(t) + \ldots + C_n X_n(t) \tag{14.30}$$

is called a *general solution* of (14.28) (on D_τ). Here $C_1, C_2, \ldots, C_n$ are numerical parameters cold *arbitrary constants*. On substituting fixed numbers c_i instead of C_i in (14.30), respectively, we obtain a matrix-function $\varphi^*(t) = c_1 X_1(t) + c_2 X_2(t) + \ldots + c_n X_n(t)$ (clearly, it is a solution to (14.28)) which is called a *partial solution* to the equation. Any solution to (14.28) may be obtained from (14.30) picking out a suitable vector $(c_1, c_2, \ldots, c_n)$. So that we may consider the general solution as a totality of all solutions to equation (14.28) on D_τ. The set of functions (14.29) specified on J_+ is said to be a *basis, fundamental system,* or *fundamental set* of solutions to system (14.28) (briefly *FSS*) on J_+ if it is an *FSS* of the equation in all D_τ for any $\tau \in [0, 1[$.

The expression

$$x = C_1(t) X_1(t) + C_2(t) X_2(t) + \ldots + C_n(t) X_n(t) \tag{14.31}$$

is called a *general solution* of (14.28) (on D_τ). Here $C_1(t), C_2(t), \ldots, C_n(t)$ are *arbitrary periodic functions with period* 1.

On substituting fixed periodic functions $c_i(t)$ with period 1 instead of $C_i(t)$ in (14.30) respectively, we obtain a matrix-function $\varphi^*(t) = c_1(t) X_1(t) + c_2(t) X_2(t) + \ldots + c_n(t) X_n(t)$ (clearly, it is a solution to (14.28)) which is called a *partial solution* to the equation. Any solution to (14.28) may be obtained from (14.30) picking out a suitable periodic functions $c_1(t), c_2(t), \ldots, c_n(t)$ with period 1. So that we may consider the general solution as a totality of all solutions to equation (14.28) on J_+.

A square matrix $\Phi(t)$ where all its columns form a basis of system (14.28) on D_τ (on J_+) is said to be a *fundamental matrix* (*FM*) on D_τ (on J_+).

Clearly

$$\Phi(t+1) = A(t)\Phi(t) \quad for \quad t \in D_\tau \ (t \in J_+) \qquad (14.32)$$

and for any solution $X(t)$ on D_τ (on J_+), there exists a constant column matrix C (a periodic column matrix-function $C(t)$ with period 1 such that $X(t) = \Phi(t)C$ $(X(t) = \Phi(t)C(t))$.

Consequently if $\Phi(t)$ is a fundamental matrix to equation (14.28) on D_τ (on J_+), then any solution $X(t)$ can be obtained from the expression $\Phi(t)C$ $(\Phi(t)C(t))$ choosing a suitable constant matrix C (periodic matrix $C(t)$). And vice versa. If any solution $X(t)$ to equation (14.28) may be obtained from the expression $\Phi(t)C$ $(\Phi(t)C(t))$ choosing a suitable constant matrix C $(C(t))$, then $\Phi(t)$ is a fundamental matrix of equation (14.28).

A square matrix $\Phi(t)$, where each its column is a solution to equation (14.28) is said to be a *matrix solution* of equation (14.28). We will designate its ith column by $\varphi_i(t)$ $(i = 1, 2, ..., n)$.

Theorem 14.13. *For a square matrix $\Phi(t)$ to be fundamental on D_τ it is necessary and sufficient to satisfy the identity (14.31) and $\det \Phi(t) \neq 0$ at least one point $t \in D_\tau$.*

PROOF. *Necessity.* If $\Phi(t)$ is fundamental then (by the definition) identity (14.32) holds and its columns form a linearly independent solutions to equation (14.28) at any point $\tau \in [0.1)$. Hence $\det \Phi(t) \neq 0$.

Sufficiency. Let

$$\Phi(t+1) = A(t)\Phi(t)$$

and $\det \Phi(t_0) \neq 0$, where t_0 is a fixed point belonging to D_τ, that is $t_0 = \tau + m_0$ where m_0 is a (fixed) natural number. Then $\det \Phi(\tau) \neq 0$ because

$$\Phi(\tau + m_0) = A(\tau)A(\tau + 1)...A(\tau + m_0 - 1)\Phi(\tau).$$

Hence

$$\det \Phi(\tau + m_0) = \det A(\tau)A(\tau + 1)... \det A(\tau + m_0 - 1) \det \Phi(\tau).$$

Consequently if $\det \Phi(\tau) = 0$ then $\det \Phi(\tau + m_0) = 0$ which is impossible. In the same way (taking into account that $\det A(t) \neq 0$) we obtain the inequality $\det \Phi(\tau + m) \neq 0$ for any m. The last means that the set of columns $\{\varphi_i(t)\}$ is linearly independent. Thus $\Phi(t)$ is a fundamental matrix. $\square$

The expression

$$x = C_1\varphi_1(t) + C_2\varphi_2(t) + ... + C_n\varphi_n(t) + \Psi(t), \qquad (14.33)$$

where $\Psi(t)$ is a solution to equation (14.27) and

$$C_1\varphi_1(t) + C_2\varphi_2(t) + ... + C_n\varphi_n(t)$$

is a general solution of equation (14.28) on D_τ and is called a *general solution* of equation (14.27) (on D_τ).

Consider (14.27) and (14.28) on J_+. If $\Phi(t)$ is an FM of matrix-equation (14.28) ($\varphi_i(t)$ are its column matrices ($i = 1, 2, ..., n$)), we say that the expression

$$x = C_1(t)\varphi_1(t) + C_2(t)\varphi_2(t) + ... + C_n(t)\varphi_n(t) + \Psi(t) \qquad (14.34)$$

is a *general solution* to equation (14.27), where $C_1(t), C_2(t), ..., C_n(t)$ are *arbitrary periodic column matrix-function* with period 1.

Remark 14.14. Expressions (14.33) and (14.34) are equivalent to the matrix expressions $X = X(t)C + \Psi(t)$ and $X = X(t)C(t) + \Psi(t)$ respectively. Here C is an arbitrary constant column matrix and $C(t)$ is an arbitrary periodic column matrix with period 1. Equation $X(t + 1) = A(t)X(t)$ is called a *homogeneous equation corresponding to the equation* $X(t + 1) = A(t)X(t) + F(t)$.

Proposition 14.15. *Let* $\Psi(t)$ *be a (partial) solution to equation* (14.27) *on* D_τ, *and let* $\Phi(t)$ *be a fundamental matrix of equation* (14.28). *Then any solution of equation* (14.27) *can be obtained from the expression*

$$X(t) = \Phi(t)C + \Psi(t) \qquad (14.35)$$

choosing a suitable constant column matrix C.

PROOF. Let us show that (14.35) is a solution to equation (14.27) for any C. To this end substitute $X(t)$ from (14.35) to (14.27). The left side of the obtained expression will be in the form

$$X(t + 1) = \Phi(t + 1)C(t + 1) + \Psi(t + 1).$$

Taking into account that $\Phi(t + 1) = A(t)\Phi(t)$, $\Psi(t + 1) = A(t)\Psi(t) + F(t)$ and $C(t + 1) = C(t)$ we have

$$\begin{aligned} X(t + 1) &= A(t)\Phi(t)C(t) + A(t)\Psi(t) + F(t) \\ &= A(t)[\Phi(t)C(t) + \Psi(t)] + F(t) \\ &= A(t)X(t) + F(t) \end{aligned}$$

which proves our assertion.   □

As a simple consequence of the last Theorem we obtain the following assertion: *The general solution* (14.33) ((14.34)) *may be considered as a totality of all solutions to equation* (14.28) *on* D_τ (*on* J_+).

Theorem 14.16. *Let* $\Phi(t)$ *be a fundamental matrix of homogeneous system* (14.28). *Then for any solution* $U(t)$ *of the equation*

$$U(t+1) = U(t) + \Phi^{-1}(t+1)F(t) \tag{14.36}$$

there exists a solution of system (14.27) *in the form* $X(t) = \Phi(t)U(t)$.

PROOF. Let us substitute $X(t) = \Phi(t)U(t)$ in (14.27). We have

$$\Phi(t+1)U(t+1) = A(t)\Phi(t)U(t) + F(t).$$

Hence

$$U(t+1) = \Phi^{-1}(t+1)A(t)\Phi(t)U(t) + \Phi^{-1}(t+1)F(t).$$

Clearly $\Phi^{-1}(t+1)A(t)\Phi(t) = E$ is a unite matrix that leads to equation (14.36).   □

Let us mark the following proposition.

Proposition 14.17. *Let the systems*

$$X(t+1) = A(t)X(t) \text{ and } X(t+1) = A^*(t)X(t) \tag{14.37}$$

possess the same set of solutions on D_τ. *Then* $A(t) = A^*(t)$ *on the set* D_τ.

Let us consider a substitution of the form

$$Y(t) = Q(t)X(t) \tag{14.38}$$

in (14.28), where $Q(t)$ is a non-singular matrix for any $t \in D_\tau$,

$$Y(t) = (y_1(t), y_2(t), ..., y_n(t))^T.$$

Substitute (14.38) in (14.28). We obtain the equation

$$Y(t+1) = B(t)Y(t), \tag{14.39}$$

where

$$B(t) = Q(t+1)A(t)Q^{-1}(t). \tag{14.40}$$

Equations (14.38) and (14.39) are equivalent on D_τ in the following sense: there is an one-to-one correspondence between the solutions to equations (14.28) and (14.39) which is given by the relation $Y(t) = Q(t)X(t)$ or (which is the same correspondence) $X(t) = Q^{-1}(t)Y(t)$.

3. SINGLE LINEAR DIFFERENCE EQUATIONS OF NTH ORDER

Here we consider a linear difference equation of nth order of the form

$$F(t, x) \equiv x(t + n) + a_1(t)x(t + n - 1) + \ldots + a_n(t)x = f(t). \quad (14.41)$$

where $a_i(t)$ and $f(t)$ are functions specified on D_τ or J_+.

Clearly the solution $x(t)$ to equation (14.41) exists and unique on D_τ if the values

$$x_{1\tau} = x(\tau), x_{2\tau} = x(\tau + 1), \ldots, x_{n\tau} = x(\tau + n - 1)$$

are given; the set $\{x_{1\tau}, x_{2\tau}, \ldots, x_{n\tau}\}$ is called *initial conditions* for the equation. On J_+ the solution exists and unique if $x(t)$ is specified at any point $t \in [0, n[$. This means that it is given an *initial function* $\varphi_0(t)$ on the interval $[0, n[$ and $x(t) = \varphi_0(t)$ on this interval. Let us put

$$x_1(t) = x(t), x_2(t) = x(t + 1), \ldots, x_n(t) = x(t + n - 1)$$

and

$$X(t) = (x_1(t), x_2(t), \ldots, x_n(t))^T,$$

then we pass from (14.41) to system (14.27) where $A(t)$ is a matrix of the form

$$A(t) = \begin{pmatrix} 0 & 1 & 0 & 0 & \ldots & 0 & 0 \\ 0 & 0 & 1 & 0 & \ldots & 0 & 0 \\ \cdot & \cdot & \cdot & \cdot & \cdot & \cdot & \cdot \\ 0 & 0 & 0 & 0 & \ldots & 0 & 1 \\ -a_n(t) & -a_{n-1}(t) & -a_{n-2}(t) & -a_{n-3}(t) & \ldots & -a_2(t) & -a_1(t) \end{pmatrix}$$

$$(14.42)$$

and $F(t) = (0, \ldots, 0, f(t))^T$. Equation (14.41) and the obtained system are equivalent on the set D_τ (on J_+) in the following sense: for every solution $x(t)$ to equation (14.41) there exists a solution

$$X(t) = (x_1(t), x_2(t), \ldots, x_n(t))^T$$

of system (14.27) such that $x_1(t) = x(t), x_2(t) = x(t + 1), \ldots, x_n(t) = x(t + n - 1)$, and for every solution

$$X(t) = (x_1(t), x_2(t), \ldots, x_n(t))^T$$

of system (14.27) there exists a solution $x(t)$ of equation (14.41) such that $x(t) = x_1(t), x(t + 1) = x_2(t), \ldots, x_r(t + n - 1) = x_n(t)$. Thus the main theorems of the general theory of equations of the form (14.41) are similar to the corresponding theorems of the general theory of linear difference systems.

If $f(t) = 0$ (14.41) is called a *homogeneous* linear difference equation of nth order. Thus it is an equation of the form

$$\Psi(t, x(t)) \equiv x(t+n) + a_1(t)x(t+n-1) + ... + a_n(t)x(t) = 0. \quad (14.1)$$

The next Theorem is proved in the same way as Theorem 14.11.

Theorem 14.18. *The set of all solutions to equation* (14.1) *on the set* D_τ *forms n-dimensional linear space over the field of all complex numbers.*

Definition 14.19. A set of n linearly independent solutions

$$\{x_1(t), x_2(t), ..., x_n(t)\} \quad (14.43)$$

to equation (14.1) on the set D_τ is called a *fundamental system* of solutions (*FSS*) or a *basis* to this equation on D_τ and the expression

$$x = C_1 x_1(t) + C_2 x_2(t) + ... + C_n x_n(t) \quad (14.44)$$

is called a *general solution* to equation (14.1) on D_τ. Here $C_1, C_2, ..., C_n$ are arbitrary constants. If (14.43) is linearly independent on the semi-axis J_+ then we say that (14.43) is a *fundamental system* of solutions (*FSS*) or a *basis* of this equation in J_+ and the expression

$$x = C_1(t)x_1(t) + C_2(t)x_2(t) + ... + C_n(t)x_n(t) \quad (14.45)$$

is called a *general solution* to equation (14.1) on the positive semi-axis where $C_1(t), C_2(t), ..., C_n(t)$ are arbitrary periodic functions with period 1.

Let $x_1(t), x_2(t), ..., x_n(t)$ be solutions to equation (14.1). Let us consider the matrix (which is called the matrix of CASORATI)

$$\begin{pmatrix} x_1(t) & x_2(t) & \cdots & x_n(t) \\ x_1(t+1) & x_2(t+1) & \cdots & x_n(t+1) \\ \cdots\cdots\cdots\cdots\cdots\cdots\cdots\cdots\cdots \\ x_1(t+n-1) & x_2(t+n-1) & \cdots & x_n(t+n-1) \end{pmatrix}. \quad (14.46)$$

This matrix is a matrix solution to system (14.28) with the matrix $A(t)$ in form (14.44). The determinant of the last matrix (which is called a *Casoratian*) we denote by $V(x_1(t), ..., x_n(t))$.

Theorems 14.20, 14.21 and Propositions 14.22, 14.23 simply follow from the corresponding part of general theory of systems of difference equations (see subsection 14.2).

Theorem 14.20. *The set of n linearly independent solutions $\{x_1(t), x_2(t), ..., x_n(t)\}$ to equation (14.1) on D_τ forms an FSS if and only if*

$$V(x_1(t), ..., x_n(t)) \neq 0$$

at least one point $t \in D_\tau$.

Theorem 14.21. *Any solution $x(t)$ to equation (14.1) on D_τ is a linear combination of functions belonging to FSS and expression (14.44) may be considered as a set of all solutions to equation (14.1).*

Proposition 14.22. *Let*

$$\Psi_1(t, x(t)) \equiv x(t+n) + a_{11}(t)x(t+n-1) + ... + a_{n1}(t)x(t) = 0$$

and

$$\Psi_2(t, x(t)) \equiv x(t+n) + a_{12}(t)x(t+n-1) + ... + a_{n2}(t)x(t) = 0$$

be two equations such that the functions $a_{i1}(t)$ and $a_{i2}(t)$ are determined on D_τ ($i = 1, 2, ..., n$) and the equations have the same set of solutions on the set D_τ. Then $\Psi_1(t, x(t)) \equiv \Psi_2(t, x(t))$, i.e. $a_{i1}(t) = a_{i2}(t)$ at any point $t \in D_\tau$ ($i = 1, 2, ..., n$).

Proposition 14.23. *Let $x_1(t), x_2(t), ..., x_n(t)$ be functions specified on D_τ and the Casoratian $V(x_1(t), ..., x_n(t)) \neq 0$ at any $t \in D_\tau$. Then there exists a unique homogeneous difference equation of nth order of the form (14.1) with coefficients defined on D_τ such that each function $x_i(t)$ ($i = 1, 2, ..., n$) is a solution of the equation. Moreover this equation is written in the form:*

$$F(t, x) \equiv (-1)^n \frac{V(x(t), x_1(t), ..., x_n(t))}{V(x_1(t), ..., x_n(t))} = 0. \qquad (14.47)$$

Here $V(x(t), x_1(t), ..., x_n(t))$ is a Casoratian of $n+1$ order for the functions $x(t), x_1(t), ..., x_n(t)$, i.e. it is a determinant where its $m+1$th row consists of the elements

$$x(t+m), x_1(t+m), ..., x_n(t+m) \quad (m = 0, 1, ..., n-1).$$

Theorem 14.24. *Let* $\{\varphi_1(t), \varphi_2(t), ..., \varphi_n(t)\}$ *be an FSS of homogeneous equation* (14.1) *and* $\Phi(t)$ *be the corresponding matrix of Casorati (see* (14.46)). *Then there exists a solution to equation* (14.41) *in the form*

$$x(t) = \varphi_1(t)u_1(t) + \varphi_2(t)u_2(t) + ... + \varphi_n(t)u_n(t) \qquad (14.48)$$

for any solution $U(t) = (u_1(t), u_2(t), ..., u_n(t))$ to equation

$$U(t+1) = U(t) + \Phi^{-1}(t+1)F(t). \qquad (14.49)$$

4.   DIFFERENCE EQUATIONS WITH CONSTANT COEFFICIENTS

We consider an equation of the form

$$L[x(t)] \equiv x(t+n) + a_1x(t+n-1) + ... + a_nx(t) = 0, \qquad (14.50)$$

where a_i are constants, $a_n \neq 0$. The algebraic equation

$$H(y) \equiv y^n + a_1y^{n-1} + ... + a_n = 0 \qquad (14.51)$$

is called the *characteristic equation* of equation (14.50).

We look for solutions to equation (14.50) in the form $x(t) = \lambda^t$ where λ is a constant.

Lemma 14.25. *The function* $x(t) = \lambda^t \neq 0$ *is a solution to equation* (14.50) *if and only if* λ *is a root of equation* (14.51).

PROOF. Since $a_n \neq 0$, clearly, any root of characteristic equation (14.51) does not equal to zero. Let us substitute $x(t) = \lambda^t$ in (14.50) where $\lambda \neq 0$. We have $x(t+k) = \lambda^k\lambda^t$ and (taking into account that $\lambda^t \neq 0$) we obtain $H(\lambda) = 0$. The last relation takes place if and only if λ is a root of equation (14.51). $\square$

The following Proposition is a simple consequence of Lemma 14.25.

Proposition 14.26. *Let characteristic equation* (14.51) *have all distinct roots* $\lambda_1, \lambda_2, ..., \lambda_n$. *Then a general solution of equation* (14.49) *on* D_τ *can be represented in the form*

$$x = C_1\lambda_1^t + C_2\lambda_2^t + ... + C_n\lambda_n^t, \qquad (14.52)$$

where C_i *are arbitrary constants* $(i = 1, 2, ..., n)$.

PROOF. Indeed any function of the form (14.51) is a solution of equation (14.49) and as it easy to show the Casoratian $V(\lambda_1^t, ..., \lambda_n^t)$ is equal to $\lambda_1^t ... \lambda_n^t \omega(\lambda_1, ..., \lambda_n)$, where $\omega(\lambda_1, ..., \lambda_n)$ is the Van der Monde determinant of the values $\lambda_1, ..., \lambda_n$. Since all the numbers λ_j are distinct ($j = 1, 2, ..., n$) we have $\omega(\lambda_1, ..., \lambda_n) \neq 0$ which leads to the required property. $\square$

The following proposition is a simple consequence of Proposition 14.26.

Proposition 14.27. *Let characteristic equation* (14.51) *have all distinct roots* $\lambda_1, \lambda_2, ..., \lambda_n$. *Then a general solution of equation* (14.50) *on* J_+ *can be represented in the form*

$$x = C_1(t)\lambda_1^t + C_2(t)\lambda_2^t + ... + C_n(t)\lambda_n^t, \qquad (14.53)$$

where $C_i(t)$ *are arbitrary periodic functions with period* 1 ($i = 1, 2, ..., n$).

Let us introduce the following notation

$$x_\lambda^{[1]}(t) = x(t+1) - \lambda x(t), \quad x_\lambda^{[m]}(t) = x^{[m-1]}(t+1) - \lambda x^{[m-1]}(t)$$

($m = 2, ...$). For $\lambda = 1$ we somewhat simplify the notation and write $x^{[m]}(t)$ instead of $x_\lambda^{[m]}(t)$. For instance $x^-(t) = x^{[1]}(t)$.

Proposition 14.28. *Let* λ *be a root of equation* (14.50). *Then the substitution*

$$u(t) = x_\lambda^{[1]}(t) \qquad (14.54)$$

in (14.50) *leads to the equation*

$$L^*[u(t)] \equiv u(t+n-1) + b_1 u(t+n-2) + ... + b_{n-1} = 0, \qquad (14.55)$$

where its characteristic polynomial

$$H^*(y) = y^n + b_1 y^{n-1} + ... + b_{n-1}$$

is connected with polynomial $H(y)$ *by the relation* $H(y) = H^*(y)(y - \lambda)$.

PROOF. To make this substitution let us rewrite (14.52) in the form $x(t+1) = u(t) + \lambda x(t)$. We have

$$x(t+2) = u(t+1) + \lambda u(t) + \lambda^2 x(t)$$

and (by induction with respect to $m = 1, 2, ...$)

$$x(t+m) = u(t+m-1) + \lambda u(t+m-2) + ... + \lambda^{m-1} u(t) + \lambda^m x(t).$$

On substituting all the relations in (14.50) we obtain $[L^*[u(t)]+H(\lambda)]x(t) = 0$, where

$$
\begin{aligned}
L^*[u(t)] = \ & u(t+n-1) + (\lambda + a_1)u(t+n-2) + ... \\
& + (\lambda^k + a_1\lambda^{k-1} + ... + a_k)u(t+n-k-1) + ... \\
& + (\lambda^{n-1}a_1\lambda^{n-2} + ... + a_{n-1})u(t).
\end{aligned}
$$

Its characteristic polynomial

$$
\begin{aligned}
& y^{n-1} + (\lambda + a_1)y^{n-2} + ... + (\lambda^k + a_1\lambda^{k-1} + ... + a_k)y^{n-k-1} + ... \\
& + (\lambda^{n-1} + a_1\lambda^{n-2} + ... + a_{n-1})
\end{aligned}
$$

as it easy to see coincides with $H^*(y)$. Moreover $H(\lambda)x(t) = 0$ which leads to the required relation. □

Lemma 14.29. *The equation $x^{[n]}(t) = 0$ has a general solution (on D_τ) in the form*

$$
x(t) = C_1 + C_2t + ... + C_nt^{n-1}, \tag{14.56}
$$

where C_i are arbitrary constants $(i = 1, 2, ..., n)$.

PROOF. Directly (on substituting t^m $(m = 0, 1, ..., n-1)$ in the left side of the considered equation) we make sure that the function is a solution to the equation. Clearly the set $\{1, t, ..., t^{n-1}\}$ is linearly independent which proves this Lemma. □

The following Lemma is a simple consequence of Lemma 14.29.

Lemma 14.30. *The general solution of the equation $x_\lambda^{[n]}(t) = 0$ (on D_τ) can be represented in the form*

$$
x(t) = \lambda^t(C_1 + C_2t + ... + C_nt^{n-1}), \tag{14.57}
$$

where $C_1, C_2, ..., C_n$ are arbitrary constants.

PROOF. Indeed substitute $x(t) = \lambda^t u(t)$ in the considered equation. We have $y^{[n]}(t) = 0$ which leads to the required relation. □

Lemma 14.31. *If characteristic equation (14.51) has a k multiple root λ then equation (14.50) (on D_τ) has any solution of the form*

$$
x(t) = \lambda^t(C_1 + C_2t + ... + C_kt^{k-1}),
$$

where $C_1, C_2, ..., C_k$ are arbitrary constants.

PROOF. Let us simultaneously make the following substitutions

$$y_1(t) = x_\lambda^{[1]}(t), y_2(t) = y_{1\lambda}^{[1]}(t), ..., y_k(t) = y_{k-1\lambda}^{[1]}(t).$$

As a result we obtain a linear difference equation of order $n - k$. Hence all the solutions of the equation $y_k(t) = 0$ or which is the same, all the solutions to equation $x_\lambda^{[k]}(t) = 0$ are solutions to equation (14.50). Hence we obtain the required relation from Lemma 14.30.   □

The following Theorem is a consequence of the previous Lemmata.

Theorem 14.32. *Let $\lambda_1, ..., \lambda_s$ be all the (distinct) roots of characteristic equation (14.51) and let the multiplicity of each root λ_j be k_j ($j = 1, 2, ..., s$) (so that $k_1 + k_2 + ... + k_s = n$). Then FSS of equation (14.49) (on D_τ and on J_+) is given by the n functions*

$$t^{m_r} \lambda_r^t \quad (m_r = 0, 1, ..., k_r - 1, \ r = 1, 2, ..., s). \tag{14.58}$$

Chapter 15

ASYMPTOTIC BEHAVIOUR OF SOLUTIONS OF LINEAR DIFFERENCE EQUATIONS

In this chapter we consider the asymptotic behavior of solutions of a linear difference homogeneous equation of the form

$$\Phi(t, x(t)) \equiv x(t+n) + a_1(t)x(t+n-1) + \ldots + a_n(t)x(t) = 0 \quad (15.1)$$

with power order growth coefficients.

First by the substitution $x(t+1) = (t)x(t)$ we arrive to the equation

$$F(t, y(t)) \equiv$$

$$y(t+n-1)y(t+n-2)\ldots y(t) + a_1(t)y(t+n-2)y(t+n-3)\ldots y(t) + \ldots + a_n(t) = 0.$$

Hence for $x(t) \neq 0$ we obtain the equation $F(t, y(t)) = 0$ and look for asymptotic solutions belonging to the class of power order growth functions The procedure consists of two stages. In the first stage we look for so called formal solutions to the equation $F(t, y(t)) = 0$. A function $g(t)$ (roughly speaking) is a formal solution to the equation if $F(t, g(t)) = O(t^{-\infty})$.

In the second stage we have to prove that the obtained formal solutions are the desired asymptotic solutions. This means that (for the examined formal solution $g(t)$) there exists an exact solution $y(t)$ such that $y(t) - g(t) = O(t^{-\infty})$. Thus the asymptotic representation of the corresponding function $x(t)$ can be obtained from the equation $x(t+1) = g(t)x(t)$.

Here we restrict our representation with the following conditions. For the first stage: the coefficients $a_1(t), \ldots, a_n(t)$ belong to a field Q of type N (or some more general, to $\{Q\}$). All the roots of a complete set of roots

$$\Lambda = \{\lambda_1(t), \ldots, \lambda_n(t)\} \quad (15.2)$$

of the characteristic equation

$$H(t,y) \equiv y^n + a_1(t)y^{n-1} + \ldots + a_n(t) = 0 \qquad (15.3)$$

have to be not equivalent in pairs for $t \to +\infty$ and no more than one of the roots $\lambda(t) \in \Lambda$ may have the estimate $\Pi\{\lambda(t)\} = -\infty$. Then for any root $\lambda_j(t) \in \Lambda$ with the estimate $\Pi\{\lambda_j(t)\} > -\infty$ there is a formal solution $g_j(t)$ to equation $F(t,y) = 0$ $(F(t,g_j(t)) = O(t^{-\infty}))$ on the positive semi-axis J_+, which is an asymptotic limit of an iterate sequence $\{s_{jm}(t)\}$ where $s_{j0}(t) = \lambda_j(t)$ and for $m = 1,2,\ldots$

$$s_{jm}(t) = \lambda_j(t) - \frac{F^*(t, s_{jm-1}(t))}{H_j^*(t, s_{jm-1}(t))}.$$

Here $F^*(t,y) = F(t,y) - H(t,y)$ and $H_j^*(t,y)$ is obtained from the identity $H(t,y) = H_j^*(t,y)(y - \lambda_j(t))$. Thus $g_j(t) \sim y_j(t)$ for $t \to +\infty$. Moreover

$$g_j(t) = \lambda_j(t) - \frac{1}{2}\lambda_j'(t)\lambda_j(t)\frac{\partial^2 H(t,\lambda_j(t))}{\partial y^2} \Bigg/ \frac{\partial H(t,\lambda_j(t))}{\partial y} + \alpha_j(t), \quad (15.4)$$

where $\Pi\{\alpha_j(t)\} \leq -2$. In particular $g_j(t) = \lambda_j(t) + O(1/t)$ for $t \to +\infty$. If $\Pi\{\lambda_j(t)\} = -\infty$ then a function $g_j(t)$ with the estimate $\Pi\{g_j(t)\} = -\infty$ is a formal solution to the equation $F(t,y) = 0$. Moreover the set $G_j(t)$ $(j = 1,2,\ldots,n)$ form a formal fundamental system of solutions to equation (15.1) where $G_j(t+1) = g_j(t)G_j(t)$.

For the second stage we suppose that $a_1(t),\ldots,a_n(t)$ belong to a field G of type M (or to $\{G\}$). If the conditions in the first stage are fulfilled then there is a fundamental system of solutions (FSS) $\{x_j(t)\}$ $(j = 1,2,\ldots,n)$ to equation (15.1) such that for any

$$y_j(t) = \frac{x_j(t+1)}{x_j(t)}, \quad y_j(t) - g_j(t) = O(t^{-\infty}).$$

Linear difference equations of the second order with power order growth coefficients are considered in detail.

In section 15.1 we consider linear difference equations of nth order where all the coefficients $a_1(t),\ldots,a_n(t)$ have finite limits for $t \to +\infty$. We prove the main important the H POINCARÉ and O. PERRON theorem which describe the asymptotic behaviour of solutions of the equation.

In conclusion of this chapter we consider a linear differential–difference equation of the first order of the form $x'(t+1) = a(t)x(t)$, where $a(t) = t^{k+\alpha(t)}$. Here k is a non-zero number and $\alpha(t) \in C_t$. And we look for its formal solution.

1.   THE POINCARÉ AND PERRON THEOREMS

First, prove several simple Lemmata. Consider an equation of the form

$$y(t + 1) = q(t)y(t) + \alpha(t), \tag{15.5}$$

where $q(t)$ and $\alpha(t)$ are functions defined for $t \gg 1$ and $\alpha(t) \to 0$ for $t \to +\infty$.

Lemma 15.1. *Let there exist a number σ such that $0 \leq \sigma < 1$ and $|q(t)| \leq \sigma$ for $t > T$ where T is a sufficiently large positive number. Then every solution $y(t)$ of equation (15.5) which is bounded in the interval $[T, T + 1]$ is infinitesimal for $t \to +\infty$.*

PROOF. Let $\tau \in [T, T + 1]$. We have

$$\begin{aligned}
y(\tau + 1) &= q(\tau)y(\tau) + \alpha(\tau), \\
y(\tau + 2) &= q(\tau + 1)y(\tau + 1) + \alpha(\tau + 1) \\
&= q(\tau + 1)[q(\tau)y(\tau) + \alpha(\tau)] + \alpha(\tau + 1) \\
&= q(\tau)q(\tau + 1)y(\tau) + q(\tau)\alpha(\tau) + \alpha(\tau + 1).
\end{aligned}$$

Reasoning in the same way by induction with respect to m we obtain the following relation

$$\begin{aligned}
y(\tau + m) &= q(\tau + m - 1)q(\tau + m - 2)...q(\tau)y(\tau) \\
&\quad + q(\tau + m - 2)...q(\tau)\alpha(\tau) + ... \\
&\quad + q(\tau)\alpha(\tau + m - 2) + \alpha(\tau + m - 1).
\end{aligned}$$

Let us estimate all the terms in the right hand side of the last relation. We have

$$|q(\tau + m - 1)q(\tau + m - 2)...q(\tau)y(\tau)| \leq A\sigma^m,$$

where

$$A = \sup_{T \leq \tau \leq T+1} |y(t)|.$$

Let us choose an arbitrary positive number ε (no matter how small) and take a number m so large that $|\alpha(\tau + k)| < \varepsilon$ for any $k > [m/2]$. Here $[m/2]$ is the nearest integer to $m/2$ such that $[m/2] \leq m/2$. We have

$$q(\tau + m - 2)...q(\tau)\alpha(\tau) + ... + q(\tau)\alpha(\tau + m - 2) + \alpha(\tau + m - 1)$$

$$= S_1(\tau, m) + S_2(\tau, m),$$

where

$$S_1(\tau, m) = q(\tau + m - 2)...q(\tau)\alpha(\tau) + ...$$

$$+ q\left(\tau + m - \left[\frac{m}{2}\right]\right)...q(\tau)\alpha\left(\tau + \left[\frac{m}{2}\right] - 1\right)$$

and

$$S_2(\tau, m) = q\left(\tau + m - \left[\frac{m}{2}\right] + 1\right)...q(\tau)\alpha\left(\tau + \left[\frac{m}{2}\right] - 2\right) + ...$$

$$+ q(\tau)\alpha(\tau + m - 2) + \alpha(\tau + m - 1).$$

Consequently

$$|S_1(t)| \leq \sigma^{[m/2]-1}B(1 + \sigma + \sigma^2 + ... + \sigma^{[m/2]}) \leq \sigma^{[m/2]-1}\frac{B}{1 - \sigma},$$

where $B = \sup_{t \geq T} |\alpha(t)| < \infty$ and

$$|S_2(\tau, m)| \leq \varepsilon(1 + \sigma + ... + \sigma^{[m/2]}) \leq \frac{\varepsilon}{1 - \sigma}.$$

Hence $|y(\tau + m)| \ll 1$ uniformly in τ for $m \gg 1$. This implies the limit $\lim_{t \to +\infty} y(t) = 0$. $\qquad\qquad\qquad\qquad\qquad\qquad\qquad\qquad\square$

Lemma 15.2. *Let there exist a number σ such that $|q(t)| > \sigma > 1$. Then there exists a unique bounded solution $y^*(t)$ to equation (15.2) and the solution is infinitesimal for $t \to +\infty$.*

PROOF. Let us choose by arbitrariness a number $\varepsilon > 0$ and take the number $T \gg 1$ such that $|\alpha(t)| < \varepsilon$ for $t \geq T$. Let us show that equation (15.5) possesses an infinitesimal solution represented in the form

$$y^*(t) = -\frac{\alpha(t)}{q(t)} - \frac{\alpha(t + 1)}{q(t)q(t + 1)} - ... - \frac{\alpha(t + p)}{q(t)...q(t + p)} - ... \ .$$

Indeed the series formally satisfy this equation and it is convergent because

$$\left|\frac{\alpha(t + p)}{q_r(t)...q_r(t + p)}\right| < \varepsilon\sigma^{-p}.$$

Consequently $|y * (t)| \leq \varepsilon/(\sigma - 1)$. Hence (taking into account the arbitrariness of ε) we conclude that $y * (t) \to 0$ for $t \to +\infty$.

All the solutions of the equation may be obtained from the general solution $y = Cg(t) + y^*(t)$ where C is an arbitrary constant, $g(t)$ is a non-trivial solution to equation (15.5). Clearly

$$|g(t+p)| \geq \sigma^{p-1}|y(t)| \to \infty \text{ for } p \to \infty.$$

Hence there is only one bounded solution to equation and clearly it is equal to $y^*(t)$.   □

Consider an equation of the form

$$y(t) = \alpha(t) + q(t)y(t)y(t+1). \tag{15.6}$$

Lemma 15.3. *Let $\alpha(t)$ be an infinitesimal function for $t \to +\infty$, $\alpha(t) \neq 0$ and $q(t)$ be a bounded function for $t \gg 1$. Then equation (15.6) has a unique infinitesimal solution $y(t)$ and it satisfies the relation $y(t) \sim \alpha(t)$ for $t \to +\infty$. As a consequence $y(t) \neq 0$ for $t \gg 1$.*

PROOF. Compose the following sequence of functions: $y_0(t) = 0$ and for $m = 1, 2, \ldots$

$$y_m(t) = \alpha(t) + q(t)y_{m-1}(t)y_{m-1}(t+1). \tag{15.7}$$

Since $q(t)$ is bounded there exists a positive number M such that $|q(t)| < M$ (for $f \gg 1$). Put $\beta(t) = \sup_{\tau \geq t} |\alpha(\tau)|$ and $y_m^*(t) = \sup_{\tau \geq t} |y_m(\tau)|$. We have $y_1^*(t) = \beta(t) < 2\beta(t)$. Let $y_{m-1}^*(t) < 2\beta(t)$ and let t be so large that $\beta(t) < 1/M$. Then

$$y_m^*(t) < \beta(t) + M\beta^2(t) < 2\beta(t).$$

Consequently (by induction) $y_m^*(t) < 2\beta(t)$ for any m. Put

$$\Delta_m(t) = \sup_{\tau \geq t} |y_m(t) - y_{m-1}(t)|.$$

Let t be so large that $|\beta(t)| < 1/(4M)$. We have

$$\Delta_2(t) = \sup_{\tau \geq t} |y_2(t) - y_1(t)|$$

$$= \sup_{\tau \geq t} |y_1(t)y_1(t+1)|$$

$$\leq M \sup_{\tau \geq t} |[\beta(t)|y_1(t) - y_0(t)|$$

$$< \frac{1}{2}\Delta_1(t).$$

Let $\Delta_{m-1} < \frac{1}{2}\Delta_{m-2}(t)$. Then

$$\Delta_m(t) < \sup_{\tau \geq t} |q(t)[y_{m-1}(t)y_{m-1}(t+1) - y_{m-2}(t)y_{m-2}(t+1)]|$$

$$\leq M \sup_{\tau \geq t} |y_{m-1}(t)[y_{m-1}(t+1) - y_{m-2}(t+1)]$$

$$+ y_{m-1}(t+1)[y_{m-1}(t) - y_{m-2}(t)]|$$

$$\leq 2M\beta(t)\Delta_{m-1}(t)$$

$$\leq \frac{1}{2}\Delta_{m-1}(t).$$

Hence (by induction) $\Delta_m(t) \leq \frac{1}{2}\Delta_{m-1}(t)$ for any m. Compose the series

$$(y_1(t) - y_0(t)) + (y_2(t) - y_1(t)) + ... + (y_m(t) - y_{m-1}(t)) + ...$$

Its partial sum is equal to $y_m(t)$. The series is convergent for $t \gg 1$ because $|y_m(t) - y_{m-1}(t)| \leq \Delta_m(t)$. This leads to the inequality

$$|y_m(t) - y_{m-1}(t)| \leq 2^{1-m}\beta(t).$$

Consequently there exists the limit $y(t) = \lim_{m \to \infty} y_m(t) \leq 2\beta(t)$ and clearly, $y(t)$ is a solution to equation (15.6). Moreover by identity (15.6) we have

$$y(t) = \alpha(t)/[1 - q(t)y(t+1)] \sim \alpha(t) \text{ for } t \to +\infty.$$

Let there be another infinitesimal solution $\tilde{y}(t)$ to the equation. Hence

$$y(t) - \tilde{y}(t) = q(t)[y(t)y(t+1) - \tilde{y}(t)\tilde{y}(t+1)].$$

Consequently

$$\sup_{\tau \geq t} |y(t) - \tilde{y}(t)| \leq M[\sup_{\tau \geq t}[|y(t)| + |\tilde{y}(t)|] \sup_{\tau \geq t} |y(t) - \tilde{y}(t)|$$

$$= o(1) \sup_{\tau \geq t} |y(t) - \tilde{y}(t)| \quad (t \to +\infty).$$

The latter is possible if $\sup_{\tau \geq t} |y(t) - \tilde{y}(t)| = 0$ and hence $y(t) - \tilde{y}(t) = 0$.

$\square$

1.1   *THE POINCARÉ THEOREM*

The Problem Statement of Poincaré's Theorem.

Equation (15.1) with constant coefficients is written in the form

$$L[x(t)] \equiv x(t+n) + a_1 x(t+n-1) + \ldots + a_n x(t) = 0. \qquad (15.8)$$

Its characteristic equation

$$H(y) \equiv y^n + a_1 y^{n-1} + \ldots + a_n = 0 \qquad (15.9)$$

has a complete set of roots $\{\lambda_1, \ldots, \lambda_n\}$. Let all the roots have distinct modulus in pairs so that we may suppose (without loss of generality) that

$$|\lambda_1| > |\lambda_2| > \ldots > |\lambda_n| \qquad (15.10)$$

(otherwise we may change the numeration of the roots). Since all the roots of the characteristic equation are different on D_τ there exists an *FSS* of equation (15.8) in the form

$$x = C_1 \lambda_1^t + C_2 \lambda_2^t + \ldots + C_n \lambda_n^t, \qquad (15.11)$$

where C_i are arbitrary constants $((i = 1, 2, \ldots, n)$. Let us fix a solution

$$x(t) = c_1 \lambda_1^t + c_2 \lambda_2^t + \ldots + c_n \lambda_n^t \neq 0,$$

where $c_1, c_2, \ldots, c_n$ are given numbers. Let their first non-zero number be c_j (that is, $c_1 = \ldots = c_{j-1} = 0$ and $c_j \neq 0$). Then (taking into account that $|\lambda_{j+s}/\lambda_j| < 1$) we obtain $(\lambda_{j+s}/\lambda_j)^t \to 0$ for $t \to +\infty$, $s = 1, 2, \ldots, n - j$, and hence

$$x(t) = c_j \lambda_j^t \left[1 + \left(\frac{c_{j+1}\lambda_{j+1}}{c_j \lambda_j} \right)^t + \left(\frac{c_{j+2}\lambda_{j+2}}{c_j \lambda_j} \right)^t + \ldots + \left(\frac{c_n \lambda_n}{c_j \lambda_j} \right)^t \right] \sim c_j \lambda_j^t$$

$(t \to +\infty)$. Clearly

$$x(t+1) \sim c_j \lambda_j^{t+1} \sim \lambda_j x(t).$$

Thus $\lim_{t \to +\infty} x(t+1)/x(t) = \lambda_j$, $t \in D_\tau$. In other words for any (fixed) non-trivial solution $x(t)$ to equation (15.8) there exists a root λ_j of equation (15.9) such that $\lim_{t \to +\infty} x(t+1)/x(t) = \lambda_j$ $(t \in D_\tau)$. The Poincaré Theorem generalizes this result.

Theorem 15.4 (Poincaré). *Let for a linear difference equation of the form* (15.1) *all the limits* $\lim_{t \to +\infty} a_i(t) = a_i$ *be finite, and let limiting characteristic equation* (15.6) *have all the roots*

$$\lambda_1, \ldots, \lambda_n$$

satisfying condition (15.10). Then for any (fixed) non-trivial solution $x(t)$ of the equation for $t \gg 1$ on D_τ there exists a root λ_j such that

$$\lim_{t \to +\infty} \frac{x(t+1)}{x(t)} = \lambda_j.$$

PROOF. Let $x(t)$ be a non-trivial solution to the considered equation. Set $u_1(t), u_2(t), ..., u_n(t)$ by means of the relations

$$\begin{cases} u_1(t) & + & u_2(t) & + & ... & + & u_n(t) & = & x(t), \\ \lambda_1 u_1(t) & + & \lambda_2 u_2(t) & + & ... & + & \lambda_n u_n(t) & = & x(t+1), \\ & & & \cdot \cdot \cdot \cdot \cdot \cdot \cdot \cdot \cdot \cdot \cdot \cdot \cdot \cdot \cdot & & & \\ \lambda_1^{n-1} u_1(t) & + & \lambda_2^{n-1} u_2(t) & + & ... & + & \lambda_n^{n-1} u_n(t) & = & x(t+n-1). \end{cases}$$

$$(15.12)$$

From the first and the second lines we have

$$x(t+1) \equiv u_1(t+1) + ... + u_n(t+1) = \lambda_1(t)u_1(t) + ... + \lambda_n(t)u_n(t)$$

which implies the relation

$$u_1{}_{\lambda_1}^{[1]}(t) + ... + u_n{}_{\lambda_n}^{[1]}(t) = 0$$

$(u_m{}_{\lambda_m}^{[1]}(t) = u_m(t+1) - \lambda_m(t))$. In the same way from the $m+1$ and $m+2$ lines we obtain

$$\lambda_1^m u_1{}_{\lambda_1}^{[1]}(t) + ... + \lambda_n^m u_n{}_{\lambda_n}^{[1]}(t) = 0, \quad m = 0, 1, ..., n-2. \qquad (15.13)$$

From the last line of (15.12) we obtain

$$x(t+n) \equiv \lambda_1^{n-1} u_1(t+1) + ... + \lambda_n^{n-1} u_n(t+1) \qquad (15.14)$$

$$= \lambda_1^{n-1} u_1{}_{\lambda_1}^{[1]}(t) + ... + \lambda_n^{n-1} u_n{}_{\lambda_n}^{[1]}(t) + \lambda_1^n u_1(t) + ... + \lambda_n^n u_n(t).$$

On substituting relations (15.12) and (15.14) in (15.1) and collecting the members containing the same multiples $u_m{}_{\lambda_m}^{[1]}(t)$, we obtain the following relation

$$H(\lambda_1)u_1(t) + ... + H(\lambda_n)u_n(t) + \lambda_1^{n-1} u_1{}_{\lambda_1}^{[1]}(t) + ... + \lambda_n^{n-1} u_n{}_{\lambda_n}^{[1]}(t)$$

$$= \alpha_1(t)u_1(t) + ... + \alpha_n(t)u_n(t),$$

where

$$\alpha_m(t) = (a_n - a_n(t)) + (a_{n-1} - a_{n-1}(t))\lambda_m(t) + ... + (a_1 - a_1(t))\lambda_m^{n-1} \to 0$$

for $t \to +\infty$ $(m = 1, 2, ..., n)$. Since $H(\lambda_m) = 0$ finally we have

$$\lambda_1^{n-1} u_{1\lambda_1}^{[1]}(t) + ... + \lambda_n^{n-1} u_{n\lambda_n}^{[1]}(t) = \sum_{m=1}^{n} \alpha_m(t)u_m(t). \qquad (15.15)$$

Equations (15.13) and (15.15) may be considered as a system of algebraic linear equations in unknowns $u_{m\lambda_m}^{[1]}(t)$ $(m = 1, 2, ..., n)$. Determinant of this system is a Van der Monde determinant. Since the roots λ_i are distinct in pairs the system has a unique solution which we may write in the following form

$$u_m(t + 1) - \lambda_m u_m(t) = \sum_{i=1}^{n} \alpha_{im}(t)u_i(t) \quad (m = 1, 2, ..., n). \qquad (15.16)$$

Consider the obtained set of functions $u_1(t), u_2(t), ..., u_n(t)$. From (15.12) (because the determinant of the system does not equal to zero) the functions are uniquely determined from the solution $x(t)$, and clearly all the functions cannot vanish simultaneously [otherwise (from (15.12)) $x(t) = x(t + 1) = ... = x(t + n - 1) = 0$ and (from (15.1)) $x(t) = 0$ for any t which is impossible]. Let

$$u(t) = \max[|u_1(t)|, ..., |u_n(t)|]. \qquad (15.17)$$

there exist a minimal number j for which $u(t) = |u_j(t)|$ for any fixed t. That is, $|u_r(t)| < |u_j(t)|$ for $r = 1, 2, ..., j - 1$, and $|u_s(t)| \le |u_j(t)|$ for $s = j + 1, ..., n$. Clearly $u(t) \ne 0$ (otherwise all $u_i(t) = 0$). Relations (15.16) may be rewritten in the form

$$u_m(t + 1) - \lambda_m u_m(t) = \beta_m(t)u_j(t) \quad (m = 1, 2, ..., n), \qquad (15.18)$$

where $\beta_m(t) \to 0$ for $t \to +\infty$. We have to suppose that j depends on t. But for $t \gg 1$, j may only decrease. This follows from the condition (15.10). Hence j achieves its minimal value at a point t_0 which is invariable if t increases. That is, $j = $ const for $t \gg 1$. We have (see (15.18))

$$u_j(t + 1) - \lambda_j u_j(t) = \beta_j(t)u_j(t)$$

hence (because $u(t) \ne 0$)

$$\lim_{t \to +\infty} \frac{u(t + 1)}{u(t)} = \lambda_j.$$

Put $v_m(t) = u_m(t)/u_j(t)$. Of cause $|v_m(t)| \le 1$. In (15.18) divide the relations by $u_j(t+1)$. Hence (for any $m = 1, 2, ..., m$)

$$v_m(t+1) - q_m(t)v_m(t) = \alpha_m(t),$$

where

$$q_m(t) = \lambda_m[u_j(t)/u_j(t+1)] = \lambda_m/[\lambda_j + \beta_j(t)].$$

Consequently $|q_r(t)| > \sigma_1 > 1$ for $r = 1, 2, ..., j-1$, and $|q_s(t)| < \sigma_2 < 1$ for $s = j+1, ..., n$ $(t \gg 1$, $\sigma_{1,2}$ are numbers independent of $t)$. Thus on the basis of Lemmata 15.1 and 15.2 for any $m \ne j$ the equation

$$v_m(t+1) - q_m(t)v_m(t) = \beta_m(t)$$

has the desired solution $v_m(t) \to 0$ for $t \to +\infty$. From the first line of (15.12)

$$x(t) = u_j(t)(1 + v_1(t) + v_2(t) + ... + v_n(t)) \sim u_j(t)$$

for $t \to +\infty$ which leads to the required relation. $\square$

Example 15.5.   This example shows that the condition (15.10) is essential in the Poincaré theorem. Given the equation

$$x(t+2) - \left(1 + \frac{(-1)^t}{t+1}\right)x(t) = 0$$

considered in D_0 (that is, t may only take the values $0, 1, ...$). Here $n = 2$, $a_1(t) = 0$ and

$$a_2(t) = -(1 + (-1)^t/(t+1)) \to -1$$

for $t \to \infty$. The limiting characteristic equation $y^2 - 1 = 0$ possesses two distinct roots $\lambda_1 = 1$ and $\lambda_2 = -1$ having equal modulus. Let us show that the Poincaré theorem does not true for this problem. Set the initial conditions $x(0) = 0$ and $x(1) = 1$. We construct the solution for the equation separately for $t = 2m+1$ and $t = 2m$ $(m = 0, 1, ...)$. If $t = 2m+1$ the equation may be written in the form

$$x(t+2) = \left(1 - \frac{1}{t+1}\right)x(t).$$

On substituting $t = 1, 3, ...$ we obtain a sequence of numbers

$$x(1) = 1, \quad x(3) = \left(1 - \frac{1}{2}\right)x(1), ..., x(2m+1) = \left(1 - \frac{1}{2m}\right)x(2m-1).$$

Hence

$$x(2m+1) = \prod_{k=1}^{m}\left(1 - \frac{1}{2k}\right).$$

If $t = 2m$ and $x(0) = 0$ we obtain $x(2m) = 0$. Thus the limit $\lim[x(t + 1)/x(t)]$ has no sense. Indeed $x(2m)/x(2m+1) = 0$ and $x(2m+1)/x(2m) = x(2m + 1)/0$, where $x(2m + 1) \neq 0$.

1.2   *PERRON'S THEOREM*

If in addition to the conditions of the Poincaré theorem $a_n(t) \neq 0$ for $t \gg 1$ then (as the The Perron theorem asserts) equation (15.1) has an FSS $\{x_j(t)\}$ $(j = 1, 2, ..., n)$ such that $x_j(t+1)/x_j(t) \to \lambda_j$ for $t \to +\infty$.

To prove this assertion we have to prove several simple propositions. Let us put

$$F_m(y(t)) = y(t)y(t+1)...y(t + m - 1) \quad (m = 1, 2, ...), \tag{15.19}$$

$$|y(t)|^* = \sup_{s \geq t}|y(s)|, \quad |F_m(y(t))|^* = \sup_{s \geq t}|F_m y(s)|,$$

$$\delta(y_1(t), y_2(t)) = \sup_{s \geq t}|y_1(s) - y_2(s)|,$$

and

$$\Delta_m(y_1(t), y_2(t)) = \sup_{s \geq t}|F_m(y_1(s)) - F_m(y_2(s))|.$$

Lemma 15.6. *Let $y(t)$, $y_1(t)$ and $y_2(t)$ be infinitesimal functions for $t \to \infty, t \in D_\tau$. Then for $m = 2, ...$ and $t \to \infty, t \in D_\tau$,*

(1) $|F_m(y(t))|^* = o(|y(t)|^*)$;

(2) $\Delta_m(y_1(t), y_2(t)) = o(\delta(y_1(t), y_2(t)))$.

PROOF. Property (1) is evident. We prove property (2) by induction with respect to $m = 2, ...$ For $m = 2$ we have $F_2(y(t)) = y(t)y(t + 1)$. Hence

$$F_2(y_1(t)) - F_2(y_2(t)) = (y_1(t) - y_2(t))y_1(t+1) + (y_1(t+1) - y_2(t+1))y_2(t).$$

Taking into account that

$$|y(t + 1)|^* \leq |y(t)|^* \quad \text{and} \quad \delta(y_1(t+1), y_2(t+1)) \leq \delta(y_1(t), y_2(t))$$

we obtain the relation

$$\Delta_2(y_1(t), y_2(t)) \leq (|y_1(t)|^* + |y_2(t)|^*)\delta(y_1(t), y_2(t)) = o(\delta(y_1(t), y_2(t)))$$

for $t \to \infty, t \to D_\tau$, which leads to relation (2). Let relation (2) be true for $m-1$ $(m > 2)$. We have

$$F_m(y_1(t)) - F_m(y_2(t)) \equiv y_1(t)y_1(t+1)...y_1(t+m-1)$$

$$- y_2(t)y_2(t+1)...y_2(t+m-1)$$

$$= (y_1(t) - y_2(t))F_{m-1}(y_1(t+1))$$

$$+ (F_{m-1}(y_1(t+1)) - F_{m-1}(y_2(t+1)))y_2(t).$$

Consequently

$$\Delta_m(y_1(t), y_2(t)) \leq \delta(y_1(t), y_2(t))|F_{m-1}(y_1(t))|^* + |y_1(t)|^* o(\delta(y_1(t), y_2(t))),$$

$(t \to \infty, t \in D_\tau)$ which implies property (2). $\qquad\square$

Consider an expression of the form

$$R(t, y(t)) = b_0(t)F_n(y(t)) + b_1(t)F_{n-}(y(t)) + ... + b_{n-2}(t)F_2(y(t)).$$

The following proposition is a simple consequence of Lemma 15.6.

Lemma 15.7. *Let* $b_j(t) \to b_j \neq \infty$ $(j = 0, 1, ..., n - 2)$ *and let* $y(t), y_1(t), y_2(t)$ *be infinitesimal functions for* $t \to \infty, t \in D_\tau$. *Then*

(1) $\sup_{s \geq t} |R(s, y(s))| = o(|y(t)|^*)$;

(2) $\sup_{s \geq t} |R(s, y_1(s)) - R(s, y_2(s))| = o(\delta(y_1(t), y_2(t)))$ $(t \to \infty, t \in D_\tau)$.

Consider an equation of the form

$$y(t) = \alpha(t) + R(t, y(t)). \tag{15.20}$$

Lemma 15.8. *Let* $R(t, y(t))$ *satisfy all the conditions of Lemma* 15.7. *Let* $\alpha(t) \to 0$ *for* $t \to \infty, t \in D_\tau$. *Then equation* (15.20) *has a unique infinitesimal solution* $y^*(t)$ $(t \to \infty, t \in D_\tau)$. *Consequently if (in addition)* $\alpha(t) \neq 0$ *for* $t \gg 1, t \in D_\tau$, *then* $y^*(t) \neq 0$ *(for* $t \gg 1, t \in D_\tau$).

PROOF. The existence of a unique infinitesimal solution $y^*(t)$ follows from the principle of attractive mappings and based of Lemma 15.7. From (15.17) we obtain the following identity

$$y^*(t) = \alpha(t) + R(t, y^*(t)).$$

Hence

$$y^*(t) = \frac{\alpha(t)}{1 - R(t, y^*(t))} \sim \alpha(t) \ (t \to \infty, t \in D_\tau).$$

□

Consider an equation of the form

$$x(t+1) - (\lambda + \gamma(t))x(t) = u(t). \tag{15.21}$$

Lemma 15.9. *Let $\gamma(t) \to 0$ for $t \to \infty, t \in D_\tau$. Let $u(t+1)/u(t) = \mu + \delta(t) \neq 0$ for $t \gg 1, t \in D_\tau$, where $\delta(t) \to 0$ for $t \to \infty, t \in D_\tau$. Let it be asserted λ and μ numbers such that $|\lambda| \neq |\mu|$. Then equation (15.21) has a non-trivial solution $x^*(t)$ such that $x^*(t+1)/x^*(t) \to \mu$ $(t \to \infty, t \in D_\tau)$.*

PROOF. Put $x(t) = z(t)u(t)$ in (15.21) where $z(t)$ is a new unknown. We have

$$z(t+1)\frac{u(t+1)}{u(t)} - (\lambda + \gamma(t))z(t) = 1$$

or (which is the same)

$$z(t+1)(\mu + \delta(t)) - z(t)(\lambda + \gamma(t)) = 1.$$

Substitute $z(t) = 1/(\mu - \lambda) + v(t)$ in the last equation. We obtain

$$v(t+1) - \frac{\lambda + \gamma(t)}{\mu + \delta(t)}v(t) = \eta(t), \tag{15.22}$$

where

$$\eta(t) = [\gamma(t) - \delta(t)]/(\mu - \lambda) \to 0 \ (t \to \infty, t \in D_\tau).$$

For the cases $|\lambda| < |\mu|$ and $|\lambda| > |\mu|$ the existence of an infinitesimal solution $v^*(t)$ to equation (15.22) follows from Lemmata 15.1 and 15.2, respectively, which leads to the required solution to equation (15.21).

□

Theorem 15.10 (Perron). *Let all the conditions of (Poincaré's) Theorem 15.4 be fulfilled and let (in addition) $a_n(t) \neq 0$ for $t \gg 1, t \in D_\tau$. Then equation (15.1) possesses an FSS $\{x_1(t), x_2(t), ..., x_n(t)\}$ such that*

$$\frac{x_i(t+1)}{x_i(t)} \to \lambda_i \ for \ (i = 1, 2, ..., n). \tag{15.23}$$

PROOF. Perron's Theorem is a consequence of the Poincaré Theorem.

We prove the theorem by induction. Being trivial for $m = 1$ we suppose that it is true for $m-1$.

Let $\lambda_n \neq 0$. By the Poincaré theorem there is a solution $x^*(t)$ of equation (15.1) such that $x^*(t+1)/x^*(t) \to \lambda \in \{\lambda_1, \lambda_2, ..., \lambda_n\}$. Thus there exists a function $\gamma(t) \to 0$ (for $t \to \infty, t \in D_r$) such that $x^*(t+1) - (\lambda + \gamma(t))x^*(t) = 0$. Let us make the substitution

$$x(t+1) = (\lambda + \gamma(t))x(t) + u(t)$$

in (15.1) where $u(t)$ is a new unknown function. We have

$$x(t+2) = (\lambda + \gamma(t+1))x(t+1) + u(t+1).$$

Hence

$$x(t+2) = u(t+1) + (\lambda + \gamma(t+1))u(t) + (\lambda + \gamma(t))(\lambda + \gamma(t+1))x(t)$$

and so on. As the result we obtain the following relation

$$u(t+n-1) + b_1(t)u(t+n-2) + ... + b_{n-1}(t)u(t) + b_n(t)x(t) = 0,$$

where

$$b_n(t) = F_n(\lambda + \gamma(t)) + a_1(t)F_{n-1}(\lambda + \gamma(t)) + ... + a_n(t).$$

Clearly $b_n(t) = 0$. Consequently the considered substitution leads to the equation

$$u(t+n-1) + b_1(t)u(t+n-2) + ... + b_{n-1}(t)u(t) = 0. \qquad (15.24)$$

Moreover as it was shown in Lemma 15.6 its limiting equation

$$H^*(y) \equiv y^{n-1} + b_1 y^{n-2} + ... + b_{n-1} = 0$$

is connected with the limiting characteristic polynomial $H(y)$ (see (15.9)) by the relation $H(y) = H^*(y)(y - \lambda)$. Thus equation (15.24) satisfies all the conditions of the Poincaré theorem. It means (by induction) that, for any root μ_j of the polynomial $H^*(y)$ there exist a non-trivial solution $u_j(t)$ of equation (15.24) such that

$$u(t+1) - (\mu_j + \delta_j(t))u(t) = 0,$$

where $\mu_j \neq \lambda$ for any $j = 1, 2, ..., n - 1$, and $\delta_j(t) \to 0$ for $t \to \infty, t \in D_r$. Equation (15.1) has all the solutions of the equation

$$x(t+1) - (\lambda + \gamma(t))x(t) = u_j(t).$$

On the basis of Lemma 15.9 (taking into account that $|\lambda| \neq |\lambda_j|$) there exist a solution $x_i(t)$ of the last equation such that

$$\frac{x_j(t+1)}{x_j(t)} \to \mu_j \text{ for } t \to \infty, t \in D_\tau.$$

Besides $\{\mu_j\} \cup \lambda = \{\lambda_i\}$ $(j = 1, 2, ..., n-1, \ i = 1, 2, ..., n)$. Consequently we proved that the required set $\{x_1(t), x_2(t), ..., x_n(t)\}$ exists. We have to prove that the set forms an *FSS* of equation (15.1). To this end consider the Casoratian of the functions. It is equal to

$$U(x_1(t), ..., x_n(t)) \sim x_1(t)...x_n(t)\omega(\lambda_1, ..., \lambda_n).$$

Here $\omega(\lambda_1, ..., \lambda_n)$ is a Van der Monde determinant. Since all the roots λ_i are simple, the Van der Monde determinant does not equal to zero which proves our assertion.

Let $\lambda_n = 0$. Then $a_n(t) \to 0$ for $t \to \infty, t \in D_\tau$ and $a_n(t) \neq 0$ for $t \gg 1, t \in D_\tau$. Substitute $x(t+1) = y(t)x(t)$ in (15.1) where $y(t)$ is a new unknown. As the result we obtain the equation

$$F(t, y(t)) \equiv F_n(y(t)) + a_1(t)F_{n-1}(y(t)) + ... + a_{n-1}(t)y(t) + a_n(t) = 0.$$
$$(15.25)$$

Clearly for any solution $y(t)$ to the last equation any solution of the equation $x(t+1) - y(t)x(t) = 0$ is a solution to equation (15.1). Of course

$$a_{n-1}(t) \sim (-1)^{n-1}\lambda_1...\lambda_{n-1} \neq 0 \ (t \gg 1, t \to \infty).$$

Equation (15.25) is transformed to equation (15.20) where $\alpha(t) = -a_n(t)/a_{n-1}(t)$ and

$$R(t, y(t)) = -[F(t, y(t)) - a_{n-1}(t)y(t) - a_n(t)]/a_{n-1}(t).$$

On the basis of Lemma 15.8 we conclude that the equation has an infinitesimal solution $y^*(t) \neq 0$ for $t \gg 1$. The last part of the proof is made in the same way as the corresponding part of the case $\lambda_n \neq 0$, where we make the substitution

$$x(t+1) = y^*(t)x(t) + u(t)$$

instead of

$$x(t+1) = (\lambda + \gamma(t))x(t) + u(t). \qquad \square$$

2.   LINEAR DIFFERENCE EQUATIONS WITH POWER ORDER GROWTH COEFFICIENTS

Beforehand let us consider the following expressions

$$F_m(y(t)) = y(t)y(t+1)...y(t+m-1) \ (m = 1, 2, ...), \qquad (15.19)$$

$F_0(y(t)) = 1$ and

$$F_m^*(y(t)) = F_m(y(t)) - y^m(t). \qquad (15.26)$$

They play an important role in the further investigation. Consider their main asymptotic properties.

Proposition 15.11. *Let* $y(t) \in A_t$. *Then*

$$F_m(y(t)) = y^m(t) + \frac{m(m-1)}{2} y'(t) y^{m-1}(t) + \alpha_m(t), \qquad (15.27)$$

where $\alpha_m(t) = O(y^m(t)/t^2)$ *for* $t \to +\infty$, *hence*

$$\Pi\{\alpha_m(t)\} \le m\Pi\{y(t)\} - 2.$$

PROOF. We have

$$y(t+s) = y(t) + sy'(t) + O(y''(t)) \quad \text{for} \quad t \to +\infty \quad (s = 1, 2, ...).$$

It follows that $y''(t) = O(y(t)/t^2)$ for $t \to +\infty$. On substituting the obtained relations in (15.16) we obtain (15.23). $\qquad \square$

Consider a determinant of the form

$$U(y_1(t), ..., y_n(t)) =$$

$$\begin{vmatrix} 1 & 1 & \cdots & 1 \\ F_1(y_1(t)) & F_1(y_2(t)) & \cdots & F_1(y_n(t)) \\ \cdot \cdot \cdot \cdot \cdot \cdot \cdot \cdot \cdot \cdot \cdot \cdot \cdot \cdot \\ F_{n-1}(y_1(t)) & F_{n-1}(y_2(t)) & \cdots & F_{n-1}(y_n(t)) \end{vmatrix}. \qquad (15.28)$$

Proposition 15.12. *Let* Q *be a field of type* N. *Let*

$$y_i(t) \in \{Q\}, \quad y_i(t) \not\sim y_j(t) \quad \text{for} \quad t \to +\infty \quad (i \neq j; \quad i, j = 1, 2, ..., n)$$

and no more than one of the functions belongs to O_t. *Then*

$$U(y_1(t), ..., y_n(t)) = \omega(y_1(t), ..., y_n(t))(1 + O(1/t)) \quad \text{for} \quad t \to +\infty, \qquad (15.29)$$

hence $U(y_1(t), ..., y_n(t)) \in A_t$ *and*

$$\Pi\{U(y_1(t), ..., y_n(t))\} \ge \frac{n(n-1)}{2} \min_{i \neq j, i, j = 1, 2, ..., n} \Pi\{y_i(t) - y_j(t)\} > -\infty. \qquad (15.30)$$

Here $\omega(y_1(t), ..., y_n(t))$ is Van der Monde's determinant for $y_1(t), ..., y_n(t)$.

PROOF.
$$\Pi\{y_i(t) - y_j(t)\} > -\infty$$

for any $i \neq j$. Since no more than one of the functions $y_j(t)$ has the estimate $O(t^{-\infty})$ we can suppose that the numeration of the functions is chosen such that if $i > j$, then

$$\lim_{t \to +\infty} \left| \frac{y_i(t)}{y_j(t)} \right| \leq 1.$$

Hence there is a number $c \neq 0$ such that for $t \to +\infty$

$$U(y_1(t), ..., y_n(t)) \sim cy_1^{n-1}(t)y_2^{n-2}(t)...y_{n-1}(t). \tag{15.31}$$

Because of (15.27) the following relation holds:

$$F_m(y_i(t)) = y_i^m(t) + O\left(\frac{y_i^m(t)}{t}\right) \quad \text{for } t \to +\infty.$$

Let us substitute them in (15.27) open the determinant ,open the brackets and collect all the terms which entered in $\omega(y_1(t), ..., y_n(t))$. Clearly they form the Van der Monde determinant. All the rest terms are

$$O\left(\frac{y_1^{n-1}(t)y_2^{n-2}(t)...y_{n-1}(t)}{t}\right) \quad \text{for } t \to +\infty$$

which leads to (15.29).

Formal Solutions

Here Q means a field of type N which will not be stipulated later on.

Let us make the substitution $x(t + 1) = y(t)x(t)$ in equation (15.1). Then (see (15.25)) $\Psi(t, x(t)) = F(t, y(t))x(t)$. Let us put

$$F^*(t, y(t)) = F(t, y(t)) - H(t, y(t)), \tag{15.32}$$

where

$$H(t, y(t)) = y^n(t) + a_1(t)y^{n-1}(t) + ... + a_n(t). \tag{15.33}$$

Let $a_i(t) \in \{Q\}$ $(i = 1, 2, ..., n)$. Consider the set of functions

$$\{y^n(t), a_1(t)y^{n-1}(t), ..., a_n(t)\}. \tag{15.34}$$

If $y(t) \in \{Q\}$ and $y(t) \notin O_t$ then all the functions of the set are compara-ble in pairs for $t \to +\infty$. Let $\psi(y(t))$ be a function of the greatest growth of the set. It means that all the limits

$$\lim_{t \to +\infty} \frac{a_j(t)y^{n-j}(t)}{\psi(y(t))}$$

are finite $(j = 0, 1, ..., n;$ here $a_0(t) = 1)$.

The following proposition is a simple consequence of Proposition 15.11.

Proposition 15.13. *Let* $a_1(t), ..., a_n(t)$ *and* $y(t)$ *belong to* $\{Q\}$ $(i = 1, 2, ..., n)$, *and* $y(t) \in A_t$. *Then (see* (15.32) *and* (15.33))

$$F^*(t, y(t)) = \frac{1}{2} y'(t) y(t) \frac{\partial^2 H(t, y(t))}{\partial y^2} + \beta(t), \qquad (15.35)$$

where

$$\beta(t) = O\left(\frac{\psi(y(t))}{t^2}\right) \quad for \ t \to +\infty,$$

hence

$$\Pi\{\beta(t)\} \leq \Pi\{\psi(y(t))\} - 2.$$

Definition 15.14. A function $G(t)$ is said to be a *simple formal solution* to equation (15.1) if $F(t, g(t)) = O(t^{-\infty})$, where $g(t) = G(t + 1)/G(t)$. The function $g(t)$ is said to be a *simple formal basis* to equation (15.1).

If $g(t) \in \Pi$ and $F(g(t), t) \asymp 0$ then the functions $G(t)$ and $g(t)$ are said to be an *analytic formal solution* and an *analytic formal basis* to equation (15.1), respectively.

Lemma 15.15. *Let* $\lambda(t)$ *be an asymptotically simple root of the poly-nomial* $H(t, y)$ *(see* (15.33)) *and* $\Pi\{\lambda(t)\} > -\infty$. *Let us introduce the designation*

$$R(y(t)) = -\frac{F^*(t, y(t))}{H^*(t, y(t))},$$

where the polynomial $H^*(t, y)$ *is determined from the identity* $H(t, y) = H^*(y)(y - \lambda(t))$. *Then* $\Pi\{R(y(t))\} \leq \Pi\{\lambda(t)\} - 1$ *for any* $y(t) \sim \lambda(t)$, $y(t) \in \Pi$ *and* $\Pi'\{R(\lambda(t))\} \leq -1$.

PROOF. Polynomial $H^*(t, y)$ has coefficients belonging to $\{Q\}$. More-over for any $y(t) \sim \lambda(t)$ for $t \to +\infty$ and $y(t) \in \Pi$ we have

$$\Pi\{H^*(t, y(t))\} = \Pi\{\psi(\lambda(t))\} - \Pi\{\lambda(t)\}.$$

Clearly $F^*(t, y)$ is of power type at the point $\lambda(t)$ with majorant

$$f^* = \Pi\{\psi(\lambda(t))\} - 1.$$

Hence the operator $-\Phi^*(t, y)/H^*(t, y)$ is of power type at the point $\lambda(t)$ with a majorant

$$r = f^* - \Pi\{\psi(\lambda(t))\} + \Pi\{\lambda(t)\} = \Pi\{\lambda(t)\} - 1$$

which leads to the required inequalities.   □

Proposition 15.16. *Let* $a_i(t) \in \{Q\}$ ($i = 1, 2, ..., n$) *and* $Pi\{a_n(t)\} > -\infty$. *Let* $\lambda(t) \in \{Q\}$ *be an asymptotically simple root of polynomial* (15.33) *for* $t \to +\infty$. *Then there exists an analytic formal solution* $G(t)$ *and an analytic formal basis* $g(t)$ *of equation* (15.1) *such that* $g(t) = G(t+1)/G(t)$, $g(t) \sim \lambda(t)$ *for* $t \to +\infty$, *and* $g(t)$ *is an analytic asymptotic limit of the sequence* $\{s_m(t)\}$, *where* $s_0(t)$ *is an arbitrary (fixed) function belonging to* $\{Q\}$ *and equivalent to* $\lambda(t)$ *for* $t \to +\infty$, *and*

$$s_m(t) = \lambda(t) - \frac{F^*(t, s_{m-1}(t))}{H^*(t, s_{m-1}(t))}, \quad m = 1, 2, ... \tag{15.36}$$

Hence

$$g(t) = \lambda(t)(1 + \eta(t) + \alpha(t)), \tag{15.37}$$

where

$$\eta(t) = -\frac{1}{2}\lambda'(t)\frac{\partial^2 H(t, \lambda(t))}{\partial y^2} \Big/ \frac{\partial H(t, \lambda(t))}{\partial y} \tag{15.38}$$

and $\Pi\{\alpha(t)\} \leq -2$. *Moreover,* $s_m(t)$ *and* $g(t)$ *belong to* $\{Q\}$. *The function* $\tilde{g}(t)$ *such that* $\tilde{g}(t) \in \Pi$ *and* $\tilde{g}(t) \sim g(t)$ *for* $t \to +\infty$*is a formal basis of the equation if and only if* $\tilde{g}(t) \asymp g(t)$.

PROOF. From the condition $\Pi\{a_n(t)\} > -\infty$ it follows that $\lambda(t) \notin O_t$, hence taking into account that $\lambda(t) \in \{Q\}$ we conclude that $\lambda(t) \in A_t$ and consequently $\Pi\{\lambda(t)\} > -\infty$. It is possible to write the equation $F(t, y(t)) = 0$ in the form

$$F^*(t, y(t)) + H^*(t, y(t))(y(t) - \lambda(t)) = 0.$$

Hence

$$y(t) = \lambda(t) - \frac{F^*(t, y(t))}{H^*(t, y(t))}. \tag{15.39}$$

Since $\Pi\{R(y(t))\} \leq \Pi\{\lambda(t)\} - 1$ for any $y(t) \sim \lambda(t)$, $y(t) \in \Pi$ and $\Pi'\{R(\lambda(t))\} \leq -1 < 0$, then all the conditions of Lemma 3.24 are fulfilled and hence there exists a formal solution $g(t)$ of equation $F(t, y(t)) = 0$ which is an asymptotic limit of sequence (15.36). Clearly $s_m(t) \in \{Q\}$ and since $\Pi\{g(t) - s_m(t)\} \ll -1$ for $m \gg 1$ we have $g(t) \in \{Q\}$. The properties of the formal solution $\tilde{g}(t)$ are also follow from Lemma 3.24.

Relation (15.37) follows from (15.35). Indeed let $s_0(t) = \lambda(t)$. We have

$$\alpha(t) = \frac{\beta(t)}{y(t) H^*(t, y(t))} = O\left(\frac{\beta(t)}{\lambda(t) H^*(t, \lambda(t))}\right) \quad \text{for} \quad t \to +\infty.$$

Hence

$$\Pi\{\alpha(t)\} \leq \Pi\{\psi(\lambda(t))\} - 2 - \Pi\{\psi(\lambda(t))\} + \Pi\{\lambda(t)\} - \Pi\{\lambda(t)\} = -2. \quad \square$$

Definition 15.17. Let $G_i(t)$ be simple (analytic) formal solutions to equation (15.1) and

$$g_i(t) = \frac{G_i(t+1)}{G_i(t)} \quad (i = 1, 2, ..., n).$$

We say that the set

$$\{G_1(t), G_2(t), ..., G_n(t)\}$$

is a *simple (analytic) formal fundamental system* of solutions $SFFS$ (or $AFFS$, respectively) if (see (15.28))

$$h\{U(g_1(t), ..., g_n(t))\} > -\infty \quad (\Pi\{U(g_1(t), ..., g_n(t))\} > -\infty).$$

The following proposition is a simple consequence of Propositions 15.11 and 15.16.

Proposition 15.18. *Let $a_i(t) \in \{Q\}$ for $i = 1, 2, ..., n$ and $h\{a_n(t)\} > -\infty$. Let characteristic polynomial (15.33) have a complete set of roots*

$$\Lambda \equiv \{\lambda_i(t)\} \quad (i = 1, 2, ..., n)$$

such that they are not equivalent in pairs for $t \to +\infty$. Then there exists an AFFS

$$\{G_1(t), G_2(t), ..., G_n(t)\}$$

of equation (15.1) such that for $i = 1, 2, ..., n$ each function

$$g_i(t) = \frac{G_i(t+1)}{G_i(t)}$$

possesses any properties obtained in Proposition 15.16 (where instead of $\lambda(t)$, $g(t)$ and $\alpha(t)$ we have to write $\lambda_i(t)$, $g_i(t)$ and $\alpha_i(t)$ respectively).

3.   ASYMPTOTIC SOLUTIONS TO LINEAR DIFFERENCE EQUATIONS

Consider a matrix equation of the form

$$V(t+1) = [Q(t) + A(t)]V(t) + B(t), \qquad (15.40)$$

where $Q(t)$ is a diagonal matrix $Q(t) = \operatorname{diag}(q_1(t), q_2(t), ..., q_n(t))$, $A(t)$ is a square matrix, $A(t) = (a_{ij}(t))_n$ and $B(t) = (b_1(t), b_2(t), ..., b_n(t))^T$. $V(t) = (v_1(t), v_2(t), ..., v_n(t))^T$ is the unknown matrix.

Lemma 15.19. *Let $q_i(t)$ be continuous functions and $|q_i(t)| \geq 1$ for $t \gg 1$ or $\mathrm{h}\{|q(t)| - 1\} = -\infty$ $i = 1, 2, ..., n$. Let $A(t)$ and $B(t)$ be continuous matrices for $t \gg 1$ and $\mathrm{h}\{\|A(t)\|\} = \mathrm{h}\{\|B(t)\|\} = -\infty$. Then equation (15.40) has for $t \gg 1$ a continuous solution $V^*(t)$ with the estimate $\mathrm{h}\{\|V^*(t)\|\} = -\infty$.*

PROOF. We shall prove only the case $|q_i(t)| \geq 1$ for any $i = 1, 2, ..., n$ because the other case is reduced to the considered case by means of the substitution $Y(t) = tV(t)$. Consider a sequence of matrices $\{V_m(t)\}$, where $V_0(t)$ is a zero matrix, and for $m = 1, 2, ...$ $V_m(t)$ is the infinitesimal solution (see Lemma 14.6) of the equation

$$X(t+1) = Q(t)X(t) + A(t)V_{m-1}(t) + B(t) \text{ for } t \to +\infty.$$

By induction with respect to m on the basis of inequality (14.20) it is easy to show that

$$\|V_m(t)\| < \frac{1}{t} \text{ for } t \geq T \text{ and } m = 1, 2, ...,$$

where T is a sufficiently large number. Let us put

$$\Delta_m(t) = V_{m+1}(t) - V_m(t).$$

The matrix $\Delta_m(t)$ is the infinitesimal solution to equation

$$X(t+1) = Q(t)X(t) + A(t)\Delta_{m-1}(t).$$

On the basis of inequality (14.20) we conclude that

$$\|\Delta_m(t)\| \leq \sup_{t \leq \tau < +\infty} \|\Delta_m(\tau)\| \leq \frac{1}{2} \sup_{t \leq \tau < +\infty} \|\Delta_{m-1}(\tau)\| \text{ for } t \geq T$$

which leads to the inequality $\|\Delta_m(t)\| < 1/2^m t$. Consequently the series

$$V_1(t) + (V_2(t) - V_1(t)) + ... + (V_{m+1}(t) - V_m(t)) + ...$$

converges for $t \geq T$ to a continuous matrix

$$V^*(t), \quad \|V^*(t)\| < \frac{1}{t}$$

and $V^*(t)$ is a solution to equation (15.40). Moreover the matrix $V^*(t)$ is a continuous (for $t \geq T$) infinitesimal (for $t \to +\infty$) solution to equation

$$X(t+1) = Q(t)X(t) + D(t),$$

where $D(t) = Q(t)V^*(t) + B(t)$. Hence

$$\mathrm{h}\{\|D(t)\|\} = -\infty \text{ and } \mathrm{h}\{\|V^*(t)\|\} = -\infty.$$

□

Lemma 15.20. *Let G be a field of type M and let $q(t)$ be represented in the form*

$$q(t) = g(t) + r(t),$$

where $g(t) \in \{G\}$, $r(t)$ is a continuous function for $t \gg 1$ and $\mathrm{h}\{r(t)\} = -\infty$. Let $\mathrm{h}\{\alpha(t)\} = -\infty$. Then equation

$$x(t+1) = q(t)x(t) + \alpha(t)$$

has a continuous solution $x^*(t)$ for $t \gg 1$ with the estimate $\mathrm{h}\{x^*(t)\} = -\infty$.

Proof. We have $g(t) \in A_t \cup O_t$. If $g(t) \in O_t$ then $\mathrm{h}\{q(t)\} = -\infty$. This case is proved in Lemma 14.8. If $g(t) \in A_t$ then (see Proposition 5.19) there exists a (finite or infinite) limit $\lim_{t \to +\infty} t \ln q(t) = p$. The case $p < +\infty$ is proved in Lemmata 14.7 and 14.8. If $p = +\infty$ then $|q(t)| > 1$ for $t \gg 1$. This case is proved in Lemma 14.6. □

Theorem 15.21. *Let $a_i(t) \in \{G\}$ for $i = 1, 2, ..., n$, where G is a field of type M, and $a_n(t) \not\equiv 0$ for $t \gg 1$. Let there be an AFFS of the form $\{G_1(t), G_2(t), ..., G_n(t)\}$ for the equation*

$$\Phi(t, x(t)) \equiv x(t+n) + a_1(t)x(t+n-1) + ... + a_n(t)x(t) = 0 \quad (15.1)$$

such that $g_i(t) = G_i(t+1)/G_i(t) \in \{G\}$. Then equation (15.1) has an FSS of the form

$$\{x_1(t), x_2(t), ..., x_n(t)\},$$

such that $x_i(t+1)/x_i(t) = g_i(t) + \delta_i(t)$. Here $\delta_i(t) = O(t^{-\infty})$ $(i = 1, 2, ..., n)$.

PROOF. Since $g_1(t) \in \{G\}$ and $\Pi\{U(g_1(t), ..., g_n(t))\} > -\infty$ there is no more than one function $g_i(t) \in \{g_1(t), g_2(t), ..., g_n(t)\}$ with the estimate $g_i(t) \asymp 0$. Therefore we may suppose that for any $j > i$ either

$$|g_j(t)/g_i(t)| \geq 1 \text{ for } t \gg 1 \text{ or } |g_j(t)| = |g_i(t)| + \theta_{ij}(t),$$

where $h\{\theta_{ij}(t)\} = -\infty$ and $h\{g_i(t)\} > -\infty$.

To find $x_1(t)$ apply to (15.1) Lagrange's method of variation of arbitrary constants and put

$$x_1(t) = G_1(t)u_1(t) + G_2(t)u_2(t) + ... + G_n(t)u_n(t), \qquad (15.41)$$

where $u_i(t)$ are the variable parameters $(i = 1, 2, ..., n)$. Let us put for $k = 1, 2, ..., n-1$

$$G_1(t+k)u_1^-(t) + G_2(t+k)u_2^-(t) + ... + G_n(t+k)u_n^-(t) = 0. \quad (15.42)$$

Hence

$$G_1(t+k)u_1^-(t) + G_2(t+k)u_2^-(t) + ... + G_n(t+n)u_n^-(t) \qquad (15.43)$$

$$= \alpha_1(t)G_1(t)u_1(t) + \alpha_2(t)G_2(t)u_2(t) + ... + \alpha_n(t)G_n(t)u_n(t),$$

where $h\{\alpha_i(t)\} = -\infty$. Equations (15.42) and (15.43) can be considered as a system with unknowns

$$u_1^-(t), u_2^-(t), ..., u_n^-(t).$$

Its determinant is equal to

$$U(g_1(t+1), ..., g_n(t+1))G_1(t+1)...G_n(t+1)$$

(see (15.28)). Taking into account that $h\{U(g_1(t+1), ..., g_n(t+1))\} > -\infty$ the system can be rewritten in the form

$$G_i(t)u_i^-(t) = \alpha_{i1}(t)G_1(t)u_1(t) + \alpha_{i2}(t)G_2(t)u_2(t) + \alpha_{i2}(t)G_n(t)u_n(t) \tag{15.44}$$

for $i = 1, 2, ..., n$, where $h\{\alpha_{ij}(t)\} = -\infty$. Let us substitute

$$u_1(t) = 1 + v_1(t) \text{ and } u_j(t) = \frac{G_1(t)}{G_2(t)}v_j(t) \text{ for } j = 2, ..., n$$

in (15.44). For

$$V(t) = (v_1(t), v_2(t), ..., v_n(t))^T$$

we obtain the system (15.40), where

$$q_1(t) = 1 \text{ and } q_j(t) = \frac{v_j(t)}{v_i(t)} \text{ for } j = 2, ..., n.$$

The system satisfies all the conditions of Lemma 15.19. Consequently there exists a solution

$$V^*(t) = (v_1^*(t), v_2^*(t), ..., v_n^*(t))^T$$

of the considered system such that $h\{\|V^*(t)\|\} = -\infty$. Hence

$$x_1(t) = G_1(t)(1 + \delta_1(t)),$$

where

$$\delta_1(t) = v_1^*(t) + v_2^*(t) + ... + v_n^*(t) = O(t^{-\infty}).$$

Suppose that it is proved that $x_k(t) = G_k(t)(1+\delta_k(t))$ for $k = 1, 2, ..., m-1$, where $h\{\delta_k(t)\} = -\infty$. Put

$$x_1(t) = G_1(t)u_1(t) + G_2(t)u_2(t) + ... + G_n(t)u_n(t). \tag{15.45}$$

Applying the well known procedure and taking into account that $x_k(t)$ are solutions to equation (15.1) we obtain the following system

$$x_k(t)u_k^-(t) = \beta_{km}(t)G_m(t)u_m(t) + \beta_{km+1}(t)G_{m+1}(t)u_{m+1}(t) + ...$$

$$+\beta_{kn}(t)G_n(t)u_n(t) \tag{15.46}$$

for $k = 1, 2, ..., m - 1$ and

$$G_p(t)u_p^-(t) = \beta_{pm}(t)G_m(t)u_m(t) + \beta_{pm+1}(t)G_{m+1}(t)u_{m+1}(t) + ...$$

$$+\beta_{pn}(t)G_n(t)u_n(t) \tag{15.47}$$

for $p = m, m + 1, ..., n$. Here $h\{\beta_{ij}(t)\} = -\infty$ $i, j = 1, 2, ..., n$. Equations (15.47) form an independent system with unknowns

$$u_m(t), u_{m+1}(t), ..., u_n(t).$$

In the same way as it is proved above the system has a solution

$$U_m^*(t) = (u_m^*(t), u_{m+1}^*(t), ..., u_n^*(t))^T,$$

where

$$u_m(t) = 1 + v_m(t) \text{ and } u_p(t) = \frac{G_m(t)}{G_p(t)}v_p(t) \text{ for } p = m + 1, ..., n$$

with the estimates $h\{v_j(t)\} = -\infty$ $(j = m, m+1, ..., n)$. Let us substitute the functions $u_j(t)$ in (15.47) and put

$$x_k(t)u_k(t) = G_m(t)v_k(t) \quad (k = 1, 2, ..., m - 1)$$

then we obtain the equations $v_k(t + 1) = q_{km}(t)v_k(t) + \gamma_k(t)$, where

$$q_{km}(t) = \frac{g_k(t) + \delta_k(t)}{g_m(t) + \delta_m(t)} \text{ and } h\{\gamma_k(t)\} = -\infty.$$

The equations have solutions $v_k(t)$ with the estimates $h\{v_k(t)\} = -\infty$. Hence $x_m(t) = G_m(t)(1 + \delta_m(t))$, where

$$\delta_m(t) \equiv v_1(t) + v_2(t) + ... + v_n(t) = O(t^{-\infty}).$$

The obtained set of functions $\{x_1(t), x_2(t), ..., x_n(t)\}$ forms an *FSS* of equation (15.1) because

$$\Pi\{U(g_1(t) + \delta_1(t), ..., g_n(t) + \delta_1(t))\} = \Pi\{U(g_1(t), ..., g_n(t))\} > -\infty.$$
$$\square$$

4.   SECOND ORDER EQUATIONS

Here we consider in detail equations of the form

$$x(t + 2) + a_1(t)x(t + 1) + a_2(t)x(t) = 0 \tag{15.48}$$

with coefficients $a_{1,2}(t)$ belonging to the space $\{G\}$ (G is a field of type M).

We begin with the most simple case when the characteristic equation

$$y^2 + a_1(t)y + a_2(t) = 0 \tag{15.49}$$

has a complete set of roots $\{\lambda_1(t), \lambda_2(t)\}$ such that at least one of the roots has an estimate of type Π more than $-\infty$, and $\lambda_1(t) \not\sim \lambda_2(t)$ for $t \to +\infty$. Let $\Pi\{\lambda_1(t)\} > -\infty$. Make the transformation $x(t + 1) = y(t)x(t)$ in (15.48). If $x(t) \neq 0$ then we obtain the equation

$$y(t + 1)y(t) + a_1(t)y(t) + a_2(t) = 0$$

which may be rewritten in the form

$$y^2(t) + a_1(t)y(t) + a_2(t) + y^-(t)y(t) = 0. \tag{15.50}$$

We rewrite the last equation in the following form

$$(y(t) - \lambda_1(t))(y(t) - \lambda_2(t)) + y^-(t)y(t) = 0$$

and

$$y(t) = \lambda_1(t) - \frac{y^-(t)y(t)}{y(t) - \lambda_2(t)}.\qquad(15.51)$$

Consider the ball

$$V_1 = \{y(t) : y(t) \in \{G\}, y(t) - \lambda_1(t) = o(\lambda_1(t))\ \text{for}\ t \to +\infty\}.\quad(15.52)$$

Thus equation (15.50) is written in the form $y(t) = A(y(t))$, where $A(y(t)) = y(t) + R(y(t))$ and

$$R(y(t)) = -\frac{y^-(t)y(t)}{y(t) - \lambda_2(t)}.$$

The operator $A(y)$ transforms any function $y(t) \in V_1$ into the function $y^*(t) \in V_1$ because

$$\Pi\left\{\frac{y^-(t)y(t)}{y(t) - \lambda_2(t)}\right\} \le \Pi\left\{\frac{\lambda_1^-(t)\lambda(t)}{2\lambda_1(t)}\right\} = \Pi\{\lambda_1(t)\} - 1.$$

Moreover this operator is of the power type at the point $\lambda_1(t)$. Hence equation (15.50) has a formal solution $G_1(t) = \lambda_1(t)(1 + \delta_1(t))$, where $\Pi\{\delta_1(t)\} < 0$. On substituting the obtained solution in the equation $y(t) = A(t, y(t))$ we obtain $y(t) = \lambda_1(t)(1 + O(1/t))$ for $t \to +\infty$. We may determine the solution more precisely. For this end, form a sequence of functions $\{s_m(t)\} : s_0(t) = \lambda_1(t)$ and for $m = 1, 2, \ldots$

$$s_m(t) = \lambda_1(t) - \frac{s_{m-1}^-(t)s_{m-1}(t)}{s_{m-1}(t) - \lambda_2(t)}.$$

So that

$$s_1(t) = \lambda_1(t) - \frac{\lambda_1^-(t)\lambda_1(t)}{\lambda_1(t) - \lambda_2(t)},$$

$$s_2(t) \equiv \lambda_1(t) - \frac{s_1^-(t)s_1(t)}{s_1(t) - \lambda_2(t)} = s_1(t) + O\left(\frac{\lambda_1(t)}{t^2}\right)$$

and so on. Any asymptotic limit $g_1(t)$ of the obtained sequence is a formal solution of the equation (15.51), and the function $G_1(t)$ which satisfy the identity $G_1(t+1) = g_1(t)G_1(t)$ is a formal solution of equation (15.51).

If $\Pi\{\lambda_2(t)\} > -\infty$ we may obtain a formal solution $g_2(t) \sim \lambda_2(t)$ in the same way as the function $g_1(t)$. If $\lambda_2(t) \asymp 0$ then any function $g_2(t) \in O_t$ is a formal solution of the equation (15.51) and the function $G_2(t)$ is any function of the form $G_2(t) = 1 + \delta(t)$, where $\delta(t) \asymp 0$. The set of functions $G_1(t), G_2(t)$ forms an $AFFS$ of equation (15.48) because

$$\Pi\{U(g_1(t), g_2(t))\} = \Pi\{\lambda_1(t) - \lambda_2(t)\} > -\infty.$$

On the basis of Theorem 15.21 there exists an *FSS* of equation (15.48) in the form

$$x_1(t) = G_1(t)(1 + \delta_1(t)), \quad x_2(t) = G_1(t)(1 + \delta_2(t)),$$

where $h\{\delta_{1,2}(t)\} = -\infty$.

Example 15.22. Consider the equation

$$x(t + 2) - 3tx(t + 1) + 2t^2 x(t) = 0.$$

Its characteristic equation has the form

$$y^2 - 3ty + 2t^2 = 0.$$

It has two roots $\lambda_1(t) = t$ and $\lambda_2(t) = 2t$. The substitution $x(t + 1) = y(t)x(t)$ leads to the equation

$$(y(t) - t)(y(t) - 2t) + y^-(t)y(t) = 0.$$

For the root $\lambda(t) = t$ we rewrite the equation in the form

$$y(t) = t - \frac{y^-(t)y(t)}{y(t) - 2t}$$

and form the sequence of functions

$$s_m(t) = t - \frac{s_{m-1}^-(t)s_{m-1}(t)}{s_{m-1}(t) - 2t}$$

Thus $s_0(t) = t$, $s_1(t) = t + 1$ and $s_2(t) = t + 1 + O(1/t)$ for $t \to +\infty$. Consequently $g_1(t) = t + 1 + O(1/t)$. On the basis of Proposition 15.16 we have

$$\ln G_1(t) = \int \ln(t + 1)dt - \frac{1}{2}\ln t + O\left(\frac{1}{t}\right).$$

Hence

$$G_1(t) = \left(1 + O\left(\frac{1}{t}\right)\right) t^{t+1/2} e^{-t}.$$

In the same way for the root $\lambda_2(t) = 2t$ we have

$$y(t) = 2t - \frac{y^-(t)y(t)}{y - t}.$$

This leads to the relations $g_2(t) = 2t - 4 + O(1/t)$ and

$$G_2(t) = \left(1 + O\left(\frac{1}{t}\right)\right) t^{t-5/2} 2^t e^{-t} \quad (t \to +\infty).$$

Thus the considered equation has an *FSS* $x_1(t), x_2(t)$, where

$$x_1(t) = \left(1 + O\left(\frac{1}{t}\right)\right) t^{t-1/2} e^{-t} \text{ and } x_2(t) = \left(1 + O\left(\frac{1}{t}\right)\right) t^{t-5/2} 2^t e^{-t}$$

for $t \to +\infty$.

Now we consider the cases when the roots of the characteristic equation are equivalent for $t \to +\infty$. That is, $\lambda_{1,2}(t) = \lambda(t) \pm \Delta(t)$, where $\Delta(t) = o(\lambda(t))$ for $t \to +\infty$. We distinguish the two cases:

(1) $\Pi\{\Delta(t)\} > \Pi\{\lambda(t)\} - 1/2$
and

(2) $\Pi\{\Delta(t)\} < \Pi\{\lambda(t)\} - 1/2$.
In the first case the function $g_1(t)$ can be find as an asymptotic limit of the sequence $\{s_m(t)\}$, where $s_0(t) = \lambda(t)$ and for $m = 1, 2...$,

$$s_m(t) = \lambda(t) + \Delta(t) - \frac{s^-_{m-1}(t) s_{m-1}(t)}{s_{m-1}(t) - \lambda(t) + \Delta}.$$

It is easy to see that for any function $y(t)$ such that $y(t) - \lambda(t) \sim \Delta(t)$
$(y(t) \in \{G\}, t \to +\infty)$

$$\Pi\left\{\frac{y^-(t)y(t)}{y(t) - \lambda(t) + \Delta(t)}\right\} \le 2\Pi\{\lambda(t)\} - 1 - \Pi\{\Delta(t)\} < \Pi\{\Delta(t)\}.$$

Hence any $s_m(t) - \lambda(t) - \Delta(t) = o(\Delta(t))$. Moreover put $\delta_m(t) = s_{m+1}(t) - s_m(t)$. Hence

$$\delta_m(t) = \frac{(s^-_{m-1}(t) + \delta^-_{m-1}(t))(s_{m-1}(t) + \delta_{m-1}(t))}{s_{m-1}(t) + \delta_{m-1}(t) - \lambda(t) + \Delta(t)}$$

$$- \frac{s^-_{m-1}(t) s_{m-1}(t)}{s_{m-1}(t) - \lambda(t) + \Delta(t)}.$$

It is easy to show that

$$\Pi\{\delta_m(t)\} \le \Pi\{\delta_{m-1}(t)\} + \Pi\{\lambda(t)\} - \Pi\{\Delta(t)\} - 1$$

$$\le \Pi\{\delta_{m-1}(t)\} - 1/2.$$

Hence $\Pi\{\delta_m(t)\} \to -\infty$ for $m \to \infty$. Consequently the considered sequence $\{s_m(t)\}$ has an asymptotic limit $g_1(t)$ such that $g_1(t) - \lambda(t) \sim \Delta(t)$ for $t \to +\infty$. More precisely

$$g_1(t) = \lambda(t) + \Delta(t) - \frac{\lambda^-(t)\lambda(t)}{2\Delta(t)} + \alpha_1(t),$$

where $\Pi\{\alpha_1(t)\} < \Pi\{\Delta(t)\} - 1/2$.

In the same way we can obtain the asymptotics for $g_2(t)$ related to the root $\lambda_2(t) = \lambda(t) - \Delta(t)$. The function $g_2(t)$ is an asymptotic limit of the sequence $\{s_m(t)\}$ where $s_0(t) = \lambda(t)$ and for $m = 1, 2...$

$$s_m(t) = \lambda(t) - \Delta(t) - \frac{s^-_{m-1}(t)s_{m-1}(t)}{s_{m-1}(t) - \lambda(t) - \Delta}.$$

Consequently

$$g_2(t) = \lambda(t) - \Delta(t) + \frac{\lambda^-(t)\lambda(t)}{2\Delta(t)} + \alpha_1(t),$$

where $\Pi\{\alpha_1(t)\} < \Pi\{\Delta(t)\} - 1/2$. Clearly equation (15.48) has an *FSS* of the form

$$x_1(t) = G_1(t)(1 + \delta_1(t)), \quad x_2(t) = G_1(t)(1 + \delta_2(t)),$$

where $h\{\delta_{1,2}(t)\} = -\infty$ and $G_{1,2}(t+1) = g_{1,2}(t)G_{1,2}(t)$.

Example 15.23. Consider the equation

$$x(t+2) - 2tx(t+1) + (t^2 - t\sqrt{t})x(t) = 0.$$

Its characteristic equation

$$y^2 - 2ty + t^2 - t\sqrt{t} = 0$$

has two roots $\lambda_1(t) = t + t^{3/4}$ and $\lambda_2(t) = t - t^{3/4}$. Here $\lambda(t) = t$, $\Pi\{\lambda(t)\} = 1$, $\Delta(t) = t^{3/4}$ and

$$\Pi\{\Delta(t)\} = 3/4 > \Pi\{\lambda(t)\} - 1/2 = 1/2.$$

For the root $\lambda_1(t)$ we have

$$y(t) = t + t^{3/4} - \frac{y^-(t)y(t)}{y(t) - t + t^{3/4}}.$$

Form the sequence

$$s_m(t) = t + t^{3/4} - \frac{s^-_{m-1}(t)s_{m-1}(t)}{s_{m-1}(t) - t + t^{3/4}}.$$

Thus

$$s_0(t) = t + t^{3/4}, \quad s_1(t) = t + t^{3/4} - (1/2)t^{1/4} + O(t^{-3/4})$$

and

$$s_2(t) = t + t^{3/4} - (1/2)t^{1/4} - 7/8 + O(t^{-1/4}) \text{ for } t \to +\infty.$$

Thus

$$g_1(t) = t + t^{3/4} - \frac{1}{2}t^{1/4} - \frac{7}{8} + O(t^{-1/4}) \ (t \to +\infty).$$

We have

$$\ln g_1(t) = \ln t + \ln(1 + t^{-1/4} - \frac{1}{2}t^{-3/4} - \frac{7}{8}t^{-1} + O(t^{-5/4})).$$

As will be recalled

$$\ln(1 + \alpha(t)) = \alpha(t) - \frac{1}{2}\alpha^2(t) + \frac{1}{3}\alpha^3(t) - \frac{1}{4}\alpha^4(t) + O(\alpha^5(t)) \text{ for } \alpha \to 0.$$

Therefore

$$\ln g_1(t) = \ln t + t^{-1/4} - \frac{1}{2}t^{-1/2} - \frac{1}{6}t^{-3/4} - \frac{5}{8}t^{-1} + O(t^{-5/4}).$$

We have $\ln G_1(t) = \int \ln g_1(t)dt - (1/2)\ln t + O(t^{-1/4})$. Hence

$$x_1(t) = (1 + O(t^{-1/4}))t^{t-9/8}e^{-t+(4/3)t^{3/4}-t^{1/2}-(2/3)t^{1/4}} \ (t \to +\infty).$$

In the same way

$$x_2(t) = (1 + O(t^{-1/4}))t^{t-9/8}e^{-t-(4/3)t^{3/4}-t^{1/2}+(2/3)t^{1/4}} \ (t \to +\infty).$$

Clearly the set $\{x_1(t), x_2(t)\}$ forms an *FSS* of the considered equation.

In the second case let $\Pi\{\lambda(t)\} \neq 0$ and $\Pi\{\lambda(t)\} > -\infty$. Then the function $g_1(t)$ can be find as an asymptotic limit of the sequence $\{s_m(t)\}$, where $s_0(t) = \lambda(t)$ and for $m = 1, 2....$

$$s_m(t) = \lambda(t) + \sqrt{\Delta^2(t) - s_{m-1}^-(t)s_m(t)}.$$

It is easy to see that for any function $y(t)$ such that $y(t) \sim \lambda(t)$ for $t \to +\infty$ $(y(t) \in \{G\}, t \to +\infty)$ we have

$$y^-(t) \sim \lambda^-(t) \sim \lambda'(t) \sim P\{\lambda(t)\}\lambda(t)/t$$

for $t \to +\infty$. We have

$$\Pi\{\sqrt{\Delta^2(t) - y^-(t)y(t)}\} = \Pi\{\lambda(t)\} - 1/2.$$

Hence $s_m(t) = \lambda(t) + \alpha_m(t)$, where $\Pi\{\alpha_m(t)\} = \Pi\{\lambda(t)\} - 1/2$. Moreover put $\delta_m(t) = s_{m+1}(t) - s_m(t)$. Hence

$$\delta_m(t) = \sqrt{\Delta^2(t) - s_m^-(t)s_m(t)} - \sqrt{\Delta^2(t) - s_{m-1}^-(t)s_{m-1}(t)}$$

$$= \frac{s_{m-1}^-(t)s_{m-1}(t) - s_m^-(t)s_m(t)}{\sqrt{\Delta^2(t) - s_m^-(t)s_m(t)}\sqrt{\Delta^2(t) - s_{m-1}^-(t)s_{m-1}(t)}}.$$

It is easy to show that

$$\Pi\{\delta_m(t)\} \leq \Pi\{\delta_{m-1}(t)\} - 1/2.$$

Hence $\Pi\{\delta_m(t)\} \to -\infty$ for $m \to \infty$. Thus the considered sequence $\{s_m(t)\}$ has an asymptotic limit $g_1(t)$ such that

$$g_1(t) - \lambda(t) \sim \sqrt{\lambda^-(t)\lambda(t)} \text{ for } t \to +\infty.$$

More precise asymptotics we can obtain from the sequence $\{s_m(t)\}$, where the asymptotics became better when m increases.

In the same way the function $g_2(t)$ can be obtained from a sequence $\{s_m(t)\}$ where $s_0(t) = \lambda(t)$ and for $m = 1, 2...$

$$s_m(t) = \lambda(t) - \sqrt{\Delta^2(t) - s_{m-1}^-(t)s_m(t)}.$$

The set $\{G_1(t), G_2(t)\}$ (where $G_{1,2}(t + 1) = g_{1,2}(t)G_{1,2}(t)$) forms an $AFFS$ because

$$U(g_1(t), g_2(t)) = g_2(t) - g_1(t) \sim -2\sqrt{\lambda^-(t)\lambda(t)},$$

hence

$$\Pi\{U(g_1(t), g_2(t))\} = \Pi\{\lambda(t)\} - 1/2 > -\infty.$$

Example 15.24. Consider the equation

$$x(t + 2) - 2tx(t + 1) + t^2 x(t) = 0.$$

Its characteristic equation

$$y^2 - 2ty + t^2 = 0$$

has only one (double) root $\lambda(t) = t$. Here $\Delta(t) = 0$. The function $g_1(t)$ is an asymptotic limit of the sequence $\{s_m(t)\}$, where $s_0(t) = t$ and

$$s_m(t) = t + \sqrt{-s_{m-1}^-(t)s_{m-1}(t)}$$

for $m = 1.2, \ldots$ So that $s_1(t) = t + i\sqrt{t}$,

$$s_2(t) = t + i\sqrt{t} - \frac{3}{4} + O(t^{-1/2})$$

for $t \to +\infty$ ($i = \sqrt{-1}$). It easy to see that the function $g_1(t)$ has its asymptotic expansion in terms of $t^{-1/2}$. Hence $g_1(t) = t + i\sqrt{t} - 3/4 + O(t^{-1/2})$ and $\ln g_1(t) = \ln t + it^{-1/2} - 1/(4t)$. Hence

$$\ln G_1(t) = t\ln t - t + 2i\sqrt{t} - \frac{3}{4}\ln t + O(t^{-3/2})$$

and

$$G_1(t) = (1 + O(t^{-1/2}))t^{t-3/4}e^{-t+2i\sqrt{t}}$$

for $t \to \infty$. In the same way

$$g_2(t) = t - i\sqrt{t} - \frac{3}{4} + O(t^{-1/2})$$

and

$$G_2(t) = (1 + O(t^{-1/2}))t^{t-3/4}e^{-t-2i\sqrt{t}} \quad (t \to +\infty).$$

Every case (2) when $\lambda(t) \to$ constant $\neq 0, \infty$ for $t \to +\infty$ or $\Pi\{\Delta(t)\} = \Pi\{\lambda(t)\} - 1/2$, has to be considered separately. If $\lambda(t) \to$ constant $\neq 0, \infty$ it is possible to recommend the following procedure: first we make the substitution $x(t + 1) = y(t)x(t)$ and pass the equation (15.50). Next substitute $y(t) = \lambda + u(t)$. And then for the last equation we try to obtain a convenient form to obtain the necessary asymptotic estimates.

Example 15.25. Consider the equation

$$x(t + 2) - 2x(t + 1) + \left(1 + \frac{1}{t}\right)x(t) = 0. \tag{15.53}$$

To solve equation (15.53) substitute $x(t + 1) = y(t)x(t)$ and $y(t) = 1 + u(t)$. Clearly

$$(y(t) - 1)^2 + y^-(t)y(t) + \frac{1}{t} = 0 \tag{15.54}$$

and

$$u^2(t) + \frac{1}{t} + u^-(t)(1 + u(t)) = 0. \tag{15.55}$$

To obtain the required asymptotic approximations we use two following representations of the last equation

$$u(t) = \frac{i}{\sqrt{t}} - \frac{u^-(t)(1 + u(t))}{u(t) + i/\sqrt{t}} \tag{15.56}$$

and

$$u(t) = -\frac{i}{\sqrt{t}} - \frac{u^-(t)(1+u(t))}{u(t) - i/\sqrt{t}}. \tag{15.57}$$

Here $i = \sqrt{-1}$. The relation (15.56) unable us form a sequence of the form $\{s_m(t)\}$ where $s_0(t) = i/\sqrt{t}$ and

$$s_m(t) = -\frac{i}{\sqrt{t}} - \frac{u_{m-1} - (t)(1 + u_{m-1}(t))}{u_{m-1}(t) - i/\sqrt{t}} \quad \text{for } m = 1, 2, \dots$$

((15.57) is considered in the same way). We have $s_1(t) = i/\sqrt{+}(1/4)t$, $s_2(t) = i/\sqrt{t} + (1/4)/t + O(1/(t\sqrt{t}))$ for $t \to +\infty$ and so on. Thus

$$\varphi_1(t) = \frac{i}{\sqrt{t}} + \frac{3}{4t} + O(t^{-3/2}) \text{ for } t \to +\infty$$

and the formal solution of equation (15.56) is in the form $g_1(t) = 1 + \varphi(t)$. The asymptotics to the formal solution $G_1(t)$ $(G_1(t+1) = g_1(t)G_1(t))$ of the equation (15.53) is obtained from the relation

$$\ln G_1(t) = \int \ln g_1(t)dt + o(1/\sqrt{t}).$$

Consequently

$$\ln g_1(t) = \ln[1 + i/\sqrt{t} + 1/(4t) + O(t^{-3/2}).$$

Clearly

$$\ln g_1(t) = i/\sqrt{t} + 3/(4t) + O(t^{-3/2}).$$

Hence $\ln G_1(t) = 2i\sqrt{t} + (3/4)\ln t + O(t^{-1/2})$. Thus

$$G_1(t) = \left(1 + O(t^{-1/2})t^{-1/4}e^{2i\sqrt{t}}\right) \quad (t \to +\infty).$$

In the same way

$$G_2(t) = \left(1 + O(t^{-1/2})t^{3/4}e^{-2i\sqrt{t}}\right) \quad (t \to +\infty).$$

Obviously the functions $G_1(t), G_2(t)$ form an *AFFS* of the considered equation. Thus for real initial conditions at the point t_0 the general solution of equation (15.53) satisfy the following estimate:

$$x(t) \approx t^{3/4}(A \cos(2\sqrt{t}) + B \sin(2\sqrt{t})) \text{ for } t \gg 1$$

on the set $D = \{t : t_0, t_0 + 1, \dots, t_0 + m, \dots\}$, where A, B are arbitrary constants.

5.   SYSTEMS OF TWO EQUATIONS

Here we consider a system of the form

$$\begin{cases} x_1(t+1) & = & a_{11}(t)x_1(t) + a_{12}(t)x_2(t) \\ \\ x_2(t+1) & = & a_{21}(t)x_1(t) + a_{22}(t)x_2(t) \end{cases} \tag{15.58}$$

where the conditions imposed upon the coefficient $a_{ij}(t)$ are the same just as for equation (15.48). System (15.58) can be written as a matrix equation of the form

$$X(t+1) = A(t)X(t), \tag{15.59}$$

where $X(t) = (x_1(t), x_2(t))^T$ and

$$A(t) = \begin{pmatrix} a_{11}(t) & a_{12}(t) \\ \\ a_{21}(t) & a_{22}(t) \end{pmatrix}.$$

We may reduce the system to one equation of second order by means of identical transformations. We use the following procedure (similar to the procedure used for differential equations): we choose a linear form

$$x(t) = b(t)x_1(t) + b_2(t)x_2(t) \tag{15.60}$$

and take its difference according to system (15.57). As the result we obtain the transformation

$$\widetilde{X}(t) = P(t)X(t), \tag{15.61}$$

where $\widetilde{X}(t) = (x(t), x(t+1))$. This leads to the matrix equation

$$\widetilde{X}(t+1) = Q(t)\widetilde{X}(t), \tag{15.62}$$

where $Q(t)$ is a matrix of the form

$$Q(t) = \begin{pmatrix} 0 & 1 \\ -q_2(t) & -q_1(t) \end{pmatrix}. \tag{15.63}$$

The matrix $P(t)$ must possess the estimate $\Pi\{\det P(t)\} > -\infty$. We have

$$\widetilde{X}(t+1) = P(t+1)X(t+1) = P(t+1)A(t)X(t) = P(t+1)A(t)P^{-1}(t)\widetilde{X}(t).$$

Hence

$$Q(t) = P(t+1)A(t)P^{-1}(t). \tag{15.64}$$

System (15.64) is equivalent to the equation

$$x(t+2) + q_1(t)x(t+1) + q_2(t)x(t) = 0 \qquad (15.65)$$

in the following sense: for a solution $x(t)$ of equation (15.65) there exists a solution $\widetilde{X}(t) = (x(t), x(t+1))^T$ of equation (15.63). And vice versa. For a solution $\widetilde{X}(t) = (x(t), x(t+1))^T$ the function $x(t)$ is a solution of equation (15.65). The solutions of the original system may be find from the relation $X(t) = P^{-1}(t)\widetilde{X}(t)$.

If $\Pi\{a_{12}(t)\} > -\infty$ the required transformation is obtained by means of the linear form $x(t) = x_1(t)$. In this case the matrix $P(t)$ is in the form

$$P(t) = \begin{pmatrix} 1 & 0 \\ a_{11}(t) & a_{12}(t) \end{pmatrix}. \qquad (15.66)$$

Clearly, $\Pi\{\det P(t)\} = \Pi\{a_{12}(t)\} > -\infty$. In the other cases we always can choose the necessary functions $b_1(t), b_2(t)$ (which may not be constants).

Example 15.26. Consider the system

$$\begin{cases} x_1(t+1) &= (t-1)x_1(t) + x_2(t), \\ x_2(t+1) &= -tx_1(t) + tx_2(t). \end{cases} \qquad (15.67)$$

Here

$$A(t) = \begin{pmatrix} t-1 & 1 \\ -t & t \end{pmatrix}. \qquad (15.68)$$

We take $x(t) = x_1(t)$. In this case the matrix $P(t)$ is in the form

$$P(t) = \begin{pmatrix} 1 & 0 \\ t-1 & 1 \end{pmatrix}, \quad P^{-1}(t) = \begin{pmatrix} 1 & 0 \\ 1-t & 1 \end{pmatrix}.$$

Clearly $\Pi\{\det P(t)\} = \Pi\{a_{12}(t)\} > -\infty$. In the other cases, we always can choose the necessary functions $b_1(t)$, $b_2(t)$. Consequently

$$Q(t) \equiv P(t+1)A(t)P^{-1}(t)$$

$$= \begin{pmatrix} 1 & 0 \\ t & 1 \end{pmatrix} \begin{pmatrix} t-1 & 1 \\ -t & t \end{pmatrix} \begin{pmatrix} 1 & 0 \\ 1-t & 1 \end{pmatrix}$$

$$= \begin{pmatrix} 1 & 0 \\ t^2 & 2t \end{pmatrix}.$$

Hence the variable $x(t) = x_1(t)$ satisfies the following equation

$$x(t+2) - 2tx(t+1) + t^2 x(t) = 0.$$

Hence (see Example 15.24) $x_1(t) \sim t^{t-1/4} e^{-t \pm 2i\sqrt{t}}$ and

$$x_2(t) = x_1(t+1) - (t-1)x_1(t)$$

$$\sim -(1 - e^{-1})x_1(t) \text{ for } t \to +\infty.$$

6.   HIGHER ORDER EQUATIONS

The Theorem 15.21 may be useful for asymptotic solution of some linear difference equations of order $n > 2$. We consider only the case when the roots of characteristic polynomial (15.33) are not equal in pairs for $t \to +\infty$.

Theorem 15.27.   *Let* $\{a_1(t), ..., a_n(t)\} \subset \{G\}$, *where* G *is a field of type* M *and let* $a_n(t) \not\equiv 0$ *for* $t \gg 1$. *Let characteristic polynomial* (15.33) *have a complete set of roots* $\Lambda = \{\lambda_i(t)\}$ $(i = 1, 2, ..., n)$ *such that they are not equivalent in pairs for* $t \to +\infty$ *and no more than one of the roots have the estimate* $O(t^{-\infty})$. *Then the equation*

$$\Phi(t, x(t)) \equiv x(t+n) + a_1(t)x(t+n-1) + ... + a_n(t)x(t) = 0 \quad (15.1)$$

has a continuous fundamental system of solutions (for $t \gg 1$) *(FSS) of the form*

$$\{x_1(t), x_2(t), ..., x_n(t)\}$$

such that for each $i \in \{1, 2, ..., n\}$ *the function*

$$y_i(t) = \frac{x_i(t+1)}{x_i(t)}$$

possesses the following properties:

(1) *if* $h\{\lambda_i(t)\} = -\infty$ *then*

$$y_i(t) = -\frac{a_n(t)}{a_{n-1}(t)}(1 + \theta_i(t)), \quad (15.69)$$

where $h\{\theta_i(t)\} = -\infty$;

(2) *if* $h\{\lambda_i(t)\} > -\infty$ *then*

$$y_i(t) \sim \lambda_i(t) \text{ for } t \to +\infty$$

and $y_i(t)$ is a simple asymptotic limit of the sequence $\{s_{im}(t)\}$, where $s_{i1}(t) = \lambda_i(t)$ and

$$s_{im+1}(t) = \lambda_i(t) - \frac{F^*(t, s_{im}(t))}{H^*(t, s_{im}(t))}, \quad m = 1, 2, ... \tag{15.70}$$

Hence

$$y_i(t) = \lambda_i(t) \left\{ 1 - \frac{1}{2} \lambda_i'(t) \frac{\partial^2 H(\lambda_i(t), t)}{\partial y^2} \middle/ \frac{\partial H(\lambda_i(t), t)}{\partial y} + O\left(\frac{1}{t^2}\right) \right\} \tag{15.71}$$

for $t \to +\infty$ and

$$x_i(t) = [\lambda_i(t)]^{-(1+q_i/2+o(1))} \exp\left[\int \ln \lambda_i(t) dt\right]. \tag{15.72}$$

Here

$$q_i = \lim_{t \to +\infty} \frac{\lambda_i(t) \partial^2 H(t, \lambda_i(t))}{\partial y^2} \middle/ \frac{\partial H(t, \lambda_i(t))}{\partial y}. \tag{15.73}$$

If in addition, all the coefficients $a_i(t)$ $(i = 1, 2, ..., n)$ have their expansions in the form of generalized power series. Then

$$x_i(t) = (1 + \alpha_i(t))t^{-(1/2)(1+q_i)P\{\lambda_i(t)\}} \exp\left[\int \ln \lambda_i(t) dt\right], \tag{15.74}$$

where $h\{\alpha_i(t)\} < 0$.

Let P be a field of type M such that $\{P\}$ contains $\{G\}$ and the set Λ. Then $y_i(t)$ $(i = 1, 2, ..., n)$ is represented in the form

$$y_i(t) = g_i(t) + r_i(t),$$

where $g_i(t) \in \{P\}$, $r_i(t)$ is a continuous function for $t \gg 1$ and $h\{r_i(t)\} = -\infty$.

The theorem is a simple consequence of Proposition 15.18 and Theorem 15.21.

On the basis of Theorem 15.27 it is possible to formulate the criteria of asymptotic stability and unstability of equation (15.1).

Theorem 15.28. *Let all the conditions of Theorem 15.27 be fulfilled. Then for $t \to +\infty$ for equation (15.1)*

(1) to be asymptotically stable it is sufficient that for any root $\lambda_i(t)$ with the estimate $h\{\lambda_i(t)\} > -\infty$ $(i = 1, 2, ..., n)$ the following inequality holds:

$$\lim_{t \to +\infty} t \ln |\lambda_i(t)| < 0; \tag{15.75}$$

(2) *to be unstable, it is sufficient that for at least one root $\lambda_i(t)$ with the estimate $h\{\lambda_i(t)\} > -\infty$ $i \in 1, 2, ..., n$ the following inequality holds:*

$$\lim_{t \to +\infty} t \ln |\lambda_i(t)| > 0. \qquad (15.76)$$

PROOF. Let us note that because G is a field of type M and $\lambda(t) \not\equiv 0$, $P\{\lambda(t)\}$ is a real number and there exists a (finite or infinite) limit $\lim_{t \to +\infty} t \ln |\lambda_i(t)|$ for any function $\lambda(t) \in \{G\}$. Let us prove the criterion of stability. Owing to condition (1) of this proposition and (15.71) it follows that there is a positive number σ such that $|y_i(t)| < (1 + 1/t)^{-\sigma}$ for $t \gg 1$ Indeed it is obvious if $P\{\lambda_i(t)\} < 0$. If $P\{\lambda_i(t)\} = 0$ then it leads from the relation

$$y_i(t) = \lambda_i(t) \left(1 + o\left(\frac{1}{t}\right)\right) \quad \text{for } t \to +\infty.$$

The case $P\{\lambda_i(t)\} > 0$ is impossible. Consequently $x_i(t)$ has the estimate $h\{x_i(t)\} \le -\sigma$ and hence $x_i(t) \to 0$ for $t \to +\infty$. The criterion of unstability follows from the inequality $|y_i(t)| > (1 + 1/t)^{\sigma}$ for $t \gg 1$ ($\sigma = \text{const}, \sigma > 0$). The last follows from condition (2) of this proposition and (15.71).

Example 15.29. Let us consider the simplest example when the coefficients of equation (15.1) become constants for $t \to +\infty$.

Let all the conditions of Theorem 15.27 be fulfilled and let

$$a_i(t) = a_i + o(1) \text{ for } t \to +\infty \ (a_i = \text{const}, \ i = 1, 2, ..., n).$$

Let all the roots of the polynomial $\lambda^n + a_1 \lambda^{n-1} + ... + a_n$ be simple. Let us designate them by $\lambda_1, \lambda_2, ..., \lambda_n$. Let $\lambda_1(t), \lambda_2(t), ..., \lambda_n(t)$ be the roots of characteristic polynomial (15.33). We have $\lambda_i(t) = \lambda_i + o(1)$ and $\lambda_i'(t) = o(1/t)$ for $t \to +\infty$. By Theorem 15.27 equation (15.1) has a fundamental system of solutions $\{x_1(t), x_2(t), ..., x_n(t)\}$ such that for

$$y_i(t) = \frac{x_i(t+1)}{x_i(t)}$$

the following relations hold:

$$y_i(t) = \lambda_i(t) + O(\lambda_i'(t)) = \lambda_i(t) \left(1 + O\left(\frac{1}{t}\right)\right) \quad \text{for } t \to +\infty.$$

For instance consider the equation

$$x(t+2) - 4\left(1 + \frac{1}{t}\right) x(t+1) + 3\left(1 + \frac{2}{t}\right) x(t) = 0.$$

Its characteristic polynomial has the following roots:

$$\lambda_1(t) = 1 + \frac{1}{t} + O\left(\frac{1}{t^2}\right), \quad \lambda_2(t) = 3\left(1 + \frac{1}{t}\right) + O\left(\frac{1}{t^2}\right)$$

and $\lambda'_{1,2}(t) = O(1/t^2)$ for $t \to +\infty$. Hence

$$y_1(t) = 1 + \frac{1}{t} + o\left(\frac{1}{t}\right) \quad \text{and} \quad y_2(t) = 3\left(1 + \frac{1}{t}\right) + o\left(\frac{1}{t}\right).$$

On the basis of (15.71) more exactly

$$y_1(t) = 1 + \frac{1}{t} + O\left(\frac{1}{t^2}\right) \quad \text{and} \quad y_2(t) = 3\left(1 + \frac{1}{t}\right) + O\left(\frac{1}{t^2}\right)$$

and hence

$$x_1(t) = t\left(1 + O\left(\frac{1}{t}\right)\right) \quad \text{and} \quad x_2(t) = 3^t t^3\left(1 + O\left(\frac{1}{t}\right)\right)$$

for $t \to +\infty$.

Example 15.30. Let $a_i(t) \in \{G\}$, where G is a field of type M, let $h\{a_n(t)\} > -\infty$ and

$$\lim_{t \to +\infty} \frac{a_j(t)}{a_n^{j/n}(t)} = 0 \quad (i = 1, 2, ..., n; j = 1, 2, ..., n - 1).$$

It is easy to show that characteristic polynomial (15.33) has simple roots

$$\lambda_j(t) \sim \varepsilon_j a_n^{1/n}(t) \quad \text{for } t \to +\infty \quad (j = 1, 2, ..., n),$$

where ε_j are the distinct roots of nth degree of -1. We have

$$\lambda'_j(t) \sim \frac{a'_n(t)}{n a_n(t)} \lambda_j(t);$$

$$y_j(t) = \lambda_j(t)\left[1 - \frac{n-1}{2n}\frac{a'_n(t)}{a_n(t)} + o\left(\frac{1}{t}\right)\right]$$

for $t \to +\infty$. Hence

$$x_j(t) = t^{o(1)}[a_n(t)]^{-1/2} \exp\left[\int \ln \lambda_j(t) dt\right]. \tag{15.77}$$

Example 15.31. Consider the equation

$$x(t+3) - 3\sqrt{t}x(t+2) + 2tx(t+1) - x(t) = 0. \tag{15.78}$$

Its characteristic polynomial $H(t, y) = y^3 - 3\sqrt{t}y^2 + 2ty - 1$ has the following roots:

$$\lambda_1(t) = \frac{1}{2t}(1 + O(t^{-3/2})), \quad \lambda_2(t) = t^{1/2}(1 + O(t^{-3/2}))$$

and $\lambda_3(t) = 2t^{1/2}(1 + O(t^{-3/2}))$ for $t \to +\infty$. We have $\lambda_1'(t) \sim -\frac{1}{2}t^{-2}$,

$$\frac{\partial H(t, \lambda_1(t))}{\partial y} \sim 2t \quad \text{and} \quad \frac{\partial^2 H(t, \lambda_1(t))}{\partial^2 y} \sim -6t^{1/2},$$

hence (see (15.71)) $y_1(t) = \frac{1}{2}t^{-1}(1 + O(t^{-3/2}))$ and

$$x_1(t) = (1 + O(t^{-1/2}))2^{-t}e^t t^{1/2-t} \quad \text{for } t \to +\infty.$$

In the same way

$$x_2(t) = (1 + O(t^{-1/2}))e^{-1/2}t^{t/2-1/4}$$

and

$$x_3(t) = (1 + O(t^{-1/2}))2^t e^{-1/2}t^{t/2-7/4} \quad \text{for } t \to +\infty.$$

7. DIFFERENTIAL–DIFFERENCE EQUATIONS

Here we consider a differential–difference equation of the first order of the form

$$x'(t+1) = a(t)x(t), \tag{15.79}$$

where the function $a(t)$ may be represented in the form $a(t) = t^{k+\alpha(t)}$. Here k is a number and $\alpha(t) \in C_t$. So kind functions form a class of functions wider than the class A_t because if $a(t) \in A_t$, then in addition $\alpha'(t)t\ln t$ must belong to C_t. We restrict our consideration for the case $k \neq 0$.

We look for solutions to the equation presented in the form

$$x(t) = e^{y(t)}. \tag{15.80}$$

Then $y'(t+1)e^{y(t+1)} = a(t)e^{y(t)}$, and (for $y'(t+1) \neq 0$)

$$\ln y'(t+1) + y(t+1) - y(t) = [k + \alpha(t)]\ln t. \tag{15.81}$$

Definition 15.32. A function $G(t) = e^{g(t)}$ is called a *formal solution* to equation (15.79) if $\ln g'(t+1) + g(t+1) - g(t) - (k + \alpha(t)) \ln t \asymp 0$.

Since we look only for the formal solutions to equation (15.81) any function $\varphi(t) \in A_t$ under consideration may be replaced by any function or formal series $\varphi^*(t)$ such that $\varphi^*(t) \asymp \varphi(t)$. We have

$$y(t+1) - y(t) \asymp y'(t+1) - \sum_{m=2}^{\infty} (-1)^m \frac{y^{(m)}(t+1)}{m!}.$$

Put $u(t) = y(t+1)$. Hence we obtain the inclusion

$$\ln u(t) + u(t) \asymp [k + \alpha(t)] \ln t + \sum_{m=1}^{\infty} (-1)^{m+1} \frac{u^{(m)}(t)}{(m+1)!}. \qquad (15.82)$$

We obtain the first formal approximation to (15.81) from the equation $\ln v(t) + v(t) = [k + \alpha(t)] \ln t$. It has a unique solution $\varphi(t) \sim k \ln t$ for $t \to +\infty$ $(\varphi(t) \in A_t)$. Let us substitute $u(t) = \varphi(t) + w(t)$ in (15.81). We have

$$w(t) + \ln\left(1 + \frac{w(t)}{\varphi(t)}\right) \asymp q(t) + \sum_{m=1}^{\infty} (-1)^{m+1} \frac{w^{(m)}(t)}{(m+1)!}, \qquad (15.83)$$

where $q(t)$ is an asymptotic sum of the series

$$\sum_{m=1}^{\infty} (-1)^m \frac{\varphi^{(m)}(t)}{(m+1)!}.$$

Clearly $q(t) \sim \varphi'(t)/2$ hence $q(t) \sim k/(2t)$ for $t \to +\infty$. To obtain the subsequent approximations we form the following iteration sequence $\{s_n(t)\}$: $s_0(t) = 0$ and for $n = 1, 2...$

$$s_n(t) + \ln\left(1 + \frac{s_n(t)}{\varphi(t)}\right) = q(t) + \sum_{m=1}^{\infty} (-1)^{m+1} \frac{s_{n-1}^{(m)}(t)}{(m+1)!}. \qquad (15.84)$$

Obviously $s_n(t) \sim k/(2t)$ for any $n = 1, 2, ...$ Let us put $\Delta_n(t) = s_{n+1}(t) - s_n(t)$. We have

$$s_{n+1}(t) - s_n(t) + \ln\left(1 + \frac{s_{n+1}(t)}{\varphi(t)}\right) - \ln\left(1 + \frac{s_n(t)}{\varphi(t)}\right) \sim \left(1 + \frac{1}{k \ln t}\right) \Delta_n(t)$$

$$\sim \Delta_n(t).$$

Besides

$$\sum_{m=1}^{\infty}(-1)^{m+1}\frac{s_n^{(m)}(t)}{(m+1)!} - \sum_{m=1}^{\infty}(-1)^{m+1}\frac{s_{n-1}^{(m)}(t)}{(m+1)!} \sim \frac{\Delta'_{n-1}(t)}{2}.$$

Consequently $\Pi\{\Delta_n(t)\} \leq \Pi\{\Delta_{n-1}(t)\} - 1$. So that the sequence $\{s_n(t)\}$ has an asymptotic limit $\varphi_1(t) \sim k/(2t)$ for $t \to +\infty$. And we may conclude that equation (15.81) has a formal solution $g(t) = \varphi(t-1)+k/(2t)+o(1/t)$ for $t \to +\infty$. Or more precisely $g(t) = \varphi(t-1) + k/(2t) + \delta(t)$, where $\Pi\{\delta(t)\} \leq -2$.

Example 15.33. Consider the equation

$$x'(t+1) = tx(t). \tag{15.85}$$

It has a formal solution $x(t) = \exp[\int g(t)dt]$, where $\ln g(t+1)+g(t+1)-g(t) = \ln t$. We have

$$g(t) = \varphi_0(t-1) + 1/2t + \delta(t),$$

where $\varphi_0(t) + \ln\varphi_0(t) = \ln t$ and $\Pi\{\delta(t)\} \leq -2$. The asymptotic approximations of the function $\varphi_0(t)$ may be obtained from the iteration sequence $\{s_n(t)\}$ where $s_0(t) = \ln t$ and for $n = 1, 2, ...,$ $s_n(t) = \ln t - \ln s_{n-1}(t)$. So that $\varphi_0(t) = s_n(t) + o(s_{n+1}(t) - s_n(t))$ for $t \to +\infty$. We have $s_1(t) = \ln t - \ln_2 t$, $s_2(t) = \ln t - \ln(\ln t - \ln_2 t)$. Hence

$$s_2(t) = \ln t - \ln_2 t + \frac{\ln_2 t}{\ln t} \quad \text{and so on.}$$

In this case it is possible to prove that the obtained formal solution is an asymptotic solution of our equation.

Chapter 16

SUPPLEMENT

1. ALGORITHMS WITH AN UNIFORMLY DISTRIBUTED THEORETICAL ERROR

At the present time when there exist high speed electronic digital computers, the algorithmic methods of representing of functions have come into importance.

A function $f(t)$ is said to be *algorithmic represented* if for any point t of its domain of definition, all the step by step operations (analytic, logical, etc.) to compute the function are given.

The number of operations can be infinitely large. In this case we suppose the following: for any positive number ε there is a number m (generally speaking, depending on t and ε) such that $|f(t) - f(t,m)| < \varepsilon$. Here $f(t,m)$ is a function determined by the part of this algorithm composed by the m first actions of the original algorithm. That is, $f(t)$ is determined as a limit of $f(t,m)$ for $m \to \infty$.

We consider algorithms without unnecessary operations which implies if $m_2 > m_1$ then $|f(t) - f(t,m_1)| \geq |f(t) - f(t,m_2)|$.

The value $\delta_m(t) = f(t) - f(t,m)$ is called the *theoretical error* of the algorithm.

As a matter of fact the total error of computation is equal to the sum of two values: $\delta_m(t) = \delta_{1m}(t) + \delta_{2m}(t)$, where $\delta_{1m}(t)$ is the theoretical error and $\delta_{2m}(t)$ is so called *instrumental*. The instrumental error arise because of errors of performing the operations of the applied device.

Here we manly consider the theoretical error. If (for any fixed t)

$$\lim_{m \to \infty} (f(t) - f(t, m)) = 0,$$

we say that the algorithm is *correct*. Otherwise, it is *incorrect*.

We distinguish two kinds of algorithms: let for the theoretical error $\delta_m(t)$ $= f(t) - f(t, m)$ in the domain D,

(1) there exists a positive number σ such that for any (fixed) m there is at least one number $t \in D$ (which, generally speaking, depends on m) such that $|\delta(t, m)| > \sigma$;

(2) for any positive number ε, there exists a number m independent of t such that $\sup_{t \in D} |\delta(t, m)| < \varepsilon$.

Algorithm (2) is called *uniform* (in D).

Uniformity is one of important qualitative characteristics of an algorithm. Asymptotic solutions give us a unique possibility to obtain uniform algorithms for computing functions having singularities in the considered domains. As a model of solution of such problem we consider the Airy equation

$$x'' + tx = 0 \tag{16.1}$$

on the entire positive semi-axis $J_+ = \{t : 0 \le t < +\infty\}$, and we try to obtain an uniform algorithm for the solution $x(t)$ with initial conditions $x(0) = 1$ and $x'(0) = 0$ (here $t = +\infty$ is the singular point of the solution).

First, we consider the possibility to calculate this solution on a fixed finite segment $[0, T]$. Since $x(t)$ is analytic in the considered domain, it can be expended into a convergent power series of the form

$$x(t) = c_0 + c_1 t + \ldots + c_m t^m + \ldots, \tag{16.2}$$

where c_m are numbers. Algorithms of the function $x(t)$ calculation by the series is correct because the series is convergent at any point $t \in J_+$. Taking into account the initial conditions, we have $c_0 = 1$ and $c_1 = 0$. To obtain the subsequent coefficients c_m of the series, substitute the series in equation (16.1) collect the like terms and equate them to zero. We obtain $c_2 = 0$ and for $m = 3, \ldots,$ $m(m-1)c_m + c_{m-3} = 0$. Hence (as it easy to see) all the numbers of the form c_{3m-2} and c_{3m-1} are zeros. Thus, we can rewrite the series in the form

$$x(t) = a_0 + a_1 t^3 + \ldots + a_m t^{3m} + \ldots, \tag{16.3}$$

where $a_0 = 1$, and

$$a_m = -\frac{a_{m-1}}{3m(3m-1)}. \tag{16.4}$$

So that $a_1 = -1/6$, $a_2 = 1/180$ and so on. The values of $x(t)$ approximately equal to

$$x(t, m) = a_0 + a_1 t^3 + \dots + a_m t^{3m}.$$

Since (16.2) is an alternating series satisfying the Leibniz test the theoretical error $x(t) - x(t, m) \approx a_m t^{3m} \to 0$ for any fixed t as $m \to \infty$, so that, the theoretical error $\delta_m(t)$ satisfies the inequality $|\delta_m(t)| \leq |a_m| T^{3m}$ for any $t \in [0, T]$. Clearly $\delta_m(t) \to 0$ for $m \to \infty$ uniformly in $t \in [0, T]$. Thus the first algorithm is uniform on any fixed interval $[0, T]$. But we have $\delta_m(t) \to \infty$ as $t \to \infty$ for any fixed m. This means that the algorithm is not uniform in the entire semi-axis J_+.

Such an algorithm is referred to as an algorithm of *first type*.

In fact the algorithm may be acceptable on an interval $[0, T]$, where T is not very large. If T is large the numbers $a_m T^{3m}$ for several m become vary large. For example to calculate the function with accuracy to within 10^{-5} (for modern computers) the number $T \approx 10$. Instead of this algorithm, other global algorithms may to increase the number T. But (since they are not uniform on J_+) they do not solve the problem to obtain the solution for $t \gg 1$.

To obtain a uniform algorithm on the entire positive semi-axis we have (in addition to first type algorithm) to use the asymptotic approximations of the function $x(t)$ for $t \gg 1$.

An algorithm, where the theoretical error may be less (by modulus) than any (fixed) number $\varepsilon > 0$ for $t \gg 1$, is referred to as an algorithm of *second type*. As a rule the algorithms of second type are based on the asymptotic methods.

We make (as usual) the substitution $x' = yx$ and pass to the equation

$$y' + y^2 + t = 0. \tag{16.5}$$

Its characteristic equation $y^2 + t = 0$ has two roots $\lambda_1(t) = i\sqrt{t}$ and $\lambda_2(t) = -i\sqrt{t}$, and correspondingly, we have two solutions $y_1(t) \sim \lambda_1(t)$ and $y_2(t) \sim \lambda_2(t)$ for $t \to +\infty$ and two solutions of the equation (16.1)

$$x_1(t) = \exp\left[\int y_1(t)dt\right], x_2(t) = \exp\left[\int y_2(t)dt\right]$$

which form an *FSS*. Thus there exist two numbers C_1, C_2 such that

$$x(t) = C_1 x_1(t) + C_2 x_2(t).$$

We look for the asymptotic representations to the function $y_1(t)$ (the function $y_2(t)$ is examined in the same way). The function $y_1(t)$ is an asymptotic sum of a series of the form

$$b_0 t^{1/2} + b_1 t^{-1} + \dots + b_m t^{-(3/2)m+1/2} + \dots \tag{16.6}$$

where b_m are integers ($m = 0, 1, ...$). Series (16.6) is divergent for any fixed $t \in J_+$, but $y_1(t) - s_m(t) = O(t^{-(3/2)m-1})$ for $t \gg 1$. The last means that for the chosen m there exists a theoretical error $\delta_m(t) \equiv y_1(t) - s_m(t) \leq A_m T^{-(3/2)m-1}$ for any $t \geq T$, and A_m is minimal. That is $A_{m-1}T > A_m$ and $A_{m+1} > A_m T$. Thus for the given T the function $y_(t)$ cannot be calculated more precisely than $A_m T^{-(3/2)m-1}$. If the accuracy is not sufficient we have to increase T. The point is that for any positive number ε we can choose numbers T and m such that the theoretical error of $x(t)$ calculation will be less then ε. That is, the combination of the two type algorithms gives us the possibility to obtain an uniform algorithm on the entire semi-axis J_+.

Usually the asymptotic methods gives some possibilities to obtain the estimates of the required number A_m and then to obtain the required number T. But it is sufficient to obtain the number T and the numbers C_1, C_2 comparing the approximate values $x(t)$ obtained by the two algorithms in an interval $[T - c, T + c]$ (c is a positive number), where the both methods give the results with the required accuracy.

In our example we shall compute the values $x(t)$ with the total error $\delta \leq 10^{-5}$ for sufficiently large interval $[0, T]$.

In fact in many cases the total error cannot be uniform. The simplest example: we cannot write down (with any accuracy) very large numbers and (which is the same problem) we cannot calculate the function t^2 with any accuracy on the entire semi-axis J_+.

(1) For $t \leq 10.5$ we compute the values $x(t)$ by means of series (16.2). The precise solution $x(t)$ is represented by the approximate value $x(t, m)$. The required number m depends on t. For $t \leq 10$ we choose $m = 40$. For example we obtain $x(1) \approx 0.83881$ and $x(10) \approx -0.19919$.

(2) For $t > 10$ we use series (16.6). For the root $\lambda_1(t) = \sqrt{t}$ in order to obtain the numbers b_m, we have to substitute the series in the left side of equation (16.5) and to collect like terms. We have $b_0^2 + 1 = 0$, $(1/2)b_0 + 2b_0 b_1 = 0$. For $m = 2$ $-a_1 + 2a_0 a_2 + a_1^2 = 0$. If m is even then we have

$$-\left(\frac{3}{2}m - 2\right)a_{m-1} + 2a_0 a_m + 2a_1 a_{m-1} + ... + 2a_{m/2-1}a_{m/2+1} + a_{m/2}^2 = 0.$$

Hence

$$a_m = \frac{(3/2)m - 2 - 2a_1 a_{m-1} + ... + 2a_{m/2-1}a_{m/2+1} - a_{m/2}^2}{2a_0}. \tag{16.7}$$

For odd m we have

$$-((3/2)m - 2)a_{m-1} + 2a_0 a_m + 2a_1 a_{m-1} + ... + 2a_{(m-1)/2}a_{(m+1)/2} = 0.$$

Hence

$$a_m = \frac{(3/2)m - 2 - 2a_1 a_{m-1} + \ldots + 2a_{(m-1)/2-1}a_{(m+1)/2+1}}{2a_0}. \qquad (16.8)$$

Since $y(t) \sim i\sqrt{t}$ we have $a_0 = i$ and consequently $a_1 = -1/4$, $a_2 = 5i/32$, $a_3 = 15/64$, $a_4 = -1105i/2048$ and so on. This means that

$$y(t) = i\sqrt{t} - \frac{1}{4t} + \frac{5i}{32t^2\sqrt{t}} + \frac{15}{64t^4} - \frac{1105i}{2048t^4\sqrt{t}} + O(1/t^6)$$

for $t \to +\infty$. Clearly $x(t) = Ae^{\int y(t)dt}$. Hence

$$x_1(t) = t^{-1/4}(1 + e^{(5/64)t^{-3}} + O(t^{-6}))e^{i[(2/3)t^{3/2} - (5/48)t^{-3/2} + (1105/9432)t^{-9/2}]}.$$

In the same way we have another solution $x_2(t)$ which corresponds to the root $\lambda_2(t) = -\sqrt{t}$:

$$x_2(t) = t^{-1/4}(1 + e^{(5/64)t^{-3}} + O(t^{-6}))e^{i[-(2/3)t^{3/2} - (5/48)t^{-3/2} + (1105/9432)t^{-9/2}]}.$$

The required solution $x(t)$ is a linear combination of the solutions $x_1(t)$ and $x_2(t)$. Taking into account that $x(t)$ is a real function, we conclude that $x(t)$ may be written in the form

$$x(t) = At^{-1/4}(1 + e^{-(5/64)t^{-3}} + O(t^{-6}))\cos((2/3)t^{3/2} - (5/48)t^{-3/2}$$

$$+ (1105/9432)t^{-9/2} + \varphi). \qquad (16.9)$$

Consequently we may choose the required approximation in the form

$$x(t) \approx At^{-1/4}(1 + e^{-(5/64)t^{-3}})\cos((2/3)t^{3/2} - (5/48)t^{-3/2}$$

$$+ (1105/9432)t^{-9/2} + \varphi). \qquad (16.10)$$

Clearly for $t \gg 1$ the approximation may be chosen in a more convenient form:

$$x(t) \approx At^{-1/4}\cos((2/3)t^{3/2} + \varphi) \qquad (16.11)$$

The constants A and φ may be obtained comparing the values obtaining from formulae (16.2) and (16.6). For $A = 0.917433$ and $\varphi = -0.261801$ the absolute theoretical error $|\delta(t)| < 0.510^{-5}$ on the interval $[9.5, 10.5]$. For example we have $x(225) \approx 0.22188$ and $x(10000) \approx -0.03936$.

Bibliography

[1] AIRY, G.B., Trans. Camb. Phil. Soc. **6** (1838), 379–402.

[2] ANIKEEVA, L.I., *About asymptotic behaviour of solutions of the equation* $y^{(4)} - a(x^\alpha y')' + bx^\beta y = \lambda y$ *for* $x \to \infty$. Vestnik MGU, Math., Mech. **6** (1976), 44–52 (in Russian).

[3] BANACH, S.S., *Course of functional analysis*. Kiev, 1948 (in Ukrainian).

[4] BANK, S., *An asymptotic analog of the Fuchs regularity theorem*. J. Math Anal. Appl. **16** (1966), 138–151.

[5] BANK, S., *On the asymptotic behaviour of solutions near an irregular singularity*. Proc. Amer. Math. Soc. **18** (1967), 15–21.

[6] BANK, S., *On the instability theory of differential polynomials*. Ann. Math. Pura Appl. **74** (3) (1966), 83–112.

[7] BANK, S., *On principal solutions of linear differential equations*. Proc. Amer. Math. Soc. **19** (3) (1968), 724–732.

[8] BANK, S., *On complex oscillation theory* Applicable Analysis **29** (3) (1988), 209–233.

[9] BELLMAN, R., *A Survey of the Theory of the Boundness, Stability and Asymptotic Behaviour of Solutions of Linear and Non-Linear Differential and Difference Equations*. Dover, N.Y., 1967.

[10] BERK, H.L., NEVINS W.M., ROBERTS K.V., *New Stokes line in WKB theory*. J. Math. Phys. **23** (6) (1982), 988–1002.

[11] BIJL, J., R., *Teopassingen van der Methode der stationare Phase (Thesis)*. Amsterdam, 1937.

[12] BIEBERBACH, L., *Funktionentheorie*. Leipzig–Berlin, 1930, **I**, VII, §8.

[13] BIRKHOFF, J.D., Trans. Amer. Math. Soc **9** (1908), 219–231, 380–382.

[14] BIRKHOFF, J.D., *Quantum Mechanics and Asymptotic Series*. Bull. Amer. Math. Soc. **32** (1933), 681–700.

[15] DU BOIS-REYMOND, P., *Sur la grandeur relative des infinis des functions*. Ann. di Math. **4** (2), (1871) 338–357.

[16] DU BOIS-REYMOND, P., *Über asymptotische Wertche, infinitäre Approximationen und infinitäre Auflösung von Gleichungen*. Math. Ann. **8** (1875), 362–414; (1871) 338–357.

[17] BOREL E., Ann. Sci, École Norm. Sup. **12** (3) (1895), 9–55.

[18] BOREL E., Ann. Sci, École Norm. Sup. **16** (3) (1895), 8–136.

[19] BOREL E., *Leçons sur le séries divergentes*. Paris, 1928.

[20] BRAAKSMA, L.J., SIAM, J., Math. Anal. **1** (1971), 1–16.

426

[21] BRIOT, J.C., BOUQUET, C.A.A., *Recherches sur les propriétés des fonctions définies par des equations differentielles.* J. École Polytech **21** (1856) 133–198.

[22] BOURBAKI, N., *Fonctions d'une variable reéle.* **IV**, Paris, 1948.

[23] BOURBAKI, N., *Integration.* Paris Hermann & Co. **VI** (5), 1956.

[24] DE BRUIJN, N.G., *Asymptotic Methods in Analysis.* North-Holland Publ. C., Amsterdam, 1958.

[25] BURCHARDT, H., München. Acad. S., **I–II**, Berlin, 1914.

[26] CARLEMAN, T.G.T., *Les fonctions quasi-analytiques.* Paris, 1926.

[27] CESARI, L., *Asymptotic Behaviour and Stability in Ordinary Differential Equations.* Springer, Berlin, 1963.

[28] CHAUDHARY, J., EVERITT, W.N., *On the square of formally self adjoint differential expressions.* J. London Math. Soc. **1** (2) (1969), 661–673.

[29] CODDINGTON, E.A., LEVINSON, N., *Uniqueness and convergence of successive approximations.* J. Indian Math. Soc. **16** (1952), 75–81.

[30] CODDINGTON, E.A., LEVINSON, N., *Theory of Ordinary Differential Equations.* McGraw-Hill, New York, 1965.

[31] COPSON, E.T., *The Asymptotic Expansion of a Function Defined by a Definite Integral of a Contour Integral.* Admirality Computing Service, London, 1946.

[32] VAN DER CORPUT, J.G., *Asymptotic Expansions.* Parts I and II. National Bureau of Standards, 1951.

[33] VAN DER CORPUT, J.G., *Asymptotic Expansions.* Part III. National Bureau of Standards, 1952.

[34] VAN DER CORPUT, J.G., Nederl. Akad. Wetensch. Amsterdam. Proc. **57** (1954), 206–217.

[35] VAN DER CORPUT, J.G., *Asymptotic Expansions I, Fundamental Theorems of Asymptotics.* Department of Mathematics, University of California, Berkeley, 1954.

[36] VAN DER CORPUT, J.G., FRANCLIN, J., Proc. Amst. Akad. **A 54** (1951), 213–219.

[37] COURANT, R., HURWITZ, A., *Functionentheorie.* Springer Verlag, 1922.

[38] DEBYE, P., Math. Ann **67** (1909), 535–558.

[39] DEVINATZ, A., *The deficiency index of certain fourth-order self-adjoint differential operators.* Quart. J. Math. Oxford **23** (91) (1972), 267–286.

[40] DEVINATZ, A., *The Deficiency Index of Certain Class of Ordinary Self-adjoint Differential Operators.* Advance in Math. **8** (1972), 434–473.

[41] DEVINATZ, A., KAPLAN, I.M., *Asymptotic estimates for solutions of linear systems of ordinary differential equations having multiple characteristic roots.* Indiana Univ. Math. J. **22** (4) (1972), 355–366.

[42] DITKIN, V.A., PRUDNIKOV, A.P., *Integral Transforms and Operational Calculus*. Pergamon Press, NY, 1965.

[43] DOETSCH, G., *Handbuch der Laplace-Transformation*. Birkhäuser, Basel, 1950.

[44] DULAC, H., *Recherches sur les points singuliers des équations différentielles*. J. École Polytech. **9** (2) (1904).

[45] DULAC, H., *Points singuliers des équations différentielles*. Mémorial des Sci. Math. Fusc. **61**, Gautier-Willars, Paris, 1934.

[46] DUNFORD, N., SCHWARTZ., J., *Linear Operators*. Part I (general Theory). Interscience, New York, 1958.

[47] EASTHAM, M.S.P., *Theory of Ordinary Differential Equations*. Van Nostrand Reinhold Company, London, 1970.

[48] ECKHAUS, W., *Asymptotic analysis of singular perturbations*. North Holand, 1979.

[49] ERDELYI, A., Proc. Edinburgh Math. Soc. **8** (2) (1947), 20–24.

[50] ERDELYI, A., *Asymptotic Expansions*. Dover Publications. Inc., N.Y., 1959.

[51] ERDELYI, A., KENNEDY, M., McGREGOR, J.L., *Parabolic cylinder functions of large order*. J. Rational Mech. and Analysis **3** (1954), 459–485.

[52] ERDELYI, A., KENNEDY, M., McGREGOR, J.L., *Asymptotic forms of Coulomb wave functions*, **I.** Tech., Rep. 5, Dept. of Math. Calif. Inst. Tech. Pasadena, 1955.

[53] ERDELYI, A., KENNEDY, M., McGREGOR, J.L., *Asymptotic forms of Coulomb wave functions*, **II.** Tech., Rep. 5, Dept. of Math. Calif. Inst. Tech. Pasadena, 1955.

[54] ERDELYI, A., SWANSON, C.A., *Asymptotic forms of Whittaker's confluent hypergeometric functions*. Mem. Amer Math. Soc. **25**, 1957.

[55] EULER, L., Comment. Acad. sci. imp. Petrop. **6** (1738), 68–97.

[56] EULER, L., Nove commentarii Ac. Sci. Petrop. **5** (1754), 205–237.

[57] FABRY, E., *Sur les intégrales des équations différentielles linéaries à coéfficients rationnels*. Thése. Paris, 1885.

[58] FEDORJUK, M.V., *Asymptotic methods in the theory of one-dimensonal singular operators*. Proc. Mosc. Math. Soc. **15** (1966), 296–345 (in Russian).

[59] FEDORJUK, M.V., *Asymptotics of eigenvalue and eigenfunctions of one-dimensional singular differential operators*. DAN U.S.S.R. **169** (2) (1966), 288–291 (in Russian).

[60] FEDORJUK, M.V., *Asymptotics of solutions of ordinary linear differential equations of n-th order*. Differential Equations **2** (4) (1966), 492–507 (in Russian).

[61] FEDORJUK, M.V., *Asymptotic methods in the theory of the one-dimensional singular operators*. Memoir. Mosc. Math. Soc. **15** (1966), 296–345 (in Russian).

[62] FEDORJUK, M.V., *Asymptotic methods in the theory of ordinary linear differential equations*. Math. Sbornik **79** (4) (1969), 477–516 (in Russian).

[63] FEDORJUK, M.V., *Asymptotic Methods for Linear Ordinary Differential Equations*. Nauka, Moscow, 1983 (in Russian).

[64] FEDORJUK, M.V., Asymptotic analysis, Springer, 1993 (translation from [64]).

[65] FORD, W.B., *The asymptotic developments of functions defined by MacLaurin series*. Univ. of Michigan Sci. **11**, 1936.

[66] FORSYTH, A.R., *Theory of differential equations*. Cambridge, 1959.

[67] FROBENIUS, G., *Über die integration der linearen Differentialgleichungen durch Reichen*. Cambridge, 1902.

[68] FUCHS, L., *Zur Theorie der linearen Differentialgleichungen mit veränderlichen Koeffizienten*. J. für Math. **66** (1866), 121–160.

[69] FUCHS, L., *Zur Theorie der linearen Differentialgleichungen mit veränderlichen Koeffizienten*, J. für Math. **68** (1868), 354–385.

[70] FULKS, W.B., Proc. Amer. Soc. **8** (2) (1951), 613–622.

[71] GEORGESCU, A., *Asymptotic treatment of differential equations*. Chapman and Hall, 1995.

[72] GILBERT, R.C., *The deficiency index of a third order operator*. Pacific J. Math. **68** (2) (1977), 369–392.

[73] GOLLWITZER, H.E., *Stokes multipliers for subdominant solutions of second order differential equations with polynomial coefficients*. Univ. Minnessota, Inst. Technology, School Math., Minneapolis, 1967, 1-67.

[74] GOURSAT, E., *Cours d'analyse mathématique*. Paris, 1924–1928.

[75] HABER, S., LEVINSON, N., *A boundary value problem for a singularity perturbed differential equation*. Proc. Amer. Math. Soc. **6** (1955), 866–872.

[76] HARDY, G.H., *Orders of infinity*. Cambridge tracts. Cambrigdge Univ. Press, **12**, 1924.

[77] HARDY, G.H., LITTLEWOOD, J.G., *Notes on the theory of series (XI): On Tauberian theorems*. Proc. London Math. Soc. **30** (1929), 23–37.

[78] HARRIS, W.A., *Singular perturbations of two-point boundary problems for systems of ordinary differential equations*. Arch Rational Mech. Analsis 5 (1960),212–225.

[79] HARRIS, W.A., *Singular perturbation problems.* Bol. Soc. Mat. Mex. **2** (2) (1960),245–254.

[80] HARRIS, W.A., *Singular perturbations of two-point boundary problems*. J. Math. Mech. **11** (1962), 371–382.

[81] HARRIS, W.A., *A boundary value problem for a singulary perturbed system of non-linear differential equation*. Intern. Sympos. Non-linear Mech., Academic Press, New York, 1963, 489–495.

[82] HARRIS, W.A., LUTZ, D.A., *On the asymptotic integration of linear differential systems* J. Math. Anal. and App. **48** (1974), 1–16.

[83] HARRIS, W.A., LUTZ, D.A., *Asymptotic integration of adiabatic oscilators* J. Math. Anal. and App. **51** (1975), 78–93.

[84] HALMOS, P.S., *Measure Theory*. Princeton, Van Nostrand, 1950.

[85] HARTMAN, F., *Ordinary differential equations*. Inc. Wiley, New York, 1969.

[86] HEADING, J., *The Stokes phenomenon and certain nth order differential equations*. Proc. Cambridge Phill. Soc. **53**, (1957) 399–441.

[87] HILLE, E., *Ordinary Differential Equations in the complex domain*. New York, Wiley, 1976.

[88] HOHEIZEL, G.K.H., Reine Angew. Math. **153** (1924), 228–248.

[89] HOLMES, M.H., *Introduction to perturbation methods*. Springer, 1997.

[90] HORN, J., *Über die Reichenentwicklung der Integrale eines Systems von Differentialgleichungen in der Umgebung gewisser singulärer Stellen*. J. Reine Angew. Math. **116** (1896), 265–306; **117** (1897), 104–128.

[91] HORN, J., *Über das Verçhalten der Integrale von Differentialgleichungen an eine Unbestimmtheitstelle*. J. Reine Angew. Math. **119** (1898), 196–209, 267–290.

[92] HORN, J., *Fakultätenreihen in der Teorie der linearen Differentialgleichungen*. Math ann. **71** (1912), 510–532.

[93] HORN, J., *Integration linear Differentialgleichungen durch Laplacesche Integrale an Facultätenreihen*. Jahresber. Deutsch. Math. Ver. **24** (1915), 309–325.

[94] HORN, J., *Integration linear Differentialgleichungen durch Laplacesche Integrale an Facultätenreihen*. Jahresber. Deutsch. Math. Ver. **25** (1915), 74–83.

[95] HORN, J., *Laplacesche Integrale, Binommialkoeffizientenreihen*. Math. Z. **21** (1924), 85–95.

[96] HORN, J., *Integration linear Differentialgleichungen durch Laplacesche Integrale, I,II*. Math. Z. **49** (1944), 359–350, 684–701.

[97] HUET, D., *Phénomènes de perturbation singulière*. C. R. Acad. Sci. Paris **246** (1958), 2096–2098.

[98] HUGHES, H.K., *The asymptotic developments of a class of entire functions*. Bull. Amer. Math. Soc. **51** (1945), 456–461.

[99] HUKUHARA, M., *Intégration formelle d'un système d'équations différentielles non-linéaires dans le voisinage d'un point singulier*. Ann. Math Pura Appl. Bologna, **19** (4) (1940), 35–44.

[100] HUKUHARA, M., *Sur les points singulaires des équations différentielles linéares III*. Mém. Fac. Sci. Kushu Univ. **2** (1942), 125–137.

[101] HUKUHARA, M., IVANO, M., *Étude de la convergence des solutions formelles d'un système différentiel ordinaire linéare*. Funccial. Ekvac. **2** (1959), 1–18.

430

[102] IMMINK, G.K., *Asymptotics of Analytic Difference Equations.* Springer Ferlag, Lecture notes in Mathematics **1085**, 1984.

[103] INCE, E.L., *Ordinary Differential Equations.* Longmans, Green and Co., London, 1927, republished in 1956 by Dover Publications.

[104] INCE, E.L., *Integration of ordinary differential equations.* Edinburgh, 1952.

[105] JEFFREYS, B.S.,JEFFREYS, H., *Methods of Mathematical Physics.* Cambridge, 1956.

[106] JEFFREYS, H., *On certain approximate solutions of linear differential equations of second order.* Proc. London Math. Soc. **23** (2) (1924), 428–436.

[107] JORDAN, C., *Traité des substitutions et des équations algébriques.* Paris, 1870.

[108] KAMKE, E., *Das Lebesgue-Stieltjes Integral.* Leipzig, 1956.

[109] KAMKE, E., *Differentialgleihungen. Lösungsmethoden und Lösungen.* Leipzig, 1956.

[110] VAN KAMPEN, E.R., *Remarks on systems of ordinary differential equations.* Amer J. Math. **59** (1937), 144–152.

[111] KARAMATA, J., *Sur certain "Tauberian theorems" de Hardy et Littlewood.* Mathematica (Cluj) **3** (1930), 34–46.

[112] KARAMATA, J., Mathematica (Cluj) **4** (1930), 38–53.

[113] KARAMATA, J., Bull. Soc. Math. France **61** (1933), 52–62.

[114] KELLY, B.J., *Admissible domains of higher order differential operators.* Studies in appl. Math. **60** (1979), 211–240.

[115] KEVORKIAN, J., COLE, J.D., *Perturbation methods in applied mathematics.* Springer, 1981.

[116] KNOPP, K., *Theory and Application of Infinite Series,* 1928.

[117] KOLMOGOROV, A.N., FOMIN, S.V., *Introductory Real Analysis.* Prentice-Hall, Englewood Cliffs, New Jersey, 1970.

[118] LANDAU, E., *Vorlesungen über Zahlentheorie,* Leipzig, **2**, 1927.

[119] LANGER, R.E., *On the asymptotic solutions of ordinary differential equations with an application to the Bessel functions of large order.* Trans. Amer. Math. Soc. **33** (1931), 23–64.

[120] LANGER, R.E., *On the asymptotic solutions of differential equations with an application to the Bessel functions of large complex order.* Trans. Amer. Math. Soc. **34** (1932), 447–480.

[121] LANGER, R.E., *The asymptotic solutions of certain linear ordinary differential equations of second order.* Trans. Amer. Math. Soc. **36** (1934), 90–106.

[122] LANGER, R.E., *The asymptotic solutions of ordinary differential equations with reference to the Stokes' phenomenon.* Bull. Amer. Math. Soc. **40** (1934), 545–582.

[123] LANGER, R.E., *The asymptotic solutions of ordinary differential equations of the second order with special reference to the Stokes' phenomenon about a singular point*. Trans. Amer. Math. Soc. **37** (1935), 397–415.

[124] LANTSMAN, M.H., *Asymptotic Properties of Solutions of Linear Differential Equations with Coefficients of the Power order of Growth*. Differential Equations **18** (35) (1982), 711–713. (in Russian)

[125] LANTSMAN, M.H., *Formal Representations of Solutions of One Class of Linear Homogeneous Differential Equations*. Differential Equations **20** (3) (1984), 398–408. (in Russian)

[126] LANTSMAN, M.H., *Asymptotic Representations of Solutions of One Class of Linear Homogeneous Differential Equations*. Differential Equations **20** (4) (1984), 958–967. (in Russian)

[127] LANTSMAN, M.H., *Asymptotic Integration of Some Linear Differential Equations*. Math. Nachr. **123** (1985), 23–37.

[128] LANTSMAN, M.H., *Asymptotic Solutions to Linear Differential Equations with Coefficients of the Power Order of Growth*. Proc. Roy. Soc. Edinburgh **100A** (1985), 301–326; **101A** (1985), 77–98.

[129] LANTSMAN, M.H., *Linear Difference Equations with Coefficients of the Power Order of Growth*. Math. Nachr. **163** (1993), 305–322.

[130] LANTSMAN, M.H., *Linear Differential Equations with Variable Coefficients in Regular and Some Singular Cases*. Math. Nachr. **186** (1997), 197–223.

[131] LANTSMAN, M.H., *The Connection between the Characteristic Roots and the Corresponding Solutions of a Single Differential Equation with Comparable Coefficients*. Math. Nachr. **204** (1999), 137–156.

[132] LAVRENTYEV, M.A., SHABAT, B.V., *Methods of the theory of functions of a comple variable*. Fizmatgiz, Moscow, 1972. (in Russian)

[133] LEFSCHETZ, S., *Differential equations: Geometric theory*. New York, 1957.

[134] LEVIN, J.J., *Singular perturbations of non-linear systems of differential equations related to conditional stability*. Duke Math. J. **23** (1956), 609–620.

[135] LEVIN, J.J., LEVINSON N., *Singular perturbations of non-linear systems of differential equations and associated boundary laer equation*. J. Rational Mech. Analysis **3** (1954), 247–270.

[136] LEVINSON, N., *The asymptotic nature of the solutions of linear systems of differential equations*. Duke Math. J. **15** (1948), 111–126.

[137] LEVINSON, N., *Perturbations of discontinuous of non-linear systems of differential equations*. Acta Math. **82** (1951), 71–106.

[138] LEVINSON, N., *A boundary value for a singularity perturbed differential equation*. Duke Math. **25** (1958), 331–342.

[139] LEBESGUE, H., *Leçons sur l'integration et la recherche de fonctions primitive*. Paris, 1928.

[140] LEVI, B., Publ. Inst. Mat. Univ. Nac. Litoral **6** (1946), 341–351.

432

[141] LIN, C.C., RABENSTEIN, A.L., *On the asymptotic solutions of a class of ordinary differential equations of fourth order.* Trans. Amer. Math. Soc. **94** (1960), 24-57.

[142] LINDELÖF, E., *Sur l'application des méthodes d'approximations succsessives à l'étude des intégrales réeles des équations différentielles ordinaries.* J. de Math. **10** (4) (1894), 117-128.

[143] LIPSCHITZ, R, J. Reine und angew. Math. **63** (1864), 296-308.

[144] MACLAURIN, C., *A treatise of fluctions.* 1-2, Edinburgh, 1742.

[145] MALMQUIST, J., *Sur létude analytique des solutions d'un système d'équations differentielles dans le voisinage d'un point singulier d'indétermination.* Acta Math. **73** (1941), 87-129; **74** (1941), 1-64; **74** (1941), 109-128.

[146] MALMQUIST, J., *Sur les points singuliers des équations différentielles.* Ark. Mat. Astr. Fys. **29A** (11,18) (1943).

[147] MARKUSHEVICH, A.I., *Theory of Functions of a Complex Variable.* 1-3, Prentice-Hall, Englewood Cliffs, New York, 1965-1967.

[148] MEIJER, C.S., Math. Ann. **108** (1933), 321-359.

[149] MILLER, K.S., *Linear Difference Equations.* W.A. Benjamin Inc., N.Y., Amsterdam, 1968.

[150] MILLER, J.C.P., Proc. Cambridge Philos. Soc. **48** (1952), 243-254.

[151] MIRANKER, W.L., *Singular perturbation eigenvalues by a method of undertermined coefficients.* J. Math. Phys. **42** (1963), 47-58.

[152] MISES, R., *Die Grezschichte in der Theorie der gevönlichen Differentialgleichungen.* Acta Univ. Szeged., Sect. Sci. Math. **12** (1950), 29-34.

[153] MOLCHANOV, A.M., *Uniform asymptotics of linear systems.* Dok. AN SSSR, **173** (3) (1967),519-522. (in Russian)

[154] MOULTON, F.R., *Differential equations.* New York, 1930.

[155] MOSER, J., *Singular perturbations of eigenvalue problems for linear differential equations of even order.* Comm. Pure Appl. Math. Z. **8** (1955), 251-258.

[156] MOSER, J., *The order of singularity in Fuchs's theory.* Math. Z. **72** (1959), 379-398.

[157] MOSER J., *Note on asymptotic expansions.* MRC Tech. Summary Rep., **104**, Math. Res. Ctr. U.S. Army, Univ. of Visconsin, 1959.

[158] MURRAY, J.D., *Asymptotic analysis.* Springer, 1981.

[159] NAIMARK M.A., *Linear differential operators.* Gostechisdat, Moscow, 1954. (in Russian)

[160] NATANSON I.P., *Theory of functions of a real variable.* New York, 1, 1955; 2, 1960.

[161] NAYFEH A.H., *Introduction to perturbation techniques.* Willey, 1981.

[162] NILSEN N., *Recherches sur les séries de factorielles*. Ann. Ecole Norm. **19** (3) (1902), 409–453.

[163] NEMITSKI, V.V., STEPANOV, V.V., *Qualitative theory of differential equations*. Moscow, Leningrad, 1949.

[164] NEVANLINNA F.E.H., *Über Riemannsche Flächen mit endlich vielen Windungspuncten*. Acta Math. **58** (1932), 295–373.

[165] NEWELL, H.E.J., *The asymptotic forms of the solution of an ordinary linear matrix equation in the complex domain*. Duke Math. J. **9** (1942), 245–258.

[166] NORLUND, N.E., *Vorlesungen über Differenzenreichung*. Berlin, 1924.

[167] OCUBO, K., *A global representation of a fundamental set of solutions and a Stokes phenomenon for a system of linear ordinary differential equations*. J. Math. Soc. Japan **15** (1963), 268–288.

[168] OLVER, F.W.J., *The asymptotic expansion of Bessel functions of large order*. Phil. Trans. Roy. Soc. London **A248** (1954), 328–368.

[169] PERRON, O., *Ein neuer Existenzbeweis für die Integrale eines Systems gewöhnlicher Differentialgleichungen*. Math Ann. **78** (1917), 378–384.

[170] PERRON, O., *Über Summengleichungen und Poincaré sone Differenzengleichungen*. Math. Ana. **84** (1921), 1–15.

[171] PERRON, O., *Über das Verhalten der Integrale einer linearen Differentialgleichungen bei grossen Werten der unabhängigen Variablen*. Math. Z., **1** (1918), 27–43.

[172] PICARD, É., *M'emorie sur la théorie des équatuons aux dérivees partielles a la méthode des approximations successives*, J. de Math., **4, 6**, (1890), 145–210.

[173] POINCARÉ, H., *Sur les èquations linnéaries aux differentielles ordinaires et aux differences finies*. Amer J. Math. **7** (1885), 203–285.

[174] POINCAR'E, H., *Sur les int'egrales irregulières des équations linéaires*. Acta Math. **8** (1886), 295–344.

[175] POINCAR'E, H., *Sur les propriétés des fonctions définies par les équations aux différentielles*. J. École Polytech. **45** (1878), 13–26.

[176] POOLE, E.G.C., *Introduction to the Theory of Linear Differential Equations*. Oxford, 1936.

[177] RANG, E.R., *Periodic solutions of singular perturbation problems*. Tech. Rept. **1**, Dept. of Math. Univ. of Minnesota, School of Thechnol., 1957.

[178] RANG, E.R., *Periodic solutions of singular perturbation problems*. Proc. of Intern. Symp. on Non-linear Dif. Equations, Academic Press, New York, 1963, 377–383.

[179] RAPOPORT, I.M., *About Some Asymptotic Methods in the Differential Equations Theory*. Ak. Nauk USSR, Kiev, 1954. (in Russian)

[180] RIEMANN, B., *Gesammelte Mathematische Werke*. Leipzig, 1892.

[181] RIESZ, F., NAGY, Sz., *Leçns d'analyse fonctionelle*. Budapest, 1953.

[182] RIESZ, F., NAGY, SZ., *The English translation: Functional analysis*. New York, 1955.

[183] RITT, F., *On the derivatives of a function at a point*. Ann. Math. **18** (1916), 18–23.

[184] ROOS, H.-G., *Die asymptotische Lösung einer linearen Differentialgleichung zweiter Ordnung mit zweisegmentigem charakteristischem Polygon*. Betr. Anal. **7**, (1975), 55–63.

[185] ROOS, H.-G., *Die asymptotische Lösung einer linearen Differentialgleichung mit dreisegmentigem charakteristischem Polygon*. Math. Nachr.**88**, (1979), 93–103.

[186] SANSONE, G., *Equazioni differenziali nel campo reale*. Bologna, 1948–1949.

[187] SCHLESENGER, L., *Handbuch der Theorie der linearen Differentialgleichungen*. Leipzig, 1895.

[188] SENETA, E., *Regularly Varying Functions*. Springer-Ferlag, Lecture notes in mathematics **508**, Berlin, Heidelberg, N.Y., 1976.

[189] SCHLÖMLICH, O., *Compendium der Höheren Analysis*. **II**, §4, Braunschweig, 1895.

[190] SIBUYA, Y., *Sur réduction analytique d'un système d'équations differentielles ordinaries linéares contenant un parmètre*. J. Fac. Sci., Univ. Tokyo, **7** (1) (1958), 527–540.

[191] SIBUYA, Y., *Simpliication of a system of linear ordinary differential equations about a singular point*. Funccial. Rkvac. **4** (1962), 29–56.

[192] SIBUYA, Y., *Some global properties of matrices of functions of one variable*. Math. Ann. **161** (1965), 67–77.

[193] SIBUYA, Y., *Global Theory of a second Order Linear Differential Equation with a Polynomial Coefficient*. North-Holland Publ.C.,Amsterdam (1975).

[194] STILTJES, TH., Ann de l'Ec. Norm. Sup. **3** (3) (1886), 201–258.

[195] STOKES, G.G., Trans. Camb. Phil. Soc. **16** (1857), 105–128.

[196] STRENBERG, W., *Über die asymptotische Integration von Differentialgleichungen*. Math. Ann. **81** (1920), 119–186.

[197] STRODT, W., Ann. Math **44** (3) (1943), 375–396.

[198] STRODT, W., Amer. J. Math. **69** (4) (1947), 717–757.

[199] STRODT, W., *Contributions to the asymptotic theory of ordinary differential equations in the complex domain*. Mem. Amer. Math. Soc. **13**, 1954.

[200] STRODT, W., *Principal solutions of ordinary differential equations in the complex domain*. Mem. Amer. Math. Soc., **26**, 1957.

[201] STRODT, W., *On the algebraic closure of certain partially ordered fields*. Trans. Amer. Math Soc. **105** (1962), 229–250.

[202] STRODT, W., WIRGHT, R.K., *Asymptotic behaviour of solutions and adjunction fields for non-linear first order differential equations.* Mem. Amer. Math Soc. **109**, 1971.

[203] TITCHMARSH, E.C., *The Theory of Functions.* Oxford University Press, London, 1939.

[204] THOMSEN, D.I.JR., Proc. Amer. Math. Soc. **5** (1954), 526–532.

[205] TRICOMI, F.C., *Equazioni differenziali.* Torino, 1953.

[206] TRJITZINSKY, W.J., *Analytic theory of linear differential equations.* Acta Math. **62** (1934), 167–226.

[207] TRJITZINSKY, W.J., *Laplace integrals and factorial series in the theory of linear differential and difference equations.* Trans. Amer. Math. Soc. **37** (1935), 80–146.

[208] TRIJITZINSKY, W.J., *Singular point problems in the theory of linear differential equations.* Bull. Amer. Math. Soc. **44** (1938), 208–223.

[209] TRJITZINSKY, W.J., *Developments in the analytic theory of algebraic differential equations.* Acta Math. **73** (1941), 1–85.

[210] TSCHEN, Y., *Über das Verhalten der Lösungen einer Folge vom Differentialgleichungen, welche im Limes ausarten.* Comp. Math. **2** (1935), 1–18.

[211] TURRITIN, H.L., *Stokes multipliers for asymptotic solutions of a certain differential equation.* Trans. Amer. Math. Soc. **68** (1950), 304–329.

[212] TURRITIN, H.L., *Convergent solutions of ordinary linear homogeneous differential equations in the neighborhood of an irregular point.* Acta Math. **93** (1955), 27–66.

[213] TURRITIN, W.J., *Asymptotic solutions of certain ordinary differential equations associated with multiple roots of the characteristic equation.* Amer. J. Math. **58** (1936), 364–376.

[214] VALIRON, G., *Équations functionalies; applications (cours d'analyse II).* Paris, 1945.

[215] DE LA VALLÉE POUSIN, C., *Cours d'analyse infinitésimale.* Paris, 1938.

[216] VAN DER WARDEN, B.L., *Algebra I.* Springer Verlag, 1971; *Algebra II.* Springer Verlag, Berlin, Heidelberg, New York, 1967.

[217] WALKER, P.W., *Deficiency indices of fourth-order singular operators.* J. Dif. Equat. **9** (1971), 133–140.

[218] WALKER, P.W., *Asymptotics of solutions to $[(ry'')' - py]' + qy = \sigma y$.* J. Dif. Equat. **9** (1971), 108–132.

[219] WASOW, W., *Asymptotic Expansions for Ordinary Differential Equations.* Interscience Publishers, N.Y., 1965.

[220] WASOW, W., *Singular perturbations of boundary value problems for non-linear differential equations of the second order.* Comm. Pure Appl. Math. **9** (1956), 93–113.

[221] WATSON, G.N., Philos. Trans. Royal Soc. **A211** (1912), 279–313.

[222] WENDEL, J.G., *Singular perturbations of a Van der Pol equation. Contribution to the theory of non-linear oscillations.* Ann. Math. Studies, Princeton **20** (1950), 243–290.

[223] WIDDER, D.V., *The Laplace Transform.* Princeton, 1941.

[224] VAN WIJNGAARDEN, A., Nederl. Akad. Wetensch., Amsterdam, Proc. **56** (1953), 522–543.

[225] WHITTAKER, E.T., WATSON, G.N., *Modern Analysis.* Cambridge U. P., 1927.

[226] WININER, A., Phis. Rev. **72** (1947), 516–517.

[227] WISWANATHAM, B., *The general uniqueness theorem and successive approximations.* J. Indian Math. **16** (1952), 69–74.

[228] WOOD, A.D., *Deficiency indices of some fourth order singular operators.* J. London Math. Soc. **3** (2) (1971), 96–106.

[229] WOOD, A.D., *On the deficiency indices of a fourth order singular operators.* Proc. Roy. Soc. Edinburgh **85A** (1980) 15–57.

[230] WRIGH, E.M., *The asymptotic expansion of integral functions and of the coefficients in their Taylor series.* Trans. Amer. Math. Soc. **64** (1948), 409–438.

List of Symbols

INDEX